Advances in Biochemistry in Health and Disease

Series Editor: Naranjan S. Dhalla
Winnipeg, Manitoba, Canada

Volume 1: S. K. Cheema, (ed), *Biochemistry of Atherosclerosis*

Volume 2: S. W. Schaffer and M-Saadeh Suleiman, (eds), *Mitochondria: The Dynamic Organelle*

Volume 3: A. K. Srivastava and M. B. Anand-Srivastava, (eds), *Signal Transduction in the Cardiovascular System in Health and Disease*

Volume 4: B. Ostadal and N. S. Dhalla, (eds), *Cardiac Adaptations-Molecular Mechanisms*

Volume 5: B. I. Jugdutt and N. S. Dhalla (eds), *Cardiac Remodeling-Molecular Mechanisms*

Volume 6: Jawahar L. Mehta and N. S. Dhalla (eds), *Biochemical Basis and Therapeutic Implications of Angiogenesis*

Volume 7: S. Chakraborti and N. S. Dhalla (eds), *Proteases in Health and Disease*

Volume 8: S. Chakraborti and N. S. Dhalla (eds), *Role of Proteases in Cellular Dysfunction*

Volume 9: B. Turan and N. S. Dhalla (eds), *Diabetic Cardiomyopathy: Biochemical and Molecular Mechanisms*

For further volumes:
http://www.springer.com/series/7064

Belma Turan • Naranjan S. Dhalla
Editors

Diabetic Cardiomyopathy

Biochemical and Molecular Mechanisms

Springer

Editors
Belma Turan
Department of Biophysics
Faculty of Medicine
University of Ankara
Ankara, Turkey

Naranjan S. Dhalla
Institute of Cardiovascular Sciences
St. Boniface Hospital Research
University of Manitoba
Winnipeg, MB, Canada

ISBN 978-1-4614-9316-7 ISBN 978-1-4614-9317-4 (eBook)
DOI 10.1007/978-1-4614-9317-4
Springer New York Heidelberg Dordrecht London

Library of Congress Control Number: 2013956435

Printed on acid-free paper

Springer is part of Springer Science+Business Media (www.springer.com)

Professor Suresh K. Gupta
Head, Department of Clinical Research
Delhi Institute of Pharmaceutical Sciences
and Research
Delhi University, New Delhi, India

This book is dedicated to Prof. S.K. Gupta for his outstanding leadership in promoting cardiovascular research and education throughout the Indian continent. In his capacity as Head of the Department of Pharmacology at the All India Institute of Medical Sciences, Dr. Gupta established the National Center for promoting health and family welfare. He has published more than 250 research papers and seven books, mostly on the role of herbal drugs in the prevention and treatment of cardiovascular diseases, and trained more than 150 postgraduate students. As Dean and Director General of The Institute of Clinical Research, he initiated and promoted a comprehensive education program for clinical research throughout India.

Dr. Gupta has been heavily engaged in promoting the scientific basis for the practice of cardiology as well as young investigators and is currently serving as President of the India Section of the International Academy of Cardiovascular Sciences. This book, presenting molecular and cellular aspects of cardiac dysfunction during the development of diabetic cardiomyopathy, pays a special tribute to Prof. Suresh Gupta for his distinguished services.

Preface

Diabetes has long been recognized as a disease of high blood sugar, and there has been a continuous search for the exact reason for its development. In the middle of the nineteenth century, autopsies of patients with diabetes mellitus showed damaged pancreases. The first evidence for the link between diabetes and the pancreas was provided in 1889 by O. Minkowski and J.V. Mering, who depancreatized a dog and demonstrated the development of polyuria, which was undistinguishable from diabetes. This disease is known as a disorder of carbohydrate metabolism resulting from insufficient production of insulin by pancreas (type 1 diabetes) or an ineffective response of cells to insulin (type 2 diabetes). In 2005, the World Health Organization had estimated that more than 180 million people worldwide suffer from diabetes mellitus and indicated that this figure is likely to double within the next 20 years. Among the 3.8 million deaths each year associated with diabetes, about two thirds are attributable to cardiovascular complications. In fact, diabetes is now considered to be a major metabolic risk factor for the occurrence of heart disease.

Metabolic syndrome has been used to describe a cluster of disorders such as diabetes and is believed to be an indicator of risk for heart disease, stroke, and other cardiovascular abnormalities. This term dates back to the late 1950s, but came into common usage in the 1970s to describe the association of various risk factors with diabetes, particularly insulin resistance, which is the cornerstone of this syndrome. Some individuals are genetically predisposed to developing insulin resistance; however, both obesity and sedentary lifestyle are conditions that promote metabolic syndrome and result in developing type 2 diabetes, heart disease, kidney disease, and stroke. It should be noted that cardiovascular dysfunction is the leading cause of mortality in diabetic individuals, primarily because of the development of cardiac abnormality, diabetic cardiomyopathy, as defined by A. Grishman and his coworkers in 1972. Diabetic cardiomyopathy is a clinical condition in which ventricular dysfunction is diagnosed in diabetic patients in the absence of coronary atherosclerosis and hypertension. Several mechanisms involved in the development of cardiomyopathy in diabetic patients have been postulated, including alterations in intracellular cation homeostasis and glucose metabolism, as well as enhanced oxidative stress. It is becoming clear that cardiovascular complications in diabetes result from multiple parameters including glucotoxicity, lipotoxicity, fibrosis, and mitochondrial uncoupling.

Diabetic cardiomyopathy is associated with both type 1 (insulin-deficient) and type 2 (insulin-resistant) diabetes and is characterized by early-onset diastolic and late-onset systolic dysfunctions, including depressions in diastolic compliance, contractility, and rate of myocardial relaxation. Diabetic cardiomyopathy was initially classified as a dilated cardiomyopathy with prominent left ventricular enlargement and depressed systolic function; however, during the past two decades, diastolic left ventricular dysfunction was also indentified as a manifestation of this disorder. Because discovering the nature of diabetic cardiomyopathy and its complications has been generally considered to help in designing effective therapeutic strategies to curb this epidemic problem in society, this book was aimed to provide a comprehensive approach to understanding diabetes and its complications as well as various strategies for its prevention and treatment. This book is a compilation of review articles devoted to the study of diabetic cardiomyopathy with respect to biochemical and molecular mechanisms of hyperglycemia. The wide range of topics covered here are of interest to basic research scientists, clinicians, and graduate students who are devoted to the study of the pathogenesis of diabetes-induced cardiovascular dysfunction. Furthermore, some chapters are directed toward increasing our understanding of novel ways for the prevention and treatment of cardiomyopathy. The 25 chapters in this book are organized into three sections.

The first section discusses general aspects of the metabolic derangements in diabetic cardiomyopathy, including metabolic alterations and substrate utilization in the heart, metabolic alterations and cardiac remodeling, role of diet in the development of metabolic syndrome in the heart, and the effects of hyperglycemia in terms of biochemical and structural alterations in the heart. This section also has chapters on the role of hyperglycemia in cardiovascular complications and their possible prevention with antioxidant treatment protocols. It is generally well accepted that diabetes is associated with increased oxidative stress. In fact, chronic hyperglycemia has been shown to cause oxidative stress, leading to cardiac complications, including hypertension, left ventricular hypertrophy, dilated cardiomyopathy, and myocardial infarction. Furthermore, significant increases in oxidants have been shown to trigger a cascade of pathological events including contractile dysfunction. Oxidative stress and imbalance between endogenous reactive oxygen species as well as antioxidant systems are involved in the etiology of diabetes-induced downregulation of heart function. There is also a close relationship between impaired insulin signaling and alteration in heart function via depressed endogenous antioxidant defense mechanisms. In addition, there is some evidence for sex-related differences in diabetes-induced cardiovascular pathologies, particularly the role of estrogen, which can exert protective effects against diabetes through modulation of altered Ca^{2+} dynamics and reduction of oxidative damage to the heart.

In the second section, several cellular mechanisms of diabetic cardiomyopathy are discussed, indicating that diabetic cardiomyopathy is a multifactorial and complex problem. Included are alterations in cardiac energy metabolism showing reduced glucose uptake and increased free fatty acid oxidation, resulting in mitochondrial uncoupling, impaired Ca^{2+} homeostasis, and depressed contractile activity. The use of PPAR-α agonist to reduce fatty acid oxidation and physical exercise to induce mitochondrial adaptation has been claimed to prevent diabetes-induced cardiac dysfunction. It should be emphasized that the diabetic heart is almost dependent on the metabolism of fatty acids as there occurs an increase in the myocardial uptake of fatty acids and an increase in its oxidation as well as a reduction in glucose oxidation. Such changes result in a decrease in ATP production per mole of oxygen and an increase in mitochondrial uncoupling, leading to an unfavorable energetic state together with an overproduction of reactive oxygen species. On the other hand, diabetes has long been reported to rapidly induce contractile dysfunctions associated with altered Ca^{2+} handling in isolated ventricular myocytes. Because miRNAs play a fundamental role in gene expression, alterations of a large number of miRNAs in chronic diabetes are closely associated with the development of heart disease. In a chronic neonatal rat model of diabetes demonstrating cardiomyopathy, 14 miRNAs were upregulated and 28 miRNAs were downregulated. In addition, miR223 was consistently upregulated in the insulin-resistant heart, and miR223 overexpression-induced GLUT4 protein expression in cardiomyocytes was found to be necessary for increased glucose uptake.

Several authors have contributed to the formulation of the seven chapters in the third section of this book, which demonstrate the prevention and treatment of diabetes using appropriate diet, proper supplements including antioxidants, angiotensin inhibitors, and some other drugs. Indeed, different epidemiological, experimental, and clinical investigations have demonstrated a close correlation between diet and increased risk of developing diabetes-induced cardiovascular complications. For example, the increased consumption of refined and simple carbohydrates, fats, red meats, and low fiber as well as low intake of specific minerals and vitamins have been shown to impair insulin response and increase plasma glucose levels and thus produce damage to the macro- and microvasculature. Some investigators have also highlighted the beneficial effects and impact of nutritional interventions on the developing fetus and described the role of epigenetics and maternal nutrition on the risk of developing diabetes and cardiovascular complications. In summary, the text covers a broad range of topics related to general aspects of diabetes, particularly diabetic cardiomyopathy, including pathophysiology, complications, management, and various treatment options.

We are taking this opportunity to offer our sincere thanks to all the eminent authors for their extraordinary contributions. The time and efforts of both Dr. Vijayan Elimban and Ms. Eva Little of the Institute of Cardiovascular Sciences at St. Boniface Hospital Research, University of Manitoba, are gratefully acknowledged. Our gratitude is also extended to Ms. Rita Beck and the staff at the Springer Media, New York, for their understanding of problems associated with the preparation of this book.

Ankara, Turkey Belma Turan
Winnipeg, MB, Canada Naranjan S. Dhalla

Contents

Part II Cellular Mechanisms of Diabetic Cardiomyopathy

Part III Therapeutic Aspects of Diabetic Cardiomyopathy

Contributors

Mariane Abdillahi Diabetes Research Program, Department of Medicine, New York University Langone Medical Center, New York, NY, USA

Adriana Adameova Department of Pharmacology and Toxicology, Faculty of Pharmacy, Comenius University, Bratislava, Slovakia

Veeranjaneyulu Addepalli Department of Pharmacology, SPP School of Pharmacy and Technology Management, SVKM's NMIMS, Mumbai, MH, India

Ernest Adeghate Department of Human Anatomy, Faculty of Medicine and Health Sciences, United Arab Emirates University, Al-Ain, United Arab Emirates

Saghir Akhtar Kuwait University, Safat, Kuwait

Takahide Arai Division of Cardiology, Department of Internal Medicine, Keio University School of Medicine, Tokyo, Japan

Murat Ayaz Department of Biophysics, Faculty of Medicine, Selcuk University, Konya, Turkey

Ibrahim F. Benter Department of Pharmacology and Toxicology, Kuwait University, Safat, Kuwait

Lokesh Kumar Bhatt Department of Pharmacology, Dr. BhanubenNanavati College of Pharmacy, Mumbai, India

Keshore Bidasee Department of Pharmacology and Experimental Therapeutics, Nebraska Medical Centre, Omaha, NE, USA

Kimberly-Ann M. Bordun Departments of Internal Medicine and Physiology, Faculty of Medicine, Institute of Cardiovascular Sciences, St. Boniface General Hospital Research Centre, University of Manitoba, Winnipeg, MB, Canada

Lindsay Brown Department of Biological and Physical Sciences, University of Southern Queensland, Toowoomba, QLD, Australia

Oscar Casis Departamento de Fisiología, Facultad de Farmacia, University of the Basque Country (UPV/EHU), Vitoria, Spain

Kai-Chun Cheng Department of Psychosomatic Internal Medicine, Kagoshima University Graduate School of Medical and Dental Sciences, Kagoshima, Japan

Juei-Tang Cheng Institute of Medical Sciences, Chang Jung Christian University, Tainan, Taiwan

Estelle R. Simo Cheyou Laboratory of Cell Signaling, Montreal Diabetes Research Centre, Centre de recherche-Centre hospitalier de l'Université de Montréal (CRCHUM) –Technopole Angus, University of Montreal, Montreal, QC, Canada

Department of Nutrition, University of Montreal, Montreal, QC, Canada

Arthur H. Cunha-Volpato Institute of Cardiovascular Sciences and Departments of Physiology, University of Manitoba, St. Boniface Hospital Research, Winnipeg, MB, Canada

Saloni Daftardar Department of Pharmacology, SPP School of Pharmacy & Technology Management, SVKM's NMIMS, Mumbai, India

Samarjit Das Department of Cardiac-Pathology, Johns Hopkins University, Baltimore, MD, USA

Department of Pathology, Cardiac Division, Johns Hopkins University, Baltimore, MD, USA

Naranjan S. Dhalla Institute of Cardiovascular Sciences and Departments of Physiology, St. Boniface Hospital Research, University of Manitoba, Winnipeg, MB, Canada

Vijayan Elimban Institute of Cardiovascular Sciences and Department of Physiology, University of Manitoba, St. Boniface Hospital Research, Winnipeg, MB, Canada

Institute of Cardiovascular Sciences, St. Boniface Hospital Research, Winnipeg, MB, Canada

Jennifer E. Enns St. Boniface Hospital Research Centre, University of Manitoba, Winnipeg, Canada

Departments of Physiology, University of Manitoba, Winnipeg, Canada

Keiichi Fukuda Department of Cardiology, Keio University School of Medicine, Tokyo, Japan

Mónica Gallego Lascaray Research Center, University of the Basque Country (UPV/EHU), Vitoria, Spain

Dipti Gatne Departments of Pharmaceutical Analysis, SPP School of Pharmacy and Technology Management, SVKM's NMIMS, Mumbai, India

Ramesh K. Goyal Institute of Life Science, Ahmedabad University, Navrangpura, Ahmedabad, India

Frank C. Howarth Department of Physiology, United Arab Emirates University, Al-Ain, United Arab Emirates

Masaki Ieda Division of Cardiology, Department of Internal Medicine, Keio University School of Medicine, Tokyo, Japan

Tahreem Iqbal School of Pharmacy and Biomedical Sciences, University of Central Lancashire, Preston, Lancashire, UK

Davinder S. Jassal Departments of Internal Medicine and Physiology, Faculty of Medicine, Institute of Cardiovascular Sciences, St. Boniface General Hospital Research Centre, University of Manitoba, Winnipeg, MB, Canada

Jasbir S. Juggi Kuwait University, Safat, Kuwait

Tengku Ain Kamalden University of Malaya Eye Research Centre, University of Malaya, Kuala Lumpur, Malaysia

Ginpreet Kaur Department of Pharmacology, SPP School of Pharmacy & Technology Management, SVKM's NMIMS, Mumbai, India

Saisudha Koka VCU Pauley Heart Center, Division of Cardiology, Department of Internal Medicine, Virginia Commonwealth University, Richmond, VA, USA

Rakesh C. Kukreja Pauley Heart Center, Division of Cardiology, Virginia Commonwealth University, Richmond, VA, USA

Ying-Xiao Li Department of Psychosomatic Internal Medicine, Kagoshima University Graduate School of Medical and Dental Sciences, Kagoshima, Japan

John H. McNeill Faculty of Pharmaceutical Sciences, University of British Columbia, Vancouver, BC, Canada

Paras K. Mishra Department of Cellular and Integrative Physiology, University of Nebraska Medical Center, Omaha, NE, USA

Mukesh Nandave SPP School of Pharmacy and Technology Management, SVKM's NMIMS, Mumbai, India

Arzu Onay-Besikci Department of Pharmacology, Faculty of Pharmacy, Ankara University, Ankara, Turkey

Isil Ozakca Department of Pharmacology, Ankara University, Ankara, Turkey

A. Tanju Ozcelikay Department of Pharmacology, Ankara University, Ankara, Turkey

Semir Ozdemir Faculty of Medicine, Department of Biophysics, Akdeniz University, Antalya, Turkey

Adriana Frota P. Pinto Institute of Cardiovascular Sciences and Department of Physiology, University of Manitoba, St. Boniface Hospital Research, Winnipeg, MB, Canada

Ravichandran Ramasamy Diabetes Research Program, Department of Medicine, New York University Langone Medical Center, New York, NY, USA

Anup Ramdhave SPP School of Pharmacy and Technology Management, SVKM's NMIMS, Mumbai, India

Vijay Sharma Department of Pathology, Aberdeen Royal Infirmary, Aberdeen, Scotland

Division of Applied Medicine, School of Medicine and Dentistry, The University of Aberdeen, Aberdeen, Scotland

BMJ Evidence Centre, BMJ Group, BMA House, London, UK

Jaipaul Singh School of Pharmacy and Biomedical Sciences, University of Central Lancashire, Preston, Lancashire, UK

Ashok K. Srivastava Laboratory of Cell Signaling, Montreal Diabetes Research Centre, Centre de recherche-Centre hospitalier de l'Université de Montréal (CRCHUM) –Technopole Angus, University of Montreal, Montreal, QC, Canada

Department of Medicine, University of Montreal, Montreal, QC, Canada

Department of Nutrition, University of Montreal, Montreal, QC, Canada

Paramjit S. Tappia Asper Clinical Research Institute, St. Boniface Hospital Research Centre, Winnipeg, Manitoba, Canada

Carla G. Taylor St. Boniface Hospital Research Centre, University of Manitoba, Winnipeg, Canada

Departments of Physiology, University of Manitoba, Winnipeg, Canada

Human Nutritional Sciences, University of Manitoba, Winnipeg, Canada

Belma Turan Department of Biophysics, Faculty of Medicine, University of Ankara, Ankara, Turkey

Suresh C. Tyagi Department of Physiology & Biophysics, University of Louisville, Louisville, KY, USA

Philip J. Welsby School of Pharmacy and Biomedical Sciences, University of Central Lancashire, Preston, Lancashire, UK

Weng-Yew Wong School of Biomedical Sciences, University of Queensland, Brisbane, QLD, Australia

Yan-Jun Xu Institute of Cardiovascular Sciences and Departments of Physiology, University of Manitoba, St. Boniface Hospital Research, Winnipeg, MB, Canada

Nazmi Yaras Faculty of Medicine, Department of Biophysics, Akdeniz University, Antalya, Turkey

Mariam H.M. Yousif Department of Pharmacology and Toxicology, Kuwait University, Safat, Kuwait

Peter Zahradka St. Boniface Hospital Research Centre, University of Manitoba, Winnipeg, Canada

Departments of Physiology, University of Manitoba, Winnipeg, Canada

Human Nutritional Sciences, University of Manitoba, Winnipeg, Canada

Molecular Physiology Laboratory, Canadian Centre for Agri-food Research in Health and Medicine, St. Boniface Hospital Research Centre, Winnipeg, Canada

Part I
Metabolic Derangements in Diabetic Cardiomyopathy

Metabolic Alterations in Diabetic Cardiomyopathy

Kimberly-Ann M. Bordun, Davinder S. Jassal, and Naranjan S. Dhalla

Abstract Diabetes mellitus causes cardiomyopathy in diabetic patients and is an important and dominant risk factor for congestive heart failure. With the growing prevalence of diabetes in Canada and throughout the world, diabetic cardiomyopathy is a significant public health issue. Diabetic cardiomyopathy has been defined as myocardial dysfunction that occurs in a diabetic milieu independent of identified causes such as coronary atherosclerosis, hypertension, or valvular heart disease. Alterations in ventricular structure as well as left ventricular systolic and diastolic dysfunctions have been reported in diabetic patients despite well-controlled glycemic levels and disease free coronary vasculature. Metabolic abnormalities such as hyperlipidemia, hyperinsulinemia, and hyperglycemia predispose the heart to cellular, structural, and functional alterations that manifest as the cardiac phenotype observed in this diabetic population. These mechanisms are likely to act synergistically and are believed to potentiate one another. Hyperglycemia is an essential factor in the development of cardiomyopathy and exerts its effects by altering protein kinase C, increasing oxidative stress and causing abnormalities in lipid

K.-A.M. Bordun
Departments of Internal Medicine and Physiology, Faculty of Medicine,
Institute of Cardiovascular Sciences, St. Boniface General Hospital Research Centre,
University of Manitoba, Winnipeg, MB, Canada

D.S. Jassal, M.D., FACC, FRCPC (✉)
Departments of Internal Medicine and Physiology, Faculty of Medicine,
Institute of Cardiovascular Sciences, St. Boniface General Hospital Research Centre,
University of Manitoba, Rm Y3531, Bergen Cardiac Care Centre,
St. Boniface General Hospital, 409 Tache Avenue, Winnipeg, MB R2H 2A6, Canada
e-mail: djassal@sbgh.mb.ca

N.S. Dhalla, Ph.D.
Institute of Cardiovascular Sciences, St. Boniface Hospital Research, University of Manitoba,
351 Tache Avenue, Winnipeg, MB R2H 2A6, Canada
e-mail: nsdhalla@sbrc.ca

B. Turan and N.S. Dhalla (eds.), *Diabetic Cardiomyopathy: Biochemical and Molecular Mechanisms*, Advances in Biochemistry in Health and Disease 9,
DOI 10.1007/978-1-4614-9317-4_1, © Springer Science+Business Media New York 2014

metabolism and calcium ion homeostasis. Regardless of the extensive information available on diabetic cardiomyopathy, translational research is scarce due to the lack of clinical trials, and therefore, much of our current knowledge is extrapolated from animal models. This review focuses on illustrating the various metabolic alterations that contribute to the development and progression of diabetic cardiomyopathy.

Keywords Diabetic cardiomyopathy • Diabetes mellitus • Diastolic function • Systolic function • Hyperlipidemia • Hyperinsulinemia • Hyperglycemia • Oxidative stress

1 Introduction

Diabetes mellitus is a global epidemic problem, affecting approximately 285 million people worldwide, with a prevalence expected to surpass 435 million individuals by 2030 [1]. In Canada, more than three million individuals are currently living with diabetes and this number is predicted to reach 3.7 million by the year 2020 [2]. Cardiovascular (CV) disease represents the major cause of morbidity and mortality in diabetic patients, accounting for nearly 70 % of related heart failure cases [3]. It is generally held that diabetes causes myocardial contractile dysfunction through accelerated atherosclerosis and hypertension [4, 5]. Diabetic patients, when compared with non-disease subjects, are two to four times more likely to experience CV events, due to micro- and macrovascular atherosclerosis, which is often exacerbated by the presence of concomitant CV risk factors including hypertension, dyslipidemia, and activation of neurohormonal and inflammatory mechanisms [6]. In the past four decades the pathologic association between diabetes and cardiac disease has been evaluated through a number of epidemiologic studies, including; (i) The Framingham Study [7]; (ii) United Kingdom Prospective Diabetes Study [8]; (iii) Cardiovascular Health Study [9]; and (iv) Euro Heart Failure Surveys [10]. Together, these trials illustrate that the presence of diabetes and impaired glycemic control, independent of atherosclerotic CV disease, increase the risk of developing incident heart failure [11]. Despite the clinical significance of these findings, the complex and multifactorial nature of the biochemical alterations that prelude myocardial dysfunction in the diabetic heart remain incompletely understood [4]. This review focuses on illustrating various metabolic alterations that lead to maladaptive myocardial structure and function, as depicted in a diabetic milieu.

2 Diabetes Mellitus

Diabetes mellitus is a multivariate metabolic condition that is distinguished by cellular dysfunction in the transport and utilization of glucose. Type 1 diabetes (T1D), previously called insulin dependent diabetes, is caused by T lymphocyte-mediated

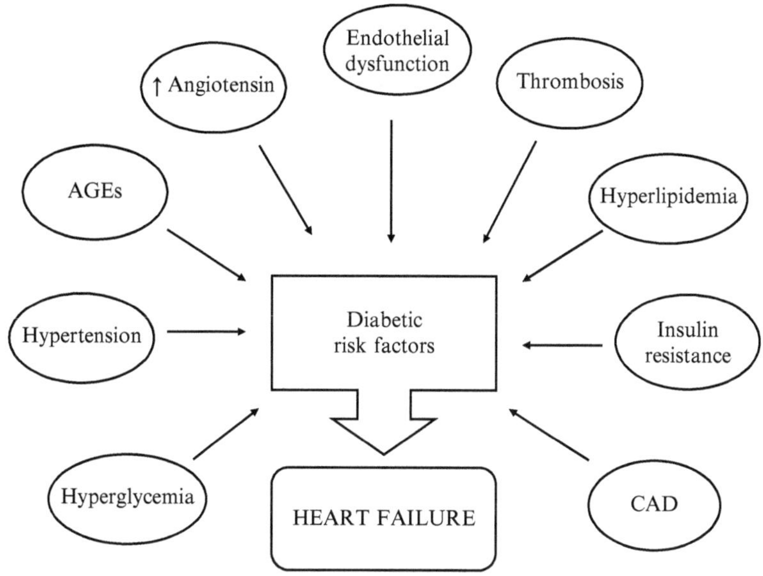

Fig. 1 Risk factors for heart failure which are associated with diabetes. *AGEs* advanced glycation end products, *CAD* coronary artery disease

autoimmune destruction of the pancreatic β-cells, which results in insufficient insulin production and reduced glucose utilization [12]. Insulin independent diabetes, commonly known as type 2 diabetes (T2D), develops due to decompensatory mechanisms invoked by insulin resistance [13]. This maladaptive response leads to hyperinsulinemia resulting in elevated blood glucose levels (hyperglycemia), impaired cellular glycolysis and pyruvate oxidation [14]. A logical explanation for these detrimental myocardial alterations is the fact that diabetes is associated with known CV risk factors including obesity, hyperlipidemia, thrombosis, myocardial infarction, hypertension, activation of multiple hormone and cytokine systems, autonomic neuropathy, endothelial dysfunction and coronary artery disease [15] (Fig. 1). Of these identified confounding factors, it remains uncertain which pathologies are most significant in the development of diabetes-mediated heart failure.

3 Diabetic Cardiomyopathy

Diabetic cardiomyopathy, a distinct clinical condition that develops in diabetic patients, is defined by the presence of myocardial dysfunction in the absence of coronary atherosclerosis, hypertension, and valvular heart disease [4, 5, 11, 16, 17]. Diabetic cardiomyopathy is an early complication, initially presenting as diastolic dysfunction with subsequent abnormalities in systolic function [18]. Rubler and colleagues [19] originally described this disease process in 1972 after diabetic patients, free of coronary artery disease (CAD), demonstrated a shortened left ventricular ejection time, longer pre-ejection period, and increased end-diastolic pressure [20]. This novel form of myocardial dysfunction was attributed to increased ventricular wall stiffness, reduced cardiac contractility, and longer isovolumic relaxation [21]. More recently, Bertoni et al. [22] evaluated 104, 395 cardiomyopathic patients in which 45, 000 cases were classified as idiopathic. These individuals were 75 % more likely to have diabetes than control subjects and this morbidity remained significantly associated with idiopathic dilated cardiomyopathy after adjustments in age, hypertension, sex, race, cholesterol level, and history of CAD [5, 22]. The presence of microvascular complications proved to be the strongest predictor of developing cardiomyopathy in this diabetic population, suggesting a possible correlation between duration of hyperglycemia and severity of myocardial injury [5].

Although diabetic cardiomyopathy is characterized by several features common to both T1D and T2D patients, murine models have recently been utilized to differentiate the mechanisms and effects caused by each disease state [23]. The manifestation of diabetes is evoked by three central metabolic disturbances: (i) *hyperlipidemia* (inclucing triglycerides and non-esterified fatty acids [NEFAs]); (ii) *hyperinsulinemia* and pancreatic β-cell failure; and (iii) *hyperglycemia* [4]. Type 1 diabetes, which is identified by early- as opposed to late-onset of hyperglycemia, is not accompanied by a period of hyperinsulinemia. A review of the literature also indicates discrepancies in contractile function, showing increased prevalence of systolic dysfunction in T1D [24]. In a pre-stage T2D condition (insulin-resistance without T2D), animals presented with collagen deposition, LV fibrosis and impaired ventricular filling, which consequently lead to early diastolic dysfunction [25]. A plausible implication for the marked differences in T1D and T2D myocardial dysfunction is the role of insulin resistance, shown to reduce protective effects in ischemic/reperfusion injury [26, 27]. Insulin resistance and hyperglycemia, both predominant defects in T2D, are considered central catalysts in the induction of adaptive and maladaptive responses contributing to heart failure [6]. These abnormalities and other pathological mechanisms including, hypertriglyceridemia, reactive oxygen species (ROS), and abnormal cardiac fuel usage are believed to act synergistically and exacerbate the cardiac phenotype present in diabetes. A clear understanding of the cellular and molecular effects of these metabolic disturbances on cardiomyocytes is essential to accurately depict the ensuing structural and functional changes associated with diabetic cardiomyopathy [4, 6, 15, 23].

4 Myocardial Alterations in Structure

Data from experimental, epidemiologic, and clinical investigations have collectively demonstrated that diabetes results in alterations in myocardial function, structure, and dimension [28]. A report from the Strong Heart Study revealed that diabetic patients, compared with non-diabetic individuals, presented with higher left ventricular mass and wall thickness, increased arterial stiffness and systolic dysfunction [29]. These cardiac alterations were independent of body mass index and blood pressure. Results from the Framingham Heart Study [7] demonstrated that LV mass and wall thickness were directly proportional to the degree of glucose intolerance and obesity, with increased prevalence among female diabetic patients. LV hypertrophy has shown to be a strong predictor of prognosis in high-risk individuals with coronary artery disease [30], diabetes [31], heart failure [32], and obesity [33]. Two recent epidemiologic trials, evaluating Japanese and Swedish populations, observed correlations between metabolic syndrome, insulin resistance and arterial stiffness and increased LV mass [34, 35]. The early stages of diabetic cardiomyopathy involve the development of concentric LV hypertrophy through cellular structural insults, which lead to abnormal LV filling and diastolic dysfunction [36]. Progression of diabetic cardiomyopathy results in eccentric hypertrophy and systolic dysfunction, frequently undetected initially due to preserved LV ejection fraction maintained by compensatory cardiac mechanisms [37]. These structural changes and vascular abnormalities have a synergistic effect in the accelerated presentation of clinical morbidities and cardiac failure [36].

5 Alterations in Cardiac Function

LV dysfunction in the setting of diabetes is identified as the earliest manifestation of diabetic cardiomyopathy and is characterized by increased diastolic stiffness and relaxation disturbances. By definition, diastole occurs during cardiac mechanical and electrophysiological inactivity, while the myocardium generates effort and energy for ensuing the systolic contraction [38]. Clinically, diastolic dysfunction refers to the echocardiographic finding of abnormal LV deceleration and relaxation time in the absence of symptomatic heart failure [39]. Diastolic dysfunction is distinguished by a delayed and extended diastolic phase accompanied by LV stiffness, which collectively comprise the passive diastolic LV compliance evident in heart failure [40]. The pathologic extent of LV dysfunction is traditionally assessed by echocardiographic analysis of the transmitral inflow velocity profile [41, 42]. Furthermore, cardiac catheterization, conventional 2D echocardiography, and M-mode Doppler imaging can also be used for the evaluation of LV isovolumic relaxation rhythm and contraction time [43]. Impaired LV diastolic function is reported to exist in 27–69 % of asymptomatic diabetic patients [39, 44] and accounts for 30 % of clinical heart failure cases in this patient population [45]. In a study conducted by Liu and colleagues [46], the degree of abnormal ventricular relaxation was found to be associated with increased glucose and hemoglobin A_{1c} levels. The coexistence of hypertension in a diabetic milieu accelerates the underlying pathophysiological process, resulting in more severe abnormal relaxation, LV hypertrophy, systolic ventricular dysfunction, and increased risk of congestive heart failure [39, 41, 42]. Diastolic heart failure is identified clinically as heart failure with preserved LV ejection fraction [39]. Patients with diastolic heart failure, when compared to healthy subjects, demonstrate a 5–8 % increased risk of cardiovascular mortality [47, 48]. A differentiation between the mechanisms responsible for increased diastolic stiffness observed in systolic and diastolic heart failure was illustrated by a recent study. The most important contributor to high LV diastolic stiffness in diabetic individuals with diastolic heart failure is elevated resting tension of hypertrophied cardiomyocytes. Comparably, fibrosis and deposition of advanced glycation end-products (AGEs) contribute most significantly to increased LV diastolic stiffness in diabetic systolic heart failure [49].

In current clinical practice, systolic dysfunction is evaluated according to the severity of depressed LV ejection fraction and is suggested to be the result of advanced diabetic cardiomyopathy leading to reduced contractile function and subsequent congestive heart failure [50]. Transthoracic echocardiography has traditionally been used for the early evaluation of cardiac performance in diabetic patients free of heart failure, to establish a relationship between prolonged pre-ejection performance, shortened ejection period and reduced resting LV ejection fraction with diminished systolic function [51]. Diabetic patients have also demonstrated to exhibit an abnormal response to exercise, expressed as decreased LV ejection fraction caused by reduced cardiac reserve [52]. Mbanya et al. [53] evaluated T2D normotensive diabetic patients and found that although 55 % of these individuals had systolic dysfunction, only 8 % had electrocardiographic changes suggestive of cardiac ischemia. Recent studies have revealed that conventional diagnostic use of 2D echocardiography may not be proficient in the detection of subtle systolic LV impairment, as it focuses on radial rather than longitudinal contraction [54]. In diabetic patients with heart failure and preserved ejection fraction, long-axis systolic dysfunction is associated with increased radial thickening and muscle mass [55]. Advanced development of sensitive techniques such as myocardial tissue Doppler imaging, strain and strain rate have allowed for measurement of regional and long-axis function in the left and right ventricles [56]. Longitudinal function of the myocardial tissue is becoming recognized as a superior prognostic indicator of ejection fraction, particularly in diabetic heart failure patients with preserved ejection fraction [57]. Several investigative groups have reported the detection of subtle abnormalities in systolic function, with the use of sensitive methods, in patients with overt diastolic dysfunction [58–60]. This discovery has led to questions of whether diastolic dysfunction exists in isolation at all [61], and of the relevance of these subtle systolic abnormalities in the presence of diastolic dysfunction [62]. Additionally, right ventricular function, although underreported in the literature, has proven to be an important independent predictor of poor cardiovascular outcome [63]. Some studies in a diabetic diseased state have shown right ventricular diastolic dysfunction [64] as well as systolic dysfunction [65].

6 Pathophysiologic Mechanisms of Diabetic Cardiomyopathy

The diagnosis of diabetic cardiomyopathy involves myocardial dysfunction in the absence of obstructive epicardial coronary artery disease and is characterized by diastolic LV dysfunction and myocardial fibrosis [66]. This pathophysiologic process is multifactorial and thought to be driven by metabolic abnormalities associated with

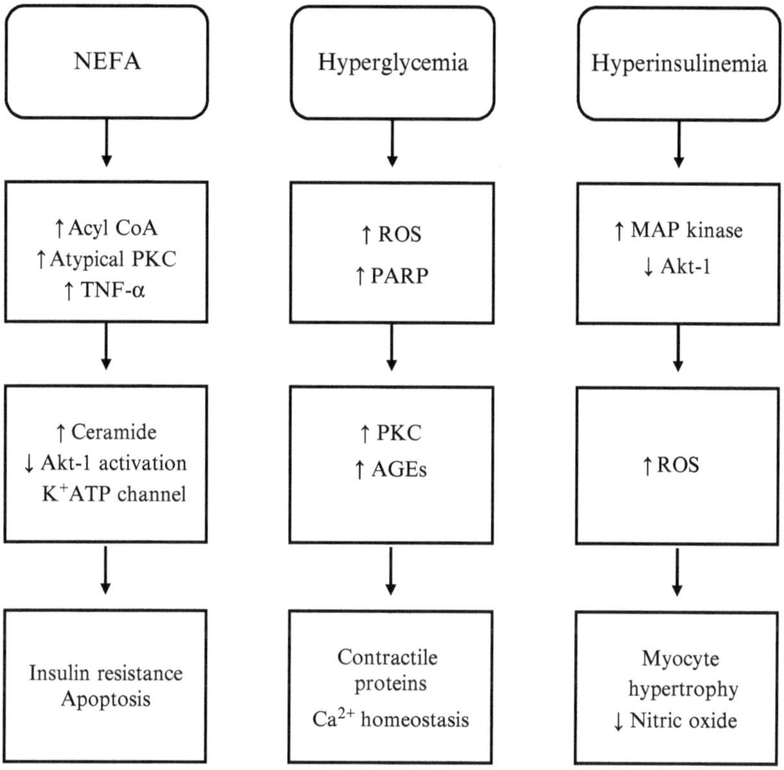

Fig. 2 Progression of diabetic cardiomyopathy leading to the abnormal cardiac phenotype. *Acyl CoA* acetyl coenzyme A, *AGEs* advanced glycation end-products, *Akt-1* a serine/threonine kinase regulated by phosphatidylinositol 3,4,5-triphosphate (*PIP3*) and phosphatidylinositol 3,4-biphosphate, *MAP* mitogen-activated protein, *NEFA* nonesterfied fatty acid, *PARP* poly adenosine diphosphate ribose polymerase, *PKC* protein kinase C, *ROS* reactive oxygen species, *TNF-α* tumor necrosis factor-α

systemic diabetes. Although the pathogenesis of diabetic cardiomyopathy remains incompletely understood, several hypotheses have been proposed including autonomic dysfunction, abnormalities in Ca^{2+} homeostasis, increased levels of circulating cytokines (tumor necrosis factor [TNF]-α), variations in structural proteins, and interstitial fibrosis [36]. In the diabetic milieu, the presence of hyperlipidemia, hyperinsulinemia, and hyperglycemia predispose the heart to cellular, structural, and functional alterations, leading to this diabetic myocardial phenotype (Fig. 2).

7 Hyperlipidemia and Increased NEFAs

One of the main abnormalities in diabetes, particularly in patients with visceral adiposity, is the increased release of triglycerides and non-esterified fatty acids (NEFAs) from the adipose tissue. In a healthy, non-diabetic cardiac model, energy is derived

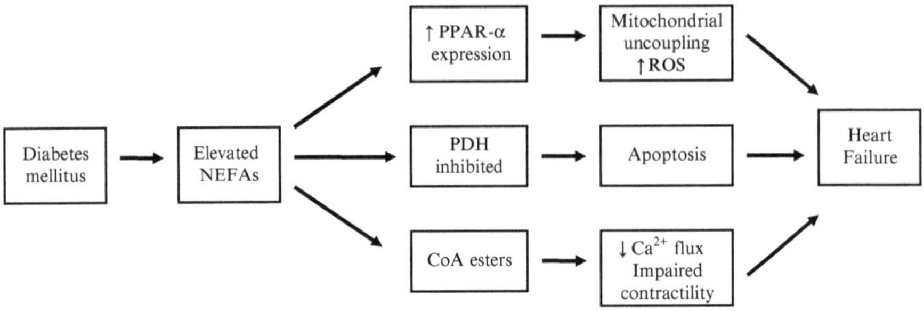

Fig. 3 Altered myocardial metabolism and its contribution to the pathogenesis of diabetic cardiomyopathy. *CoA* coenzyme A, *NEFAs* non-esterified fatty acids, *PDH* pyruvate dehydrogenase, *PPAR-α* peroxisome proliferator-activated receptor alpha, *ROS* reactive oxygen species

approximately in equivalent proportions from glucose metabolism and free fatty acids [67]. It is well known in the literature that impaired myocardial energy metabolism occurs as a result of elevated concentrations of circulating NEFA, inducing the maladaptive shift from glucose to NEFA uptake and utilization in the myocardium [68–70]. By using positron emission tomography, [71] increased myocardial FA use and reduced glucose oxidation in T1D diabetes have been demonstrated, while another study comprising of insulin resistant obese women has revealed an increase in FA oxidation and myocardial oxygen consumption with decreased cardiac efficiency [72]. In the diabetic myocardium, decreased glucose use results from depleted glucose transporter proteins, glucose transporter-1 (GLUT-1) and GLUT-4. Cardiac myocytes, in the presence of high circulating levels of NEFA, up-regulate the expression of enzymes necessary for their disposal through mitochondrial β-oxidation. As compared with glucose oxidation, β-oxidation yields more ATP but requires large amounts of oxygen, therefore resulting in reduced cardiac efficiency [73, 74]. Consequently, NEFA myocardial uptake may exceed NEFA β-oxidation capacity, in which lipids in the form of triglyceride accumulate in the myocytes [75, 76]. Lipotoxicity, which describes the accumulation of NEFAs in non-adipose tissues, initially acts as a cardiac protective mechanism, providing fuel storage for subsequent oxidation and impeding the toxic effects of lipid metabolites [77, 78]. Eventually, this chronic imbalance between lipid storage and lipid oxidation leads to morphological changes [79, 80] and impaired myocardial function [81, 82]. Unger [83] was first to introduce the concept that only adipocytes are competent for extensive storage of lipids, suggesting that all other cell types are vulnerable to lipotoxic injury. Furthermore, increased NEFAs inhibit pyruvate dehydrogenase and may directly contribute to cell death through the generation of the sphingolipid ceramide (Fig. 3). This toxic lipid product is generated by the reaction between palmitoyl-CoA and serine, and is facilitated by the cytokine, TNF-α [84]. As intracellular FA content increases, the proportion of intracellular FA derivatives such as, acyl-CoA, diacylglycerol, and ceramide also increase [28]. Some studies have demonstrated that increased ceramide levels can induce cellular apoptosis by the induction of nuclear factor κB, caspase 3 activation, and cytochrome c release [85]. Furthermore, ceramide inhibits DNA repair by blocking poly (ADP ribose) polymerase and is associated with increased oxidative stress and reduced contractile function [86].

Peroxisome proliferator-activated receptors (PPARs), which are transcriptional factors activated by increased FA levels, play an active role in cardiac lipid metabolism due to the fact that their target genes participate in lipid metabolism. The ligand-activated transcription factor PPAR-α is a key regulator of FA metabolism and is primarily expressed in tissues that rely on fatty acid oxidation for energy, such as heart, liver, kidney, and skeletal muscle. Various target genes such as heart-type fatty acid-binding protein, lipoprotein lipase and CD36 are regulated by PPAR-α and participate in lipid metabolism. Additionally, PPAR-α increases the expression of pyruvate dehydrogenase kinase 4, thus reducing glucose oxidation, and stimulating mitochondrial fatty acid uptake. Increased FA oxidation leads to augmented oxygen consumption, generation of reactive oxygen species (ROS), and the promotion of mitochondrial uncoupling, which contributes to reduced myocardial high-energy reserves and contractile dysfunction [87]. Transgenic mice overexpressing PPAR-α show increased myocardial fatty acid oxidation rates, decreased glucose uptake and oxidation, and LV abnormalities [88]. Metabolic myocardial alterations, as present in the diabetic milieu, are mainly caused by malfunctions of acetyl-CoA carboxylase, CPT1, and pyruvate dehydrogenase. Through the overexpression of LPL, stimulation of PPAR-α expression, or synthesis of long-chain CoA, FA metabolism is enhanced, manifesting as the cardiac phenotype observed in diabetic cardiomyopathy [88–90].

The accumulation of triglycerides and NEFAs are critical precursors in the development of cellular insulin resistance and also has been seen to contribute to myocardial contractile dysfunction. Alterations in the contractile state occur due to increasing NEFA influx into the myocardium [91], causing increased fatty acyl coenzyme A (CoA) esters to activate KATP channels [92]; this results in a shortened action potential, reduced trans-sarcolemmal calcium flux, and consequent impairment in myocardial contractility. Increased secretion and activation of pro-inflammatory adipokines and cytokines from inflamed adipose tissue are also known to result in low-grade inflammation [93]. Due to the high levels of circulating NEFAs, insulin-mediated uptake of glucose is impaired through inhibition of insulin receptor substrate and protein kinase B. Atypical protein kinase C (PKC) θ, a serine/threonine kinase, is activated by NEFAs and through its phosphorylation, activates IκB kinase [94]. Serine residues of insulin receptor substrate-1 are phosphorylated by IκB kinase, which inhibits the binding of SH2 domains of the p85 regulatory subunit of phosphatidylinositol 3-kinase (PI3K), therefore, inhibiting insulin signal transduction [95, 96]. Additionally, increased intracellular NEFAs, independent from IRS-1/PI3K activation, can alter insulin signaling by inhibiting Akt-1. Akt-1 activation is dependent on the generation of phosphatidylinositol 3,4,5-triphosphate (PtdIns(3,4,5)P_3) to bind the N-terminal pleckstrin domain and activate membrane bound kinases that phosphorylate serine and threonine residues [97, 98]. NEFAs serve as natural ligands for PPAR-γ, which allows this nuclear receptor to induce the upregulation of phosphatase and tensin homolog, leading to the dephosphorylation of PtdIns(3,4,5)P_3 and subsequent prevention of Akt-1 activation [99]. Therefore, these findings indicate that NEFAs are a key contributor to the induction of cellular insulin resistance, reduced myocardial contractility and cardiomyocyte apoptosis.

8 Insulin Resistance and Hyperinsulinemia

Cellular insulin resistance precedes the clinical diagnosis of diabetes and involves the compensatory increase in plasma insulin levels in order to maintain glucose homeostasis, particularly in skeletal muscle and liver [100]. Myocardial insulin resistance when evaluated in vivo, is characterized by reduced myocardial glucose uptake in the absence of coronary artery disease [101]. Bonora et al. [102] have demonstrated the presence of reduced insulin sensitivity, despite well-controlled conditions of T2D free of other co-morbidities. In diabetic cardiomyopathy, cardiac hypertrophy is suggested to result due to systemic hyperinsulinemia that accentuates cellular insulin action in the myocardium, which is free of apparent cellular insulin resistance [103–105]. Previous studies have illustrated the pathophysiologic significance of insulin resistance and its involvement in the development of myocardial contractile dysfunction [106] and LV diastolic abnormalities, independent of hypertension, obesity and LV hypertrophy [107, 108]. Hyperinsulinemia mediates cardiomyocyte hypertrophy through three chief cellular mechanisms. Acutely, insulin stimulates the growth of cardiomyocytes through the same PI3Kα/Akt-1 pathway in which the uptake of glucose is mediated. Glycogen synthases kinase-3β, an inhibitor of nuclear transcription that controls the hypertrophic process, is phosphorylated and subsequently inactivated by Akt-1 [109]. Akt-1 also activates the mammalian target of rapamycin (mTOR) which activates the p70 ribosomal subunit S6kinase-1, resulting in increased protein synthesis [110, 111]. Activation of this mammalian target of rapamycin complex 1 (mTORC1)/S6 kinase 1 pathway, due to chronic insulin resistance and inflammation-induced oxidative stress, is known as one of the prime causes of insulin resistance in diabetes [112]. In chronic conditions, hyperinsulinemia augments myocardial Akt-1 activation indirectly by increasing sympathetic nervous system (SNS) activity [113, 114]. The overstimulation of the SNS, evoked by elevated insulin levels and insulin resistance, also leads to increased hypertension [81]. It has been shown that protein kinase A and Ca^{2+}-calmodulin dependent kinase (CaMK) activate β_2 -adrenergic receptors, which leads to Akt-1 activation in cardiac myocytes [115]. Furthermore, there are various insulin-mediated pathways, apart from Akt-1, that may be active in the development of myocardial hypertrophy, such as the extracellular signal-regulated kinase (ERK)/mitogen-activated protein (MAP) kinase pathways [116]. In both a murine [117] and clinical model [118], leptin has been shown to stimulate myocyte hyperplasia through PI3K- and ERK1/2-dependent mechanisms. Wang and colleagues [119] demonstrated significant cellular evidence for insulin-induced activation of the p38 MAP kinase pathway, as well as prenylation of Rho and Ras in a setting of hyperinsulinemia, manifesting as myocyte hypertrophy and expansion of the extracellular matrix.

Despite the vast literature that exists on insulin resistance and hyperinsulinemia, the specific defect in insulin-stimulated GLUT4 translocation, which manifests as muscle insulin resistance, remains unknown. Therefore, further investigations are required to evaluate the mechanisms of NEFAs and their involvement in insulin-mediated glucose transport [120], as well as to discover novel pathways involved in this complex insulin-signaling cascade [121].

9 Hyperglycemia

Blood glucose concentration is an independent and continuous risk factor for cardiovascular disease, with hyperglycemia serving as an essential trigger of metabolic alterations. The progressive relationship between glucose levels and cardiovascular risk begins prior to achieved glucose threshold conventionally required for clinical diagnosis of diabetes [122]. A cohort study performed by Iribarren et al. [48] evaluated approximately 50,000 diabetic patients and found that each 1 % increase in glycosylated hemoglobin resulted in an 8 % increase in heart failure risk. An investigation encompassing two clinical trials for the evaluation of plasma glucose levels reported similar results. The study population, which included 31,546 high CV risk patients, demonstrated a modest increase in risk of heart failure hospitalization with each 1mmol/L increase in baseline fasting plasma glucose [22]. Increased glycemic levels have been associated with adaptive and maladaptive cardiac responses [39, 123], with resultant cardiomyopathy occurring due to the combination of increased insulin resistance and/or low insulin levels and reduced GLUT 1 and 4 proteins. Current knowledge suggests that hyperglycemia and poor glycemic control promote the progression of heart failure by accelerating atherosclerosis and excessive interstitial myocardial collagen accumulation, resulting in impaired systolic and diastolic function [124]. Hyperglycemia contributes to the development of cardiomyopathy by altering PKC, increasing oxidative stress, and up-regulating the activity of the angiotensin-converting enzyme (ACE) and angiotensin II (Fig. 4) [125, 126].

10 Oxidative Stress and ROS

Hyperglycemia leads to an increase in oxidative stress by exacerbating glucose oxidation and mitochondrial generation of ROS, which causes DNA damage and contributes to accelerated apoptosis. ROS encompasses a range of highly reactive oxygen base molecules, consisting of both free radicals (superoxide) and hydroxyl radicals as well as oxidizing agents such as hydrogen peroxide. In the setting of chronic diabetes, oxidative stress occurs when the production of ROS is greater than their degradation by antioxidant defenses. The resultant elevation of ROS has numerous harmful effects on the cardiovascular system via cellular damage by oxidation, disruption of vascular hemostasis through interference with nitric oxide

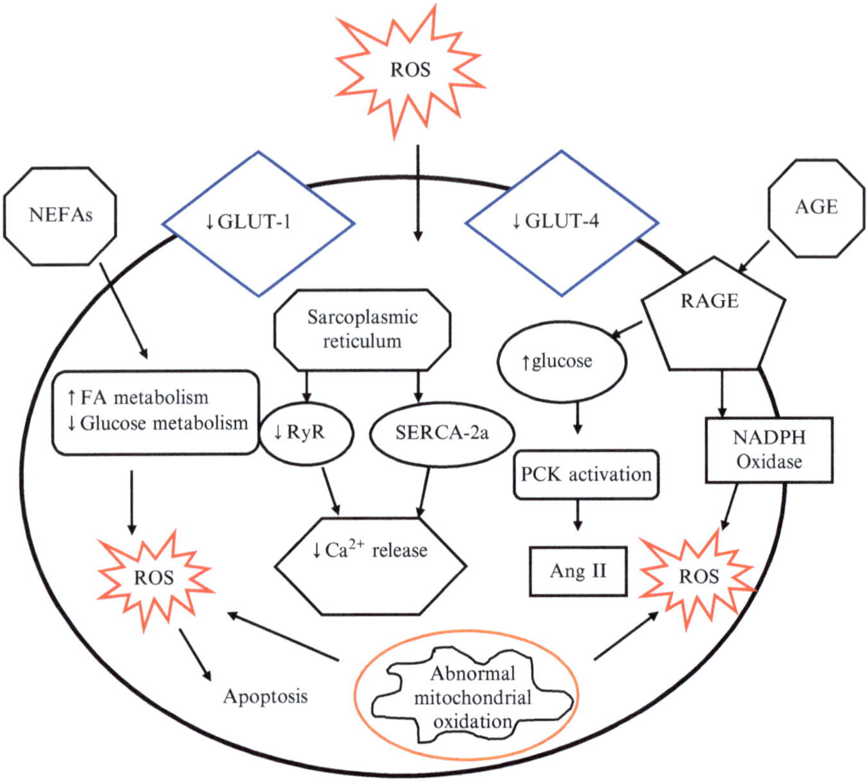

Fig. 4 Cellular alterations in diabetic cardiomyopathy. *AGE* advanced glycation end-products, *Ang II* angiotensin II, *GLUT* glucose transporters, *NADPH* nicotinamide adenine dinucleotide phosphate, *NEFAs* non-esterified fatty acids, *PKC* protein kinase C, *RAGE* advanced glycation end-products receptor, *ROS* reactive oxygen species, *RyR* ryanodine receptor, *SERCA-2a* sarcoplasmic reticulum Ca^{2+}-ATPase

(NO), and most recently, by the modulation of detrimental intracellular signaling pathways, also known as redox signaling [127]. ROS reacts with NO to form peroxynitrite species, which triggers a range of proatherosclerotic pathogenic events including vasoconstriction, enhanced leukocyte adherence, platelet activation, mitogenesis, oxidation, pro-trombotic state, impaired coagulation and vessel inflammation [128]. In diseased conditions, ROS is produced by multiple sources, rather than exclusively from the mitochondria. The group of nicotinamide adenine dinucleotide phosphate (NADPH) oxidase enzymes has recently been recognized as a contributor to ROS and for its involvement in redox signaling [129]. These enzymes act as catalysts for electron transfer from NADPH molecular oxygen, in which the generation of free radicals results. Through the interaction with various transcription factors, redox signaling influences the expression of growth-related genes and in turn affects contractile function [130]. Greater amounts of DNA are damaged due to augmented ROS production, which leads to activation of reparative enzyme poly (ADP ribose) polymerase (PARP) [131]. However, PARP has been shown to promote endothelial damage through the activation of NFκB and to induce the overexpression of vasoconstrictor endothelin 1 and its receptors [132]. PARP also mediates the ribosylation and inhibition of glyceraldehype phosphate dehydrogenase, which acts to divert glucose from its glycolytic pathway and into alternative biochemical pathways that are considered the mediators of hyperglycemia induced cellular injury. Several of these pathologic abnormalities include increases in AGEs, increased hexosamine and polyol flux, and activation of classical isoforms of PKC.

11 Alterations in Protein Kinase C (PKC)

PKC is a family of serine/threonine kinases that consists of approximately 12 members [133], and can be categorized into three groups: (i) conventional PKC; (ii) novel PKC; and (iii) typical PKC. Phospholipase C-induced intracellular Ca^{2+} and diacylglycerol levels activate PKC, in which triggers several transduction pathways involved in cardiovascular function such as increased endothelial cell permeability, apoptosis, angiogenesis, smooth muscle cell and cardiomyocyte growth and augmented contraction, cytokine activation and increased leukocyte adhesion [134]. The activation of PKC may also mediate several important biochemical alterations, including reduced blood flow and increased vascular permeability, basement membrane thickening, and extracellular matrix deposition [135, 136]. Previous in vitro studies have demonstrated that increased activation of PKC (specifically isoforms β and δ) are characterized by a reduction in the PI3K/Akt pathway, decreased NO bioavailability vascular endothelial growth factor, and enhanced MAPK [137]. Therefore, PKC is proven to affect endothelial homeostasis resulting in endothelial dysfunction and increased permeability [134]. Additionally, the glucose-medicated activation of PKC may directly contribute to increased ACE activity. ACE plays an active role in the conversion of inactive angiotensin I to angiotensin II (Ang II), which is a potent vasoconstrictor and hypertrophic peptide. Ang II contributes to several alterations in diabetes including fibrosis by stimulating extracellular matrix component synthesis, apoptosis/proliferation, vascular inflammation and oxidative damage [138]. A model of streptozotocin-induced diabetic rats has revealed the generation of fibrosis through the overexpression of connective tissue growth factor and PKC activation [139]. Furthermore, in transgenic PKC-β2 mice, hyperglycemic-induced PKC activation was shown to contribute to cardiac fibrosis by stimulating the expression of connective tissue growth factor [140]. In this transgenic murine model, the resultant cardiac phenotype was reminiscent of that seen in diabetic cardiomyopathy, characterized by early diastolic dysfunction, small vessel disease, myocardial hypertrophy and loss of cardiomyocytes [141]. The suggested mechanism responsible for these alterations is the PKC-β2-mediated phosphorylation of troponin I, which may decrease myofilament Ca^{2+} responsiveness [142]. Recent investigations have focused on the PKC-β inhibitor ruboxistaurin mesylate (LY333531) [135], as it has been reported to reverse cardiac hypertrophy, improve fractional shortening and diminish cardiac injury in a transgenic PKC-β cardiac murine model [141]. These findings suggest that PKC-β inhibition may represent a novel therapeutic strategy for the prevention of diabetes-associated cardiac dysfunction [143].

12 Advanced Glycation Endproducts (AGEs)

Experimental models of T1D [144] and T2D [145] diabetes have provided extensive evidence of the vital role of hyperglycemia in the pathogenetic development of cardiomyopathy. Hyperglycemia activates the myocardial renin-angiotensin-aldosterone system (RAAS) and endothelin systems [146], contributing to myocyte necrosis and fibrosis [147]. Within the myocardium, collagen type I and II are distributed predominately in the epicardial and perivascular regions, in myocardial hypertrophy, interstitial fibrosis, and capillary basal laminae thickening upon pathologic examination [148]. Glucose has the ability to react with collagen, proteins, lipids, and DNA in a process known as the Maillard reaction [149, 150], which yields Schiff's base (early glycation product), Amadori products (intermediate glycation products), and AGEs. During the Maillard reaction, reactive intermediate products, methylglyoxoal, 3-deoxyglucosone, and glyoxal are produced and believed to participate in the development of carbonyl stress and AGE formation [151]. AGEs have been implicated in the development of diabetic complications including cardiomyopathy [152] and affect myocardial function via two chief mechanisms. Firstly, AGEs covalently bind other AGEs forming cross-links between ECM proteins such as collagen, laminin, and elastin [152, 153]. These modifications impair the process of collagen degradation, which leads to increased collagen deposition and fibrosis [154], reduced compliance and flexibility of the tissue [155], and subsequent myocardial dysfunction. AGEs may also produce adverse effects by binding to their receptors (RAGEs), a member of the immunoglobulin superfamily of cell surface molecules expressed on endothelial cells, monocytes, and neurons [156]. The AGE/RAGE pathway primarily governs the up-regulation of transforming growth factor-β and NADPH oxidase, which produces copious levels of superoxide that when combined with NO, forms highly reactive oxygen species [15]. In diabetic murine models, increased expression of RAGE was associated with the up-regulation of connective tissue growth factor and collagen deposition [152]. Furthermore, overexpression of myocardial RAGE was reported to influence Ca^{2+} metabolism and correlated with reduced systolic and diastolic intracellular Ca^{2+} concentrations as well as a significant delay in Ca^{2+} uptake [157]. Although several recent studies have evaluated the relationship between the AGE-RAGE pathway and heart failure [49, 158], further investigations are warranted.

13 Abnormalities in Calcium Homeostasis

Intracellular calcium serves as a pivotal ionic regulator in the heart. Abnormalities of Ca^{2+} handling in the cardiomyocyte indicate impaired cardiac function and epitomize the cardiac phenotype seen in diabetic cardiomyopathy. Shifts in myosin isoenzyme composition (from V1 to V3 isoforms) and predominant foetal β myosin heavy chain (MHC) expression, as compared with α MHC, are seen to be associated with reduced contractile protein Ca^{2+}-ATPase activity and shortened velocity, in a diabetic milieu [159]. In a model of streptozotocin-induced diabetic rats, cardiac dysfunction was caused by a reduction in the release of Ca^{2+} from the sarcoplasmic reticulum by the ryanodine receptor, resulting in a decreased upstroke phase of Ca^{2} transient [160]. A similar murine model exhibited the reduced ability of sarcoplasmic reticulum to sequester Ca^{2+} with accompanied reduced Ca^{2+} efflux via the sarcolemmal Na^{+}/Ca^{2+} exchanger [161]. There is extensive literature reporting on the metabolic abnormalities present in T1D and T2D types of diabetic murine models, which demonstrates the altered expression, activity, and function of numerous transporters involved in excitation-contraction coupling; these include, sarcoplasmic reticulum Ca^{2+}-pump ATPase (SERCA) [162], sarcolemmal Ca^{2+}-pump ATPase [163], and dysfunctional intracellular calcium signaling [164]. In diabetic-induced cardiac dysfunction, alterations affecting the activity of SERCA-2a and its inhibitor phospholamban (PLB) appear to be primary variables in the pathogenesis of this disease. Tang et al. [165] demonstrated abnormal Ca^{2+} signaling in hyperglycemic cardiomyocytes due to decreased activity of SERCA evoked by oxidative stress. Additionally, clinical investigations have confirmed the findings of Ca^{2+} dysregulation in diabetic cardiomyocytes, due to reduced myofilament Ca^{2+} sensitivity [166]. Although recent investigations have enhanced our knowledge of metabolic changes resulting from disrupted Ca^{2+} homeostasis, further studies are needed to understand the potential clinical consequences of these alterations.

14 Conclusions

Diabetic cardiomyopathy represents a distinct entity that encompasses a diverse spectrum from subclinical disease to the syndrome of end-stage heart failure. The epidemic rise in diabetes mellitus worldwide has made diabetic cardiomyopathy an increasing health concern with major medical and economic impacts. It is hypothesized that the metabolic perturbations that are associated with diabetic cardiomyopathy may be caused by the magnitude and distribution of increasing NEFAs, which subsequently lead to hyperinsulinemia, β-cell failure, and hyperglycemia. The maladaptive shift from glucose to NEFA utilization seen in diabetic myocardium leads to lipotoxicity, apoptosis, and the generation of fibrotic tissue. A myriad of metabolic derangements are involved in the development of reduced myocardial contractility in diabetic cardiomyopathy including increased oxidative stress, altered substrate metabolism, impaired homeostasis and mitochondrial dysfunction. These pathological defects act synergistically to exacerbate the cardiac phenotype seen in this diabetic population. Clinically, diabetic cardiomyopathy first manifests as asymptomatic diastolic dysfunction, which progresses to symptomatic combined diastolic and systolic dysfunction (Fig. 5) [162]. The key issue is that diabetic individuals have proven to be at significant risk of cardiac failure, regardless of whether this results from diabetes itself or from the combination of various cardiovascular risks. Due to the complex and multifactorial etiology of diabetic cardiomyopathy, as well as the lack of clinical intervention trials performed on this population, this pathologic condition remains

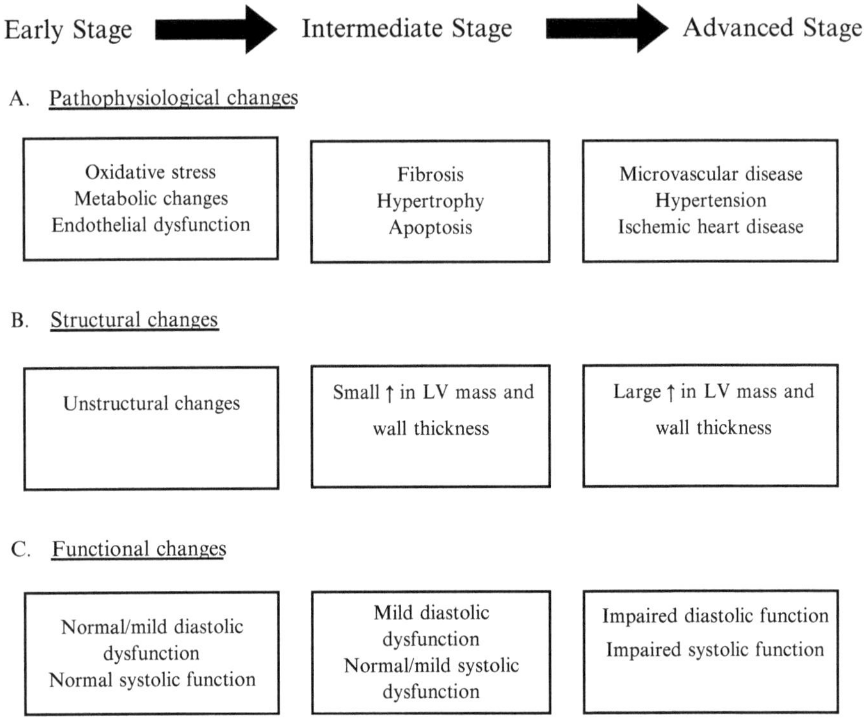

Fig. 5 Maladaptive stages of diabetic cardiomyopathy

underdiagnosed and inadequately treated. As future investigations continue to elucidate the mechanisms of this cardiomyopathy in diabetics, it is trusted that they will provide the incentive needed for generating novel therapies designed to reduce the risk of heart failure in patients with diabetes mellitus.

Acknowledgments The infrastructure for the work in this article was provided by the St. Boniface Hospital Research Foundation.

References

1. International Diabetes Federation (2009) IDF Diabetes Atlas, 4th edin
2. Public Health Agency of Canada (2011) Diabetes in Canada (2011): Facts and figures from a public health perspective
3. Department of Health and Human Services CfDCaP (2011) National Diabetes Fact Sheet, National Estimates and General Information on Diabetes and Prediabetes in the United States, Atlanta. GA
4. Poornima IG, Parikh P, Shannon RP (2006) Diabetic cardiomyopathy: the search for a unifying hypothesis. Circ Res 98:596–605
5. Karnik AA, Fields AV, Shannon RP (2007) Diabetic cardiomyopathy. Curr Hypertens Rep 9:467–473

6. Cas AD, Spigoni V, Ridolfi V et al (2013) Diabetes and chronic heart failure: from diabetic cardiomyopathy to therapeutic approach. Endocr Metab Immune Disord Drug Targets 13:38–50
7. Kannel WB, Hjortland M, Castelli WP (1974) Role of diabetes in congestive heart failure: the Framingham study. Am J Cardiol 34:29–34
8. Stratton IM, Adler AI, Neil HA et al (2000) Association of glycemia with macrovascular and microvascular complications of type 2 diabetes (United Kingdom Prospective Diabetes Study 35): prospective observational study. BMJ 321:405–412
9. Gottdiener JS, Arnold AM, Aurigemma GP et al (2000) Predictors of congestive heart failure in the elderly: the Cardiovascular Health Study. J Am Coll Cardiol 35:1628–1637
10. Follath F (2007) University Hospital Zürich, Switzerland: ESC Congress 2007 Press Release
11. Aneja A, Tang WH, Bansilal S et al (2008) Diabetic cardiomyopathy: insights into pathogenesis, diagnostic challenges, and therapeutic options. Am J Med 121:748–757
12. Shao CH, Rozanski GJ, Patel KP et al (2007) Dyssynchronous (non-uniform) Ca^{2+} release in myocytes from streptozotocin-induced diabetic rats. J Mol Cell Cardiol 42:234–246
13. Pereira L, Matthes J, Schuster I et al (2006) Mechanisms of $[Ca^{2+}]i$ transient decrease in cardiomyopathy of db/db type 2 diabetic mice. Diabetes 55:608–615
14. An D, Rodrigues B (2006) Role of changes in cardiac metabolism in development of diabetic cardiomyopathy. Am J Physiol Heart Circ Physiol 291:H1489–H1506
15. Wang J, Song Y, Wang Q et al (2006) Causes and characteristics of diabetic cardiomyopathy. Rev Diabet Stud 3:108–117
16. Asghar O, Al-Sunni A, Khavandi K et al (2009) Diabetic cardiomyopathy. Clin Sci 116:741–760
17. Murarka S, Movahed MR (2010) Diabetic cardiomyopathy. J Card Fail 16:971–979
18. Watanabe K, Thandavarayan RA, Harima M et al (2010) Role of differential signaling pathways and oxidative stress in diabetic cardiomyopathy. Curr Cardiol Rev 6:280–290
19. Rubler S, Dlugash J, Yuceoglu YZ et al (1972) A new type of cardiomyopathy associated with diabetic glomerulosclerosis. Am J Cardiol 30:595–602
20. Adameova A, Dhalla NS (2013) Role of microangiopathy in diabetic cardiomyopathy. Heart Fail Rev. doi:10.1007/s10741-013-9378-7
21. Shapiro LM (1982) Echocardiographic features of impaired ventricular function in diabetes mellitus. Br Heart J 47:439–444
22. Bertoni AG, Tsai A, Kasper EK et al (2003) Diabetes and idiopathic cardiomyopathy: a nationwide case–control study. Diabetes Care 26:2791–2795
23. Isfort M, Stevens SC, Schaffer S et al (2013) Metabolic dysfunction in diabetic cardiomyopathy. Heart Fail Rev. doi:10.1007/s10741-013-9377-8
24. Wold LE, Dutta K, Mason MM et al (2005) Impaired SERCA function contributes to cardiomyocyte dysfunction in insulin resistant rats. J Mol Cell Cardiol 39:297–307
25. Mizushige K, Yao L, Noma T et al (2000) Alteration in left ventricular diastolic filling and accumulation of myocardial collagen at insulin-resistant prediabetic stage of a type II diabetic rat model. Circulation 101:899–907
26. Greer JJ, Ware DP, Lefer DJ (2006) Myocardial infarction and heart failure in the db/db diabetic mouse. Am J Physiol Heart Circ Physiol 290:H146–H153
27. Hoshida S, Yamashita N, Otsu K et al (2000) Cholesterol feeding exacerbates myocardial injury in Zucker diabetic fatty rats. Am J Physiol Heart Circ Physiol 278:H256–H262
28. Stratmann B, Gawlowski T, Tschoepe D (2010) Diabetic cardiomyopathy—to take a long story serious. Herz 35:161–168
29. Devereux RB, Roman MJ, Paranicas M et al (2000) Impact of diabetes on cardiac structure and function: the Strong Heart Study. Circulation 101:2271e6
30. Yotsukura M, Suzuki J, Yamaguchi T (1998) Prognosis following acute myocardial infarction in patients with ECG evidence of left ventricular hypertrophy prior to infarction. J Electrocardiol 31:91–99
31. Boner G, Cooper ME, McCarroll K et al (2005) Adverse effects of left ventricular hypertrophy in the reduction of endpoints in NIDDM with the angiotensin II antagonist losartan (RENAAL) study. Diabetologia 48:1980–1987

32. Quinones MA, Greenberg BH, Kopelen HA et al (2000) Echocardiographic predictors of clinical outcome in patients with left ventricular dysfunction enrolled in the SOLVD registry and trials: significance of left ventricular hypertrophy. Studies of left ventricular dysfunction. J Am Coll Cardiol 35:1237–1244
33. de Simone G, Gottdiener JS, Chinali M et al (2008) Left ventricular mass predicts heart failure not related to previous myocardial infarction: the Cardiovascular Health Study. Eur Heart J 29:741–747
34. Ozasa N, Furukawa Y, Morimoto T et al (2008) Relation among left ventricular mass, insulin resistance, and hemodynamic parameters in type 2 diabetes. Hypertens Res 31:425–432
35. Sundstrom J, Arnlov J, Stolare K et al (2008) Blood pressure-independent relations of left ventricular geometry to the metabolic syndrome and insulin resistance: a population-based study. Heart 94:874–878
36. Falcao-Pires I, Leite-Moreira AF (2012) Diabetic cardiomyopathy: understanding the molecular and cellular basis to progress in diagnosis and treatment. Heart Fail Rev 17:325–344
37. Schilling JD, Mann DL (2012) Diabetic cardiomyopathy: bend to bedside. Heart Fail Clin 8:619–631
38. Brutsaert DL, Housmans PR, Goethals MA (1980) Dual control of relaxation. Its role in the ventricular function in the mammalian heart. Circ Res 47:637–652
39. Voulgari C, Papadogiannis D, Tentolouris N (2010) Diabetic cardiomyopathy: from the pathophysiology of the cardiac myocytes to current diagnosis and management strategies. Vasc Health Risk Manag 6:883–903
40. Zile MR, Baicu CF, Gaasch WH (2004) Diastolic heart failure: abnormalities in active relaxation and passive stiffness of the left ventricle. N Engl J Med 350:1953–1959
41. Giles TD, Sander GE (2004) Diabetes mellitus and heart failure: basic mechanisms, clinical features, and therapeutic considerations. Cardiol Clin 22:553–568
42. Sander GE, Giles TD (2003) Diabetes mellitus and heart failure. Am Heart Hosp J 1:273–280
43. Ommen SR, Nishimura RA (2003) A clinical approach to the assessment of left ventricular diastolic function by Doppler echocardiography: update. Heart 89:18–23
44. Raev DC (1994) Which left ventricular dysfunction is impaired earlier in the evolution of diabetic cardiomyopathy? An echocardiographic study of young type 1 diabetic patients. Diabetes Care 17:633–639
45. Riggs TW, Transue D (1990) Doppler echocardiographic evaluation of left ventricular diastolic dysfunction in adolescents with diabetes mellitus. Am J Cardiol 65:899–902
46. Li JE, Palmieri V, Roman MJ et al (2001) The impact of diabetes on left ventricular filling pattern in normotensive and hypertensive adults: the Strong Heart Study. J Am Coll Cardiol 37:1943–1949
47. Brogan WC III, Hillis LD, Flores ED et al (1992) The natural history of isolated left ventricular diastolic dysfunction. Am J Med 92:627–630
48. Iribarren C, Karter AJ, Go AS (2001) Glycaemic control and heart failure among adult patients with diabetes. Circulation 103:2668–2673
49. van Heerebeek L, Hamdani N, Handoko ML et al (2008) Diastolic stiffness of the failing diabetic heart: importance of fibrosis, advanced glycation end products, and myocyte resting tension. Circulation 117:43–51
50. Ashrafian H, Frenneaux MP, Opie LH (2007) Metabolic mechanisms in heart failure. Circulation 116:434–448
51. Zarich S, Nesto R (1989) Diabetic cardiomyopathy. Am Heart J 118:1000e12
52. Mildenberger RR, Bar-Shlomo B, Druck MN et al (1984) Clinically unrecognized dysfunction in young diabetic patient. J Am Coll Cardiol 4:234e8
53. Mbanya JC, Sobngwi E, Mbanya DS et al (2001) Left ventricular mass and systolic function in African diabetic patients: association with microalbuminuria. Diabetes Metab 27: 378–382
54. Petrie MC, Caruana L, Berry C et al (2002) "Diastolic heart failure" or heart failure caused by subtle left ventricular systolic dysfunction? Heart 87:29–31

55. Fang ZY, Leano R, Marwick TH (2004) Relationship between longitudinal and radial contractility in subclinical diabetic heart disease. Clin Sci 106:53–60
56. Ha JW, Lee HC, Kang ES et al (2007) Abnormal left ventricular longitudinal functional reserve in patients with diabetes mellitus: implication for detecting subclinical myocardial dysfunction using exercise tissue Doppler echocardiography. Heart 93:1571–1576
57. Svealy BG, Olofsson EL, Andersson B (2008) Ventricular long-axis function is of major importance for long-term survival in patients with heart failure. Heart 94:284–289
58. Fang ZY, Schull-Meade R, Leano R et al (2005) Screening for heart disease in diabetic subjects. Am Heart J 149:349e54
59. Fang ZY, Yuda S, Anderson V et al (2003) Echocardiographic detection of early diabetic myocardial disease. J Am Coll Cardiol 41:611e7
60. Yu CM, Lin H, Yang H et al (2002) Progression of systolic abnormalities in patients with "isolated" diastolic heart failure and diastolic dysfunction. Circulation 105:1195e203
61. Maciver DH, Townsend M (2008) A novel mechanism of heart failure with normal ejection fraction. Heart 94:446–449
62. Aurigemma GP, Zile MR, Gaasch WH (2006) Contractile behavior of the left ventricle in diastolic heart failure: with emphasis on regional systolic function. Circulation 113: 296–304
63. Zornoff LA, Skali H, Pfeffer MA et al (2002) Right ventricular dysfunction and risk of heart failure and mortality after myocardial infarction. J Am Coll Cardiol 39:1450–1455
64. Karamitsos TD, Karvounis HI, Dalamanga EG et al (2007) Early diastolic impairment of diabetic heart: the significance of right ventricle. Int J Cardiol 114:218e23
65. Movahed MR, Milne N (2007) Presence of biventricular dysfunction in patients with type II diabetes mellitus. Congest Heart Fail 13:78e80
66. Shehadeh A, Regan TJ (1995) Cardiac consequences of diabetes mellitus. Clin Cardiol 18:301–305
67. Acar E, Ural D, Bildirici U et al (2011) Diabetic cardiomyopathy. Anadolu Kardiyol Derg 11:732–737
68. Boudina S, Abel ED (2007) Diabetic cardiomyopathy revisited. Circulation 115:3213–3223
69. Lopaschuk GD (2002) Metabolic abnormalities in the diabetic heart. Heart Fail Rev 7:149–159
70. Carroll R, Carley AN, Dyck JR et al (2005) Metabolic effects of insulin on cardiomyocytes from control and diabetic db/db mouse hearts. Am J Physiol Endocrinol Metab 288:E900–E906
71. Herrero P, Peterson LR, McGill JB et al (2006) Increased myocardial fatty acid metabolism in patients with type 1 diabetes mellitus. J Am Coll Cardiol 47:598–604
72. Peterson LR, Herrero P, Schechtman KB et al (2004) Effect of obesity and insulin resistance on myocardial substrate metabolism and efficiency in young women. Circulation 109:2191–2196
73. Mazumder PK, O'Neill BT, Roberts MW et al (2004) Impaired cardiac efficiency and increased fatty acid oxidation in insulin resistant ob/ob mouse hearts. Diabetes 53:2366–2374
74. Dirkx E, Schwenk RW, Glatz JF et al (2011) High fat diet induced diabetic cardiomyopathy. Prostaglandins Leukot Essent Fatty Acids 85:219–225
75. Sharma S, Adrogue JV, Golfman L et al (2004) Intramyocardial lipid accumulation in the failing human heart resembles the lipotoxic rat heart. FASEB J 18:1692–1700
76. Szczepaniak LS, Dobbins RL, Metzger GJ et al (2003) Myocardial triglycerides and systolic function in humans: in vivo evaluation by localized proton spectroscopy and cardiac imaging. Magn Reson Med 49:417–423
77. Murray AJ, Anderson RE, Watson GC et al (2004) Uncoupling proteins in human heart. Lancet 364:1786–1788
78. Opie LH (1970) Effect of fatty acids on contractility and rhythm of the heart. Nature 227: 1055–1056

79. Nakayama H, Morozumi T, Nanto S et al (2001) Abnormal myocardial free fatty acid utilization deteriorates with morphological changes in the hypertensive heart. Jpn Circ J 65:783–787
80. Yazaki Y, Isobe M, Takahashi W et al (1999) Assessment of myocardial fatty acid metabolic abnormalities in patients with idiopathic dilated cardiomyopathy using 123i bmipp spect: correlation with clinicopathological findings and clinical course. Heart 81:153–159
81. Abe T, Ohga Y, Tabayashi N et al (2002) Left ventricular diastolic dysfunction in type 2 diabetes mellitus model rats. Am J Physiol Heart Circ Physiol 282:H138–H148
82. Malhotra A, Reich D, Nakouzi A et al (1997) Experimental diabetes is associated with functional activation of protein kinase C epsilon and phosphorylation of troponin I in the heart, which are prevented by angiotensin II receptor blockade. Circ Res 81:1027–1033
83. Unger RH (2002) Lipotoxic diseases. Annu Rev Med 53:319–336
84. Halse R, Pearson SL, McCormack JG et al (2001) Effects of tumor necrosis factor-alpha on insulin action in cultured human muscle cells. Diabetes 50:1102–1109
85. Zhang DX, Fryer RM, Hsu AK et al (2001) Production and metabolism of ceramide in normal and ischemicreperfused myocardium of rats. Basic Res Cardiol 96:267–274
86. Young ME, McNulty P, Taegtmeyer H (2002) Adaptation and maladaptation of the heart in diabetes: Part II. Potential mechanisms. Circulation 109:121
87. Boudina S, Abel ED (2006) Mitochondrial uncoupling: a key contributor to reduced cardiac efficiency in diabetes. Physiology 21:250–258
88. Finck BN, Lehman JJ, Leone TC (2002) The cardiac phenotype induced by PPARα overexpression mimics that caused by diabetes mellitus. J Clin Invest 109:121–130
89. Chiu HC, Kovacs A, Blanton RM et al (2005) Transgenic expression of fatty acid transport protein 1 in the heart causes lipotoxic cardiomyopathy. Circ Res 96:225–233
90. Yagyu H, Chen G, Yokoyama M et al (2003) Lipoprotein lipase (LpL) on the surface of cardiomyocytes increases lipid uptake and produces a cardiomyopathy. J Clin Invest 111: 419–426
91. Fang ZY, Prins JB, Marwick TH (2004) Diabetic cardiomyopathy: evidence, mechanisms, and therapeutic implications. Endocr Rev 25:543–567
92. Liu GX, Hanley PJ, Ray J et al (2001) Long-chain acylcoenzyme A esters and fatty acids directly link metabolism to K (ATP) channels in the heart. Circ Res 88:918–924
93. Calle MC, Fernandez ML (2012) Inflammation and type 2 diabetes. Diabetes Metab 38:183–191
94. King KL, Okere IC, Sharma N et al (2005) Regulation of cardiac malonyl-CoA content and fatty acid oxidation during increased cardiac power. Am J Physiol Heart Circ Physiol 289:H1033–H1037
95. Kim JK, Kim YJ, Fillmore JJ et al (2001) Prevention of fat-induced insulin resistance by salicylate. J Clin Invest 108:437–446
96. Rui L, Aguirre V, Kim JK et al (2001) Insulin/IGF-1 and TNF-alpha stimulate phosphorylation of IRS-1 at inhibitory Ser307 via distinct pathways. J Clin Invest 107:181–189
97. Brazil DP, Hemmings BA (2001) Ten years of protein kinase B signalling: a hard Akt to follow. Trends Biochem Sci 26:657–664
98. Lawlor MA, Alessi DR (2001) PKB/Akt: a key mediator of cell proliferation, survival and insulin responses? J Cell Sci 114:2903–2910
99. Schwartzbauer G, Robbins J (2001) The tumor suppressor gene PTEN can regulate cardiac hypertrophy and survival. J Biol Chem 276:35786–35793
100. Shulman GI (2000) Cellular mechanisms of insulin resistance. J Clin Invest 106:171–176
101. Iozzo P, Chareonthaitawee P, Dutka D et al (2002) Independent association of type 2 diabetes and coronary artery disease with myocardial insulin resistance. Diabetes 51:3020–3024
102. Bonora E, Targher G, Alberiche M et al (2002) Predictors of insulin sensitivity in type 2 diabetes mellitus. Diabet Med 19:535–542
103. Ilercil A, Devereux RB, Roman MJ et al (2002) Associations of insulin levels with left ventricular structure and function in Am Indians: the strong heart study. Diabetes 51: 1543–1547

104. Iacobellis G, Ribaudo MC, Zappaterreno A (2003) Relationship of insulin sensitivity and left ventricular mass in uncomplicated obesity. Obes Res 11:518–524
105. McNulty PH (2003) Insulin resistance and cardiac mass: the end of the beginning? Obes Res 11:507–508
106. Liang Q, Carlson EC, Donthi RV et al (2002) Overexpression of metallothionein reduces diabetic cardiomyopathy. Diabetes 51:174–181
107. Galderisi M, Paolisso G, Tagliamonte MR et al (1997) Is insulin action a determinant of left ventricular relaxation in uncomplicated essential hypertension? J Hypertens 15:745–750
108. Guida L, Celentano A, Iannuzzi R et al (2001) Insulin resistance, ventricular mass and function in normoglycaemic hypertensives. Nutr Metab Cardiovasc Dis 11:306–311
109. O'Neill BT, Abel ED (2005) Akt1 in the cardiovascular system: friend or foe? J Clin Invest 115:2059–2064
110. Khamzina L, Veilleux A, Bergeron S et al (2005) Increased activation of the mammalian target of rapamycin pathway in liver and skeletal muscle of obese rats: possible involvement in obesity-linked insulin resistance. Endocrinology 146:1473–1481
111. Manning BD (2004) Balancing Akt with S6K: implications for both metabolic diseases and tumorigenesis. J Cell Biol 167:399–403
112. Pulakat L, Demarco VG, Whaley-Connell A et al (2011) The impact of overnutrition on insulin metabolic signaling in the heart and the kidney. Cardiorenal Med 1:102–112
113. Kern W, Peters A, Born J et al (2005) Changes in blood pressure and plasma catecholamine levels during prolonged hyperinsulinemia. Metabolism 54:391–396
114. Grassi G (2004) Counteracting the sympathetic nervous system in essential hypertension. Curr Opin Nephrol Hypertens 13:513–519
115. Morisco C, Condorelli G, Trimarco V et al (2005) Akt mediates the cross-talk between beta-adrenergic and insulin receptors in neonatal cardiomyocytes. Circ Res 96:180–188
116. Naito Z, Takashi E, Xu G et al (2003) Different influences of hyperglycemic duration on phosphorylated extracellular signal-regulated kinase 1/2 in rat heart. Exp Mol Pathol 74:23–32
117. Tajmir P, Ceddia RB, Li RK et al (2004) Leptin increases cardiomyocyte hyperplasia via extracellular signal-regulated kinaseand phosphatidylinositol 3-kinase-dependent signaling pathways. Endocrinology 145:1550–1555
118. Paolisso G, Tagliamonte MR, Galderisi M (1999) Plasma leptin level is associated with myocardial wall thickness in hypertensive insulin-resistant men. Hypertension 34:1047–1052
119. Wang CC, Goalstone ML, Draznin B (2004) Molecular mechanisms of insulin resistance that impact cardiovascular biology. Diabetes 53:2735–2740
120. Schmitz-Peiffer C (2000) Signalling aspects of insulin resistance in skeletal muscle: mechanisms induced by lipid oversupply. Cell Signal 12:583–594
121. Baumann CA, Ribon V, Kanzaki M et al (2000) Cap defines a second signalling pathway required for insulin-stimulated glucose transport. Nature 407:202–207
122. Coutinho M, Gerstein HC, Wang Y et al (1999) The relationship between glucose and incident cardiovascular events: a metaregression analysis of published data from 20 studies of 95,783 individuals followed for 12.4 years. Diabetes Care 22:233–240
123. Mytas DZ, Stougiannos PN, Zairis MN et al (2009) Diabetic myocardial disease: pathophysiology, early diagnosis and therapeutic options. J Diabetes Complications 23:273–282
124. Suskin N, McKelvie RS, Burns RJ et al (2000) Glucose and insulin abnormalities relate to functional capacity in patients with congestive heart failure. Eur Heart J 21:1368–1375
125. Giles TD, Ouyang J, Kerut EK et al (1998) Changes in protein kinase C in early cardiomyopathy and in gracilis muscle in the BB/Wor diabetic rat. Am J Physiol 274(1 pt 2): H295–H307
126. Giles TD (2001) Angiotensin I-converting enzyme/kinase II dysregulation in cardiovascular complications of diabetes mellitus. In: Giles TD (ed) Angiotensin converting enzyme: clinical and experimental insights. Health Care Communications, Fort Lee, pp 135–144
127. Turan B (2010) Role of antioxidants in redox regulation of diabetic cardiovascular complications. Curr Pharm Biotechnol 11:819–836

128. Imrie H, Abbas A, Kearney M (2010) Insulin resistance, lipotoxicity and endothelial dysfunction. Biochim Biophys Acta 1801:320–326
129. Seddon M, Looi YH, Shah AM (2007) Oxidative stress and redox signalling in cardiac hypertrophy and heart failure. Heart 93:903–907
130. Gao WD, Liu Y, Marban E (1996) Selective effects of oxygen free radicals on excitation-contraction coupling in ventricular muscle. Implications for the mechanism of stunned myocardium. Circulation 42:2597e604
131. Du X, Matsumura T, Edelstein D (2003) Inhibition of GADPH activity by poly (ADP-ribose) polymerase activates three major pathways of hyperglycemic damage in endothelial cells. J Clin Invest 112:1049e57
132. Szabo C (2002) PARP as a drug target for the therapy of diabetic cardiovascular dysfunction. Drug News Perspect 15:197e205
133. Mellor H, Parker PJ (1998) The extended protein kinase C super-family. Biochem J 332 (Pt 2):281–292
134. Geraldes P, King GL (2010) Activation of protein kinase C isoforms and its impact on diabetic complications. Circ Res 106:1319–1331
135. Way KJ, Katai N, King GL (2001) Protein kinase C and the development of diabetic vascular complications. Diabet Med 18:945–959
136. Hayat SA, Patel B, Khattar RS et al (2004) Diabetic cardiomyopathy: mechanisms, diagnosis and treatment. Clin Sci (Lond) 107:539–557
137. Sheetz MJ, King GL (2002) Molecular understanding of hyperglycemia's adverse effects for diabetic complications. JAMA 288:2579–2588
138. Pratt RE (1999) Angiotensin II and the control of cardiovascular structure. J Am Soc Nephrol 10:S120–S128
139. He Z, Way KJ, Arikawa E et al (2005) Differential regulation of angiotensin II-induced expression of connective tissue growth factor by protein kinase C isoforms in the myocardium. J Biol Chem 280:15719–15726
140. Way KJ, Isshiki K, Suzuma K et al (2002) Expression of connective tissue growth factor is increased in injured myocardium associated with protein kinase C beta2 activation and diabetes. Diabetes 51:2709–2718
141. Wakasaki H, Koya D, Schoen FJ et al (1997) Targeted overexpression of protein kinase C beta2 isoform in myocardium causes cardiomyopathy. Proc Natl Acad Sci USA 94: 9320–9325
142. Takeishi Y, Chu G, Kirkpatrick DM et al (1998) In vivo phosphorylation of cardiac troponin I by protein kinase Cbeta2 decreases cardiomyocyte calcium responsiveness and contractility in transgenic mouse hearts. J Clin Invest 102:72–78
143. Connelly KA, Kelly DJ, Zhang Y et al (2009) Inhibition of protein kinase C-beta by ruboxistaurin preserves cardiac function and reduces extracellular matrix production in diabetic cardiomyopathy. Circ Heart Fail 2:129–137
144. Retnakaran R, Zinman B (2008) Type 1 diabetes, hyperglycaemia, and the heart. Lancet 371:1790–1799
145. Ginsberg BJ, Mazze R (1994) Clinical consequences of the Diabetes Control and Complications Trial. N J Med 91:221–224
146. Brunner F, Bras-Silva C, Cerdeira AS et al (2006) Cardiovascular endothelins: essential regulators of cardiovascular homeostasis. Pharmacol Ther 111:508–531
147. Frustaci A, Kajstura J, Chimenti C et al (2000) Myocardial cell death in human diabetes. Circ Res 87:1123–1132
148. Shimizu M, Umeda K, Sugihara N et al (1993) Collagen remodelling in myocardia of patients with diabetes. J Clin Pathol 46:32–36
149. John WG, Lamb EJ (1993) The Maillard or browning reaction in diabetes. Eye 7:230–237
150. Raj DS, Choudhury D, Welbourne TC et al (2000) Advanced glycation end products: a nephrologist's perspective. Am J Kidney Dis 35:365–380
151. Miyata T, Kurokawa K, Van Ypersele De Strihou C (2000) Advanced glycation and lipoxidation end products: role of reactive carbonyl compounds generated during carbohydrate and lipid metabolism. J Am Soc Nephrol 9:1744–1752

152. Candido R, Forbes JM, Thomas MC et al (2003) A breaker of advanced glycation end products attenuates diabetes-induced myocardial structural changes. Circ Res 18:785–792
153. Brownlee M, Cerami A, Vlassara H (1988) Advanced glycosylation end products in tissue and the biochemical basis of diabetic complications. N Engl J Med 318:1315–1321
154. Aronson D (2003) Cross-linking of glycated collagen in the pathogenesis of arterial and myocardial stiffening of aging and diabetes. J Hypertens 21:3–12
155. Dyer DG, Dunn JA, Thorpe SR et al (1993) Accumulation of Maillard reaction products in skin collagen in diabetes and aging. J Clin Invest 91:2463–2469
156. Bierhaus A, Hofmann MA, Ziegler R et al (1998) The AGE/RAGE pathway in vascular disease and diabetes mellitus. Part I. The AGE-concept. Cardiovasc Res 37:586–600
157. Petrova R, Yamamoto Y, Muraki K et al (2002) Advanced glycation endproduct-induced calcium handling impairment in mouse cardiac myocytes. J Mol Cell Cardiol 34:1425–1431
158. Koyama Y, Takeishi Y, Niizeki T et al (2008) Soluble receptor for advanced glycation end products (RAGE) is a prognostic factor for heart failure. J Card Fail 14:133–139
159. Malhotra A, Sanghi V (1997) Regulation of contractile proteins in diabetic heart. Cardiovas Res 34:34–40
160. Teshima Y, Takahashi N, Saikawa T et al (2000) Diminished expression of sarcoplasmic reticulum $Ca^{(2+)}$-ATPase and ryanodine sensitive $Ca^{(2+)}$ channel mRNA in streptozotocin-induced diabetic rat heart. J Mol Cell Cardiol 32:655–664
161. Choi KM, Zhong Y, Hoit BD et al (2002) Defective intracellular $Ca^{(2+)}$ signaling contributes to cardiomyopathy in Type 1 diabetic rats. Am J Physiol Heart Circ Physiol 283: H1398–H1408
162. Trost SU, Belke DD, Bluhm WF et al (2002) Overexpression of the sarcoplasmic reticulum $Ca^{(2+)}$-atpase improves myocardial contractility in diabetic cardiomyopathy. Diabetes 51:1166–1171
163. Golfman L, Dixon IM, Takeda N et al (1998) Cardiac sarcolemmal $Na^{(+)}$-Ca^{2+} exchange and $Na^{(+)}$-K^{+} ATPase activities and gene expression in alloxaninduced diabetes in rats. Mol Cell Biochem 188:91–101
164. Vetter R, Rehfeld U, Reissfelder C et al (2002) Transgenic overexpression of the sarcoplasmic reticulum Ca^{2+} ATPase improves reticular Ca^{2+} handling in normal and diabetic rat hearts. FASEB J 16:1657–1659
165. Tang WH, Cheng WT, Kravtsov GM et al (2010) Cardiac contractile dysfunction during acute hyperglycemia due to impairment of SERCA by polyol pathway-mediated oxidative stress. Am J Physiol Cell Physiol 299:C643–C653
166. Jweied EE, McKinney RD, Walker LA et al (2005) Depressed cardiac myofilament function in human diabetes mellitus. Am J Physiol Heart Circ Physiol 289:H2478–H2483

Metabolic and Contractile Remodelling in the Diabetic Heart: An Evolutionary Perspective

Vijay Sharma and John H. McNeill

Abstract The application of evolutionary biology to the study of human disease has given rise to the idea that disease can result from inappropriate adaptations to a change in environment. This concept can also be applied to the function of organs and responding to their local environments within the human body. The heart is an omnivorous organ which can use any substrate it is supplied with. The metabolic machinery of the heart is exquisitely attuned both to its metabolic needs and to the available energy substrates in its local environment. Diabetic cardiomyopathy is a disease process which arises as a result of the inability of the heart to adapt to a diabetic metabolic milieu. The heart becomes locked into a progressively maladaptive state from which it cannot escape by its own devices; due to the phenomenon of hyperglycemic memory, even restoration of a normal milieu may not be sufficient to completely reverse the remodeling. The pathways which initiate, progress and perpetuate this downward spiral are the same pathways which normally allow the heart to sense and respond to its local metabolic environment. These include metabolite-sensitive transcriptional regulatory pathways and, most probably, epigenetic and miRNA regulatory pathways. Overall, the application of evolutionary concepts provides a valuable framework for understanding the origins and importance of metabolic and contractile disturbances in the diabetic heart, and a strong rationale for the use of metabolic therapy as a treatment.

V. Sharma
Department of Pathology, Aberdeen Royal Infirmary, Foresterhill, Aberdeen, Scotland

Division of Applied Medicine, School of Medicine and Dentistry, The University of Aberdeen, Aberdeen, Scotland

BMJ Evidence Centre, BMJ Group, BMA House, Tavistock Square, London, UK

J.H. McNeill (✉)
Faculty of Pharmaceutical Sciences, University of British Columbia,
Vancouver, BC V6T 1Z3, Canada
e-mail: jmcneill@mail.ubc.ca

B. Turan and N.S. Dhalla (eds.), *Diabetic Cardiomyopathy: Biochemical and Molecular Mechanisms*, Advances in Biochemistry in Health and Disease 9,
DOI 10.1007/978-1-4614-9317-4_2, © Springer Science+Business Media New York 2014

Keywords Diabetic cardiomyopathy • Diabetes • Heart • Evolution • Mitochondria • Cardiac metabolism • Contractile remodeling • Epigenetics • MicroRNA • Fetal gene program

1 Introduction

Nothing in biology makes sense except in the light of evolution. Theodosus Dobzhansky, 1973

The application of evolutionary biology to the study of human disease has given rise to the idea that disease can result from adaptations which, although appropriate for the environment in which humans originally evolved, are inappropriate for the environment in which humans live today. This concept can also be applied to the function of cells, tissues, organs and systems responding to their local environments within the human body. In this context, disease results from a maladaptive response of the organ to a change in its local environment. The concept of the reaction norm describes the pattern of phenotypes that a single genotype can produce across a range of environments (Fig. 1). This pattern provides what can be regarded as a map of all possible adaptations, or "a map of plasticity". In the heart, reaction norms include the heart's ability to undergo metabolic and contractile remodeling, which it will do in response to a wide range of stressors. These include metabolic dysregulation (most notably diabetes), pressure overload, unloading, ischemia, hypobaric hypoxia and hypothyroidism. These stressors change the phenotype of the heart, and when

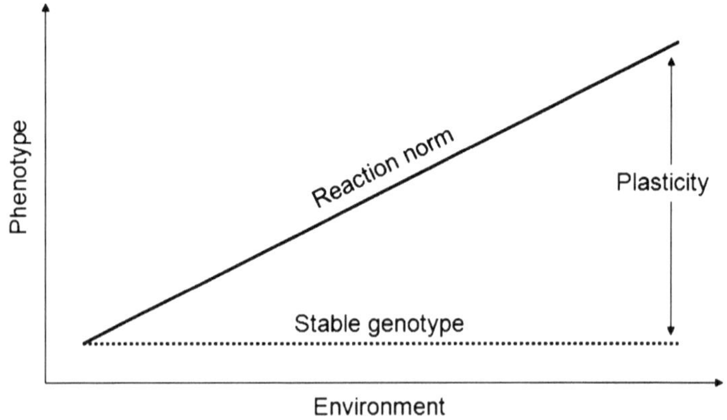

Fig. 1 A simplified representation of a reaction norm. Genotypes which produce the same phenotype over a range of environments are stable, whereas those which produce a range of phenotypes in different environments produce plasticity. Plasticity refers to the range of phenotypes which a genotype can produce. The reaction norm is the relationship between the environment and the phenotype

the phenotype moves from a point on the reaction norm which is adaptive to one which is maladaptive, function is impaired. Diabetic cardiomyopathy can be understood, from an evolutionary perspective, by understanding how and to what extent the phenotype of the heart becomes maladaptive in response to the metabolic milieu it is exposed to following the onset of diabetes. To put it simply, the heart has not evolved to be able to cope with a diabetic metabolic milieu. Before exploring this concept further, it is worthwhile to consider how the heart, and its metabolic machinery, first evolved.

2 An Overview of the Evolution of the Heart

The heart probably first emerged in the Bilateria, a group of organisms whose root is the flatworm and who lived 600–700 million years ago. The Bilateria, in turn, were ancestors of three main groups:

1. The Ecdysoa, which gave rise to the insects
2. The Lophotrochozoa, which gave rise to annelids and mollusks
3. The Deuterosomes, or "two-ended animals", which gave rise to the Echinodermata, whose members include sea urchins and starfish, the hemichordates, also known as acorn worms, and the Chordates [1].

The first heart is believed to have been a tubular pulsatile structure, devoid of chambers, septae and valves, which pumped fluid through a peri-cellular interstitium [1]. The features of the myocytes of which this tube was comprised probably varied, with features of striated muscle, vertebrate cardiomyocytes and myoepithelium all emerging in different organisms. This basic structure then underwent subsequent morphological and genetic modifications.

With the development of Deuterostomes, genes encoding the contractile proteins began to increase in number and diversify in their function. The Deuterostomes, in turn, gave rise to the Chordates, defined by the presence of a dorsal nerve cord and its supporting notochord. Two of the current branches of this group, namely the Urochordates (tunicates) and the Cephalocordates, retained a simple tubular heart. It is in the third branch, the vertebrates that the heart continued to evolve. Hagfish are at the base of the Vertebrates, and are followed, in order, by the Elasmobranchs, Teleost fish, Amphibians, Reptiles, Birds and Mammals. Hagfish possess a closed vascular system which has contractile elements at several points. The systemic heart of the hagfish is equivalent to the left-sided heart chambers in birds and mammals, whereas the portal vein heart, which pumps blood to the gill vasculature, is equivalent to the right-sided chambers. As we move upwards through the taxa, the Elasmobranchs are the first vertebrates to show vagal innervation of the heart. In the Teleost fish, the heart is a two-chambered organ, separated by an atrioventricular valve, which pumps blood to both the gill and the systemic vasculatures. Between the development of the teleost fish, and their separation into the amphibians during

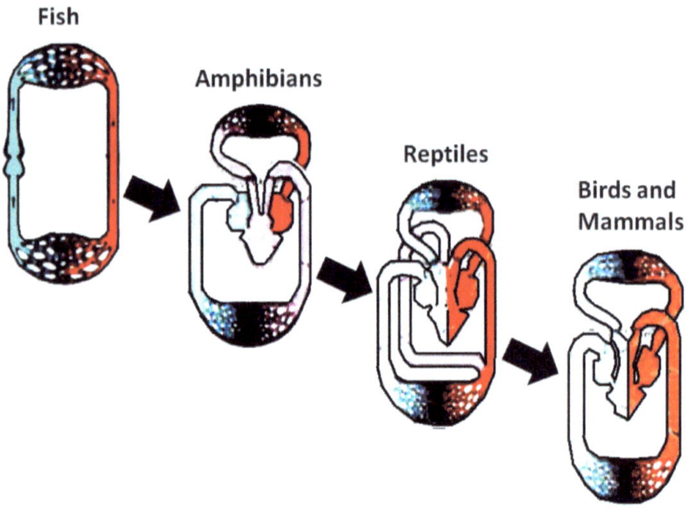

Fig. 2 The heart and vasculature of fish, amphibians, reptiles, birds and mammals, illustrating the key developments

the transition to terrestrial life, separation of the systemic and pulmonary vasculatures began. The amphibian heart has two atria and a single ventricle. In reptiles, a muscular ridge running from the base to the apex of the heart creates a partial division of the ventricular chamber, allowing oxygenated and deoxygenated blood to be separately directed. The degree of mixing which occurs depends on the size of the ridge; in turtles, the ridge is small, but in lizards and snakes, it is larger. Birds and mammals have a complete interventricular septum, enabling complete separation of the pulmonary and systemic circulations (Fig. 2).

The evolution of the heart from a tubular to a four-chambered structure was associated with a myriad of adaptations. In this brief treatise, a few key adaptations are worthy of mention. The first is the specialization of the contractile elements. The initial radiation occurred, as mentioned above, with the evolution of the Deuterostomes during the Cambrian explosion. Separation and specialization of striated and smooth muscle isoforms appeared at this time. Separation between skeletal and cardiac isoforms occurred later, initially with the separation of cardiac actin and troponin C before the frog-mammal divergence, followed by the separation of the cardiac, slow skeletal and fast skeletal isoforms before the bird-mammal divergence [1]. The second is the ability to replace myocytes, which was mostly lost during the development of endothermy, although a small pool of undifferentiated precursor cells does exist in human myocardium [1]. The third is the development of connexin-based gap junctions and a specialized conduction system. Gap junctions allowed rapid communication and enabled the transition from the peristaltic contractions of the early tubular heart to the rapid contractions of the vertebrate heart.

The development of a specialized conduction system further promoted synchronous contraction.

In terms of energy metabolism, the fundamental biochemical machinery of the cell has been conserved by evolution. Basic problems of transport created by an increase in vertebrates' size have been addressed by reducing the metabolic rate in larger organisms. Likewise, many of the adaptations seen during the evolution of the vertebrate heart are adaptations to optimize energy consumption and storage on the background of conserved biochemical machinery. These include the development of an efficient vasculature, sophisticated neural and endocrine regulatory systems, and plasticity of the heart itself.

It has been observed that ontogeny can recapitulate phylogeny. In other words, evolutionary stages in the development of an organism can be observed during the stages of embryonic and fetal development of that organism. During fetal development of the mammalian heart, the heart begins as a tubular heart, develops into a trabeculated heart and thence into the fully septated heart. Full separation of the pulmonary and systemic circulations does not occur until after birth. The tubular heart is mainly dependent on anaerobic glycolysis. Once this develops into the trabeculated heart, metabolism transitions to oxidative metabolism, but oxidative glucose metabolism predominates, with only very low rates of fatty acid oxidation (see [2] for review). In the newborn period, fatty acid oxidation rates increase dramatically, and the preference of the adult mammalian heart for fatty acids becomes established.

To a certain extent, this ontogenic pattern recapitulates the phylogeny of cardiac energy metabolism. Examination of enzyme activities and high-energy phosphate concentrations show that, across the taxa, carbohydrate metabolism plateaus at low levels of cardiac demand, and the increased requirement for ATP production necessitated by the increased need for power development by the heart is met by increasing the capacity for fatty acid oxidation [3]. This pattern is also supported by studies looking directly at substrate utilisation; fatty acid use tends to increase in higher taxa [4–7] whereas glycolysis decreases [8]. However, superimposed on this broad trend is a considerable degree of interspecies variation, particularly in fish, which exhibit marked differences in their substrate preferences [9]. The evolutionary advantage of fatty acids is that they generate more ATP per gram of fuel than carbohydrates. Despite the fact that fatty acid oxidation is less efficient in terms of oxygen consumption, expansion of fatty acid oxidation to meet increased energy demands is the adaptation which seems to have been preferred by evolution.

The mammalian heart is an omnivorous organ which can use any substrate it is supplied with. To cope with transient changes in substrate supply, this is clearly an advantageous adaptation. However, the ability of the heart to cope with sustained shifts in substrate preference, away from its normal balance, is limited. The balance of substrate utilization by the heart has been slowly established through its evolution, and the pathways which deliver substrates to their myriad cellular fates have evolved with this balance 'in mind'. Although short term shifts in preference are easily handled, sustained shifts lead to the cytoplasmic accumulation of metabolic intermediates, some of which are toxic, and to an increase in flux of substrates through alternate, potentially maladaptive, pathways. The main exception is the adaptation to hypoxia and hypobaria; the hearts of peoples who live at high altitude, such as the Sherpas, Tibetans and Andean natives, use 60 % glucose and 40 % fatty acids [10–13]. This appears to be a beneficial adaptation because glucose oxidation is more efficient in a hypobaric environment. However, most instances of a sustained shift in cardiac substrate preference, either towards glucose oxidation or towards fatty acid oxidation, are associated with disease. This phenomenon has a clear evolutionary origin.

3 The Pathogenesis of Diabetic Cardiomyopathy

The diabetic state immediately changes the delivery and utilization of metabolic substrates by the heart. Myocardial glucose transport is all but lost as a result of decreased Glut-4 translocation to the membrane, and decreased total Glut-4 protein and mRNA levels [14]; this is a direct result of the loss of insulin signalling. In contrast, fatty acid delivery to the heart from the coronary lumen by the enzyme lipoprotein lipase (LPL), and its subsequent uptake by fatty acid transporters, is increased to such an extent that it swamps the capacity of the heart to utilize fatty acids, even though cardiac fatty acid oxidation is increased [15]. Cytoplasmic lipids therefore accumulate, mainly in the form of long-chain acyl-CoA's [15]. These are converted into the toxic substance ceramide, which induces reactive oxygen species (ROS) and cardiomyocyte apoptosis [16]. This process has been named 'lipotoxicity' [15, 17]. Fatty acids bind and activate peroxisome proliferator-activated receptors (PPAR's) of which PPAR-α is the key isoform in the heart. It acts as a 'lipostat' which induces genes involved in fatty acid metabolism [18, 19]. PPAR-α-overexpression induces a phenotype similar to that seen in diabetic cardiomyopathy, and worsens the diabetic phenotype [15]. Conversely, deletion of PPAR-α protects against diabetic cardiomyopathy [15]. The phenotype of PPAR-α overexpression is rescued by knocking out cardiac lipoprotein lipase [20]. PPAR-α is therefore essential to the pathogenesis of diabetic cardiomyopathy, and seemingly requires the cytoplasmic accumulation of fatty acid intermediates.

High rates of fatty acid oxidation increase the ratios of $NADH/NAD^+$ and acetyl CoA/free CoA, both of which feed back into and inhibit glucose oxidation by decreasing flux through pyruvate dehydrogenase (PDH) [15]. High rates of fatty acid oxidation also increase citrate production, which in turn nhibits glycolysis by inhibiting the key glycolytic enzyme phosphofructokinase (PFK). This is referred to as the Randle Cycle, and it results in an intracellular accumulation of intermediate products of glucose metabolism [2].

As mentioned above, inhibition of the main pathways of glucose metabolism increases the flux of glucose through alternate, potentially maladaptive, pathways. The first of these is the polyol pathway. Aldose reductase is an enzyme which catalyses the conversion of glucose to sorbitol, which is subsequently converted into fructose by sorbitol dehydrogenase (SDH) using NAD^{+29} [21]. Excess glucose enters the polyol pathway and selectively stimulates aldose reductase without affecting sorbitol dehydrogenase; leading to an accumulation of sorbitol [21]. This has several consequences: the $NADH/NAD^+$ ratio is increased, leading to inhibition of glycolysis, and the consumption of NADPH increases. Reduction in the availability of NADPH compromises the ability of myocytes to regenerate reduced glutathione [22], potentially interfering with the heart's ability to cope with oxidative stress. Effects on myocardial gene expression have been postulated but not extensively studied [23]. The second pathway is the hexosamine biosynthetic pathway, which converts glucose-6-phosphate to hexosamine-6-phosphate which, in turn, is converted into uridine-5′-diphosphate-N-acetylglucosamine (UDP-GlcNAc). Approximately 5 % of the glucose which enters cardiac myocytes is metabolized by this route. UDP-GlcNAc serves as an important substrate for protein glycosylation which is thought to play a role in decreased expression of several key components of the mitochondrial oxidative phosphorylation complexes as well as increased expression of pro-inflammatory proteins [22]. An important mechanism of this effect is the glycosylation of key transciption factors such as sp-1, leading to an increase in their binding affinity for DNA [24]. One of the consequences of this may be activation of the fetal gene program [25].

The shift toward exclusive use of fatty acid oxidation may be energetically detrimental, partly because fatty acid oxidation requires more oxygen per mole of ATP produced, and partly because it is associated with increased mitochondrial uncoupling. It also, at a global level, restricts the reaction norm; the heart is unable to use the range of substrates it can normally use (glucose being the key example). The heart is therefore effectively locked into a maladaptive state from which it cannot escape by its own devices.

In parallel with this metabolic remodeling, contractile remodeling also occurs. It is possible that contractile remodeling is a consequence of metabolic remodeling. Induction of the so called 'fetal gene program' results in a shift from the fast V_1 isomyosin pattern seen in the normal heart to a predominantly V_3 pattern in the diabetic heart [26]. The fetal gene program involves the re-expression within the heart of genes that are expressed during fetal development (e.g. skeletal muscle actin, beta myosin heavy chain, atrial natriuretic peptide within the ventricle) along with blunting of the expression of genes expressed in the adult heart (e.g. alpha-myosin heavy chain, cardiac actin, sarcoplasmic reticulum calcium ATPase-2 (SERCA-2)). The decrease in SERCA-2 is one of the changes which causes disturbances in calcium handling. The debate continues as to whether activation of this program is adaptive or maladaptive. In the setting of mechanical overload and pathological hypertrophy, re-expression of the fetal gene program allows the stretched fibres to contract at their usual energy cost, because Vmax is slowed. The adaptation is therefore helpful for individual fibres. Likewise, the fetal gene program may also be beneficial at the level of the cell, because it supports pro-survival signalling pathways, including the Akt pathway [27, 28]. However, at the level of the heart itself, the resulting slowing of Vmax is the initial deleterious change that eventually leads to heart failure.

Changes to the ultrastructure are also observed, including alterations in myofibrillar arrangements, disrupted mitochondria and an increased cytoplasmic area [29]. It is unclear whether these are causes or consequences of metabolic remodeling. The metabolic pathways within the cardiomyocyte are spatially organized, and although the majority of the cell's energy is derived from fatty acid oxidation, there are particular ATPases which rely more heavily on glycolytic ATP production (see [30] for review); these include the ATPases involved in sarcoplasmic reticulum function, and it therefore makes sense that sarcoplasmic reticulum function, and therefore calcium handling, would be susceptible to both the ultrastructural and the metabolic changes which occur in the diabetic heart.

When myocardium starts to fail in the non-diabetic heart, the reaction norms aim to restore normal economy (calories/g of tension). Two of the adaptations which permit this are reexpression of the fetal phenotype (with the consequent slowing of Vmax), and a decrease in the heat produced per gram of tension, achieved through shifts in metabolism towards glucose oxidation [31]. The former adaptation occurs in the diabetic heart, but the latter adaptation is clearly not available to the diabetic heart. One intriguing question, very difficult to answer in practice, is whether a switch towards glucose oxidation, is more cardioprotective, or at least less maladaptive, than a switch to fatty acid oxidation, because of a need to limit the heat produced per gram of tension. A third mechanism for restoring cardiac economy is provided by cardiac hypertrophy, which increases muscle mass to restore wall stress. These processes are beneficial at the level of the cardiomyocyte, because they improve economy, but they become maladaptive at the level of the heart because a slowing of Vmax contributes to decreased contractile function and remodelling of the heart impairs its mechanical function. The basic problem is that the adaptive responses seem to favour restoration of economy over restoration of inotopy. In other words, they have evolved to serve the needs of the cells but not of the organ they form.

The changes discussed so far induce a mild diastolic dysfunction which is too subtle to be detected by routine echocardiography and is only appreciable on Doppler echocardiography. The superimposition of ventricular hypertrophy, cardiomyocyte apoptosis and necrosis, and fibrosis causes progression of the diastolic dysfunction and induces a mild systolic dysfunction sufficient to activate the renin-angiotensin system. Autonomic neuropathy also occurs and exacerbates the dysfunction. As the disease reaches its most advanced stage, microangiopathy develops, and the superimposition of hypertension and ischemic heart disease leads to combined systolic and diastolic function, severe enough to activate the sympathetic nervous system. Within these responses, a range of other maladaptive responses can be observed. Activation of the renin-angiotensin and sympathetic nervous systems are maladaptive responses of systems which have not evolved to discern and respond differentially to cardiogenic causes of low blood pressure. The superimposition of pressure overload may trigger mechanosensor-mediated contractile remodeling. It is impossible to dissect the relative importance of metabolic, hormonal and mechanical stimuli to a particular reaction norm. However, the fact that metabolic and contractile remodeling predates any appreciable pressure overload in the pathogenesis of diabetic cardiomyopathy strongly suggests that the early triggers are biochemical and hormonal rather than mechanical.

4 The Mechanisms of the Reaction Norms

There are two major mechanisms by which the metabolic milieu could change the cardiac phenotype, thereby creating the reaction norms. The first involves molecular transcriptional regulators such as PPAR-α or the effectors of the hexosamine biosynthetic pathway which have already been discussed. However, an additional feature of this response is worthy of brief consideration. One of the most crucial adaptations a cell, tissue, organ system or organism requires is the ability to anticipate and prepare for predictable cyclical changes in the environment. If the appearance of a stimulus follows a circadian pattern, then the reaction norm would be well served by linking the phenotype not just to the environmental signal, but to an internal clock which can anticipate the appearance of that signal. Such a molecular clock exists, and it is intricately interwoven with a range of metabolic loops [32]. Intriguingly, diabetes induces a phase advance in the clock, which could exacerbate an already maladaptive state. There is a difference in the disposition of glucose and fatty acids between the awake and sleep phases [33, 34]. During the awake phase, glucosyl units are channelled towards complete oxidation, whereas during the sleep phase, exogenous fatty acids are channelled into synthetic pathways for phospholipids, DAG and TAG and oxidation of fatty acids is reduced. The susceptibility of the heart to cytoplasmic accumulation of fatty acids and depression of cardiac function is normally greatest during the sleep phase. This has implications for the pathogenesis of the metabolic syndrome and type 2 diabetes which are beyond the scope of the present discussion. What is relevant is that the phase shift would be expected to alter the periods during which the diabetic heart is susceptible to depression of cardiac function so that they occur during the awake phase (see [35] for review).

The second potential mechanism is epigenetic regulation. Epigenetics refers to heritable information other than the DNA sequence. Two of the most common and best characterized mechanisms are DNA methylation and acetylation. In general, genes are inactivated by the attachment of methyl groups to cytosines or the detachment of acetyl groups from lysines on histones. Methylation of histones can either activate or deactivate a gene. Epigenetic regulation is known to be intricately linked to calorie availability through mitochondrial energetics (see Fig. 3). This nuclear-mitochondrial interaction has an evolutionarily ancient origin [36]. The original proto-nucleus-cytosol had a limited metabolic capacity which was relieved when a symbiotic relationship developed between the proto-nucleus-cytosol and an oxidative protobacterion, which became the proto-mitochondrion, giving rise to the first eukaryotic cells. The limiting factor for the growth and replication of these cells was the amount of energy the proto-mitochondria could produce, which in turn was limited by the calorie availability. It was therefore necessary for growth and replication to be regulated so that it could respond to calorie availability. Epigenetic mitochondrial-nuclear interactions therefore evolved. Given the fact that diabetes is associated with fundamental alterations in metabolic fluxes, it is highly likely that there are associated alterations in the mitochondrial-nuclear interactions which lead to altered gene expression. However, this is a virtually untapped area of research.

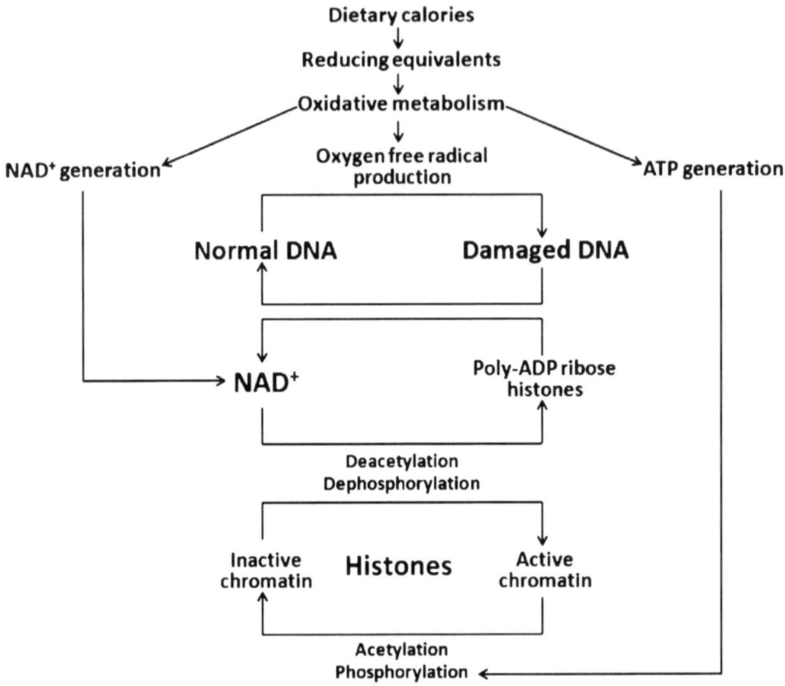

Fig. 3 Simplified schematic providing an overview of the regulation of the epigenome by energetics. Oxidative metabolism of reducing equivalents leads to the production of NAD^+, ATP and oxygen free radicals. Oxygen free radicals produce DNA damage, the repair of which is coupled via NAD^+ to epigenetic regulation. NAD^+ itself drives deacetylation reactions, while ATP fuels phosphorylation reactions, including chromatin phosphorylation

There is emerging evidence that vascular complications of diabetes can develop even with intensive glycemic control. One possible explanation for this is the phenomenon of hyperglycemic memory, in which a transient exposure to hyperglycemia produces permanent cellular effects which persist long after the restoration of normoglycemia. There is evidence which suggests that DNA methylation induced by transient hyperglycemia could be the underlying mechanism of this effect [37]. Furthermore, a wide range of genes involved in the pathogenesis of diabetic cardiomyopathy have been shown to be subject to epigenetic regulation. These include the fetal gene program, endothelin-1, glut-4, the angiotensin-1 receptor, transforming growth factor-β, matrix metalloproteinases, PPAR-γ and a range of pro-inflammatory genes (see [38, 39] for review). A few such genes have also been shown to be epigenetically regulated in the heart itself, either by hyperglycemia or chronic diabetes [40–49] (Table 1).

The final stage in the development of diabetic cardiomyopathy is associated with the superimposition of ischemic heart disease, and epigenetic regulation, in addition to mediating the primary pathogenesis of the disease, could conceivably increase

Table 1 A summary of currently described examples of pathologically-relevant epigenetic regulation in the diabetic heart

Gene	Cell type or organ	Effect	Reference
NF-Kb p65 subunit	Vascular cells	Activation by hyperglycemia	[41, 42]
Interleukin-6 and monocyte chemotactic protein-1	H9C2 cells	Activation by hyperglycemia	[43]
IGF-1 receptor	Cardiomyocytes	Repressed by hyperglycemia	[48]
Hypertrophic genes	Mouse heart	Activation in type 2 diabetes, increased by renal failure	[40]
Angiogenic factors	Human cardiac tissue	Differential expression in cardiomyopathy	[44]
SERCA-2	Cardiomyocytes	Repression in diabetes, mediated by tumour necrosis factor-α	[45]
p21 (cell cycle control protein)	Cardiomyocytes	Up-regulation in streptozotocin-diabetes	[47]
Cyclin D1	Cardiomyocytes	Repression in streptozotocin-diabetes	[46]
Liver X receptor-α	Rat heart	Activation in streptozotocin-diabetes	[49]

the susceptibility of the diabetic heart to ischemic damage. The Reperfusion Injury Salvage Kinase (RISK) pathway (which includes Akt and Erk 1/2), the Survivor Activating Factor Enhancement pathway (which includes STAT-3) and the Sirtuin-1 (Sirt-1)-p53 pathway are activated in the setting of myocardial infarction and are believed to be cardioprotective in the short-term, and to be involved in cardiac remodelling in the long-term [50, 51]. There is some evidence that the expression and activity of the proteins in these pathways is inhibited in the diabetic heart; the mechanism appears to partly involve decreased acetylation of a transcription factor involved in regulating Sirtuin-1 expression, which hints at a possible link to epigenetic regulation [52].

An additional form of gene regulation exists in the form of micro-RNAs (miRNAs), which are endogenous short non-coding RNAs that inhibit gene expression either by repressing translation or by enhancing degradation of their target mRNAs. Micro-RNAs are evolutionarily conserved, and their evolution and development has been linked to the emergence of organismal complexity (see [53] for review). In the context of heart disease, miRNA-mediated mechanisms have, to date, mostly been identified in the pathways of cardiac hypertrophy, fibrosis and apoptosis. Intriguingly, there is preliminary evidence of bi-directional modulation between miRNAs and epigenetic regulatory pathways. A subgroup of miRNAs called epi-miRNAs target the proteins involved in the epigenetic machinery. Conversely, the first evidence for epigenetic regulation of miRNA expression is beginning to emerge (see [39] for review).

5 Conclusion

The heart is an omnivorous organ which can use any substrate it is supplied with. The metabolic machinery of the heart is exquisitely attuned both to its metabolic needs and to the available energy substrates in its local environment. Diabetic cardiomyopathy is a disease process which arises as a result of the inability of the heart to adapt to a diabetic metabolic milieu. The heart becomes locked into a progressively maladaptive state from which it cannot escape by its own devices; due to the phenomenon of hyperglycemic memory, even restoration of a normal milieu may not be sufficient to completely reverse the remodeling. The pathways which initiate, progress and perpetuate this downward spiral are the same pathways which normally allow the heart to sense and respond to its local metabolic environment. These include metabolite-sensitive transcriptional regulatory pathways and, most probably, epigenetic and miRNA regulatory pathways. Overall, the application of evolutionary concepts provides a valuable framework for understanding the origins and importance of metabolic and contractile disturbances in the diabetic heart, and a strong rationale for the use of metabolic therapy as a treatment.

References

1. Bishopric NH (2005) Evolution of the heart from bacteria to man. Ann N Y Acad Sci 1047: 13–29
2. Stanley WC, Chandler MP (2002) Energy metabolism in the normal and failing heart: potential for therapeutic interventions. Heart Fail Rev 7:115–130
3. Driedzic WR, Sidell BD, Stowe D et al (1987) Matching of vertebrate cardiac energy demand to energy metabolism. Am J Physiol 252:R930–R937
4. Clark AJ, Gaddie R, Stewart CP (1932) The carbohydrate metabolism of the isolated heart of the frog. J Physiol 75:311–320
5. Neely JR, Morgan HE (1974) Relationship between carbohydrate metabolism and energy balance of heart muscle. Annu Rev Physiol 36:413–459
6. Sidell BD, Stowe DB, Hansen CA (1984) Carbohydrate is the preferred metabolic fuel of the hagfish (*Myxine glutinosa*) heart. Physiol Zool 7:266–273
7. Moyes CD (1996) Cardiac metabolism in high performance fish. Comp Biochem Physiol 113A:69–75
8. Wu JJ, Chang I (1948) The glycolytic activity of the hearts of vertebrates. Q J Exp Physiol Cogn Med Sci 34:91–95
9. Moyes CD, Suarez RK, Hochachka PW (1989) A comparison of fuel preferences of mitochondria from vertebrates and invertebrates. Can J Zool/Rev Can Zool 68:1337–1349
10. Beall CM (2007) Detecting natural selection in high-altitude human populations. Respir Physiol Neurobiol 158:161–171
11. Moore LG (2001) Human genetic adaptation to high altitude. High Alt Med Biol 2:257–279
12. Rupert JL, Hochachka PW (2001) The evidence for hereditary factors contributing to high altitude adaptation in Andean natives: a review. High Alt Med Biol 2:235–256
13. Wu T, Kayser B (2006) High altitude adaptation in Tibetans. High Alt Med Biol 7:193–208
14. Camps M, Castello A, Munoz P et al (1992) Effect of diabetes and fasting on GLUT-4 (muscle/fat) glucose-transporter expression in insulin-sensitive tissues. Heterogeneous response in heart, red and white muscle. Biochem J 282(Pt 3):765–772

15. Severson DL (2004) Diabetic cardiomyopathy: recent evidence from mouse models of type 1 and type 2 diabetes. Can J Physiol Pharmacol 82:813–823
16. Bielawska AE, Shapiro JP, Jiang L et al (1997) Ceramide is involved in triggering of cardiomyocyte apoptosis induced by ischemia and reperfusion. Am J Pathol 151:1257–1263
17. Chiu HC, Kovacs A, Blanton RM et al (2005) Transgenic expression of fatty acid transport protein 1 in the heart causes lipotoxic cardiomyopathy. Circ Res 96:225–233
18. Brandt JM, Djouadi F, Kelly DP (1998) Fatty acids activate transcription of the muscle carnitine palmitoyltransferase I gene in cardiac myocytes via the peroxisome proliferator-activated receptor alpha. J Biol Chem 273:23786–23792
19. Djouadi F, Brandt JM, Weinheimer CJ et al (1999) The role of the peroxisome proliferator-activated receptor alpha (PPAR alpha) in the control of cardiac lipid metabolism. Prostaglandins Leukot Essent Fatty Acids 60:339–343
20. Duncan JG, Bharadwaj KG, Fong JL et al (2010) Rescue of cardiomyopathy in peroxisome proliferator-activated receptor-alpha transgenic mice by deletion of lipoprotein lipase identifies sources of cardiac lipids and peroxisome proliferator-activated receptor-alpha activators. Circulation 121:426–435
21. Narayanan S (1993) Aldose reductase and its inhibition in the control of diabetic complications. Ann Clin Lab Sci 23:148–158
22. Wold LE, Ceylan-Isik AF, Ren J (2005) Oxidative stress and stress signaling: menace of diabetic cardiomyopathy. Acta Pharmacol Sin 26:908–917
23. Liu TP, Juang SW, Cheng JT et al (2005) The role of sorbitol pathway and treatment effect of aldose reductase inhibitor ONO2235 in the up-regulation of cardiac M2-muscarinic receptors in streptozotocin-induced diabetic rats. Neurosci Lett 383:131–135
24. Jiang T, Che Q, Lin Y et al (2006) Aldose reductase regulates TGF-beta1-induced production of fibronectin and type IV collagen in cultured rat mesangial cells. Nephrology (Carlton) 11: 105–112
25. Young ME, Yan J, Razeghi P et al (2007) Proposed regulation of gene expression by glucose in rodent heart. Gene Regul Syst Bio 1:251–262
26. Scognamiglio R, Avogaro A, Negut C et al (2004) Early myocardial dysfunction in the diabetic heart: current research and clinical applications. Am J Cardiol 93:17A–20A
27. Sack MN, Yellon DM (2003) Insulin therapy as an adjunct to reperfusion after acute coronary ischemia: a proposed direct myocardial cell survival effect independent of metabolic modulation. J Am Coll Cardiol 41:1404–1407
28. Jonassen AK, Mjos OD, Sack MN (2004) p70s6 kinase is a functional target of insulin activated Akt cell-survival signaling. Biochem Biophys Res Commun 315:160–165
29. Searls YM, Smirnova IV, Fegley BR et al (2004) Exercise attenuates diabetes-induced ultrastructural changes in rat cardiac tissue. Med Sci Sports Exerc 36:1863–1870
30. Weiss JN, Yang L, Qu Z (2006) Systems biology approaches to metabolic and cardiovascular disorders: network perspectives of cardiovascular metabolism. J Lipid Res 47:2355–2366
31. Swynghedauw B, Delcayre C, Samuel JL et al (2010) Molecular mechanisms in evolutionary cardiology failure. Ann N Y Acad Sci 1188:58–67
32. Durgan DJ, Young ME (2010) The cardiomyocyte circadian clock: emerging roles in health and disease. Circ Res 106:647–658
33. Meng QJ, Logunova L, Maywood ES et al (2008) Setting clock speed in mammals: the CK1 epsilon tau mutation in mice accelerates circadian pacemakers by selectively destabilizing PERIOD proteins. Neuron 58:78–88
34. Um JH, Yang S, Yamazaki S et al (2007) Activation of 5′-AMP-activated kinase with diabetes drug metformin induces casein kinase Iepsilon (CKIepsilon)-dependent degradation of clock protein mPer2. J Biol Chem 282:20794–20798
35. Bray MS, Young ME (2008) Diurnal variations in myocardial metabolism. Cardiovasc Res 79:228–237
36. Wallace DC, Fan W (2010) Energetics, epigenetics, mitochondrial genetics. Mitochondrion 10:12–31

37. Villeneuve LM, Natarajan R (2010) The role of epigenetics in the pathology of diabetic complications. Am J Physiol Renal Physiol 299:F14–F25
38. Singh GB, Sharma R, Khullar M (2011) Epigenetics and diabetic cardiomyopathy. Diabetes Res Clin Pract 94:14–21
39. Asrih M, Steffens S (2012) Emerging role of epigenetics and miRNA in diabetic cardiomyopathy. Cardiovasc Pathol 22:117–125
40. Gaikwad AB, Sayyed SG, Lichtnekert J et al (2010) Renal failure increases cardiac histone h3 acetylation, dimethylation, and phosphorylation and the induction of cardiomyopathy-related genes in type 2 diabetes. Am J Pathol 176:1079–1083
41. El-Osta A, Brasacchio D, Yao D et al (2008) Transient high glucose causes persistent epigenetic changes and altered gene expression during subsequent normoglycemia. J Exp Med 205: 2409–2417
42. Brasacchio D, Okabe J, Tikellis C et al (2009) Hyperglycemia induces a dynamic cooperativity of histone methylase and demethylase enzymes associated with gene-activating epigenetic marks that coexist on the lysine tail. Diabetes 58:1229–1236
43. Yu XY, Geng YJ, Lin QX et al (2010) High glucose leads to increased inflammatory gene expression via epigenetic histone H3 lysine 9 methylation in cardiomyocyte. Circulation 122: 8891–8898
44. Movassagh M, Choy MK, Goddard M et al (2010) Differential DNA methylation correlates with differential expression of angiogenic factors in human heart failure. PLoS ONE 5:e8564
45. Kao YH, Chen YC, Cheng CC et al (2010) Tumor necrosis factor-alpha decreases sarcoplasmic reticulum Ca^{2+}-ATPase expressions via the promoter methylation in cardiomyocytes. Crit Care Med 38:217–222
46. Kuan CJ, al-Douahji M, Shankland SJ (1998) The cyclin kinase inhibitor p21WAF1, CIP1 is increased in experimental diabetic nephropathy: potential role in glomerular hypertrophy. J Am Soc Nephrol 9:986–993
47. Kaneto H, Kajimoto Y, Fujitani Y et al (1999) Oxidative stress induces p21 expression in pancreatic islet cells: possible implication in beta-cell dysfunction. Diabetologia 42:1093–1097
48. Yu XY, Geng YJ, Liang JL et al (2010) High levels of glucose induce apoptosis in cardiomyocyte via epigenetic regulation of the insulin-like growth factor receptor. Exp Cell Res 316: 2903–2909
49. Cheng Y, Liu G, Pan Q et al (2011) Elevated expression of liver X receptor alpha in myocardium of streptozotocin induced diabetic rats. Inflammation 34:698–706
50. Hausenloy DJ, Yellon DM (2004) New directions for protecting the heart against ischaemia-reperfusion injury: targeting the Reperfusion Injury Salvage Kinase (RISK)-pathway. Cardiovasc Res 61:448–460
51. Fuglesteg BN, Suleman N, Tiron C et al (2008) Signal transducer and activator of transcription 3 is involved in the cardioprotective signalling pathway activated by insulin therapy at reperfusion. Basic Res Cardiol 103:444–453
52. Vahtola E, Louhelainen M, Forsten H et al (2010) Sirtuin1-p53, forkhead box O3a, p38 and post-infarct cardiac remodeling in the spontaneously diabetic Goto-Kakizaki rat. Cardiovasc Diabetol 9:5
53. Berezikov E (2011) Evolution of microRNA diversity and regulation in animals. Nat Rev Genet 12:846–860

Induction of Metabolic Syndrome by Excess Fructose Consumption

Weng-Yew Wong and Lindsay Brown

Abstract Fructose is an important nutritive component of foods such as honey and fruit, but this easily available sweetener may contribute to increased caloric consumption from overeating. Fructose is now a major component of the Western diet, with increased consumption associated with obesity, metabolic syndrome, and cardiovascular disorders in observational and short-term intervention studies, mainly in animal models. Rodent studies have identified possible mechanisms for the adverse effects of fructose when ingested in large amounts. Fructose promoted de novo lipogenesis, inflammation, and increased sympathetic tone. These mechanisms induced hepatic insulin resistance, increased total and visceral fat mass with accumulation of ectopic fat in the liver and skeletal muscle, and dyslipidemia. Fructose reduced leptin and insulin signals for satiety, caused structural and functional damage to the heart and blood vessels, and disrupted the diversity of the gut microbiota. These early effects may initiate the development of the metabolic syndrome. Despite this evidence from rodents, there are few long-term intervention studies in humans, especially at a moderate dose. The definition of prudent fructose consumption is needed, but this will require carefully controlled dose–response studies in humans.

Keywords Fructose • Metabolic syndrome • Leptin • Insulin • Lipogenesis • Hypertension • Inflammation • Gut microbiota • Cardiac hypertrophy

---

W.-Y. Wong
School of Biomedical Sciences, The University of Queensland, Brisbane, QLD 4072, Australia

L. Brown (✉)
School of Health, Nursing and Midwifery, University of Southern Queensland, Toowoomba, QLD 4350, Australia
e-mail: Lindsay.Brown@usq.edu.au

B. Turan and N.S. Dhalla (eds.), *Diabetic Cardiomyopathy: Biochemical and Molecular Mechanisms*, Advances in Biochemistry in Health and Disease 9,
DOI 10.1007/978-1-4614-9317-4_3, © Springer Science+Business Media New York 2014

1 Introduction

Fructose, a polyhydroxyketone with chemical formula $C_6H_{12}O_6$, is a naturally occurring monosaccharide isomeric with glucose and galactose. In contrast to fructose with a ketone group at position 2, glucose and galactose have an aldehyde group at position 1 of the carbon chain. In solution, fructose can be present as α- or β-pyranoside and furanoside rings [1]. In nature, fructose is found in honey and fruits, hence the name fruit sugar. In addition, fructose occurs as sucrose, a disaccharide of one glucose molecule linked through an α1–4 glycosidic bond to one fructose molecule. Sugar consumption in the United States (USA) was predominantly sucrose until about 1970 with the introduction of fructose in high-fructose corn syrup (HFCS), increasing fructose intake to approximately 49 g/day but decreasing the sucrose intake in a total sugar intake around 110 g/day in the NHANES 1999–2004 results [1]. Sugar consumption in Australia was approximately 42 kg/person/year in 2011 (approximately 115 g/person/day), down from 57 kg/person/year in 1951 [2]. Fructose intake in Australia has not been reported, but intake in 14-year-old adolescents in the Raine cohort in Western Australia was 48.3 ± 20.2 g/day for girls and 58.8 ± 26.6 g/day for boys [3].

Fructose, considered to be about 1.7 times as sweet as glucose with a more rapid response, is produced commercially from corn through starch isolation and hydrolysis to glucose, followed by enzymatic isomerization of part of the glucose into fructose using enzymes grown in bacteria. Fructose is then diluted, providing the commercially available mixture known as HFCS, which contains either 42 % or 55 % fructose [4–6]. HFCS appears to be a cheaper sweetener than other sugar products as it is derived from renewable and abundant corn; its consumption is progressively increasing [6–8]. The escalating use of HFCS in food and beverages has generated interest in both fructose metabolism and the possible detrimental effects during chronic consumption. Increased dietary fructose contributes to caloric overconsumption from overeating and the associated energy imbalance because it does not stimulate insulin and leptin secretion that promotes satiety. In addition, fructose intake is a potential risk factor for obesity, diabetes, dyslipidemia, and cardiovascular morbidity, collectively known as metabolic syndrome [1, 7, 9]. There are many animal studies showing that chronic high fructose consumption induced insulin resistance [10–13], obesity [14, 15], dyslipidemia [16–19], hypertension [11, 20–22], fatty liver [19, 23–26], and ventricular hypertrophy [11] as signs of the metabolic syndrome. In humans, the association of fructose consumption with metabolic syndrome is less clear, with disagreement regarding the role of chronic fructose consumption in the development of metabolic syndrome. High doses of fructose (17–25 % of total energy for 6 days to 10 weeks) caused insulin resistance [27, 28], hyperlipidemia [27, 29, 30], and intraabdominal fat accumulation [28] in some human studies but not in others [31–33]. High intakes of pure fructose (60 % of diet) have been utilized in short-term studies in rats to study the biological pathways of fructose, but these doses are unlikely in humans. It is clear that the evidence from animal studies has raised serious concerns about chronic fructose consumption and

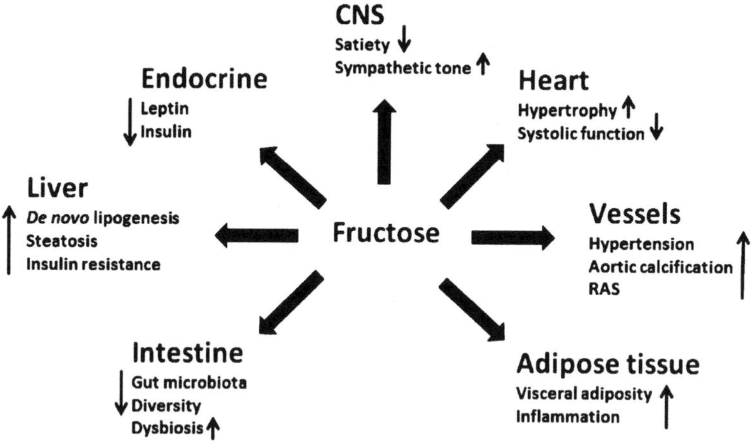

Fig. 1 Detrimental effects of excess fructose consumption in rats

metabolic syndrome in humans. It is generally agreed that human data on adverse metabolic effects of fructose consumption are insufficient. However, increased fructose consumption in human subjects has been linked to each of the characteristic features of the metabolic syndrome.

Several mechanisms have been suggested linking fructose intake to the changes in the metabolic syndrome. Fructose activated genes involved in de novo lipogenesis, generating fatty acids that cause dyslipidemia, fatty liver, and insulin resistance [30, 34, 35]. Fructose impaired the anorexic action of leptin [36] and the satiety response of insulin [7], causing obesity. Fructose induced inflammation and oxidative stress in liver [37]. In response to fructose overload, gut microbiota diversity was disturbed, favoring the *Firmicutes*, which cause intestinal inflammation that later develops as endotoxemia [38]. Figure 1 depicts the detrimental effects of excess fructose consumption, at least in rats.

2 Metabolic Syndrome

Metabolic syndrome was recognized by the World Health Organization (1998) as the clustering of interrelated risk factors for cardiovascular disease and type 2 diabetes, including hypertension, dyslipidemia, hyperglycemia, insulin resistance, and abdominal obesity [39–41]. The metabolic syndrome is a global phenomenon. Based on National Health and Nutrition Examination Survey (NHANES) 2003–2006 data, 35.1 % of US men and 32.6 % of US women 20 years of age and older met the criteria for metabolic syndrome [42]. The prevalence of metabolic syndrome ranged between 13.4 % and 30.7 % in Australian adults [43]. In the Japanese population, 51 % of male and 38 % of female subjects met World Health Organization

(WHO) criteria for metabolic syndrome [44]. In Chinese men, the prevalence of metabolic syndrome associated with hypertension was 32.9 % and 53.1 % by the National Cholesterol Education Program (NCEP) Adult Treatment Panel III definition [45]. In an urban Indian population, the NCEP definition included 31.6 % with metabolic syndrome, with higher prevalence in women (39.9 %) than in men (22.9 %) [46].

Metabolic syndrome is associated with increased abdominal adipose tissue together with increased plasma concentrations of reactive oxidant species, nonesterified fatty acids, and oxidized low density lipoproteins (LDL) [39, 47]. Adipose tissue includes lipid-filled adipocytes, endothelial cells, pericytes, fibroblasts, preadipocytes, mast cells, and immune cells such as resident macrophages and T cells to function not only as an energy store but also as a dynamic organ [48]. Adipocytes and immune cells within adipose tissue secrete many bioactive molecules, known as adipokines and pro-inflammatory cytokines, to initiate vascular dysfunction and impaired glucose metabolism [47, 49]. Increased production of nonesterified fatty acids by adipocytes inhibited carbohydrate metabolism via substrate competition and impaired insulin signaling [50].

Inflammatory lipid mediators such as prostaglandins, thromboxanes, and leuko-trienes are involved in increased cardiac fibrosis and chronic inflammatory diseases [51]. These mediators can act as ligands for immune receptors such as class A G protein-coupled receptors and Toll-like receptors [51]. Lipid mediators localized in the circulation and adipose tissue may bind to these immune receptors and induce low-grade tissue inflammation, causing metabolic dysfunction [52]. Fat accretion may alter synthesis and action of metabolic hormones, with deposition in liver, heart, muscle, and pancreatic beta cells causing lipotoxicity, or initiate specialized extracellular and intracellular signaling through lipid-derived mediators, thereby causing systemic inflammation and insulin resistance [53].

In metabolic syndrome, inflammation causes both insulin resistance and vascular dysfunction [52, 53]. Tumor necrosis factor-α (TNF-α) expression in adipose tissue induced by obesity can cause systemic insulin resistance [54]. In addition, adipocytes secrete proinflammatory cytokines that are detrimental to the vasculature as they cause endothelial dysfunction [55]. Production of proinflammatory cytokines is linked to nuclear factor-κB (NF-κB) activation [56]. Activation of NF-κB initiates TNF-α-induced insulin resistance [57].

3 Fructose Absorption and Metabolism

In nature, fructose exists mainly as free fructose, the monosaccharide, or as sucrose, the disaccharide. The body can absorb free fructose directly from the gastrointestinal tract. However, the major source of fructose is sucrose, which upon ingestion is hydrolzsed into equimolar amounts of fructose and glucose in the small intestine by sucrase [58]. After absorption, fructose enters the liver through the hepatic portal vein, where it is later metabolized [59]. Transport of fructose from intestine into enterocytes can occur via Na^+-independent transport using the GLUT5 transporter. Fructose can then be transported out of the enterocyte across the basolateral membrane by either GLUT2 or GLUT5 [60]. The presence of fructose rapidly upregulates both GLUT2 and GLUT5 mRNA [61]. GLUT5 knockout in mice induces the symptoms of fructose malabsorption, including distension of the cecum and proximal colon with increased cecal contents, including fluid and gas [62]. Fructose absorption in humans is consistent with the limited absorption capacity of a facilitative transport system [61]. Higher doses of fructose induce fructose malabsorption with gastrointestinal symptoms including chronic diarrhea and abdominal pain, possibly the result of decreased expression of GLUT5, as in infants and toddlers [61]. Chronic fructose increased salt absorption in the small intestine and kidney tubules mediated by upregulation of chloride/base exchangers and the Na^+/H^+-exchanger; this could be the basis of fructose-induced hypertension [63].

The metabolism of fructose in the liver differs from glucose because it is insulin independent. The fructose metabolism pathway involves three enzymes, fructokinase, aldolase B, and triokinase [59]. Although these enzymes are present in liver and kidney in hamster, guinea pig, dog, rat, and human, they can also be found in the small intestine of hamster, guinea pig, and dog [59]. Absence of these enzymes in the intestine of both rats and humans makes rats a good model for human fructose consumption [59]. On entering the liver, fructose is phosphorylated by adenosine triphosphate (ATP) catalyzed by the enzyme fructokinase to form fructose-1-phosphate; this is further converted into the three-carbon phosphate intermediates, dihydroxyacetone phosphate and glyceraldehyde, members of the glycolysis sequence of intermediates, via aldolase B [7]. Triokinase catalyzes the phosphorylation of glyceraldehyde by ATP to form glyceraldehyde-3-phosphate, another intermediate of the glycolytic pathway. A portion of these intermediates enters the gluconeogenic pathway and is converted to glucose or glycogen, or is metabolized to form lactate or pyruvate and later triglycerides. Hence, fructose metabolism bypasses the vital checkpoint of glycolysis, the rate-limiting step, phosphofructokinase [12]. This reaction is negatively regulated by citrate and ATP, which allosterically inhibit phosphofructokinase, thereby preventing further glucose uptake into the liver [64]. As fructose uptake by the liver is not regulated by phosphofructokinase, excess fructose consumption results in larger increases in the triglyceride pool than the consumption of a comparable amount of glucose. Figure 2 summarizes fructose metabolism and glycolysis.

4 Chronic Fructose Consumption and Organ Changes

4.1 Fructose and Fat Accumulation

Increased fructose consumption over the past decades has raised major concerns worldwide as it has been implicated in increased energy intake and obesity [1, 9]. Increased energy intake, body weight, and adiposity occurred in some studies with

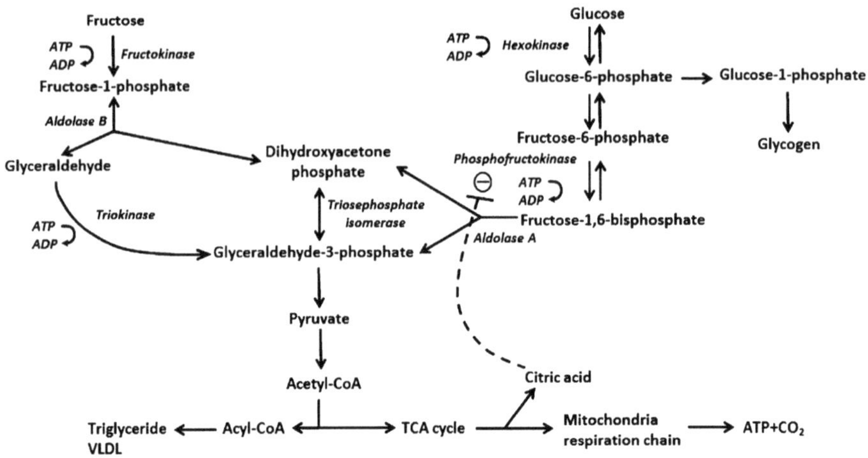

Fig. 2 Fructose metabolism and glycolysis. Fructose bypasses the vital checkpoint of phosphofructokinase, which is negatively regulated by citric acid

animals fed with high fructose [14, 15], but not in others [11, 20, 22, 65]. A study on humans showing different regional fat accumulation between fructose and glucose consumption may explain the detrimental effect of fructose not confounded by body weight difference [28]. Fructose consumption increased visceral adipose tissue deposition whereas consumption of glucose resulted primarily in subcutaneous adipose tissue deposition [28], possibly because of the activation of lipoprotein lipase (LPL), the rate-limiting enzyme involved in the uptake of triglyceride from the circulation for storage in the adipose depots [66]. LPL of subcutaneous adipose tissue is more sensitive to activation by insulin than LPL associated with visceral adipose tissue [28, 67]. In contrast to glucose, fructose does not stimulate insulin secretion from the pancreas because pancreatic β cells lack the transporter GLUT5 [7]. Therefore, it was hypothesized that activation of LPL in subcutaneous fat would be reduced in subjects consuming high fructose. Conversely, high fructose intake would increase the availability of triglycerides for uptake in visceral adipose tissue [28]. Rats fed a 60 % w/w fructose diet for 20 weeks had comparable body weight with controls but exhibited an apparent increase in abdominal fat mass size [68]. Visceral obesity is defined as fat accumulation around the viscera and inside the intraabdominal solid organs. Visceral obesity, or abdominal obesity, is strongly associated with pathogenesis of type 2 diabetes, hypertension, and cardiovascular changes such as atherosclerosis, coronary calcification, and endothelial dysfunction [69]. In addition, abdominal obesity is associated with elevated serum concentrations of small dense LDL cholesterol particles, apo-B [70] and triglycerides, and reduced HDL cholesterol concentrations [71, 72]. Studies on fructose and regional adipose distribution are scarce and future studies looking at the role of fructose in abdominal adiposity are warranted.

Patients with Cushing's syndrome develop central obesity, together with hypertension and insulin resistance, indicating that lipid uptake into adipocytes may be enhanced by enhanced glucocorticoid responses within the adipose tissue. Fructose feeding in rats increased the glucocorticoid-activated enzyme, 11-beta-hydroxysteroid dehydrogenase type 1, increased corticosterone concentrations within the adipose tissue, and enhanced nuclear glucocorticoid receptor accumulation, before the onset of obesity [73]. Increased plasma nonesterified fatty acids suggest a stimulation of lipolysis by this fructose-initiated glucocorticoid action, rather than an increased lipogenesis [73].

4.2 *Fructose and Nervous System*

Insulin- and leptin-induced stimulation of the central nervous system may be one of the mechanisms linked to increased body weight and calorie intake with increased fructose consumption [7, 9]. Fructose does not stimulate secretion of insulin from the pancreas; hence, suppression of food intake by insulin in the central nervous system may not occur after fructose consumption [74]. Insulin secreted from the pancreas enters the central nervous system, where it acts as a humoral feedback regulator of food intake and energy balance [75–77], providing a negative feedback signal and hence lower food intake [7]. Insulin receptors [78] and related intracellular signaling molecules such as insulin receptor substrate 1, which are concentrated in hypothalamic areas such as the arcuate nucleus [79], may mediate the effect of insulin on food intake. The responses of insulin on peripheral tissues and the brain appear to be different. Systemic infusion of insulin in diabetic subjects induced excess body weight gain [80], whereas in a separate study, central infusion of insulin in diabetic rats produced weight loss [75].

In addition, there is consistent evidence for a role of the sympathetic nervous system and the renin-angiotensin system in fructose-induced cardiovascular and renal changes. In humans, infusion of fructose amplified the release of adrenaline in response to hypoglycemia [81]. Fructose-fed mice had increased mean arterial pressure and heart rate during the dark period, when mice are active and sympathetic activity was at the highest. Alpha-adrenoceptor antagonism with prazosin in fructose-fed mice greatly decreased mean arterial pressure [82]. In addition, blood pressure variability, a detector for early abnormalities in autonomic modulation in diabetes, was increased in fructose-fed AT1 wild-type mice during dark phase, suggesting a high-fructose diet activated sympathetic input to the circulation [82]. Suppression of sympathetic activity by adrenal medullectomy, followed by weekly 6-hydroxydopamine injections, abrogated the development of both hyperinsulinemia and hypertension in fructose hypertensive rats without affecting these parameters in control rats [83]. This result suggests that sympathetic excitation may play an early and integral role in the final expression of elevated plasma insulin concentrations and blood pressure in rats fed a high-fructose diet [83]. Fructose feeding in C57/BL mice increased plasma angiotensin II concentrations, accompanied by increased expression of AT1a receptors and tyrosine hydroxylase mRNA in the brainstem locus coeruleus [84]. Treatment with AT1 receptor antisense DNA in fructose-fed rats inhibited AT1 receptors and prevented development of an increased blood pressure [85, 86]. Losartan normalized blood pressure, NADPH oxidase activity, endothelial function, and angiotensin II-induced vasoconstriction in fructose-fed rats. Expression of AT1a receptor mRNA was enhanced in fructose-fed mice [87]. Also, rats with fructose overload had increased blood pressure and insulin resistance combined with reduced cardiac vagal tone [88]. In healthy humans, acute ingestion of fructose increased heart rate, blood pressure, and systolic arterial pressure variability, although cardiovagal baroreflex sensitivity was decreased [89]. Similarly, rats fed a fructose-rich diet for 4 weeks had an impaired vagal component of baroreflex response while the sympathetic component remained intact [90]. These studies suggested that fructose overload may induce autonomic imbalance, which triggers hemodynamic and metabolic changes, as seen in increased sympathetic modulation to the cardiovascular system, and consequently reduced parasympathetic modulation. It is therefore possible that impaired vagal responses would attenuate the satiety signal to the brain, hence increasing body weight or food intake. Adipose, gut, and hepatic tissues that secrete metabolically important hormones are heavily innervated by sympathetic and parasympathetic fibers [91]. Gastric distension, sensors in the portal blood vessels for cholecystokinin, glucose, osmolality, and pH during meal ingestion activate vagal afferents, sending signals from the stomach to brainstem and hypothalamus that result in the perception of fullness and satiety [92]. Thus, autonomic imbalance caused by fructose overload may reduce satiety signaling, leading to obesity. Impairment of the parasympathetic nervous system could be an etiological factor in the pathological process, rather than just a consequence of diabetes [93], which predates the overt development of type 2 diabetes mellitus by many years [90].

The satiety effects of fructose consumption may be related to the impaired anorexic responses of leptin [36]. Leptin derived from adipocytes plays a crucial role in energy balance and body weight regulation through interactions with hypothalamic nuclei to reduce food intake and increase energy expenditure. Leptin, along with insulin, regulates food intake and energy metabolism via neuropeptide systems including neuropeptide-Y and melanocortins [66]. Patients with leptin deficiency exhibited increased hunger and impaired satiety [94]. Magnetic resonance imaging (MRI) scans on leptin-deficient patients injected with leptin after food consumption showed that leptin-responsive neurons at the ventral striatal region provided signals that modulate physiological responses to energy restriction [95]. There appears to be both a short-term and a long-term system for controlling feeding behavior and energy balance [96]. Leptin and insulin provide the signals to the central nervous system in the long-term regulation of the food consumption quantity relative to energy expenditure [97]. In addition, evidence from animals suggested that leptin is necessary for the biosynthesis and secretion of thyrotropin-releasing hormone [98, 99]. Complete leptin deficiency is associated with a moderate degree of hypothalamic hypothyroidism characterized by low free thyroxine and high serum TSH concentrations, which is relatively bio-inactive [100]. It was hypothesized that a reduction in circulating leptin concentrations during prolonged consumption of diets high in energy from fructose could lead to increased energy intake or decreased energy expenditure and weight gain [66].

Leptin resistance ensues when leptin fails to produce metabolic responses despite high plasma concentrations. Leptin failed to induce an anorexic response in rats fed a high-fat diet, predisposing to increased weight gain [36]. The onset of leptin resistance may predate adiposity, as leptin resistance induced by fructose occurred in the absence of an increased body weight or elevated serum insulin concentrations [36]. Two hypotheses were proposed for the mechanism underlying fructose-induced leptin resistance: (1) reduced activation of JAK-mediated STAT3 phosphorylation and (2) impaired leptin transport across the blood–brain barrier. Reduced phosphoSTAT3 appeared to cause defective insulin receptor signaling in the brain, and subsequently failure of downstream signaling events such as anorexic action and regulation of liver glucose fluxes [36, 101]. In addition, fructose is lipogenic because of its ability to induce de novo lipogenesis in the liver, increasing plasma triglyceride concentrations [30, 59]. Serum triglycerides may impair the ability of the blood–brain barrier to transport leptin, an action probably through direct binding of triglycerides to leptin in the circulation or direct action on leptin transporter via a regulatory site controlled by triglycerides [102]. The impaired transport can cause leptin resistance [102].

4.3 Fructose and De Novo Lipogenesis

Animal studies on fructose feeding have shown hyperlipidemia [17–19] with high fructose feeding in mice for 16 weeks increasing plasma and muscle triglycerides and plasma free fatty acids, similar to high-fat-fed mice [17]. In livers of mice chronically fed with a 30 % fructose solution, triglycerides concentrations were approximately 2-fold higher than in glucose-fed mice and 5.3-fold higher than in controls [19]. Rats fed with a cholesterol-free high-fructose diet had increased serum total cholesterol, free cholesterol, and triglyceride and phospholipid concentrations. Increased serum total cholesterol concentrations showed enhancement of its synthesis in the liver because no cholesterol was given in the diet [18]. Hypertriglyceridemia reflects an imbalance between the production and clearance rates of very low density lipoprotein–triglycerides. The main factor influencing hepatic triglyceride secretion is fatty acid availability [103]. As already described, fructose-derived intermediates rapidly enter the glycolytic pathway, providing an unrestricted source of acetyl coenzyme A (acetyl CoA) for de novo lipogenesis and formation of long-chain fatty acids, which can be esterified to form triglycerides in the liver [59]. The initial step of de novo lipogenesis is the synthesis of malonyl-CoA, which acts as a potent inhibitor of carnitine palmitoyl transferase-1, the mitochondrial enzyme that mediates the transport of long-chain fatty acids across the mitochondrial membrane for β-oxidation. Therefore, de novo lipogenesis and lipid oxidation are simultaneously and inversely regulated by fructose in liver cells [30]: this makes fructose highly lipogenic compared with glucose.

Fig. 3 Lipogenic effect of high fructose consumption

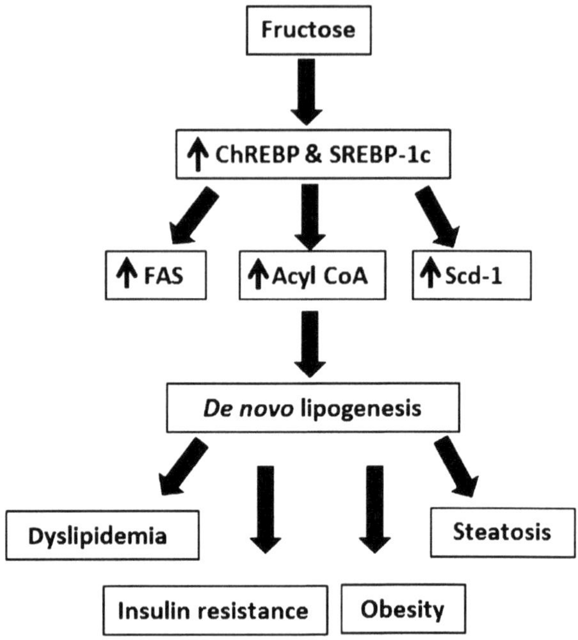

Also, high fructose intake can enhance fatty acid availability by increasing hepatic de novo lipogenesis through the activation of transcription factors, including the carbohydrate regulatory element-binding protein (ChREBP)[34] and sterol regulatory element-binding protein-1c (SREBP-1c) [104]. ChREBP signals the induction of lipogenic genes by carbohydrates [34], whereas SREBP-1c is responsible for the insulin-mediated induction of lipogenic enzymes in liver [65]. In rats and mice, fructose feeding activated ChREBP or SREBP-1c, increasing the expression of lipogenic genes such as fatty acid synthase (FAS), acyl coenzyme-A (Acyl CoA) carboxylase (ACC), and stearoyl coenzyme-A desaturase-1 (Scd-1) [65, 104]. Whether the activation of these lipogenic genes is a direct effect of ChREBP or a secondary response mediated by SREBP-1c [34], these transcriptional factors are induced during fructose feeding and they could perhaps work in concert because activation of both factors is required for normal expression of lipogenic genes [34, 105]. Fructose may activate SREBP-1c, independent of insulin [104]. The increased rate of de novo lipogenesis contributes fatty acids that are incorporated into hepatic triglycerides; this is associated with increased very low density lipoprotein (VLDL) synthesis and secretion from the liver. Hyperlipidemia appears to be a major effect of high-fructose feeding. In rats, increased dietary fructose rapidly leads to high plasma triglyceride concentrations and more specifically to increased VLDL triglycerides [16], as summarized in Fig. 3.

Decreased extrahepatic clearance of VLDL triglycerides could contribute to hypertriglyceridemia. Fructose does not stimulate insulin required to increase the activity of LPL in peripheral tissues. The lower postprandial LPL reduced triglyceride clearance, hence contributing to fructose-induced postprandial hypertriglyceridemia, increasing circulating and portal concentrations of free fatty acids [106].

High carbohydrate intake may promote the development of nonalcoholic fatty liver disease (NAFLD) [19]. Steatosis is the initial change in NAFLD and is thought to be the benign state of liver injury. As steatosis develops further, it renders the liver more vulnerable to injuries, progresses to steatohepatitis, and increases the chances of further liver damage [107]. Animal studies have shown that fructose and sucrose play a critical role in the pathogenesis of NAFLD, with increased accumulation of lipids in the liver accompanied by insulin resistance, elevated plasma triglycerides, and oxidative stress [19, 23–26]. Fructose, having bypassed the vital glycolytic checkpoint, rapidly increases hepatic triglyceride synthesis and deposition, causing fatty liver in rats [23]. In addition to these hepatic effects, fructose may alter many nonconventional pathways in other organs, including the gut, to cause fatty liver [19, 25]: this is discussed later.

Fructose-induced and high fat-induced insulin resistance share some commonalities that could be attributed to altered lipid metabolism at the level of skeletal muscle [9]. Diabetes mellitus is a condition in which the blood glucose concentration is persistently high because of insufficient insulin production or insulin resistance. In obesity, diabetes mellitus is commonly caused by insulin resistance, where insulin is unable to produce responses such as increasing glycogen synthesis and decreasing gluconeogenesis. Chronic consumption of a high-fructose diet may induce insulin resistance, as evident in many animal studies [10–13], although fructose does not stimulate insulin secretion in the short term [108]. Increased exposure to nonesterified fatty acids may reduce insulin sensitivity by increasing the intramyocellular lipid content [35]. This lipid accumulation in muscle might be associated with increased expressions of proteins related to lipid transportation and lipid synthesis such as FAS and translocase (FAT/CD36) [17], hence inducing lipotoxicity in muscle.

Recent progress on insulin resistance suggested that translocation of GLUT4-containing vesicles to the cell surface may involve a group of complex proteins called the SNARE proteins [109–112]. Interaction of SNARE proteins with GLUT4 translocation may be disrupted by high lipid content in skeletal muscles, hence prompting insulin resistance [109]. Accumulation of lipid in skeletal muscles attributed to unregulated fructose metabolism may disrupt SNARE proteins–GLUT4 interaction, leading indirectly to insulin resistance [9, 109, 110, 112]. SNARE proteins consist of SNAP23, syntaxin-5, and VAMP4. SNAP23 is involved in the insulin-dependent translocation of GLUT4 to the plasma membrane and has an important role in the development of insulin resistance. It was proposed that the SNARE proteins catalyzed fusion of newly synthesized triglyceride droplets with the larger old droplets [109]. Further, it was proposed that SNAP23 combines with syntaxin-4 to form a t-SNARE complex in the plasma membrane, which is essential for the insulin-dependent translocation of the glucose transporter GLUT4 to the plasma membrane. Treatment of muscle cells with fatty acids decreased the levels of SNAP23 in the plasma membrane and increased the amount in the interior of the cell, including on lipid droplets. Lipid accumulation in the cell shifts SNAP23 from the glucose uptake mechanism in the plasma membrane to the interior of the cell. Insulin signals for GLUT4-storing vesicles to bind to the membrane and take up glucose were reduced as binding of GLUT4 to membranes required SNAP23; this caused insulin resistance, and fructose, being lipogenic, may enhance the process [111].

5 Fructose and Inflammation

Inflammation plays a vital role in metabolic syndrome, causing both insulin resistance and vascular dysfunction [52, 53]. TNF-α expression in adipose tissue is induced by obesity, hence contributing to systemic insulin resistance [54]. Production of pro-inflammatory cytokines is linked to NF-κB activation [56]. Activation of NF-κB leads to TNF-α-induced insulin resistance [57]. Peroxisome proliferator-activated receptors (PPARs) may attenuate inflammatory signaling pathways and, as such, interfere with cardiac remodeling via inhibition of NF-κB [113].

Fructose feeding in animals increased TNF-α mRNA in plasma [19, 114] and TNF-α concentrations [114, 115] in hepatic tissue. The activation of inflammatory pathways by fructose feeding may influence directly the hepatic and intestinal secretion of lipoproteins [106]. Fructose feeding for 2 weeks increased hepatic NF-κB activity and fatty acid oxidation in rats following reduced activity of hepatic PPAR-α [37]. In another study, fructose feeding in mice for 8 weeks increased mRNA expression of TNF-α [19], the inflammatory genes regulated by NF-κB [116]. Fructose induced inflammation, particularly in the liver, as it is an essential organ for the maintenance of lipid, glucose, and hormonal homeostasis. As such, the liver is at the crossroads of metabolic health and disease. As discussed earlier, fructose induced de novo lipogenesis as it provides unregulated amounts of the substrate, acetyl CoA, for the formation of long-chain fatty acids and triglycerides, increasing hepatic fat deposition [30], hence steatosis. Prolonged insult to the liver induced oxidative stress. High dietary fructose consumption for 19 days in rats induced hepatic stress response through the c-Jun amino-terminal kinase (JNK)/AP-1 pathway, which is similar to that observed for the inflammatory cytokine TNF-α, suggesting that hepatic inflammation and lipid dysregulation are linked [117].

Inflammation and hepatic stress induced by increased fructose intake may partly be caused by the metabolism of fructose itself during glycolysis [117]. Fructose bypasses the vital checkpoint, phosphofructokinase, producing excess (hypothetically during high-fructose diet) dihydroxyacetone phosphate (DHAP) and D-glyceraldehyde, which can be phosphorylated to glyceraldehyde-3-phosphate (G3P). The excess supply of unused DHAP and G3P, as during the rested state, leads to methylglyoxal accumulation. Mcthylglyoxal, a highly reactive ketoaldehyde, arises from nonenzymatic phosphate elimination from DHAP and G3P [118]. Methylglyoxal reacts with free amino groups of lysine and arginine and with thiol groups of cysteine, forming advanced glycation end products (AGEs) [119]. Methylglyoxal has been associated with NF-κB activation and diabetic complications [120].

Oxidative stress and inflammation converge in shared pathways connected to insulin resistance, such as JNK1, making both stimuli important factors to consider with fructose exposure [106]. Insulin resistance in a fructose-fed model may be caused by inflammation. TNF-α is an important mediator of insulin resistance in obesity and diabetes through its ability to decrease the tyrosine kinase activity of the insulin receptor. Binding of insulin stimulated the intrinsic tyrosine kinase of the insulin receptor, which results in autophosphorylation of the subunits on tyrosine residues and subsequent phosphorylation of insulin receptor substrate 1 (IRS-1), activating the phosphatidylinositol-3 kinase/PDK/AKT signaling and hence glucose uptake [54]. However, TNF-α inhibited these proximal steps in insulin receptor signaling in both cultured cells and whole animals [54], through activation of JNK [121].

6 Fructose and Cardiovascular Diseases

6.1 Hypertension

Chronic hypertension is a key risk factor for stroke, myocardial infarction, and heart failure [122]. Fructose feeding produced systolic hypertension in dogs [10] and rats [11, 20–22]. High-fructose-fed rats developed high systolic blood pressure over 16 weeks but without endothelial dysfunction [11]. However, in a study with 40 % fructose feeding for 2 weeks, rats had increased blood pressure accompanied by reduced acetylcholine-induced relaxation in the aortic rings, suggesting endothelial damage [22]. In this study, insulin resistance induced endothelial dysfunction, resulting in hypertension. Insulin, apart from stimulating the uptake of glucose in muscle via GLUT4 transporter, dilates the skeletal muscle vasculature in humans [123, 124]. Insulin activates endothelial nitric oxide synthase (eNOS) and stimulates the production of vasodilator NO in the endothelium [124]. The activation of eNOS is via the PI3K/PDK/AKT pathway, triggered by insulin [124]. In addition, insulin vasodilation contributes, in part, to an inhibition of voltage-operated Ca^{2+} channels [125]. A fructose diet impaired the vasodilator effects of insulin in precontracted aortas [22, 126]. Further, fructose induced autonomic imbalance, triggering the increased blood pressure [91].

In addition, excess fructose feeding alters glucose metabolism by shunting glucose through the polyol pathway, which leads to a buildup of intracellular glyceraldehyde and dihydroxyacetone phosphate. Aldehydes are able to react non-enzymatically with sulfhydryl groups of protein, thus altering their function [68]. Aldehydes can impair the function of L-type calcium channels, and this may possibly lead to an increased intracellular calcium concentration in vascular smooth muscle and to an increase in peripheral vascular resistance [1, 127, 128].

Although some studies reported hypertension in fructose-fed rats [11, 20–22], others did not [129, 130]. Further, chronic consumption of fructose for 24 weeks in rats increased blood pressure, which was not maintained after 40 weeks of feeding [14]. In a separate study, rats developed hypertension during the first 3 months of

12 months of high-fructose feeding [16]. These conflicting data may be attributed to differences in the composition of the diets and to variable blood pressure responses to fructose among different ages and strains [129] or to different methods used for blood pressure measurements [131].

6.2 *Cardiac Function and Hypertrophy*

Eccentric hypertrophy, defined as a progressive and proportional increase in chamber volume and wall thickness in response to volume overload, has been commonly described in obese individuals with altered diastolic function and prone to heart failure [132, 133]. A fructose-rich diet in a rat model of chronic volume overload, caused by aortic valve regurgitation, worsened left ventricular eccentric hypertrophy, as shown by increased left ventricular hypertrophy, atrial and brain natriuretic factor mRNA, and increased estimated left ventricular mass [134]. In addition, left ventricular function was decreased, shown by lower systolic ejection fraction [134]. Fructose feeding for 16 weeks induced left ventricular dilation with hypertrophy and decreased contractile function, and increased inflammatory cell infiltration into the ventricular myocardium, resulting in excess collagen deposition and increased stiffness of the left ventricle [11, 20]. Increases in collagen deposition are linked to the excess production of AGEs, as products of nonenzymatic glycosylation in hyperglycemic states [135].

Vascular calcification is a well-recognized pathological entity associated with cardiovascular morbidity and mortality [136]. In type 2 diabetes, medial arterial calcification of peripheral arteries is a strong independent predictor of total cardiovascular mortality and a predictor of future coronary heart disease events, stroke, and amputation [137]. Aortic calcification was observed in rats fed with high fructose for 12 weeks, accompanied by mild hypertension, insulin resistance, hyperinsulinemia, hyperglycemia, and hypertriglyceridemia [138]. It was suggested that fructose increased oxidative stress by increasing lipid peroxidation and inhibition of activity of enzymes such as superoxide dismutase and catalase, and oxidative stress may be involved in the pathogenesis of vascular calcification [138].

7 Fructose and Gut Microbiota

The prevalence of obesity is attributed to the so-called Western diet with highly refined carbohydrates and fat but reduced complex plant polysaccharides [38]. A concomitant increase in fructose intake has mirrored this trend, changing the regulation and homeostatic maintenance of host energy balance produced by hormones from the gut and adipocytes [38]. The gastrointestinal tract is a complex digestive organ including genetic interactions between host and microbes, modulating a multitude of processes related to digestion, absorption, and metabolism of dietary compounds. Hence, there is a probable link between commensal gut microflora and obesity [139].

It was hypothesized that obesity is associated with reduced gut microflora diversity and altered genes, hence manifestation of metabolic syndrome [140, 141]. The *Firmicutes:Bacteroidetes* ratio appears to be critical [140]. The gut in the Western diet includes *Firmicutes* class *Mollicutes* with genes encoding pathways related to the phosphotransferase system, fructose and mannose metabolism, and glycolysis/gluconeogenesis, but depleted of genes required for starch and sucrose metabolism [142]. This Western diet is detrimental to *Bacteroidetes*, as these microbes metabolize polysaccharides and glycans and so prefer complex plant materials [143]. Excess exposure to fructose may cause dysbiosis with loss of microbial genetic and phylogenic diversity and perturbed energy regulation [38]. The innate and adaptive immune system may sense these differences in microbial composition and metabolic activity, causing intestinal inflammation and then endotoxemia [38]. The combination of these processes can exacerbate metabolic disorders associated with obesity. It was postulated that bacterial endotoxins may cause NAFLD because increased translocation of bacterial endotoxins was found in obese humans and animals [25]. Chronic fructose consumption can increase intestinal translocation of bacterial endotoxins, inducing hepatic TNF-α, and subsequently induce liver steatosis in mice [25]. Furthermore, the bacterial flora of rodents can ferment carbohydrates to alcohol when intestinal stasis permits their overgrowth in the upper parts of the gastrointestinal tract [144]. This finding further supports the hypothesis that elevated intake of simple sugars such as fructose and sucrose can contribute to the development of obesity and liver damage through mechanisms involving alterations in the intestinal microflora [19].

8 Fructose Consumption in Humans

In contrast to the animal studies, there are relatively few data from humans with chronic excess fructose intake. Short-term studies with fructose overfeeding (25 % of total calories) for 6 days of seven healthy, normal-weight males increased plasma triglycerides concentration by 79 % and impaired the suppression of nonesterified fatty acids, indicating insulin resistance [27]. A 6-week diet providing 17 % of energy from fructose increased postprandial triglyceride concentrations compared with an isocaloric glucose diet in healthy men but not in healthy women [29]. A study on the effects of fructose on intrahepatocellular lipids and insulin sensitivity in healthy offspring of patients with type 2 diabetes concluded that 7 days of high-fructose diet (20 % solution in three meals) increased ectopic lipid deposition in liver and muscle and fasting VLDL-triacylglycerols and decreased hepatic insulin sensitivity [30]. Overweight and obese patients consuming a high fructose-sweetened beverage (25 % of total calories) had increased visceral adiposity and lipids with decreased insulin sensitivity [28]. In contrast, dietary fructose (20 % of carbohydrate calories, 45–65 g daily for 4 weeks) for ten type 2 diabetic patients lowered plasma glucose concentrations and improved insulin sensitivity without affecting the lipid profile [145]. Fructose consumption (63–99 g/day) for 2 weeks by 11 subjects with normal oral glucose tolerance and body weight had no adverse effects on triglyceride, pyruvate, lactate, or uric acid metabolism [146]. A review article on the effects of dietary fructose on lipid metabolism pointed out that there is strong evidence that fructose consumed at approximately 20 % of total energy increases total and LDL cholesterol concentrations, but the effect of dietary fructose on triacylglycerol concentrations is less clear [147].

Fructose had no overall effect on body weight in a review of the effects of fructose on body weight in controlled feeding trials lasting 7 or more days, comparing free fructose and nonfructose carbohydrate in diets providing similar calories (isocaloric trials) or of diets supplemented with free fructose to provide excess energy and usual or control diets (hypercaloric trials) [31]. High doses of fructose in hypercaloric trials lead to significant increases in weight but fructose does not seem to cause weight gain when it is substituted for other carbohydrates in diets providing similar calories. Free fructose at high doses that provided excess calories modestly increased body weight, an effect that may be caused by the extra calories rather than the fructose [31]. Although fructose induces insulin resistance and affects metabolic processes in animal models, the existing data for humans are unclear [7]. Fructose did not differentially affect satiety when compared with glucose [32, 33], although different effects have also been reported [148, 149].

9 Conclusions

In summary, findings from human studies are inconsistent; however, animal studies have raised serious concerns regarding increased dietary fructose in humans. Animal experiments have relied on the importance of fructose overloads (up to 60 % total energy intake), which by far exceeded the consumption of fructose in humans. However, the animal studies identify possible mechanisms for the adverse effects of fructose, although they do not provide clear dietary recommendations for humans. Finally, although fructose appears to alter lipid metabolism independently of any change in body composition [28, 150], increased whole-body and visceral fat mass may play a significant role in other metabolic disturbances [9, 28]. There are, therefore, several important questions to be addressed regarding metabolic and cardiovascular short- and long-term effects of fructose consumption and the mechanisms underlying these effects. Also, future studies in rodents should address the responses to chronic moderate fructose consumption to be more realistic and relevant to humans.

To date, there is evidence from short-term studies suggesting consumption of sweetened beverages is associated with obesity and metabolic syndromes. Further, pure fructose caused metabolic changes at high concentrations, especially when fed as the sole carbohydrate source. However, there is little evidence from human studies that strongly suggest common fructose-glucose (HFCS or sucrose) sweeteners would do the same. In addition, there is no solid evidence to support that moderate intake of fructose, or fructose from fruits or honey, is unsafe. The fructose that comes from whole fruits and vegetables, dairy products, and 100 % fruit juices is associated with vitamins, minerals, and fiber [6] whereas honey contains phenolic acids, flavonoids, and minerals [151]. There are no specific recommendations for fructose intake. The 2009 American Heart Association recommended that the intake of added sugar for women should be no more than 100 calories per day and that for men no more than 150 calories per day [152]. Obesity and diabetes rates were low when total fructose intake was in the range of 25–40 g/day [153]. Conclusions as to the safe and prudent amounts of fructose consumption are needed but will require carefully controlled dose–response studies in humans.

References

1. Tappy L, Le KA (2010) Metabolic effects of fructose and the worldwide increase in obesity. Physiol Rev 90:23–46
2. Sugar consumption in Australia: a statistical update. http://greenpoolcommodities.com/files/8113/4932/3223/121004_Sugar_Consumption_in_Australia_-_A_Statistical_Update_-_Public_Release_Document.pdf. 1 Feb 2013
3. Bobridge KS, Haines GL, Mori TA et al (2012) Dietary fructose in relation to blood pressure and serum uric acid in adolescent boys and girls. J Hum Hypertens 27:217–224
4. Wolf A, Bray GA, Popkin BM (2008) A short history of beverages and how our body treats them. Obes Rev 9:151–164
5. White JS (2008) Straight talk about high-fructose corn syrup: what it is and what it ain't. Am J Clin Nutr 88:1716S–1721S
6. Bray GA (2010) Fructose: pure, white, and deadly? Fructose, by any other name, is a health hazard. J Diabetes Sci Technol 4:1003–1007
7. Elliott SS, Keim NL, Stern JS et al (2002) Fructose, weight gain, and the insulin resistance syndrome. Am J Clin Nutr 76:911–922
8. Johnson RJ, Segal MS, Sautin Y et al (2007) Potential role of sugar (fructose) in the epidemic of hypertension, obesity and the metabolic syndrome, diabetes, kidney disease, and cardiovascular disease. Am J Clin Nutr 86:899–906
9. Le KA, Tappy L (2006) Metabolic effects of fructose. Curr Opin Clin Nutr Metab Care 9: 469–475
10. Martinez FJ, Rizza RA, Romero JC (1994) High-fructose feeding elicits insulin resistance, hyperinsulinism, and hypertension in normal mongrel dogs. Hypertension 23:456–463
11. Patel J, Iyer A, Brown L (2009) Evaluation of the chronic complications of diabetes in a high fructose diet in rats. Indian J Biochem Biophys 46:66–72
12. Tran LT, Yuen VG, McNeill JH (2009) The fructose-fed rat: a review on the mechanisms of fructose-induced insulin resistance and hypertension. Mol Cell Biochem 332:145–159
13. Hashikawa-Hobara N, Hashikawa N, Inoue Y et al (2012) Candesartan cilexetil improves angiotensin II type 2 receptor-mediated neurite outgrowth via the PI3K-Akt pathway in fructose-induced insulin-resistant rats. Diabetes 61:925–932
14. Takagawa Y, Berger ME, Hori MT et al (2001) Long-term fructose feeding impairs vascular relaxation in rat mesenteric arteries. Am J Hypertens 14:811–817
15. Ishimoto T, Lanaspa MA, Le MT et al (2012) Opposing effects of fructokinase C and A isoforms on fructose-induced metabolic syndrome in mice. Proc Natl Acad Sci U S A 109: 4320–4325
16. Park OJ, Cesar D, Faix D et al (1992) Mechanisms of fructose-induced hypertriglyceridaemia in the rat. Activation of hepatic pyruvate dehydrogenase through inhibition of pyruvate dehydrogenase kinase. Biochem J 282:753–757

17. Song GY, Ren LP, Chen SC et al (2012) Similar changes in muscle lipid metabolism are induced by chronic high-fructose feeding and high-fat feeding in C57BL/J6 mice. Clin Exp Pharmacol Physiol 8:1440–1681
18. Okazaki M, Zhang H, Yoshida Y et al (1994) Correlation between plasma fibrinogen and serum lipids in rats with hyperlipidemia induced by cholesterol free-high fructose or high cholesterol diet. J Nutr Sci Vitaminol 40:479–489
19. Haub S, Kanuri G, Volynets V et al (2010) Serotonin reuptake transporter (SERT) plays a critical role in the onset of fructose-induced hepatic steatosis in mice. Am J Physiol Gastrointest Liver Physiol 298:G335–G344
20. Patel J, Matnor NA, Iyer A et al (2011) A regenerative antioxidant protocol of vitamin E and α-lipoic acid ameliorates cardiovascular and metabolic changes in fructose-fed rats. Evid Based Complement Altern Med Article 120801, doi:10.1155/2011/120801
21. Dai S, McNeill JH (1995) Fructose-induced hypertension in rats is concentration- and duration-dependent. J Pharmacol Toxicol Methods, pp 101–107
22. Uchida T, Okumura K, Ito T et al (2002) Quinapril treatment restores the vasodilator action of insulin in fructose-hypertensive rats. Clin Exp Pharmacol Physiol 29:381–385
23. Sanchez-Lozada LG, Mu W, Roncal C et al (2010) Comparison of free fructose and glucose to sucrose in the ability to cause fatty liver. Eur J Nutr 49:1–9
24. Armutcu F, Coskun O, Gurel A et al (2005) Thymosin alpha-1 attenuates lipid peroxidation and improves fructose-induced steatohepatitis in rats. Clin Biochem 38:540–547
25. Bergheim I, Weber S, Vos M et al (2008) Antibiotics protect against fructose-induced hepatic lipid accumulation in mice: role of endotoxin. J Hepatol 48:983–992
26. Jurgens H, Haass W, Castaneda TR et al (2005) Consuming fructose-sweetened beverages increases body adiposity in mice. Obes Res 13:1146–1156
27. Faeh D, Minehira K, Schwarz JM et al (2005) Effect of fructose overfeeding and fish oil administration on hepatic de novo lipogenesis and insulin sensitivity in healthy men. Diabetes 54:1907–1913
28. Stanhope KL, Schwarz JM, Keim NL et al (2009) Consuming fructose-sweetened, not glucose-sweetened, beverages increases visceral adiposity and lipids and decreases insulin sensitivity in overweight/obese humans. J Clin Invest 119:1322–1334
29. Bantle JP, Raatz SK, Thomas W et al (2000) Effects of dietary fructose on plasma lipids in healthy subjects. Am J Clin Nutr 72:1128–1134
30. Le KA, Ith M, Kreis R et al (2009) Fructose overconsumption causes dyslipidemia and ectopic lipid deposition in healthy subjects with and without a family history of type 2 diabetes. Am J Clin Nutr 89:1760–1765
31. Sievenpiper JL, de Souza RJ, Mirrahimi A et al (2012) Effect of fructose on body weight in controlled feeding trials: a systematic review and meta-analysis. Ann Intern Med 156:291–304
32. Spitzer L, Rodin J (1987) Effects of fructose and glucose preloads on subsequent food intake. Appetite 8:135–145
33. Warwick ZS, Weingarten HP (1994) Dynamics of intake suppression after a preload: role of calories, volume, and macronutrients. Am J Physiol 266:R1314–R1318
34. Iizuka K, Bruick RK, Liang G et al (2004) Deficiency of carbohydrate response element-binding protein (ChREBP) reduces lipogenesis as well as glycolysis. Proc Natl Acad Sci U S A 101:7281–7286
35. Virkamaki A, Korsheninnikova E, Seppala-Lindroos A et al (2001) Intramyocellular lipid is associated with resistance to in vivo insulin actions on glucose uptake, antilipolysis, and early insulin signaling pathways in human skeletal muscle. Diabetes 50:2337–2343
36. Shapiro A, Mu W, Roncal C et al (2008) Fructose-induced leptin resistance exacerbates weight gain in response to subsequent high-fat feeding. Am J Physiol Regul Integr Comp Physiol 295:R1370–R1375
37. Roglans N, Vila L, Farre M et al (2007) Impairment of hepatic Stat-3 activation and reduction of PPARalpha activity in fructose-fed rats. Hepatology 45:778–788

38. Payne AN, Chassard C, Lacroix C (2012) Gut microbial adaptation to dietary consumption of fructose, artificial sweeteners and sugar alcohols: implications for host-microbe interactions contributing to obesity. Obes Rev 13:799–809
39. Chew GT, Gan SK, Watts GF (2006) Revisiting the metabolic syndrome. Med J Aust 185: 445–449
40. Eckel RH, Grundy SM, Zimmet PZ (2005) The metabolic syndrome. Lancet 365: 1415–1428
41. Moller DE, Kaufman KD (2005) Metabolic syndrome: a clinical and molecular perspective. Annu Rev Med 56:45–62
42. American Heart Association (2010) Heart disease and stroke statistics—2010 update. American Heart Association, Dallas
43. Cameron AJ, Magliano DJ, Zimmet PZ et al (2007) The metabolic syndrome in Australia: prevalence using four definations. Diabetes Res Clin Pract 777:471–478
44. Sone H, Mizuno S, Fujii H et al (2005) Is the diagnosis of metabolic syndrome useful for predicting cardiovascular disease in Asian diabetic patients? Diabetes Care 28:1463–1471
45. Li WJ, Xue H, Sun K et al (2008) Cardiovascular risk and prevalence of metabolic syndrome by differing criteria. Chin Med J (Engl) 121:1532–1536
46. Gupta R, Deedwania PC, Gupta A et al (2004) Prevalence of metabolic syndrome in an Indian urban population. Int J Cardiol 97:257–261
47. Van Gaal LF, Mertens IL, De Block CE (2006) Mechanisms linking obesity with cardiovascular disease. Nature 444:875–880
48. Lee MJ, Wu YY, Fried SK (2010) Adipose tissue remodeling in pathophysiology of obesity. Curr Opin Clin Nutr Metab Care 13:371–376
49. Haffner SM (2006) The metabolic syndrome: inflammation, diabetes mellitus, and cardiovascular disease. Am J Cardiol 97:3–11
50. Shulman GI (2000) Cellular mechanisms of insulin resistance. J Clin Invest 106:171–176
51. Shimizu T (2009) Lipid mediators in health and disease: enzymes and receptors as therapeutic targets for the regulation of immunity and inflammation. Annu Rev Pharmacol Toxicol 49: 123–150
52. Iyer A, Brown L (2010) Lipid mediators and inflammation in glucose intolerance and insulin resistance. Drug Discov Today Dis Mech 7:e191–e197
53. Iyer A, Fairlie DP, Prins JB et al (2010) Inflammatory lipid mediators in adipocyte function and obesity. Nat Rev Endocrinol 6:71–82
54. Hotamisligil GS, Peraldi P, Budavari A et al (1996) IRS-1-mediated inhibition of insulin receptor tyrosine kinase activity in TNF-alpha- and obesity induced insulin resistance. Science 271:665–668
55. Tesauro M, Cardillo C (2011) Obesity, blood vessels and metabolic syndrome. Acta Physiol 203:279–286
56. Kaileh M, Sen R (2010) Role of NFkB in the anti-inflammatory effects of tocotrienols. J Am Coll Nutr 29:334–339
57. Tilg H, Moschen AR (2006) Adipocytokines: mediators linking adipose tissue, inflammation and immunity. Nat Rev Immunol 6:772–783
58. Heman RH, Zakim D (1968) Fructose metabolism: 1. The fructose metabolic pathway. Am J Clin Nutr 21:245–249
59. Mayes PA (1993) Intermediary metabolism of fructose. Am J Clin Nutr 58:754S–765S
60. Buchs AE, Sasson S, Joost HG et al (1998) Characterization of GLUT5 domains responsible for fructose transport. Endocrinology 139:827–831
61. Jones HF, Butler RN, Brooks DA (2011) Intestinal fructose transport and malabsorption in humans. Am J Physiol Gastrointest Liver Physiol 300:G202–G206
62. Barone S, Fussell SL, Singh AK et al (2009) Slc2a5 (Glut5) is essential for the absorption of fructose in the intestine and generation of fructose-induced hypertension. J Biol Chem 284: 5056–5066

63. Soleimani M (2011) Dietary fructose, salt absorption and hypertension in metabolic syndrome: towards a new paradigm. Acta Physiol (Oxf) 201:55–62
64. Havel PJ (2005) Dietary fructose: implications for dysregulation of energy homeostasis and lipid/carbohydrate metabolism. Nutr Rev 63:133–157
65. Rodriguez-Calvo R, Barroso E, Serrano L et al (2009) Atorvastatin prevents carbohydrate response element binding protein activation in the fructose-fed rat by activating protein kinase A. Hepatology 49:106–115
66. Stanhope KL (2012) Role of fructose-containing sugars in the epidemics of obesity and metabolic syndrome. Annu Rev Med 63:329–343
67. Fried SK, Russell CD, Grauso NL et al (1993) Lipoprotein lipase regulation by insulin and glucocorticoid in subcutaneous and omental adipose tissues of obese women and men. J Clin Invest 92:2191–2198
68. Abdullah MM, Riediger NN, Chen Q et al (2009) Effects of long-term consumption of a high-fructose diet on conventional cardiovascular risk factors in Sprague–Dawley rats. Mol Cell Biochem 327:247–256
69. Hamdy O, Porramatikul S, Al-Ozairi E (2006) Metabolic obesity: the paradox between visceral and subcutaneous fat. Curr Diabetes Rev 2:367–373
70. Despres JP, Moorjani S, Lupien PJ et al (1990) Regional distribution of body fat, plasma lipoproteins, and cardiovascular disease. Arteriosclerosis 10:497–511
71. Despres JP (1998) The insulin resistance-dyslipidemic syndrome of visceral obesity: effect on patients' risk. Obes Res 6:8S–17S
72. Pouliot MC, Despres JP, Nadeau A et al (1992) Visceral obesity in men. Associations with glucose tolerance, plasma insulin, and lipoprotein levels. Diabetes 41:826–834
73. Bursac BN, Djordjevic AD, Vasiljevic AD et al (2012) Fructose consumption enhances glucocorticoid action in rat visceral adipose tissue. J Nutr Biochem 24:1166–1172
74. Schwartz MW, Baskin DG, Kaiyala KJ et al (1999) Model for the regulation of energy balance and adiposity by the central nervous system. Am J Clin Nutr 69:584–596
75. Sipols AJ, Baskin DG, Schwartz MW (1995) Effect of intracerebroventricular insulin infusion on diabetic hyperphagia and hypothalamic neuropeptide gene expression. Diabetes 44:147–151
76. Brief DJ, Davis JD (1984) Reduction of food intake and body weight by chronic intraventricular insulin infusion. Brain Res Bull 12:571–575
77. Woods SC, Porte DJ (1976) Insulin and the set-point regulation of body weight. In: Novin D, Bray GA, Wyrwichka W (eds) Hunger: basic mechanisms and clinical implications. Raven Press, New York, pp 273–280
78. Baskin DG, Wilcox BJ, Figlewicz DP et al (1988) Insulin and insulin-like growth factors in the CNS. Trends Neurosci 11:107–111
79. Baskin DG, Sipols AJ, Schwartz MW et al (1993) Immunocytochemical detection of insulin receptor substrate-1 (IRS-1) in rat brain: colocalization with phosphotyrosine. Regul Pept 48: 257–266
80. Purnell JQ, Zinman B, Brunzell JD (2013) The effect of excess weight gain with intensive diabetes mellitus treatment on cardiovascular disease risk factors and atherosclerosis in type 1 diabetes mellitus: results from the Diabetes Control and Complications Trial/Epidemiology of Diabetes Interventions and Complications Study (DCCT/EDIC) study. Circulation 127: 180–187
81. Gabriely I, Hawkins M, Vilcu C et al (2002) Fructose amplifies counterregulatory responses to hypoglycemia in humans. Diabetes 51:893–900
82. Farah V, Elased KM, Morris M (2007) Genetic and dietary interactions: role of angiotensin AT1a receptors in response to a high-fructose diet. Am J Physiol Heart Circ Physiol 293:20
83. Verma S, Bhanot S, McNeill JH (1999) Sympathectomy prevents fructose-induced hyperinsulinemia and hypertension. Eur J Pharmacol 373:R1–R4
84. Farah V, Elased KM, Chen Y et al (2006) Nocturnal hypertension in mice consuming a high fructose diet. Auton Neurosci 130:41–50

85. Mohuczy D, Gelband CH, Phillips MI (1999) Antisense inhibition of AT1 receptor in vascular smooth muscle cells using adeno-associated virus-based vector. Hypertension 33:354–359
86. Katovich MJ, Reaves PY, Francis SC et al (2001) Gene therapy attenuates the elevated blood pressure and glucose intolerance in an insulin-resistant model of hypertension. J Hypertens 19:1553–1558
87. Shinozaki K, Ayajiki K, Nishio Y et al (2004) Evidence for a causal role of the renin-angiotensin system in vascular dysfunction associated with insulin resistance. Hypertension 43:255–262
88. Brito JO, Ponciano K, Figueroa D et al (2008) Parasympathetic dysfunction is associated with insulin resistance in fructose-fed female rats. Braz J Med Biol Res 41:804–808
89. Brown CM, Dulloo AG, Yepuri G et al (2008) Fructose ingestion acutely elevates blood pressure in healthy young humans. Am J Physiol Regul Integr Comp Physiol 294:R730–R737
90. Miller AW, Sims JJ, Canavan A et al (1999) Impaired vagal reflex activity in insulin-resistant rats. J Cardiovasc Pharmacol 33:698–702
91. De Angelis K, Senador DD, Mostarda C et al (2012) Sympathetic overactivity precedes metabolic dysfunction in a fructose model of glucose intolerance in mice. Am J Physiol Regul Integr Comp Physiol 302:R950–R957
92. Wang GJ, Tomasi D, Backus W et al (2008) Gastric distention activates satiety circuitry in the human brain. Neuroimage 39:1824–1831
93. Carnethon MR, Jacobs DR Jr, Sidney S et al (2003) Influence of autonomic nervous system dysfunction on the development of type 2 diabetes: the CARDIA study. Diabetes Care 26: 3035–3041
94. Morton GJ, Cummings DE, Baskin DG et al (2006) Central nervous system control of food intake and body weight. Nature 443:289–295
95. Farooqi IS, O'Rahilly S (2009) Leptin: a pivotal regulator of human energy homeostasis. Am J Clin Nutr 89:980S–984S
96. Friedman JM, Halaas JL (1998) Leptin and the regulation of body weight in mammals. Nature 395:763–770
97. Stanhope KL, Havel PJ (2008) Endocrine and metabolic effects of consuming beverages sweetened with fructose, glucose, sucrose, or high-fructose corn syrup. Am J Clin Nutr 88: 1733S–1737S
98. Harris M, Aschkenasi C, Elias CF et al (2001) Transcriptional regulation of the thyrotropin-releasing hormone gene by leptin and melanocortin signaling. J Clin Invest 107:111–120
99. Legradi G, Emerson CH, Ahima RS et al (1998) Arcuate nucleus ablation prevents fasting-induced suppression of ProTRH mRNA in the hypothalamic paraventricular nucleus. Neuroendocrinology 68:89–97
100. Nillni EA, Vaslet C, Harris M et al (2000) Leptin regulates prothyrotropin-releasing hormone biosynthesis. Evidence for direct and indirect pathways. J Biol Chem 275:36124–36133
101. Buettner C, Pocai A, Muse ED et al (2006) Critical role of STAT3 in leptin's metabolic actions. Cell Metab 4:49–60
102. Banks WA, Coon AB, Robinson SM et al (2004) Triglycerides induce leptin resistance at the blood–brain barrier. Diabetes 53:1253–1260
103. Lewis GF (1997) Fatty acid regulation of very low density lipoprotein production. Curr Opin Lipidol 8:146–153
104. Matsuzaka T, Shimano H, Yahagi N et al (2004) Insulin-independent induction of sterol regulatory element-binding protein-1c expression in the livers of streptozotocin-treated mice. Diabetes 53:560–569
105. Dentin R, Girard J, Postic C (2005) Carbohydrate responsive element binding protein (ChREBP) and sterol regulatory element binding protein-1c (SREBP-1c): two key regulators of glucose metabolism and lipid synthesis in liver. Biochimie 87:81–86
106. Dekker MJ, Su Q, Baker C et al (2010) Fructose: a highly lipogenic nutrient implicated in insulin resistance, hepatic steatosis, and the metabolic syndrome. Am J Physiol Endocrinol Metab 299:E685–E694

107. Adams LA, Lymp JF, St. Sauver J et al (2005) The natural history of nonalcoholic fatty liver disease: a population-based cohort study. Gastroenterology 129:113–121
108. Curry DL (1989) Effects of mannose and fructose on the synthesis and secretion of insulin. Pancreas 4:2–9
109. Olofsson S-O, Andersson L, Håversen L et al (2011) The formation of lipid droplets: possible role in the development of insulin resistance/type 2 diabetes. Prostaglandins Leukot Essent Fatty Acids 85:215–218
110. Pessin JE, Saltiel AR (2000) Signaling pathways in insulin action: molecular targets of insulin resistance. J Clin Invest 106:165–169
111. Bostrom P, Andersson L, Rutberg M et al (2007) SNARE proteins mediate fusion between cytosolic lipid droplets and are implicated in insulin sensitivity. Nat Cell Biol 9:1286–1293
112. Bostrom P, Andersson L, Vind B et al (2010) The SNARE protein SNAP23 and the SNARE-interacting protein Munc18c in human skeletal muscle are implicated in insulin resistance/type 2 diabetes. Diabetes 59:1870–1878
113. Álvarez-Guardia D, Palomer X, Coll T et al (2011) PPARβ/δ activation blocks lipid-induced inflammatory pathways in mouse heart and human cardiac cells. Biochim Biophys Acta 1811:59–67
114. Spruss A, Kanuri G, Wagnerberger S et al (2009) Toll-like receptor 4 is involved in the development of fructose-induced hepatic steatosis in mice. Hepatology 50:1094–1104
115. Tsai J, Zhang R, Qiu W et al (2009) Inflammatory NF-kappaB activation promotes hepatic apolipoprotein B100 secretion: evidence for a link between hepatic inflammation and lipoprotein production. Am J Physiol Gastrointest Liver Physiol 296:G1287–G1298
116. Weng-Yew W, Brown L (2011) Nutrapharmacology of tocotrienols for metabolic syndrome. Curr Pharm Des 17:2206–2214
117. Kelley GL, Allan G, Azhar S (2004) High dietary fructose induces a hepatic stress response resulting in cholesterol and lipid dysregulation. Endocrinology 145:548–555
118. Wyllie S, Fairlamb AH (2011) Methylglyoxal metabolism in trypanosomes and leishmania. Semin Cell Dev Biol 22:271–277
119. Shinohara M, Thornalley PJ, Giardino I et al (1998) Overexpression of glyoxalase-I in bovine endothelial cells inhibits intracellular advanced glycation endproduct formation and prevents hyperglycemia-induced increases in macromolecular endocytosis. J Clin Invest 101:1142–1147
120. Hammes HP, Du X, Edelstein D et al (2003) Benfotiamine blocks three major pathways of hyperglycemic damage and prevents experimental diabetic retinopathy. Nat Med 9:294–299
121. Hotamisligil GS (2006) Inflammation and metabolic disorders. Nature, pp 860–867
122. Aggarwal BB, Sundaram C, Prasad S et al (2010) Tocotrienols, the vitamin E of the 21st century: its potential against cancer and other chronic diseases. Biochem Pharmacol 80:1613–1631
123. Anderson EA, Hoffman RP, Balon TW et al (1991) Hyperinsulinemia produces both sympathetic neural activation and vasodilation in normal humans. J Clin Invest 87:2246–2252
124. Chaudhuri A, Dandona P, Fonseca V (2012) Cardiovascular benefits of exogenous insulin. J Clin Endocrinol Metab 97:3079–3091
125. Goud C, Pitt B, Webb RC et al (1998) Synergistic actions of insulin and troglitazone on contractility in endothelium-denuded rat aortic rings. Am J Physiol 275:E882–E887
126. Verma S, Bhanot S, Yao L et al (1997) Vascular insulin resistance in fructose-hypertensive rats. Eur J Pharmacol 322:R1–R2
127. Vasdev S, Longerich L, Gill V (2004) Prevention of fructose-induced hypertension by dietary vitamins. Clin Biochem 37:1–9
128. Sorrell MF, Tuma DJ (1987) The functional implications of acetaldehyde binding to cell constituents. Ann N Y Acad Sci 492:50–62
129. Bezerra RM, Ueno M, Silva MS et al (2001) A high-fructose diet induces insulin resistance but not blood pressure changes in normotensive rats. Braz J Med Biol Res 34:1155–1160
130. Song J, Hu X, Shi M et al (2004) Effects of dietary fat, NaCl, and fructose on renal sodium and water transporter abundances and systemic blood pressure. Am J Physiol Renal Physiol 287:F1204–F1212

131. Axelsen LN, Lademann JB, Petersen JS et al (2010) Cardiac and metabolic changes in long-term high fructose-fat fed rats with severe obesity and extensive intramyocardial lipid accumulation. Am J Physiol Regul Integr Comp Physiol 298:R1560–R1570
132. Paunovic K, Jakovljevic B, Stojanov V (2006) Left ventricular hypertrophy in hypertensive obese women. Acta Cardiol 61:623–629
133. Messerli FH (1982) Cardiovascular effects of obesity and hypertension. Lancet 319: 1165–1168
134. Bouchard-Thomassin AA, Lachance D, Drolet MC et al (2011) A high-fructose diet worsens eccentric left ventricular hypertrophy in experimental volume overload. Am J Physiol Heart Circ Physiol 300:H125–H134
135. Brownlee M (2001) Biochemistry and molecular cell biology of diabetic complications. Nature 414:813–820
136. Shaw LJ, Raggi P, Berman DS et al (2006) Coronary artery calcium as a measure of biologic age. Atherosclerosis 188:112–119
137. Lehto S, Niskanen L, Suhonen M et al (1996) Medial artery calcification. A neglected harbinger of cardiovascular complications in non-insulin-dependent diabetes mellitus. Arterioscler Thromb Vasc Biol 16:978–983
138. Zhou YB, Zhang J, Cai Y et al (2012) Insulin resistance induces medial artery calcification in fructose-fed rats. Exp Biol Med 237:50–57
139. Bjorneklett A, Viddal KO, Midtvedt T et al (1981) Intestinal and gastric bypass. Changes in intestinal microecology after surgical treatment of morbid obesity in man. Scand J Gastroenterol 16:681–687
140. Ley RE, Backhed F, Turnbaugh P et al (2005) Obesity alters gut microbial ecology. Proc Natl Acad Sci U S A 102:11070–11075
141. Turnbaugh PJ, Ley RE, Mahowald MA et al (2006) An obesity-associated gut microbiome with increased capacity for energy harvest. Nature 444:1027–1031
142. Turnbaugh PJ, Backhed F, Fulton L et al (2008) Diet-induced obesity is linked to marked but reversible alterations in the mouse distal gut microbiome. Cell Host Microbe 3:213–223
143. Sonnenburg ED, Zheng H, Joglekar P et al (2010) Specificity of polysaccharide use in intestinal bacteroides species determines diet-induced microbiota alterations. Cell 141: 1241–1252
144. Baraona E, Julkunen R, Tannenbaum L et al (1986) Role of intestinal bacterial overgrowth in ethanol production and metabolism in rats. Gastroenterology 90:103–110
145. Koivisto VA, Yki-Jarvinen H (1993) Fructose and insulin sensitivity in patients with type 2 diabetes. J Intern Med 233:145–153
146. Crapo PA, Kolterman OG (1984) The metabolic effects of 2-week fructose feeding in normal subjects. Am J Clin Nutr 39:525–534
147. Hollenbeck CB (1993) Dietary fructose effects on lipoprotein metabolism and risk for coronary artery disease. Am J Clin Nutr 58:800S–809S
148. Stevens HC (2012) Fructose impacts on gut microbiota and obesity by A.N. Payne et al. Obes Rev 13:1182–1183, 1 author response 1184–1185
149. Teff KL, Elliott SS, Tschop M et al (2004) Dietary fructose reduces circulating insulin and leptin, attenuates postprandial suppression of ghrelin, and increases triglycerides in women. J Clin Endocrinol Metab 89:2963–2972
150. Lingelbach LB, McDonald RB (2000) Description of the long-term lipogenic effects of dietary carbohydrates in male Fischer 344 rats. J Nutr 130:3077–3084
151. Alvarez-Suarez JM, Giampieri F, Battino M (2013) Honey as a source of dietary antioxidants: structures, bioavailability and evidence of protective effects against human chronic diseases. Curr Med Chem 20:621–638
152. Johnson RK, Appel LJ, Brands M et al (2009) Dietary sugars intake and cardiovascular health: a scientific statement from the American Heart Association. Circulation 120: 1011–1020
153. Sánchez-Lozada LG, Le M, Segal M et al (2008) How safe is fructose for persons with or without diabetes? Am J Clin Nutr 88:1189–1190

Substrate Metabolism in the Diabetic Heart

Arzu Onay-Besikci

Abstract Diabetic cardiomyopathy, which is defined as cardiac disease independent of vascular complications, is considered one of the consequences of the altered metabolic milieu during diabetes. Constant requirement for energy in the form of ATP is fulfilled mainly by utilizing carbohydrates (glucose and lactate) and fatty acids in the heart. Only minor differences exist between species, and the healthy adult heart relies on the oxidation of fatty acids for ATP production. Utilization of energetic substrates depends on many factors and hormones play a major role in the process. Insulin deficiency, for example, affects the levels of circulating glucose as well as fatty acids and, most certainly, these alterations contribute to the utilization of these substrates. In the past few decades, adipose tissue-originated hormones, such as leptin and adiponectin with major effects on metabolism, have been identified. Not only the amounts of hormones or substrate supply but also subcellular modifications seem to determine the heart's preference for certain substrates during physiological and pathological transitions. Among them, in diabetes, the preference of the heart changes, or perhaps the heart becomes obligated to adapt to dramatic shifts in hormones, substrate supply, and subcellular alterations. This chapter summarizes the contribution of energetic substrate metabolism to the development of diabetic cardiomyopathy.

Keywords Cardiac glucose oxidation • Cardiac fatty acid oxidation • Adiponectin • Leptin • Carvedilol

---

A. Onay-Besikci (✉)
Department of Pharmacology, Faculty of Pharmacy, Ankara University, Tandogan, Ankara 06100, Turkey
e-mail: abesikci@ankara.edu.tr

B. Turan and N.S. Dhalla (eds.), *Diabetic Cardiomyopathy: Biochemical and Molecular Mechanisms*, Advances in Biochemistry in Health and Disease 9,
DOI 10.1007/978-1-4614-9317-4_4,

1 Introduction

The heart has a very high energy demand because of the continuous work performed by the cardiac muscle. Under physiological conditions, the main substrates for energy production are fatty acids (60–90 % of ATP production), glucose, and lactate. Under nonischemic conditions, more than 95 % of the energy requirement is supplied by the oxidation of fatty acids and carbohydrates, and myocardial energy expenditure can be estimated from oxygen consumption. The contribution of these substrates to overall energy production is a dynamic process, and physiological adaptations such as fetal-to-newborn transition [1–3], and alterations related to disease states have been well established [4–8]. The heart exerts a metabolic flexibility, and myocardial substrate utilization depends on substrate availability, nutritional status, and exercise level. With glucose as the more energetically efficient substrate, the healthy heart is able to switch to glucose under stress conditions, such as ischemia, pressure overload, or in heart failure. Interestingly, interventions such as increasing fatty acid uptake [9, 10], or fatty acid oxidation [11, 12], result in alterations that resemble diabetic cardiomyopathy, and normalization of substrate metabolism in diabetic models reverses these alterations [13]. These studies indicate an important contributing role for substrate metabolism in the development of diabetic cardiomyopathy.

Although excessively available during diabetes, utilization of carbohydrates is compromised. Thus, the myocardium is forced to use alternative sources for ATP production such as fatty acids, which are high in diabetes as well [14–18]. The consequences of this pathological switch occur at multiple levels, as is discussed here.

From the mechanical efficiency point of view, an increase in fatty acid usage at the expense of carbohydrates results in ATP hydrolysis for noncontractile purposes and is attributed to a lower phosphate/oxygen (P/O) ratio for fatty acid metabolism, increased uncoupling of mitochondrial oxidative phosphorylation, and futile cycling (reviewed in [4]).

High levels of fatty acids further lower glucose usage as first defined by Randle et al. [19]. We now know that intracellular changes occur during diabetes and contribute to the aforementioned switch as well as development of diabetic cardiomyopathy.

2 Alterations in Cardiac Carbohydrate Utilization

Two glucose transporters, GLUT1 and GLUT4, are involved in basal and insulin-mediated glucose uptake. GLUT1 shows sarcolemmal localization and represents basal cardiac uptake. GLUT4, on the other hand, is located in the intracellular pool, and insulin facilitates the localization of this transporter to the sarcolemma [20]. More recently, AMP-dependent protein kinase (AMPK)-mediated and insulin-independent uptake of glucose by this transporter has been documented [21]. A decrease in the number and translocation of GLUT4 to the sarcolemma has been suggested to play a major role in the reduction of glucose metabolism in diabetes. A reduction in both glycolysis and glucose oxidation was reported in db/db mice in which cardiac dysfunction has been already defined [13]. Because both metabolic parameters and cardiac function were normalized in transgenic mice that overexpressed GLUT4, it has been concluded that there was a causative relationship between impaired substrate metabolism and diabetic cardiomyopathy [13]. A key enzyme in the regulation of glycolysis is phosphofructokinase (PFK)-I, the enzyme that catalyzes the phosphorylation of fructose-6-phosphate to produce fructose-1,6-bisphosphate. PFK-I activity is inhibited by citrate and acetyl CoA and activated by a low ATP/ADP ratio [22]. An increase in citrate levels as a result of increased fatty acid oxidation in the diabetic heart probably contributes to the inhibition of PFK-I and therefore glycolysis. At the transcriptional level of glucose uptake and metabolism, Finck et al. reported that both GLUT4 and PFK expression were lower and that PDK4 expression was higher in transgenic mice that overexpress peroxisome proliferator-activated receptor (PPAR)-α [12]. The inhibition of GLUT4 and PFK was probably not a direct result of PPAR-α overexpression, but associated with alterations in substrate metabolism that were mediated by PPAR-α. The increase in PDK4, on the other hand, was likely linked to PPAR-α overexpression because PPAR-α ligands were previously shown to activate this enzyme [23]. Another member of the PPAR family of transcription factors is PPAR-δ. PPAR-δ is the predominant form in the heart and regulates cardiac substrate metabolism [24]. A decrease in PPAR-δ expression was reported in diabetic heart [25]. However, another similar study reported that mice overexpressing PPAR-β/δ did not accumulate myocardial lipid and had normal cardiac function [26]. Conversely, cardiac glucose transport and glycolytic enzymes were activated in PPAR-β/δ transgenics.

Another limiting step in myocardial glucose metabolism is pyruvate dehydrogenase complex (PDH), which catalyzes the irreversible conversion of pyruvate to acetyl CoA. The amount of active dephosphorylated PDH is reduced when phosphorylated by PDH kinase (PDK) and induced also by PDH phosphatase. The rate of pyruvate oxidation depends not only on the phosphorylation state but also on the concentrations of its substrates (pyruvate, NAD^+, and CoA) and products (NADH and acetyl CoA). Thus, an increase in mitochondrial acetyl CoA, through an increase in fatty acid oxidation, for example, inhibits pyruvate oxidation. Indeed, the active dephosphorylated form of the PDH is reduced in diabetes models [7]. Moreover, PDK-4 is one of the targets of PPAR-α, and upregulation of PDK-4 in mice overexpressing PPARα is associated with reduced glucose oxidation [27].

Inhibition of pyruvate conversion into acetyl CoA results in accumulation and diversion of glycolytic intermediates into diacylglycerol biosynthesis, which contributes to the activation of diacylglycerol-sensitive protein kinase C (PKC) isoforms. Recently, inhibition of one of the PKC isoforms, PKC-beta, was shown to preserve cardiac function in a transgenic mice model of diabetic diastolic failure [28].

The reports on carbohydrate utilization in the human diabetic heart are controversial. Studies in type 1 diabetic patients reported lower [29, 30] or unchanged [31] uptake of carbohydrates by the myocardium. In type 2 diabetes, GLUT4 protein levels in diabetics were about 30 % lower compared to control patients [32].

However, other studies reported that cardiac glucose uptake was not compromised in type 2 diabetes [33–35] and reduced only in type 2 diabetics with hypertriglyceridemia [36] and increased plasma fatty acids. Because glucose can still enter the cell through mass action, as evidenced by a high glucose pool in the type 1 diabetic heart [22], glucose metabolism is unlikely to be regulated at the level of uptake in diabetes despite a deficiency in or resistance to insulin action.

Lactate is another potential substrate for myocardial ATP production in vivo [37], but the information on diabetes-related alterations in lactate oxidation is relatively scant. A greater decrease in lactate oxidation relative to glucose oxidation in hearts from diabetic rats was observed when lactate and glucose were the only substrates in the perfusate for ATP production. Under these conditions, a specific inhibition of lactate oxidation independent of pyruvate dehydrogenase was suggested [38]. Hearts from ZDF rats also showed lower lactate oxidation [39, 40]. The contribution of lactate in diabetic cardiomyopathy clearly needs further study.

3 Alterations in Cardiac Fatty Acid Utilization

An increase of fatty acids that are supplied as free acids bound to albumin and as esters in chylomicrons and very low density lipoproteins have been reported in diabetes [7]. The effects of elevated levels of lipoproteins on myocardial fatty acid metabolism are not clear, and neither is the relative contribution of cardiac lipoprotein lipase (LPL) activity to the delivery of free fatty acids to the diabetic heart. Unchanged, increased, and decreased levels of LPL protein and activity were reported in the diabetic heart, and this discrepancy was suggested to be related to the diversity in rat strains, the dosage of diabetogenic agent, and the duration of diabetes [41].

Free fatty acids enter the cardiac myocyte either by passive diffusion or via a protein carrier-mediated pathway. These protein carriers include fatty acid translocase (FAT)/CD36, the plasma membrane isoform of fatty acid-binding protein (FABPpm), and fatty acid transport protein (FATP) 1/6. FAT/CD36 plays a major role in the translocation of fatty acid across the sarcolemmal membrane of cardiac myocytes as this protein was shown to mediate 50–60 % of fatty acid and transport of the heart. Also, FAT/CD36 is able to translocate between intracellular endosomes and the sarcolemmal membrane and thereby regulate fatty acid uptake [4].

Fatty acid uptake is increased in diabetes and leads to increased fatty acid oxidation and triacylglycerol (TAG) storage. In the streptozotocin (STZ)-induced model of type 1 diabetes, this increase is facilitated by fatty acid translocase (FAT/CD36) [42]. In type 2 diabetic models, an increase in FAT/CD36 and fatty acid-binding protein (FATP1) [43] and a permanent relocation of FAT/CD36 to the cardiomyocyte membrane was shown to increase fatty acid uptake [44, 45]. Interestingly, insulin was suggested to upregulate FAT/CD36 and translocate it to the sarcolemma [46].

The majority (70–90 %) of the fatty acids that enter cardiomyocyte is oxidized for energy production; the rest is converted to TAGs [47]. Excessive accumulation of lipids, or lipotoxicity, within nonadipose tissue provides substrates for nonoxidative processes such as ceramide and diacylglycerol synthesis, which can lead to apoptosis [48, 49]. Accumulation of TAG within the myocardium of insulin-resistant rats was associated with contractile dysfunction [50]. We also have shown that insulin-resistant rats have increased TAG accumulation, which reduced insulin-stimulated glucose metabolism [51]. Although the exact mechanism of lipotoxicity-induced cardiac dysfunction is not known, it seems to be related to a combination of apoptotic cell death and impaired substrate metabolism.

The most important step in the regulation of fatty acid oxidation is the transport of fatty acids into the mitochondria for further metabolism. The activation of short- and medium-chain fatty acids occurs in the matrix and does not require carnitine. However, long-chain fatty acids are shuttled into the mitochondria by three carnitine-dependant enzymes. Carnitine palmitoyltransferase (CPT)-I catalyzes the conversion of long-chain acyl CoA to long-chain acylcarnitine. Carnitine:acylcarnitine translocase (CAT) transports long-chain acylcarnitine across the inner mitochondrial membrane, and CPT-II regenerates long-chain acyl-CoA in the mitochondrial matrix [52]. Of these, CPT-I is the master regulator of mitochondrial uptake of fatty acids and is allosterically inhibited by malonyl CoA [53]. The turnover of malonyl CoA in the heart is fast. Therefore, myocardial malonyl CoA concentrations are dependent on the balance between its synthesis from acetyl CoA via acetyl CoA carboxylase (ACC) and its degradation via malonyl CoA decarboxylase (MCD) [4]. A good correlation has been established between levels of malonyl CoA and rates of fatty acid oxidation, and a reduction in malonyl CoA levels is almost consistent in situations in which fatty acid oxidation is increased [18]. The decrease in malonyl CoA levels seems to be related to increased degradation of malonyl CoA by MCD [18]. MCD is transcriptionally regulated by PPAR-α [54, 55]. In addition to diabetes, activity and expression of cardiac MCD was increased in fasting, high-fat-fed, and newborn hearts [56–59]. Moreover, PPAR-α null mice had increased rates of glucose oxidation and decreased expression and activity of MCD [56].

Increase in circulating fatty acids directly modifies the enzymes in substrate metabolism because fatty acids and their derivatives are ligands for the PPAR family of nuclear receptors, of which PPAR-α and its coactivator peroxisome proliferator-activated receptor-gamma coactivator (PGC)-1 are particularly important in the heart [60–62]. PPAR-α signaling was increased in 15-week-old ob/ob and db/db mice [60]. Other studies reported an increase in the expressions of PPAR-α, PGC-1, and their targets in models of insulin resistance and type 2 diabetes [12, 63–65].

Once in the mitochondrial matrix, long-chain fatty acyl CoAs pass through the β-oxidation enzyme system to produce one acetyl CoA at each cycle, one NADH, and one FADH. The key enzyme in the β-oxidation pathway is β-hydroxyacyl-CoA dehydrogenase. The activity of this enzyme was shown to be normal [66] or high [67] in diabetic rat mitochondria. Higher expression of another enzyme, 3-ketoacyl-CoA thiolase, has also been shown in streptozotocin-diabetic rat hearts [8]. Thus, a combination of high circulating levels of fatty acids, decreased inhibition of fatty acid uptake by the mitochondria, and a normal or accelerated β-oxidation pathway results in a large proportion of acetyl CoA for the tricarboxylic acid (TCA) cycle being supplied from fatty acid oxidation.

4 Alterations in Diabetic Mitochondria

Acetyl CoA derived from both β-oxidation of fatty acids and PDH enters the TCA cycle. The myocardial TCA cycle does not appear to be altered in diabetes because measurements of TCA cycle enzyme activities are similar in diabetic and control hearts [7].

Mitochondrial proteins are produced in the mitochondria and are regulated under the mitochondrial transcriptional factor A (TFAM) by mitochondrial DNA. A recent study reported that overexpression of TFAM improved ATP content and SERCA2a content, both of which were deteriorated by exposing neonatal rat myocytes to hyperglycemic conditions [68].

5 Ketone Body Metabolism

Plasma ketone bodies (β-hydroxybutyrate and acetoacetate) are formed from fatty acids in the liver, and their plasma concentration is normally very low. Thus, ketone bodies are normally a minor substrate for the myocardium. When plasma levels increase during extreme conditions, such as poorly controlled diabetes and starvation, the heart extracts and oxidizes ketone bodies [29]. Ketone bodies inhibit the uptake and metabolism of other energetic substrates [69–74]. The biochemical mechanisms responsible for the inhibition of glucose, lactate, or fatty acid oxidation is not clearly understood.

6 Potential Drugs with Metabolic Effects in the Treatment of Diabetes

Our knowledge of the adipose tissue-originated hormones—adipokines—and their biological effects has increased tremendously since the discovery of the OB gene product, leptin, in 1994 [75]. Two of these, leptin and adiponectin, have direct effects on substrate metabolism as well as insulin secretion. Despite the presence of some conflicting reports, leptin seems to regulate pancreatic cell function and thus insulin secretion [76–79]. Acute exposure of myocytes [80] and isolated working hearts [51] to leptin stimulates fatty acid oxidation. In contrast, type 2 diabetic models, ob/ob and db/db mice, also display increased myocardial fatty acid β-oxidation [60, 81, 82], which is probably related to other alterations in these genetic models. Moreover, leptin administration to subjects with lipodystrophy, abnormal distribution of adipose tissue in which leptin levels are usually decreased [83, 84], has been shown to improve metabolic abnormalities such as hypertriglyceridemia and impaired glucose control, which often are resistant to maximum doses of insulin sensitizers or very high doses of insulin [85–87]. Taken together, these studies suggest that leptin has the potential to become an effective antidiabetic agent with effects on substrate metabolism.

Described by Scherer et al. in 1995 [88], adiponectin is another antidiabetic adipokine. Adiponectin possibly affects β-cell function as adiponectin receptors are expressed in β cells [89–91]. More recently, adiponectin was shown to protect against caspase-8-mediated apoptosis in β cells [92]. As for the effects on substrate metabolism, globular head domain of adiponectin (gAd) stimulates fatty acid oxidation in isolated working 1-day-old rabbit heart [93]. Similar effects on fatty acid oxidation with both the high molecular weight hexameric form of adiponectin and gAd have been reported for cardiac myocytes [94].

Finally, it is worth mentioning other drugs that have beneficial effects on substrate metabolism. Beta-adrenergic receptor antagonists (β-blockers) are used in the treatment of many disease states such as hypertension, ischemic heart disease, arrhythmia, heart failure, glaucoma, and migraine, as well as to reduce the symptoms related to anxiety and hyperthyroidism. Depending on their receptor subtype specificity, β-blockers have different effects on substrate usage. As clearly shown by clinical trials, long-term use of nonselective β-blockers such as proporanolol increases the incidence of diabetes [95, 96], and cardioselective β-blockers such as atenolol and metoprolol reduce insulin sensitivity [97, 98], whereas carvedilol improves serum lipid profile, reduces insulin resistance, and decreases the risk for diabetes [99, 100]. Based on these reports, nonselective and cardioselective β-blockers are accepted as "pro-insulin resistant" whereas carvedilol, the drug that also blocks α1-adrenergic receptors, is known to have opposite effects (reviewed in [101, 102]). The beneficial properties of carvedilol seem to be related to its direct effects on substrate metabolism because carvedilol was shown to reduce the myocardial uptake of long-chain fatty acids and improved left ventricular ejection fraction in heart failure patients [103]. More recently, carvedilol treatment reduced fatty acid oxidation and increased glycolysis in C2C12 cells [104]. The possible effects of carvedilol on diabetic cardiomyopathy need future investigations. However, the fact that carvedilol has beneficial effects on insulin resistance warrants an improvement in substrate utilization and hence diabetic cardiomyopathy.

7 Conclusion

Diabetes is a complex disorder. Its contributors exist at multiple levels. Hormones predominantly determine the levels of energetic substrates, and their major impact on energetic substrate utilization has been well documented. However, modulation of subcellular steps that are involved in ATP-producing processes seems to be just as important in the regulation of substrate metabolism and is expected to have beneficial effects on the mechanical function of the heart. Future studies will reveal such interventions that improve the hormonal milieu or deteriorated substrate metabolism, and leptin and adiponectin agonists are good candidates.

References

1. Onay-Besikci A (2006) Regulation of cardiac energy metabolism in newborn. Mol Cell Biochem 287:1–11
2. Lopaschuk GD, Spafford MA (1990) Energy substrate utilization by isolated working hearts from newborn rabbits. Am J Physiol 258:H1274–H1280
3. Onay-Besikci A, Campbell FM, Hopkins TA et al (2003) Relative importance of malonyl CoA and carnitine in maturation of fatty acid oxidation in newborn rabbit heart. Am J Physiol Heart Circ Physiol 284:H283–H289
4. Lopaschuk GD, Ussher JR, Folmes CD et al (2010) Myocardial fatty acid metabolism in health and disease. Physiol Rev 90:207–258
5. Paolisso G, Gambardella A, Galzerano D et al (1994) Total-body and myocardial substrate oxidation in congestive heart failure. Metabolism 43:174–179
6. Taylor M, Wallhaus TR, Degrado TR et al (2001) An evaluation of myocardial fatty acid and glucose uptake using PET with ^{18}F-fluoro-6-thia-heptadecanoic acid and ^{18}F-FDG in patients with congestive heart failure. J Nucl Med 42:55–62
7. Stanley WC, Lopaschuk GD, McCormack JG (1997) Regulation of energy substrate metabolism in the diabetic heart. Cardiovasc Res 34:25–33
8. Onay-Besikci A, Guner S, Arioglu E et al (2007) The effects of chronic trimetazidine treatment on mechanical function and fatty acid oxidation in diabetic rat hearts. Can J Physiol Pharmacol 85:527–535
9. Yagyu H, Chen G, Yokoyama M et al (2003) Lipoprotein lipase (LpL) on the surface of cardiomyocytes increases lipid uptake and produces a cardiomyopathy. J Clin Invest 111: 419–426
10. Chiu HC, Kovacs A, Blanton RM et al (2005) Transgenic expression of fatty acid transport protein 1 in the heart causes lipotoxic cardiomyopathy. Circ Res 96:225–233
11. Chiu HC, Kovacs A, Ford DA et al (2001) A novel mouse model of lipotoxic cardiomyopathy. J Clin Invest 107:813–822
12. Finck BN, Lehman JJ, Leone TC et al (2002) The cardiac phenotype induced by PPAR-alpha overexpression mimics that caused by diabetes mellitus. J Clin Invest 109:121–130
13. Belke DD, Larsen TS, Gibbs EM et al (2000) Altered metabolism causes cardiac dysfunction in perfused hearts from diabetic (db/db) mice. Am J Physiol Endocrinol Metab 279: E1104–E1113
14. Randle PJ, Garland PB, Hales CN et al (1963) The glucose fatty-acid cycle. Its role in insulin sensitivity and the metabolic disturbances of diabetes mellitus. Lancet 1:785–789
15. Goodale WT, Olson RE, Hackel DB (1959) The effects of fasting and diabetes mellitus on myocardial metabolism in man. Am J Med 27:212–220
16. Wall SR, Lopaschuk GD (1989) Glucose oxidation rates in fatty acid-perfused isolated working hearts from diabetic rats. Biochim Biophys Acta 1006:97–103
17. Reaven GM, Hollenbeck C, Jeng CY et al (1988) Measurement of plasma glucose, free fatty acid, lactate, and insulin for 24 h in patients with NIDDM. Diabetes 37:1020–1024
18. Sakamoto J, Barr RL, Kavanagh KM et al (2000) Contribution of malonyl-CoA decarboxylase to the high fatty acid oxidation rates seen in the diabetic heart. Am J Physiol Heart Circ Physiol 278:H1196–H1204
19. Randle PJ, Garland PB, Hales CN et al (1966) Interactions of metabolism and the physiological role of insulin. Recent Prog Horm Res 22:1–48
20. Luiken JJ, Coort SL, Koonen DP et al (2004) Regulation of cardiac long-chain fatty acid and glucose uptake by translocation of substrate transporters. Pflugers Arch 448:1–15
21. Li J, Hu X, Selvakumar P et al (2004) Role of the nitric oxide pathway in AMPK-mediated glucose uptake and GLUT4 translocation in heart muscle. Am J Physiol Endocrinol Metab 287:E834–E841
22. Isfort M, Stevens SC, Schaffer S et al (2013) Metabolic dysfunction in diabetic cardiomyopathy. Heart Fail Rev

23. Wu P, Inskeep K, Bowker-Kinley MM et al (1999) Mechanism responsible for inactivation of skeletal muscle pyruvate dehydrogenase complex in starvation and diabetes. Diabetes 48: 1593–1599
24. Cheng L, Ding G, Qin Q et al (2004) Cardiomyocyte-restricted peroxisome proliferator-activated receptor-delta deletion perturbs myocardial fatty acid oxidation and leads to cardiomyopathy. Nat Med 10:1245–1250
25. Yu BC, Chang CK, Ou HY et al (2008) Decrease of peroxisome proliferator-activated receptor delta expression in cardiomyopathy of streptozotocin-induced diabetic rats. Cardiovasc Res 80:78–87
26. Burkart EM, Sambandam N, Han X et al (2007) Nuclear receptors PPAR-beta/delta and PPAR-alpha direct distinct metabolic regulatory programs in the mouse heart. J Clin Invest 117:3930–3939
27. Hopkins TA, Sugden MC, Holness MJ et al (2003) Control of cardiac pyruvate dehydrogenase activity in peroxisome proliferator-activated receptor-alpha transgenic mice. Am J Physiol Heart Circ Physiol 285:H270–H276
28. Connelly KA, Kelly DJ, Zhang Y et al (2009) Inhibition of protein kinase C-beta by ruboxistaurin preserves cardiac function and reduces extracellular matrix production in diabetic cardiomyopathy. Circ Heart Fail 2:129–137
29. Avogaro A, Nosadini R, Doria A et al (1990) Myocardial metabolism in insulin-deficient diabetic humans without coronary artery disease. Am J Physiol 258:E606–E618
30. Doria A, Nosadini R, Avogaro A et al (1991) Myocardial metabolism in type 1 diabetic patients without coronary artery disease. Diabet Med 8(spec no):S104–S107
31. Nuutila P, Knuuti J, Ruotsalainen U et al (1993) Insulin resistance is localized to skeletal but not heart muscle in type 1 diabetes. Am J Physiol 264:E756–E762
32. Armoni M, Harel C, Bar-Yoseph F et al (2005) Free fatty acids repress the GLUT4 gene expression in cardiac muscle via novel response elements. J Biol Chem 280:34786–34795
33. Jagasia D, Whiting JM, Concato J et al (2001) Effect of non-insulin-dependent diabetes mellitus on myocardial insulin responsiveness in patients with ischemic heart disease. Circulation 103:1734–1739
34. Maki M, Nuutila P, Laine H et al (1997) Myocardial glucose uptake in patients with NIDDM and stable coronary artery disease. Diabetes 46:1491–1496
35. Utriainen T, Takala T, Luotolahti M et al (1998) Insulin resistance characterizes glucose uptake in skeletal muscle but not in the heart in NIDDM. Diabetologia 41:555–559
36. Monti LD, Landoni C, Setola E et al (2004) Myocardial insulin resistance associated with chronic hypertriglyceridemia and increased FFA levels in type 2 diabetic patients. Am J Physiol Heart Circ Physiol 287:H1225–H1231
37. Chatham JC (2002) Lactate – the forgotten fuel! J Physiol 542:333
38. Chatham JC, Gao ZP, Bonen A et al (1999) Preferential inhibition of lactate oxidation relative to glucose oxidation in the rat heart following diabetes. Cardiovasc Res 43:96–106
39. Chatham JC, Seymour AM (2002) Cardiac carbohydrate metabolism in Zucker diabetic fatty rats. Cardiovasc Res 55:104–112
40. Wang P, Lloyd SG, Zeng H et al (2005) Impact of altered substrate utilization on cardiac function in isolated hearts from Zucker diabetic fatty rats. Am J Physiol Heart Circ Physiol 288:H2102–H2110
41. An D, Rodrigues B (2006) Role of changes in cardiac metabolism in development of diabetic cardiomyopathy. Am J Physiol Heart Circ Physiol 291:H1489–H1506
42. Luiken JJ, Arumugam Y, Bell RC et al (2002) Changes in fatty acid transport and transporters are related to the severity of insulin deficiency. Am J Physiol Endocrinol Metab 283: E612–E621
43. Luiken JJ, Arumugam Y, Dyck DJ et al (2001) Increased rates of fatty acid uptake and plasmalemmal fatty acid transporters in obese Zucker rats. J Biol Chem 276:40567–40573
44. Coort SL, Hasselbaink DM, Koonen DP et al (2004) Enhanced sarcolemmal FAT/CD36 content and triacylglycerol storage in cardiac myocytes from obese Zucker rats. Diabetes 53:1655–1663

45. Coort SL, Luiken JJ, van der Vusse GJ et al (2004) Increased FAT (fatty acid translocase)/CD36-mediated long-chain fatty acid uptake in cardiac myocytes from obese Zucker rats. Biochem Soc Trans 32:83–85
46. Chabowski A, Coort SL, Calles-Escandon J et al (2004) Insulin stimulates fatty acid transport by regulating expression of FAT/CD36 but not FABPpm. Am J Physiol Endocrinol Metab 287:E781–E789
47. Stanley WC, Recchia FA, Lopaschuk GD (2005) Myocardial substrate metabolism in the normal and failing heart. Physiol Rev 85:1093–1129
48. Unger RH, Orci L (2000) Lipotoxic diseases of nonadipose tissues in obesity. Int J Obes Relat Metab Disord 24(suppl 4):S28–S32
49. Zhou YT, Grayburn P, Karim A et al (2000) Lipotoxic heart disease in obese rats: implications for human obesity. Proc Natl Acad Sci USA 97:1784–1789
50. Finck BN, Han X, Courtois M et al (2003) A critical role for PPARalpha-mediated lipotoxicity in the pathogenesis of diabetic cardiomyopathy: modulation by dietary fat content. Proc Natl Acad Sci USA 100:1226–1231
51. Atkinson LL, Kozak R, Kelly SE et al (2003) Potential mechanisms and consequences of cardiac triacylglycerol accumulation in insulin-resistant rats. Am J Physiol Endocrinol Metab 284:E923–E930
52. McGarry JD, Brown NF (1997) The mitochondrial carnitine palmitoyltransferase system. From concept to molecular analysis. Eur J Biochem/FEBS 244:1–14
53. McGarry JD, Mills SE, Long CS et al (1983) Observations on the affinity for carnitine, and malonyl-CoA sensitivity, of carnitine palmitoyltransferase I in animal and human tissues. Demonstration of the presence of malonyl-CoA in non-hepatic tissues of the rat. Biochem J 214:21–28
54. Kim SK, Zhao ZS, Lee YJ et al (2003) Left-ventricular diastolic dysfunction may be prevented by chronic treatment with PPAR-alpha or -gamma agonists in a type 2 diabetic animal model. Diabetes Metab Res Rev 19:487–493
55. Lee GY, Kim NH, Zhao ZS et al (2004) Peroxisomal-proliferator-activated receptor alpha activates transcription of the rat hepatic malonyl-CoA decarboxylase gene: a key regulation of malonyl-CoA level. Biochem J 378:983–990
56. Campbell FM, Kozak R, Wagner A et al (2002) A role for peroxisome proliferator-activated receptor alpha (PPAR-alpha) in the control of cardiac malonyl-CoA levels: reduced fatty acid oxidation rates and increased glucose oxidation rates in the hearts of mice lacking PPAR-alpha are associated with higher concentrations of malonyl-CoA and reduced expression of malonyl-CoA decarboxylase. J Biol Chem 277:4098–4103
57. Dyck JR, Berthiaume LG, Thomas PD et al (2000) Characterization of rat liver malonyl-CoA decarboxylase and the study of its role in regulating fatty acid metabolism. Biochem J 350 (Pt 2):599–608
58. Goodwin GW, Taegtmeyer H (1999) Regulation of fatty acid oxidation of the heart by MCD and ACC during contractile stimulation. Am J Physiol 277:E772–E777
59. Young ME, Goodwin GW, Ying J et al (2001) Regulation of cardiac and skeletal muscle malonyl-CoA decarboxylase by fatty acids. Am J Physiol Endocrinol Metab 280:E471–E479
60. Buchanan J, Mazumder PK, Hu P et al (2005) Reduced cardiac efficiency and altered substrate metabolism precedes the onset of hyperglycemia and contractile dysfunction in two mouse models of insulin resistance and obesity. Endocrinology 146:5341–5349
61. Gilde AJ, van der Lee KA, Willemsen PH et al (2003) Peroxisome proliferator-activated receptor (PPAR) alpha and PPAR-beta/delta, but not PPAR-gamma, modulate the expression of genes involved in cardiac lipid metabolism. Circ Res 92:518–524
62. Murray AJ, Panagia M, Hauton D et al (2005) Plasma free fatty acids and peroxisome proliferator-activated receptor alpha in the control of myocardial uncoupling protein levels. Diabetes 54:3496–3502
63. Duncan JG, Fong JL, Medeiros DM et al (2007) Insulin-resistant heart exhibits a mitochondrial biogenic response driven by the peroxisome proliferator-activated receptor-alpha/PGC-1alpha gene regulatory pathway. Circulation 115:909–917

64. Bernal-Mizrachi C, Weng S, Feng C et al (2003) Dexamethasone induction of hypertension and diabetes is PPAR-alpha dependent in LDL receptor-null mice. Nat Med 9:1069–1075
65. Sharma S, Adrogue JV, Golfman L et al (2004) Intramyocardial lipid accumulation in the failing human heart resembles the lipotoxic rat heart. FASEB J 18:1692–1700
66. Glatz JF, van Breda E, Keizer HA et al (1994) Rat heart fatty acid-binding protein content is increased in experimental diabetes. Biochem Biophys Res Commun 199:639–646
67. Chen V, Ianuzzo CD, Fong BC et al (1984) The effects of acute and chronic diabetes on myocardial metabolism in rats. Diabetes 33:1078–1084
68. Suarez J, Hu Y, Makino A et al (2008) Alterations in mitochondrial function and cytosolic calcium induced by hyperglycemia are restored by mitochondrial transcription factor A in cardiomyocytes. Am J Physiol Cell Physiol 295:C1561–C1568
69. Bing RJ, Siegel A, Ungar I et al (1954) Metabolism of the human heart. II. Studies on fat, ketone and amino acid metabolism. Am J Med 16:504–515
70. Stanley WC, Meadows SR, Kivilo KM et al (2003) Beta-hydroxybutyrate inhibits myocardial fatty acid oxidation in vivo independent of changes in malonyl-CoA content. Am J Physiol Heart Circ Physiol 285:H1626–H1631
71. Sultan AM (1992) Effects of diabetes and insulin on ketone bodies metabolism in heart. Mol Cell Biochem 110:17–23
72. Chen V, Wagner G, Spitzer JJ (1984) Regulation of substrate oxidation in isolated myocardial cells by beta-hydroxybutyrate. Horm Metab Res 16:243–247
73. Forsey RG, Reid K, Brosnan JT (1987) Competition between fatty acids and carbohydrate or ketone bodies as metabolic fuels for the isolated perfused heart. Can J Physiol Pharmacol 65:401–406
74. Hasselbaink DM, Glatz JF, Luiken JJ et al (2003) Ketone bodies disturb fatty acid handling in isolated cardiomyocytes derived from control and diabetic rats. Biochem J 371:753–760
75. Zhang Y, Proenca R, Maffei M et al (1994) Positional cloning of the mouse obese gene and its human homologue. Nature 372:425–432
76. Ahren B, Larsson H (1997) Leptin–a regulator of islet function? Its plasma levels correlate with glucagon and insulin secretion in healthy women. Metabolism 46:1477–1481
77. Kulkarni RN, Wang ZL, Wang RM et al (1997) Leptin rapidly suppresses insulin release from insulinoma cells, rat and human islets and, in vivo, in mice. J Clin Invest 100:2729–2736
78. Brown JE, Thomas S, Digby JE et al (2002) Glucose induces and leptin decreases expression of uncoupling protein-2 mRNA in human islets. FEBS Lett 513:189–192
79. Tanizawa Y, Okuya S, Ishihara H et al (1997) Direct stimulation of basal insulin secretion by physiological concentrations of leptin in pancreatic beta cells. Endocrinology 138: 4513–4516
80. Palanivel R, Eguchi M, Shuralyova I et al (2006) Distinct effects of short- and long-term leptin treatment on glucose and fatty acid uptake and metabolism in HL-1 cardiomyocytes. Metabolism 55:1067–1075
81. Mazumder PK, O'Neill BT, Roberts MW et al (2004) Impaired cardiac efficiency and increased fatty acid oxidation in insulin-resistant ob/ob mouse hearts. Diabetes 53:2366–2374
82. Carley AN, Semeniuk LM, Shimoni Y et al (2004) Treatment of type 2 diabetic db/db mice with a novel PPAR-gamma agonist improves cardiac metabolism but not contractile function. Am J Physiol Endocrinol Metab 286:E449–E455
83. Pardini VC, Victoria IM, Rocha SM et al (1998) Leptin levels, beta-cell function, and insulin sensitivity in families with congenital and acquired generalized lipoatropic diabetes. J Clin Endocrinol Metab 83:503–508
84. Nagy GS, Tsiodras S, Martin LD et al (2003) Human immunodeficiency virus type 1-related lipoatrophy and lipohypertrophy are associated with serum concentrations of leptin. Clin Infect Dis 36:795–802
85. Musso C, Cochran E, Javor E et al (2005) The long-term effect of recombinant methionyl human leptin therapy on hyperandrogenism and menstrual function in female and pituitary function in male and female hypoleptinemic lipodystrophic patients. Metabolism 54: 255–263

86. Oral EA, Simha V, Ruiz E et al (2002) Leptin-replacement therapy for lipodystrophy. N Engl J Med 346:570–578
87. Chong AY, Lupsa BC, Cochran EK et al (2010) Efficacy of leptin therapy in the different forms of human lipodystrophy. Diabetologia 53:27–35
88. Scherer PE, Williams S, Fogliano M et al (1995) A novel serum protein similar to C1q, produced exclusively in adipocytes. J Biol Chem 270:26746–26749
89. Kharroubi I, Rasschaert J, Eizirik DL et al (2003) Expression of adiponectin receptors in pancreatic beta cells. Biochem Biophys Res Commun 312:1118–1122
90. Brown JE, Conner AC, Digby JE et al (2010) Regulation of beta-cell viability and gene expression by distinct agonist fragments of adiponectin. Peptides 31:944–949
91. Wijesekara N, Krishnamurthy M, Bhattacharjee A et al (2010) Adiponectin-induced ERK and Akt phosphorylation protects against pancreatic beta cell apoptosis and increases insulin gene expression and secretion. J Biol Chem 285:33623–33631
92. Holland WL, Miller RA, Wang ZV et al (2011) Receptor-mediated activation of ceramidase activity initiates the pleiotropic actions of adiponectin. Nat Med 17:55–63
93. Onay-Besikci A, Altarejos JY, Lopaschuk GD (2004) gAd-globular head domain of adiponectin increases fatty acid oxidation in newborn rabbit hearts. J Biol Chem 279:44320–44326
94. Palanivel R, Fang X, Park M et al (2007) Globular and full-length forms of adiponectin mediate specific changes in glucose and fatty acid uptake and metabolism in cardiomyocytes. Cardiovasc Res 75:148–157
95. Samuelsson O, Hedner T, Berglund G et al (1994) Diabetes mellitus in treated hypertension: incidence, predictive factors and the impact of non-selective beta-blockers and thiazide diuretics during 15 years treatment of middle-aged hypertensive men in the Primary Prevention Trial Goteborg, Sweden. J Hum Hypertens 8:257–263
96. Gress TW, Nieto FJ, Shahar E et al (2000) Hypertension and antihypertensive therapy as risk factors for type 2 diabetes mellitus. N Engl J Med 342:905–912
97. Pollare T, Lithell H, Morlin C et al (1989) Metabolic effects of diltiazem and atenolol: results from a randomized, double-blind-study with parallel groups. J Hypertens 7:551–559
98. Pollare T, Lithell H, Selinus I et al (1989) Sensitivity to insulin during treatment with atenolol and metoprolol: a randomized, double-blind study of effects on carbohydrate and lipoprotein metabolism in hypertensive patients. Br Med J 298:1152–1157
99. Jacob S, Rett K, Wicklmayr M et al (1996) Differential effect of chronic treatment with two beta-blocking agents on insulin sensitivity: the carvedilol-metoprolol study. J Hypertens 14:489–494
100. Torp-Pederson C, Cleland J, Di Lenarda A et al (2005) Carvedilol reduces the risk for new onset of diabetes related adverse events in heart failure compared to metoprolol: results of the COMET study. J Am Coll Cardiol 45:187a
101. Ashrafian H, Frenneaux MP, Opie LH (2007) Metabolic mechanisms in heart failure. Circulation 116:434–448
102. Kostis JB, Sanders M (2005) The association of heart failure with insulin resistance and the development of type 2 diabetes. Am J Hypertens 18:731–737
103. Wallhaus TR, Taylor M, DeGrado TR et al (2001) Myocardial free fatty acid and glucose use after carvedilol treatment in patients with congestive heart failure. Circulation 103:2441–2446
104. Onay-Besikci A, Suzmecelik E, Ozcelikay AT (2012) Carvedilol suppresses fatty acid oxidation and stimulates glycolysis in C2C12 cells. Can J Physiol Pharmacol 90:1087–1093

Effects of Diabetes-Induced Hyperglycemia in the Heart: Biochemical and Structural Alterations

Tahreem Iqbal, Philip J. Welsby, Frank C. Howarth, Keshore Bidasee, Ernest Adeghate, and Jaipaul Singh

Abstract Hyperglycemia (HG) plays a major role in the development of diabetes mellitus (DM) and its complications. HG induces numerous maladaptations at the cellular level and moreover it is an independent risk factor to worsen cardiac performance and cell survival. The heart is a major target organ for damage with hyperglycemia. Alterations as a result of HG can lead to the development of a diabetic cardiomyopathy, resulting in changes to cardiac structure and function. Mechanisms damaging the heart are similar to those that damage the vasculature, but are more widespread in the myocardium. Four major pathways are implicated in HG-induced cardiac and vascular damage, including increases in advanced glycation end products (AGEs), enhanced hexosamine and polyol flux, and activation of classical isoforms of protein kinase C (PKC). These changes lead to abnormalities such as increased ventricular stiffness, cardiac fibrosis, derangement in cellular calcium ion homeostasis, and reduced myocyte contractility, resulting in heart failure (HF) over time. These pathways reflect upon a single HG-induced process of overproduction of superoxide by the mitochondrial electron-transport chain, which is

T. Iqbal • P.J. Welsby • J. Singh (✉)
School of Pharmacy and Biomedical Sciences,
University of Central Lancashire, Preston, Lancashire PR1 2HE, UK
e-mail: jsingh3@uclan.ac.uk

F.C. Howarth
Department of Physiology, United Arab Emirates University,
Al-Ain, United Arab Emirates

K. Bidasee
Department of Pharmacology and Experimental Therapeutics,
Nebraska Medical Centre, Omaha, NE 68198, USA

E. Adeghate
Department of Human Anatomy, Faculty of Medicine and Health Sciences,
United Arab Emirates University, Al-Ain, United Arab Emirates

B. Turan and N.S. Dhalla (eds.), *Diabetic Cardiomyopathy: Biochemical and Molecular Mechanisms*, Advances in Biochemistry in Health and Disease 9,
DOI 10.1007/978-1-4614-9317-4_5, © Springer Science+Business Media New York 2014

responsible for the changes occurring in the heart. This chapter discusses the HG-induced pathways, focusing on their effects on the structure of the diabetic heart, as well as examining the downstream signaling whereby oxidative stress leads to myocardial fibrosis and impaired contractile function. In addition, this review highlights the role of endothelin-1 (ET-1) in endothelial dysfunction and the effects of humoral factors, angiotensin II and transforming growth factor-β, in evoking multiple signaling pathways in cardiac fibroblasts or fibrosis that leads to cardiac remodelling. How these signaling pathways mediated by HG contribute to the pathophysiological alterations in the heart is also discussed in this review.

Keywords Hyperglycemia • Diabetes mellitus • Diabetic cardiomyopathy • Heart structure • Signaling pathways • Endothelial cell dysfunction • Cardiac remodelling

1 Introduction

Diabetes mellitus (DM) is universally characterised by microangiopathic complications, consisting of microvascular (nephropathy, retinopathy, neuropathy) and macrovascular (ischemic heart disease, cerebrovascular diseases, peripheral vascular diseases) complications [1]. DM is a heterogeneous metabolic disorder characterized by chronic HG resulting from malfunctioning of insulin secretion, its action, or both [2]. DM is classified into insulin-dependent diabetes mellitus (T1DM) and non-insulin-dependent diabetes mellitus (T2DM) on the basis of the history, etiology, and clinical presentation of the patient [3]. T1DM is a lifelong metabolic disorder characterized mainly by a deficiency in endogenous insulin production caused by destruction of insulin-producing β-cells of the endocrine pancreas. Consequently, the patients become dependent on exogenous insulin administration [4]. T1DM is the most common childhood disease in the developed world and is treated with a complex regime of insulin injections, diet, and exercise [5]. On the other hand, T2DM is characterized by insulin resistance (IR), which plays a major role in metabolic abnormalities such as dyslipidemia and hypertension [6]. T2DM is a polygenic disorder as is T1DM that may be triggered by genetic and environmental factors, including lifestyle habits [4, 7]. Currently, more than 250 million people worldwide suffer from T1DM and T2DM, and this could increase to 350 million within the next 20 years [8].

It was previously estimated [9, 10] that the number of adults with diabetes worldwide would increase from 135 million in 1995 to 320 million by 2025. It was also claimed that patients with DM are likely to have increased risk of cardiovascular diseases (CVD). In fact, the UK was reported to have more than 2.9 million people diagnosed with DM, with a further 850,000 people still underdiagnosed, which has almost doubled in 2012. In terms of expense, it costs about £850 billion to diagnose, treat, and care for DM patients worldwide. It is now apparent that more women have been diagnosed with diabetes in the age range of 45 to 65 years [11]. The prevalence of diabetes is higher in the developed countries. However, the major increase in people with diabetes is occurring in developing countries such as China, India, and the United States. Throughout the developing world, the increasing rates of diabetes represent an emerging epidemic with marked pathological consequences that are likely to drive previously unpredicted rates of vascular target organ complications [11, 12]. Adults with diabetes are at a two- to four-fold increased risk of cardiovascular events relative to those without diabetes [13]. CVD account for up to 80 % of premature excess mortality in diabetic patients [14]. In a recent study [15], it has been suggested that diabetes may now be considered a CVD because of the high level of morbidity and mortality attributed to CVD resulting from DM. It has been established that CVD are the leading cause of morbidity and mortality in less-developed countries as well as in developed countries [9, 14, 16].

Although medications such as insulin, and many others including hypoglycemic drugs, can control many aspects of diabetes, an assortment of complications affects the vascular system, heart, kidney, and peripheral nerves in the body. These complications are very common, and the cost is extremely high in terms of quality of life for diabetic patients. Unsurprisingly, DM has a profound influence on cardiac metabolism in terms of altered substrate supply, insulin action, and metabolic adaptations [10].

2 Diabetic Cardiomyopathy (DCM)

A high proportion of diabetic patients (both T1DM and T2DM) develop a unique cardiomyopathy. This "diabetic cardiomyopathy," as the name suggests, initiates with asymptomatic left ventricular diastolic dysfunction (slowing of relaxation kinetics) and, on progression of the disease, systolic function becomes compromised, which results in an increase in the incidence of morbidity and mortality. Diabetic cardiomyopathy (DCM) develops as a result of changes to cardiac structure and function in the absence of blood pressure, hypertension, and coronary artery disease [9, 17, 18]. Initially, the term DCM was introduced based upon the findings of postmortem patients with heart failure in the absence of other complications [19]. It has recently been reported that 56 % of diabetic patients have had DCM [20]. The etiology and pathophysiology of DCM are poorly understood and multifactorial.

Diabetic macrovasculopathy is most commonly coupled with structural (glycation of wall components) and functional changes (endothelial cell dysfunction and increased arterial stiffness) [1], vasoconstriction, and inflammation via specific cytokines [transforming growth factor (TGF)-β1]. In turn, these ultimately promote the progression of left ventricular hypertrophy, a self-governing risk factor for cardiovascular (CV) mortality and important mechanisms in the development of DCM [14, 18]. These studies have shown that conventional cardiac risks such as hypertension, atherosclerosis, and dyslipidemia are more frequent in diabetic patients and further aggravate cardiac function. The most prominent histopathological finding in diabetic patients is fibrosis, which is either perivascular, interstitial, or both. Other studies have also claimed that, as the disease progressed, an increase in myocyte loss and replacement fibrosis were also identified [18].

Diabetic patients are prone to the development of left ventricular (LV) dysfunction, which is different from seen in chronic diseases such as hypertension and chronic ischemia, in which the ventricle is not prominently hypertrophied [21]. Numerous mechanisms have been proposed to contribute to this clinical situation, including oxidative stress [22], microvascular abnormalities, and decreased sarcoplasmic reticular calcium ion uptake. However, the molecular mechanisms that cause this cardiac dysfunction are still largely undefined. It has been suggested that diastolic dysfunction may be caused by myocellular hypertrophy and myocardial fibrosis. In turn, at the cellular level these effects are associated with defects in calcium ion transport, myocardial contractile protein collagen formation, and fatty acid metabolism [23]. More recently, it has been suggested that DCM is a very common condition whose etiology is largely the result of HG, which in turn causes LV hypertrophy. Therefore, the occurrence of DCM in DM patients is found to be elevated, and this DCM is the result of diastolic dysfunction, which is caused by myocardial fibrosis that occurs in response to HG [24]. Present evidence suggests that persistent HG-induced mitochondrial oxidative stress (MOS) is a significant contributor to DCM [25, 26]. Certainly, the activity of several antioxidant enzymes is decreased in the diabetic heart in both rats and humans [27].

3 Hyperglycemia (HG)

HG is a hallmark of DM and is most commonly recognized as the causal link between diabetes and diabetic complications [10, 22, 28, 29]. Brief episodes of HG induce numerous maladaptations at cellular, subcellular, and molecular levels of vascular tissues as an independent risk factor and also worsen cardiac performance and cell survival [30]. HG is the most common factor leading to the development of chronic diabetic complications that are of high risk and long term [31]. HG also has adverse effects on left ventricular function. An experimental study demonstrated that acute HG is not merely a marker of severe myocardial damage but a potential direct cause of abnormal left ventricular function [32].

Numerous mechanisms of HG are claimed to be responsible for the generation of diabetes-induced heart disease, including metabolic abnormalities such as cellular overload of fatty acid metabolites, defective glucose transport, structural alterations in the form of microangiopathy, and altered calcium metabolism in cardiomyocytes [28, 33–35]. Increased reactive oxidative stress (ROS) and depleted antioxidant defences together with raised levels of ROS are well established in diabetes. The role of ROS in relationship to HG and diabetic complications has been well described [35–37]. Previous reports have shown that an increase in the production of ROS in the mitochondria can result from an HG-induced increase in the proton gradient across the mitochondrial membrane [36, 37]. Electron transfer is blocked when the gradient exceeds the threshold, leading to the leakage of electrons and the formation of the superoxide ion. Through ROS-induced DNA damage and poly-ADP ribose polymerase (PARP) activation, ADP-ribose polymers attach and inhibit

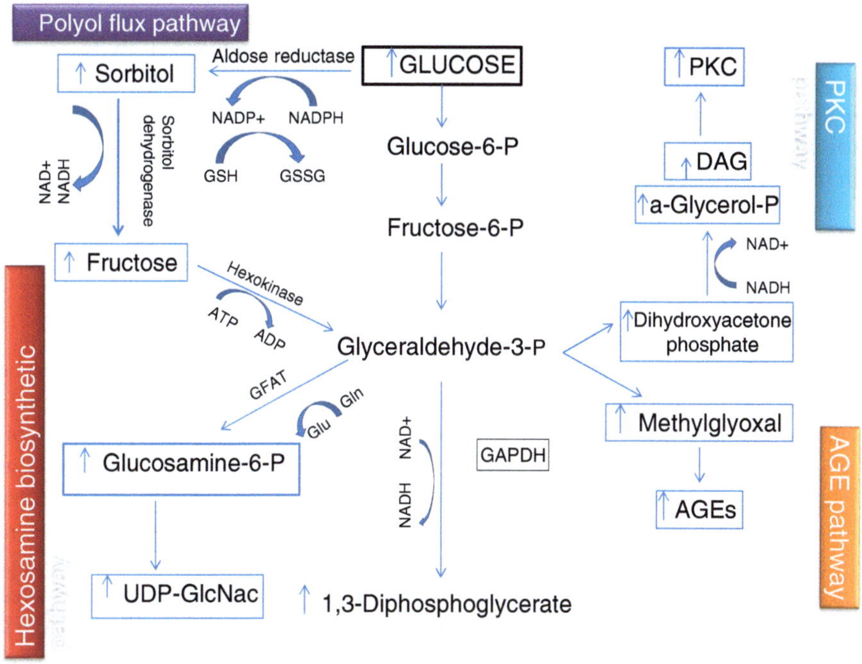

Fig. 1 Schematic representation of major mechanisms of hyperglycemia-induced damage. *AEGs* advanced glycation end products, *DAG* diacylglycerol, *PKC* protein kinase C, *GFAT* glutamine:fructose-6-phosphate amidotransferase, *GAPDH* glyceraldehyde-3-phosphate dehydrogenase (Modified from Rolo and Palmeira [10])

the cytosolic enzyme glyceraldehyde-3-phosphate dehydrogenase (GAPDH). It is this inhibition that mediates the activation of pathways leading to HG-associated tissue damage [38]. The Diabetes Control and Complications Trial (DCCT) and the U.K. Prospective Diabetes Study (UKPDS) have established that HG is the initiating cause of diabetic complications, relative to tissue damage [39]. It has been shown that these processes are modified by both genetic determinants and independent factors such as hypertension. The mechanisms that mediate the tissue-damaging effects of HG are of great interest. DM selectively damages cells such as endothelial and mesengial cells whose glucose transport rate does not decline as a result of HG, leading to an increase in glucose inside the cell. There is much evidence that these complications must involve mechanisms that occur inside these cells.

Diabetes and insulin resistance (failure of insulin to stimulate glucose uptake by fat and muscle cells) cause glucose concentrations in the blood to remain high. Consequently, glucose uptake by insulin-independent tissues increases. Increased glucose flux enhances oxidant production and impairs antioxidant defenses by multiple interacting nonenzymatic, enzymatic, and mitochondrial pathways [13, 28, 32, 36, 40–42]. Figure 1 illustrates the four major metabolic pathways as being the most important contributors for the pathological alterations in the diabetic vasculature as a consequence of HG-induced cell damage: these include increased glucose flux through the polyol pathway, increased formation of advanced glycation end products (AGEs) (discovered in the 1970s), the activation of protein kinase C (PKC) isoforms via de novo synthesis of the lipid second messenger diacylglycerol (DAG) (discovered in the early 1990s), and increased hexosamine pathway flux, discovered in the late 1990s. Each pathway is now discussed in detail.

4 Increased Polyol Pathway Flux

The polyol pathway is based on a family of aldo-keto reductase enzymes that can be utilized as substrates for a variety of carbonyl compounds, reduced via nicotinamide adenine dinucleotide phosphate (NADPH) to respective sugar alcohols (polyol) [38]. This pathway involves the HG-induced increase in enzymatic conversion of glucose to polyalcohol sorbitol with associated decreases in NADPH and glutathione. Aldose reductase carries out the function of reducing toxic aldehydes present in cells to inactive alcohols. When there is an increase in glucose concentration in the cell, aldose reductase reduces glucose to sorbitol. This sorbitol is later oxidized to fructose by sorbitol dehydrogenase (SDH), which increases the ratio of NADH/ NAD^+. This reduction is made possible by the cofactor NADPH, which also plays an essential role in being the cofactor for regenerating intracellular antioxidant such as glutathione. As a consequence, the polyol pathway increases its vulnerability to intracellular oxidative stress [36]. Aldose reductase is found in tissues such as cardio vascular cells and nerve cells. Much consideration has been proposed to ascertain how HG-induced increase observed in the polyol pathway flux could cause damage to tissues. Most papers have cited an increase in redox stress as the causative process for the consumption of NADPH, because NADPH cofactor is required to regenerate reduced GSH, which is an important scavenger of ROS. Hence, this could either induce or aggravate intracellular oxidative stress [38].

5 Advanced Glycation End Products (AGEs)

The most damaging effects of DM-induced HG are caused by the reactions of a nonenzymatic nature that occur between sugars and free amino groups on proteins, nucleic acids, and lipids leading to the development of molecular dysfunction through formation of advanced glycation end products (AGEs) [36, 43]. Studies have shown AGEs can play a key role in the development of DCM. AGEs formulate a broad range of cellular, chemical, and tissue effects through changes in their solubility and confirmation and also they interact with specific receptors and binding proteins that influence the expression of growth factors including TGF-β1 and CTGF, thus regulating proliferation and growth [43].

The production of AGEs is a physiological process that is enhanced by HG [44]. Previous studies have described the formation of AGEs resulting from a sequence of chemical reactions between glucose and proteins, most commonly known as the Millard or Browning reaction. The first step involves a condensation reaction between the amino and carboxyl group (sugar), leading to the formation of a Schiff base. The early glycation products that then undergo molecular rearrangement are known formally as the Amadori rearrangement products. These form stable, irreversible AGEs that have biological effects such as vasodilation, matrix deposition, and cytokine production [45]. AGEs are formed by intracellular oxidation of glucose to glyoxal decomposition of the Amadori product (glucose-derived 1-amino-1-deoxyfructose lysine adducts) to 3-deoxyglucosone (accelerated by an amadoriase) and fragmentation of glyceraldehyde-3-phosphate and dihydroxyacetone phosphate to methylglyoxal [46]. AGEs are derived from these intracellular dicarbonyls (glyoxal, methylglyoxal (detoxified by the glutathione-dependent glyoxalase system) and 3-deoxyglucosone by reacting with amino groups of intracellular and extracellular proteins. Methylglyoxal reacts with collagen and thus interferes with the crucial cell–matrix as well as matrix–matrix interactions. Methylglyoxal, a highly reactive alpha-dicarbonyl by-product of glycolysis, is increased in DM; it readily reacts with arginine, lysine, and sulfhydryl groups of proteins as well as nucleic acids, leading to the induction of various structurally identified AGEs in target cells and plasma [47]. Methylglyoxal also has an inhibitory effect on mitochondrial respiration as well as methylglyoxal-induced modifications that specifically target mitochondrial proteins. This observation is of particular importance because for the first time a study has discovered a direct relationship between formation of intracellular AGEs on mitochondrial proteins and the decline in mitochondrial function and exceeded formation of ROS [48].

The precursors of AGEs are known to damage target cells by three main mechanisms: these include (a) intracellular proteins being modified by AGEs, resulting in altered functions; (b) extracellular matrix components modified by AGE precursors, which interact abnormally with other matrix components and also with the receptors for matrix proteins (integrins) on cells (change signalling between matrix and the cell and cause cellular dysfunction); and lastly (c) plasma proteins that are modified by AGE precursors to bind to AGE receptors on endothelial cells, causing the production of inflammatory cytokines and growth factors. In turn, these cause vascular pathology and receptor-mediated production of ROS [36, 38, 49, 50].

An increasing number of AGE receptors have been identified including RAGE, AGE-R1, AGE-R2, galectin-3 (AGE-R3), and the macrophage scavenger receptors type II. By binding to AGE-modified proteins, RAGE initiates multiple cascades of signaling that activate protein kinase C (PKC), which induces pro-inflammatory cytokines and consequently inflammation, growth factor release, and fibrosis [38, 51]. These proteins are expressed on a wide range of cells including smooth muscle cells and endothelial cells. In DM, it has been postulated that an increase in AGEs activates the AGE receptors. DM has been associated with significant increases in left ventricular RAGE protein. Ligation of RAGE brings about activation of transcription factor nuclear factor-kappa beta (NF-κB) that coordinates the inflammatory responses [51].

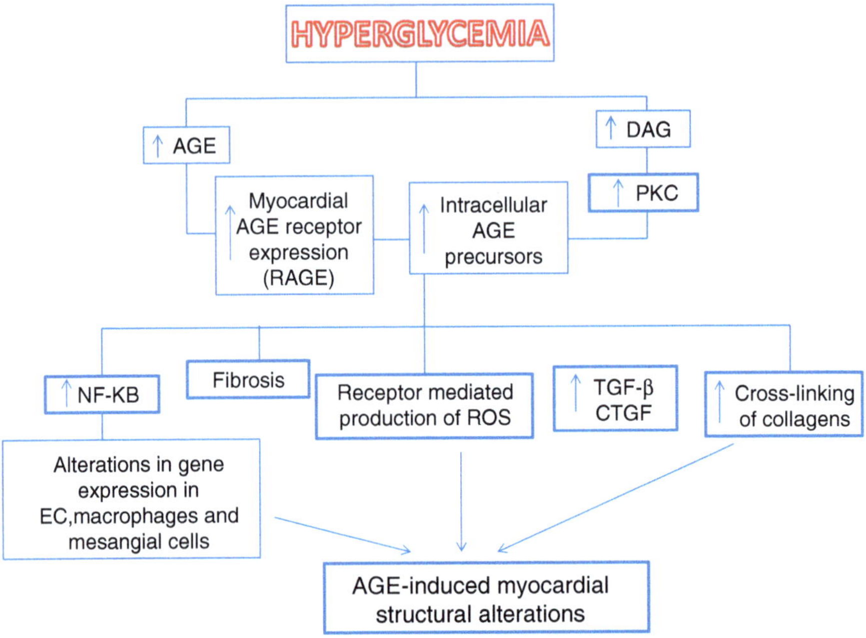

Fig. 2 Hyperglycemia-induced intracellular production of AGE precursors leading to AGE-induced myocardial structural alterations. *AEGs* advanced glycation end products, *DAG* diacylglycerol, *PKC* protein kinase C, *NF-κB* nuclear transcription factor-kappa B, *TGF-β* transforming growth factor-β

An increase in RAGE protein has been identified in the LV of diabetic animals, implying a role for this receptor in mediating AGE-induced myocardial structural alterations. Over expression of AGE-R3 receptor can mediate events by modifying the function of the AGE receptor complex. The AGE-R3 receptor has also been involved in exerting direct effects on cardiac remodeling, independently on AGE ligands, by its adhesive and growth-regulating properties [52]. The whole process is illustrated as a flow diagram in Fig. 2.

Altered cardiac performance in DM involves alterations in myocardial collagen structure. Multiple mechanisms have been discovered whereby chronic HG contributes to pathological cardiac remodeling: these involve the direct effects of elevated glucose on cell nonenzymatic glycation as well as oxidative stress. Studies have shown that AGEs can affect the structural components of the extracellular matrix such as collagen, which is elevated in tissues of diabetic patients [53]. The Strong Heart Study reported that accumulation of AGEs in the myocardium is involved in the frequency and extent of diastolic dysfunction and this was directly proportional to HBA_{1c} levels. As mentioned earlier, it is well known that DM not only increases myocardial AGE receptor expression, but it can also increase cross-linking of collagen and myocardial fibrosis. It was discovered that the lysis of these collagen crosslinks decreases myocardial fibrosis and improves diastolic dysfunction [54]. Previous studies have demonstrated that inhibitors of AGE formation such as aminoguanidine or ALT-946 are able to retard the development of diabetic nephropathy [55]. Treatment of diabetic rats with aminoguanidine prevents the increase in myocardial stiffness by decreasing the formation of myocardial collagen AGEs [56].

6 Protein Kinase C

PKC activation is one of the sequelae of HG and is thought to play a major role in the development of diabetic complications such as retinopathy [57, 58], nephropathy [59], neuropathy, cardiomyopathy, and atherosclerosis [60]. Many vascular abnormalities in cardiovascular tissues in the diabetic state including regulation of endothelial permeability, extracellular membrane (ECM) synthesis/turnover, cytokine activation, and cell growth angiogenesis are known to be associated with the activation of the diacylglycerol-protein kinase C (DAG-PKC) pathway [59, 61, 62]. There are multiple mechanisms for its activation in the diabetic state and multiple downstream effects attributable to DAG-PKC pathway activation. The essential role of PKC activation in the development of cardiovascular and microvascular complications as well as the mechanisms by which HG causes vascular cell damage through activation of PKC are now discussed.

It has been suggested that an elevation in the levels of the lipid diacylglyceride (DAG) induced by either HG or free fatty acids can act as a stimulus to activate PKC pathways, particularly the PKC-β or -δ isoforms [61] (Fig. 3). The metabolic byproducts of HG (glycation products and oxidants) are also proposed to increase DAG levels and activate the PKC system. However, some reports have suggested that chronic activation of PKC activities in diabetes could also be caused by increased expression of the various PKC isoforms. Once the PKC isoforms are activated, they possess the ability to alter signalling pathways such as inhibition of PI3 kinase Akt and the activation of MAP kinase systems [62]. The overexpression of the PKC-β isoforms in the myocardium and vascular tissue can cause cardiac hypertrophy, myocardium necrosis, and fibrosis, leading to the development of cardiac failure [63]. This observation indicates that the activation of DAG and PKC pathways induced by HG, oxidants, and glycation products can result in biological and pathological changes in the micro-cardiovascular systems to mimic many of the changes observed in DM. The biological consequence of isoform activation includes changes in enzyme activation and expression, such as endothelial nitric oxide synthase (eNOS), Na^+/K^+-ATPase, and increased expression of endothelin [59, 64].

PKC isoforms comprise at least 12 isoforms of serine threonine kinases, which are further classified into subfamilies based on their enzymatic, structural, and biochemical properties as well as tissue distribution, subcellular localization, and substrate specificity [64]. For DAG activation to take place, PKC must undergo a series of phosphorylations at three conserved sites by a series of important kinases. Initially, phosphatidylinositol (PI) is phosphorylated to PIP and phosphatidylinositol-4,5-bisphosphate (PIP_2), leading to the generation of phosphatidylinositol-3,4,5-trisphosphate (PIP_3). PIP_2 stimulates the activation of phospholipase C (PLC), which hydrolyzes PIP_2 to produce DAG and inositol-1,4,5-trisphosphate (IP_3). The IP_3 causes

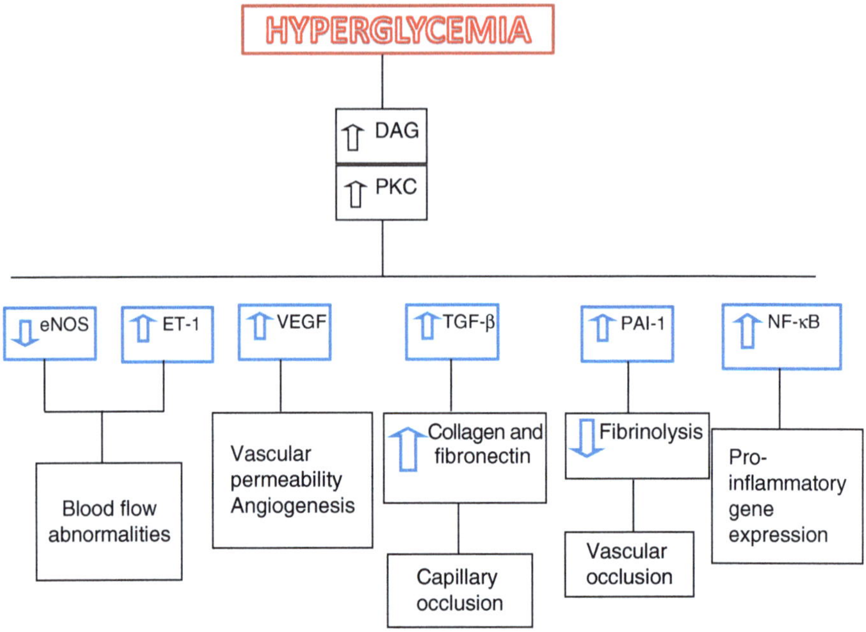

Fig. 3 Flowchart highlighting the consequences of hyperglycemia-induced activation of PKC. *eNOS* endothelial nitric oxide (NO) synthase, *ET-1* endothelin-1, *TGF-β* transforming growth factor-β, *PAI-1* plasminogen activator inhibitor-1, *NF-κB* nuclear transcription factor-kappa B (Modified from Brownlee [43])

the release of endogenous Ca^{2+} that binds to the cytosolic PKC and exposes the phospholipid-binding site. The binding of Ca^{2+} translocates PKC to the membrane, where it interacts with DAG and is transformed into a fully active enzyme [65]. HG can also activate the PKC isoforms indirectly through ligation of AGE receptors and also through increased activity of the polyol pathway. presumably by increasing ROS [43].

Several previous studies have shown that vitamin C and LY333531 (PKC isoform S-specific inhibitor) can inhibit PKC activation either directly or indirectly by decreasing levels of DAG. Intraperitoneal injection of vitamin E can prevent increases in DAG levels and PKC activity in the heart of STZ-induced diabetic rats, and in turn they can also improve functional abnormalities [59, 61]. These findings strongly suggest that the observed increase in DAG levels is responsible for PKC activation induced by either diabetes or HG and also that PKC activation is responsible for the functional abnormalities seen in DM.

Vascular alterations at the histological, cellular, and functional levels have been observed, and these changes have been linked to DAG-PKC activation. Vascular alterations have included basement membrane thickening, extracellular matrix expansion, vascular permeability, and enzymatic alterations such as Na^{+}/

K^+-ATPase, and MAP kinase. Between the various fibrotic factors, the expression of TGF-β1 and connective tissue growth factor (CTGF) have been found to play key roles in the growth of the basement membrane thickening and increasing extracellular membrane (ECM) in DM. Both these factors are known regulators of ECM accumulation, and in turn they stimulate the production of matrix components including IV collagen and fibronectin [61]. Increased TGF-β_1 and CTGF levels have been reported in the kidney, heart, and many other organs of diabetic animals and patients. Hence, it is suggested that activation of PKC may also be involved in glucose-induced increases of CTGF and TGF-β1 expression, leading to the synthesis of ECM. The activation of PKC by HG also induces expressions of permeability-enhancing factor vascular endothelial growth factor (VEGF) in vascular smooth muscle cells [38]. PKC activation is also known to induce synthesis of type IV collagen and fibronectin. It is now believed that the increased levels of basement membranes could to some extent be causing the vascular dysfunction observed in DM [10, 61, 65]. HG-induced activation of PKC has also been implicated in the overexpression of fibrinolytic inhibitor, plasminogen activator inhibitor (PAI)-1, and also in the activation of NF-κB in endothelial cells and smooth muscle cells, resulting in the development of multiple pathological changes in gene expression [38, 62, 66].

7 Hexosamine Biosynthetic Pathway

Diversion of fructose-6-phosphate from glycolysis takes place via the rate-limiting enzyme of this pathway involving glutamine:fructose 6-phosphate amido-transferase (GFAT) [67]. GFAT converts fructose-6-phosphate to glucosamine 6-phosphate, which is then converted to UDP-*N*-acetylglucosamine, the end product of the hexosamine biosynthetic pathway (HBP). Specific *O*-GlcNAc transferases use this for posttranslational modification of specific serine and threonine residues on cytoplasmic and nuclear proteins by *O*-GlcNAc [38, 43]. Inhibition of GFAT blocks HG-induced increases in the transcription of TGF-α, TGF-β1, and PAI-1. It has been proposed that HG results in a fourfold increase in *O*-GlcNAcylation of the transcriptor factor SP 1, which is involved in mediating HG-induced activation of PAI-1 promoter in vascular smooth muscle cells and TGF-β1 in endothelial cells. In parallel to transcription factors, various other nuclear and cytoplasmic proteins are modified by *O*-GlcNAcylation [67]. An example would be the inhibition of eNOS activity by HG-induced *O*-acetyl-glucosaminylation at the Akt site of the endogenous nitric oxide synthase (eNOS) protein. Previous studies have shown that HG can also impair cardiomyocyte Ca^{2+} cycling via increased nuclear O-GlcNAcylation, which decreases sarcoplasmic reticulum Ca^{2+}-ATPase 2a (SERCA2a) mRNA and protein expression and also reduced SERCA2a promoter activity, leading to prolonged Ca^{2+} transients and also impaired relaxation [49, 68]. Hence, activation of the HBP by HG may cause many changes in gene expression as well as protein functions, which in conjunction contribute to the pathogenesis of diabetic complications.

8 Single Hyperglycemia-Induced Process Responsible for Pathogenic Mechanisms: Mitochondrial Superoxide Production

It is unclear whether the processes just described (AGE formation, RAGE ligand binding, PKC activation, hexosamine pathway flux, aldose reductase inhibitors, etc.) and various other diabetes-induced abnormalities are interconnected or whether they have a common cause. HG-induced oxidative stress can lead to modifications in intracellular proteins that can cause altered functions: these include DNA damage, activation of NF-κB causing abnormal changes in gene expression, decreased production of NO, and increased expressions of growth factors, cytokines, and pro-coagulant and pro-inflammatory molecules [38, 43, 49, 61]. Many researchers believe that the four different pathogenic mechanisms reflect a single HG-induced process and that overproduction of the superoxide by mitochondrial electron-transport chain is the causal link between high glucose and the major pathways responsible for HG-induced damage in the heart and blood vessels [10, 63, 69–71]. In actual fact, DM is naturally accompanied by increased production of free radicals or depleted antioxidant defence capabilities, which indicate a vital role for reactive oxygen species ROS in the onset, development, and pathological consequences of DM. Most studies have reported mitochondrial overproduction of the superoxide ion as the "unifying hypothesis" for the development of diabetic complications [38, 72]. Hence, it is rationale to assume that mitochondrial superoxide production is a vital player in this process. However, before elaborating on how HG increases the production of ROS in the endothelial cells, it is helpful to briefly explain glucose metabolism.

9 Overview of Glucose Metabolism and the Electron-Transport Chain (ETC)

The process of glycolysis is the starting point for glucose oxidation in which NADH and pyruvate are the products of this process. NADH has the ability to donate reducing equivalents to the mitochondrial electron-transport chain via two shuttle systems, as well as by reduction of pyruvate to lactate. Pyruvate can also be transported into the mitochondria where oxidation by the tricarboxylic acid (TCA) cycle produces CO_2 and H_2O, four NADH molecules, and one molecule of $FADH_2$. Mitochondrial NADH and $FADH_2$ provide the energy for production of ATP through the oxidative phosphorylation by the electron transport chain (ETC).

ETC consists of a series of complexes localized within the inner mitochondrial membrane, which include four important inner membrane-associated complexes, with the help of cytochrome *c* and the moving electron, known as ubiquinone carry-out electron transport through the mitochondrial ETC (see flow diagram in Fig. 4). First, NADH (derived from cytosolic glucose oxidation and mitochondrial TCA cycle activity) donates electrons to complex I (NADH:ubiquinone oxido-reductase).

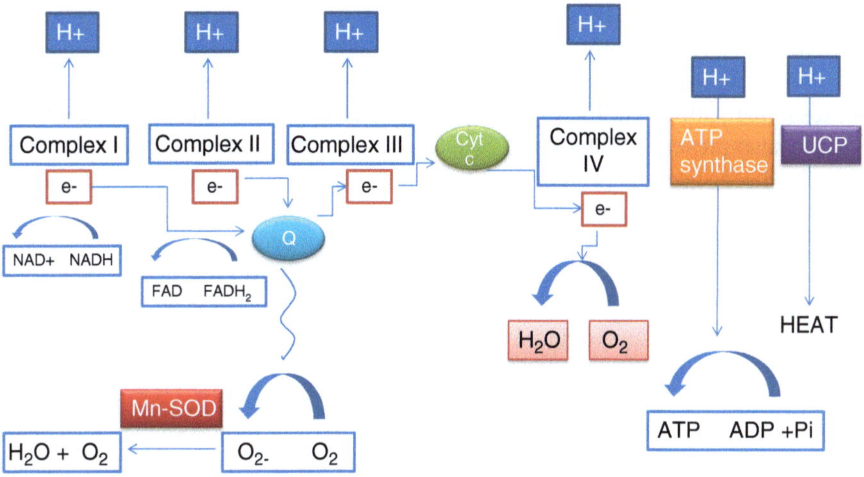

Fig. 4 Production of the superoxide via the mitochondrial electron-transport chain. Increases in hyperglycemia (HG)-derived electron donors from the tricarboxylic acid (TCA) cycle (NADH and $FADH_2$) generate a high mitochondrial membrane potential by pumping protons across the mitochondrial inner membrane; this inhibits electron transport at complex III, increasing the half-life of ubiquinone, which reduces O_2 to superoxide. H ions pass back across the inner membrane via ATP synthase, which produces ATP, or via UCPs that give off the energy as heat (Modified from Giacco and Brownlee [38])

Ubiquinone is also reduced by electrons donated from several $FADH_2$-containing dehydrogenases such as succinate:ubiquinone oxido-reductase (complex II) and glycerol-3-phosphate dehydrogenase. Electrons from reduced ubiquinone are then transferred to ubiquinol:cytochrome *c* oxido-reductase (complex III). Electron transport then proceeds through complex IV, the cytochrome *c* oxidase, and finally, molecular oxygen. Electron transfer through complexes I, III, and IV generates a proton gradient that drives ATP synthase (complex V). However, this process is disturbed in the diabetic state because diabetic cells have a higher intracellular glucose concentration and consequently more glucose-derived pyruvate is being oxidized in the TCA cycle, which increases the flux of NADH and $FADH_2$ in the ETC. Consequently, the voltage gradient across the membrane increases until a threshold is reached. This threshold causes an electron transfer blockage inside complex III of the ETC, which causes the electrons to back up to coenzyme Q that donates electrons one by one to molecular oxygen, leading to increased production of the superoxide ion [38].

Manganese superoxide dismutase (MnSOD) (the mitochondrial form of superoxide dismutase; SOD overexpression) abolishes the signal generated by ROS and also the overexpression of uncoupling protein-1 (UCP-1). Together these collapse the proton gradient that prevents HG-induced overproduction of ROS. Dynamic changes in mitochondrial morphology have been demonstrated that are associated with HG overproduction of ROS. However, studies have shown that the inhibition of either MnSOD or UCP-1 of HG-induced overproduction of mitochondrial superoxide can completely prevent the increases in AGE formation, PKC activation as well as the increase observed in the polyol pathway flux and the activity of the hexosamine pathway in EC [36, 38, 43, 70]. Overexpression of either MnSOD or UCP-1 also prevents inhibition of eNOS activity by hyperglycemia [73]. MnSOD has also been shown to prevent morphological changes in diabetic hearts and completely normalize contractility in diabetic cardiomyocytes [74, 75].

Animals and patients with diabetes have decreased activity of the key glycolytic enzyme glyceraldehyde-3-phosphate dehydrogenase (GAPDH) in cell types that develop intracellular HG. Inhibition of GAPDH does not occur when mitochondrial overproduction of superoxide is prevented by either UCP-1or MnSOD [75]. However, when activity of GAPDH is inhibited, it increases all the glycolytic intermediates that are upstream of GAPDH. As a result, increased flux of the pathways mentioned earlier takes place. Inhibition of GAPDH was discovered to be a consequence of poly (ADP)-ribosylation of GAPDH by PARP [poly (ADP-ribose) polymerase], which is known to be activated by breakage of DNA, produced by mitochondrial superoxide overproduction [76]. The AGE pathway is activated by the major intracellular AGE precursor methylglyoxal, which has been shown to increase expressions of RAGE and also its activating ligands [77]. Increase in glyceraldehyde-3-phosphate also leads to enhanced activity of the de novo DAG synthetic pathway, which causes further activation of the PKC pathway and activation of NADPH oxidase. It is also similar in both the polyol biosynthetic and the hexosamine pathways where the enzyme aldose reductase reduces it by consumption of NADPH in the process. Having revised these complex mechanical mechanisms of HG-induced complications in DM, the evidence so far indicates that these pathways are interconnected in many ways and can potentiate each other [36, 38, 43, 77].

Under normal conditions, poly (ADP-ribose) polymerase (PARP) is present in the nucleus in an inactive form awaiting DNA damage for its activation. An increase in intracellular glucose generates an increase in ROS, causing the DNA strands to break, and this in turn leads to the activation of PARP, which separates the NAD molecule into its component parts, nicotinic acid and ADP-ribose. PARP then continues to make polymers of ADP-ribose that accumulate on GAPDH, which is mainly found in the cytosol. However, it does shuttle in and out of the nucleus where it plays an essential role in DNA repair. Decrease in GAPDH can lead to the pathways described above (see flow diagram in Fig. 5) [73, 77].

10 HG-Induced Cardiac Fibrosis

Cardiac remodeling is caused by changes in shape, size, and function of the heart [78]. Histopathologically, cardiac remodeling is characterized by structural rearrangement of normal chamber components, which involve cardiomyocyte hypertrophy, cardiac fibroblast proliferation, cardiac fibrosis, and cell death [79]. Cardiac fibrosis is a widespread outcome of numerous pathological causes, including ischemia and inflammation.

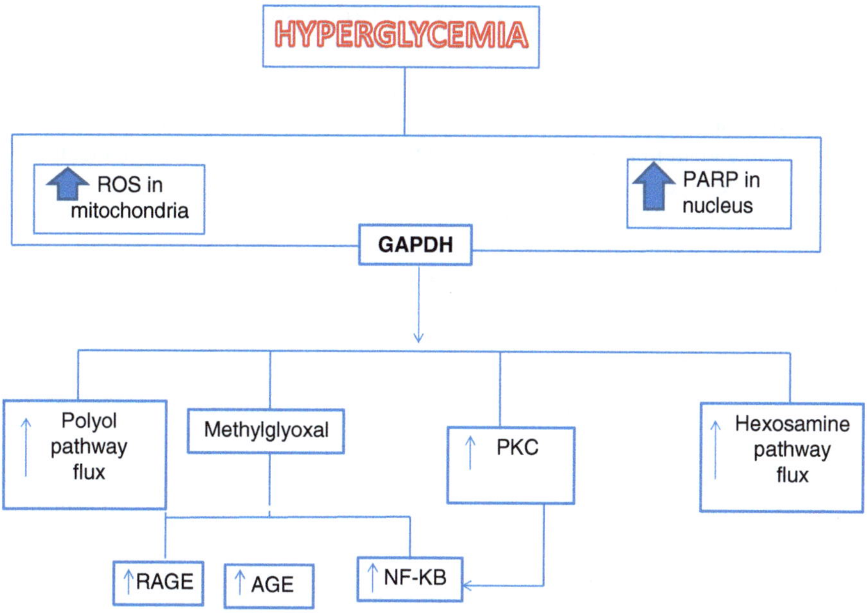

Fig. 5 Elements of the unifying mechanism of hyperglycemia-induced cellular damage. *GAPDH* glyceraldehyde 3-phosphate dehydrogenase, *ROS* reactive oxygen species, *PARP* poly (ADP-ribose) polymerase, *NF-κB* nuclear transcription factor-kappa B

It is also a major feature of DCM [80]. In this condition, overproduction of extracellular matrix (ECM) proteins leads to increased myocardial stiffness, resulting in cardiac dysfunction, impairment of cardiac performance, and ultimately cardiac failure [81]. Cardiac fibrosis represents a disproportionate accumulation of fibrillar collagen, and it is a fundamental element of remodeling characteristic of the failing heart. Type I collagen is the fundamental collagen found in cardiac fibrosis, resulting in stiffening of the ventricles, and moreover it impedes contraction as well as relaxation of the heart [79]. The underlying mechanisms and signaling pathways involved in ECM remodeling are of great importance. Oxidative stress, which is a major cause for all HG-induced diabetic complications, is associated with abnormal gene expression, altered signal transduction, and also the activation of secondary messenger pathways leading to the development of myocyte cell death and cardiac fibrosis [28, 82, 83]. The physiological alterations associated with muscle tissue and cells include the presence of myocardial hypertrophy and impaired contraction. However, alterations associated with the ECM include excess deposition of collagens, abnormal glycosylation/crosslinking via the AGE pathway, as well as the alterations in diastolic compliance [83]. Fibrosis impairs electrical coupling of cardiomyocytes by separation of myocytes and ECM proteins. In addition, fibrosis leads to reduced capillary density and increased oxygen diffusion distances, which eventually lead to myocyte hypoxia [84]. Therefore, fibrosis profoundly alters myocyte metabolism and performance and eventually ventricular function.

11 Extracellular Matrix (ECM) Regulation

The myocardial ECM forms an organized network that consists of a group of fibril collagens, a basement membrane, proteoglycans, and also glycosaminoglycans [85]. The most abundant fibril collagens include type I and type III and less-abundant collagens include types IV, V, and VI, elastin as well as laminin [86]. The distribution of collagens I and III varies in different regions of the heart. The ECM network of collagens plays a supportive role for the myocardial cellular functions in addition to being responsible for the distribution of mechanical forces and signal transduction throughout the cell via ECM receptors on the cell surface. Fibril collagen plays an important role in maintaining cardiac shape, size, and function as already mentioned because of its rigid architecture, which is in close contact with cellular and noncellular components of the myocardium. ECM homeostasis is maintained by coordinated action of the stimulators and inhibitors. Conversely, in the failing heart, activation of numerous humoral autocrine and paracrine pathways determine the regulation of the ECM metabolism and also the extent of myocardial remodeling. As a consequence, changes in ECM synthesis and degradation lead to disturbance of the composition of the collagen network in the heart. This disproportionate increase in synthesis/inhibition of the ECM proteins may result in cardiac fibrosis [87].

Fibrosis has been classified as reparative (replacement) and reactive fibrosis. Reparative fibrosis/scarring accompanies myocyte death. Reactive fibrosis, however, appears as "interstitial" or "perivascular" fibrosis and does not directly lead to myocyte death. Collagen in interstitial fibrosis appears in intramuscular spaces. However, collagen in perivascular fibrosis accumulates within the adventitia of intramyocardial coronary arteries and arterioles. Words such as "focal" and "diffuse" are also used widely to describe the distribution of fibrosis. There are apparent differences between reparative and reactive fibrosis, and many factors work in conjunction to control fibroblast function [87, 88].

Degradation of collagens occurs by means of metalloproteinases (MMPs), which are capable of enzymatically digesting various ECM proteins. The activity of these MMPs is controlled at transcriptional level and also via activation and inhibition by other proteins including tissue inhibitors of MMPs (TIMPs). Cardiac fibroblasts play a crucial role in the production of both ECM proteins and MMPs, hence playing a role in regulating the ECM [81].

12 Cardiac Fibroblasts and Myofibroblasts

Cardiac fibroblasts (CFs) and cardiomyocytes are the two major cells of the mammalian heart, which together account for 90 % of cells in the myocardium. Normal cardiac function is regulated via coordinated and dynamic interactions of these cell types. CFs account for about 60–70 % of the cells and are the key source of components of the ECM. Regulation of the structure of the heart, as well as the mechanical, chemical, and electrical signals between cellular and noncellular components, are performed by the CFs [87]. Cardiomyocytes, however, are fewer in number, but they occupy the bulk volume of the myocardium and hold the position as contractile cells that provide mechanical force transmission involved as the principal function of the ECM network. Fibroblasts synthesize the main components of the ECM, including collagen I and collagen II, comprising approximately 80 % and 10 % of the ECM, respectively. These collagens are the major stress-bearing element of the ECM that forms a three-dimensional (3D) network around bundles of myocytes to generate a stress-tolerant network. CFs are hence described as the most important cells of the ECM. CFs are important in the remodeling process of the heart in response to the pathological changes following such diseased conditions as hypertension, myocardial infarction (MI), diabetes-induced cardiomyopathy (DCM), and heart failure (HF) [88].

CFs differentiate into myofibroblasts (myoFbs) to mediate their functional effects, expressing contractile proteins, including the alpha-smooth muscle actin (α-SMA), and exhibiting increased migratory proliferative and secretory properties [89]. Cardiac myoFbs predominantly respond to pro-inflammatory cytokines including tumour necrosis factor-alpha (TNF-α), transforming growth factor-beta (TGF-β), interleukin-1 (IL-1), and interleukin-6 (IL-6), vasoactive peptides, including angiotensin II (Ang II), endothelin-1 (ET-1), and A and B type natriuretic peptides (ANP, BNP), and hormones including noradrenaline (NA), all of which are increased in the remodeling heart [90]. Cardiac myoFbs respond to these stimuli via alteration of their proliferative as well as their migration properties, thus becoming highly proliferative and invasive as they actively reform the cardiac interstitium. They do this by modifying ECM turnover through production of ECM proteins and matrix MMPs and secretion of the cytokines and vasoactive peptides along with growth factors such as vascular endothelial growth factor (VEGF). Hence, CFs are acknowledged for their ability to maintain the structural integrity of the heart through controlled proliferations and ECM turnover [90, 91].

13 Autocrine and Paracrine Factors Controlling Cardiac Fibroblast

A common histopathological finding in cardiac fibrosis is the accumulation and structural remodeling of the fibrillar collagen matrix of the heart. In ventricular tissue, fibrosis is responsible for a decrease in rate of relaxation, diastole suction, and myocardial stiffness. Various factors have been discovered to contribute to this abnormal accumulation of fibrillar matrix [92].

Humoral factors affecting the phenotype as well as the function of CFs include angiotensin II (Ang II), TGF-β1, as well as insulin-like growth factor-1 (IGF-1) and basic fibroblast growth factor (bFGF/FGF-2) [93]. Among these factors, the most important are Ang II and TGF-β1, which regulate cardiac fibrosis and remodeling [92].

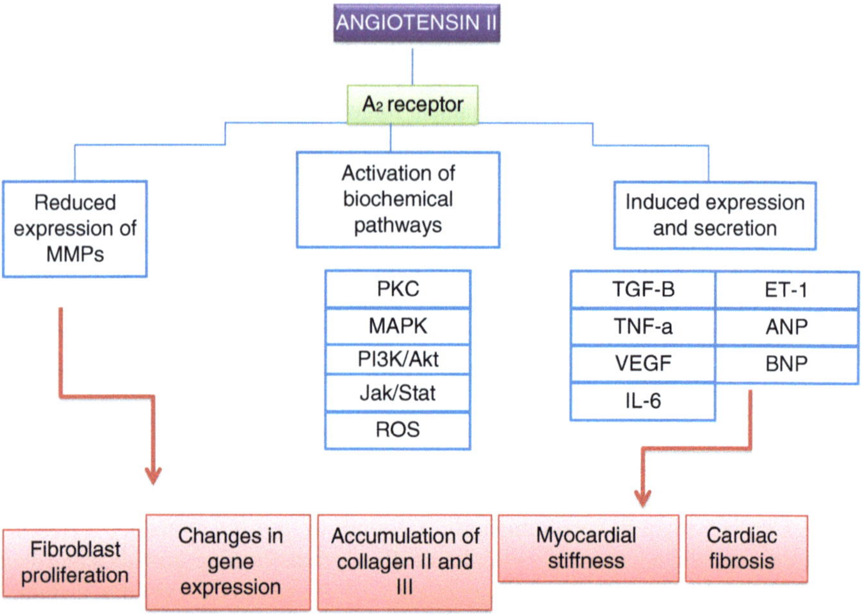

Fig. 6 Angiotensin (Ang) II stimulation evokes multiple signaling pathways in cardiac fibroblasts. Binding of Ang II activates G proteins via the AT_1 receptor, which triggers numerous pathways including PKC, MAPK, and PI3K/Akt; Jak/Stat and ROS are also involved in Ang II signaling. Activation of these pathways results in changes in gene expression

Several studies have supported a role for the rennin-angiotensin-aldosterone (RAA) system to be responsible for myocardial fibrosis. Angiotensin II encourages fibroblast proliferation and alteration of the fibrillar collagen turnover, as well as stimulation of aldosterone leading to the accumulation of collagen type I and III fibers, and hence the development of cardiac fibrosis. In turn, this results in tissue structure distortion that may be partly responsible for the increase in myocardial stiffness, leading to diastolic as well as systolic dysfunctions [93, 94]. Ang II acts as an autocrine/paracrine factor whereby Ang II stimulates CFs to secrete growth-promoting substances that affect the progression of hypertrophy and remodeling [95].

Ang II acts via two types of receptors, namely, type 1 (AT_1) and type 2 (AT_2) which are members of the G protein-coupled receptor (GPCR) superfamily [96]. Ang II mostly mediates its physiological effects through the AT_1 receptor (see flow diagram in Fig. 6). The binding of Ang II to AT1 receptors results in the activation of well-defined G protein-linked pathways such as phospholipase C (PLC), causing the release of Ca^{2+} and subsequent activation of the PKC pathway. Previous studies [97, 98] have suggested that Ang II can act on CFs to elicit pro-fibrotic effects on the heart via numerous mechanisms, including elevated ECM protein synthesis, decreased MMP activity, and increased TIMP activity. In particular, Ang II can stimulate collagen I and III as well as fibronectin synthesis through AT_1 receptor activation in human CFs. Ang II is a potent stimulus for the expression of TGF-β while being a strong inducer of the myoFb phenotype. However, it is not yet clear whether Ang II is able to directly induce the myoFb phenotype or whether it does this through increased TGF-β expression [99].

In some cases, the AT_2 receptor has been reported to counterregulate the AT_1 receptor effects [100]. AT_2 receptor stimulation has been shown to suppress the growth of cardiomyocytes and CFs. AT_2 receptor expression was found to be unregulated in the failing heart, mainly in the CFs. Hence, it has been postulated that selective blocking of the AT_1 signal might have a beneficial effect in heart failure via the AT_2 receptors [101]. Ang II can also reduce expression of collagenases such as MMP-1 in rat CFs, and it can also downregulate MMP-2 expression in human CFs [102]. Inhibitory effects of Ang II on MMP activity have been demonstrated by an increase in the activity of TIMPs (endogenous inhibitors of MMPs, in particular, increases in TIMP-1 and TIMP-2) in rat CF by Ang II. This may be the case for human CFs, findings that are supported by Sato et al. [102] and Chao et al. [103] that Ang II decreases the mRNA expressions of TIMP-1, TIMP-2, and TIMP-3. Ang II can also cause secretion of several important bioactive molecules from CFs indirectly, to regulate cellular function in an autocrine/paracrine manner. Ang II (via activation of AT1 receptor) can induce both the expression and secretion of TGF-β1, TNF-α, IL-6, ET-1, and the natriuretic peptides (ANP and BNP), and VEGF in a number of CF and myoFb models. Many studies have shown that this indirect effect may be the major mechanism by which Ang II controls CFs [104, 105].

14 Transforming Growth Factor-Beta (TGF-β)

TGFs are members of a superfamily of cytokines, consisting of three isoforms, of which TGF-β1 is the most prevalent. In their active form, all the isoforms are dimers of 12-kDa polypeptides that arise from a larger precursor through proteolytic processing. TGF-β1s are involved in a wide assortment of cell functions, including regulation of inflammation, ECM deposition, cell proliferation, and differentiation as well as tissue growth [28]. TGF-β1 acts to simultaneously stimulate the synthesis of ECM to inhibit the actions of proteases that cause matrix degradation to increase the expression of cell-surface integrins that interact with matrix components. Through these particular effects, TGF-β1 rapidly causes the deposition of an exuberant ECM [106]. TGF-β1 plays a crucial role in the development of fibrosis. It stimulates fibroblast chemotaxis along with the production of collagen and fibronectin while inhibiting collagen degradation [107]. It can also induce expressions of alpha-smooth muscle actin (α-SMA) in fibroblasts, hence it is considered to be involved in the factors held responsible for the formation of myofibroblasts [95]. In a previous study, Wu and Derynck [108] reported a central role of TGF-β1 signaling in the ability of HG to induce hypertrophy in fibroblasts and epithelial cells. It was implied that either blocking or enhancing the expression of the kinase activity of TGF-βRI receptor can prevent the hypertrophic effects of HG. Exposure of cells to HG has been shown to elicit a rapid increase in cell-surface levels of TGF-β1 receptors (TGF-βRI/TGF-βRII) along with the activation of a TGF-β1 ligand by MMP2 and MMP9. This autocrine signaling of TGF-β1 in response to HG leads to the activation of the pro-hypertrophic Akt-TOR pathway [99, 108].

15 Role of Micro-RNA-21 and TGF-βRIII in Cardiac Fibrosis

A recent article by Liang et al. [109] suggested that TGF-β1 signaling pathways play crucial roles in modulating cell function and in the progression of fibrosis. The study hypothesised that micro-RNA-21 (miR-21, which belongs to a class of endogenous, noncoding small RNAs) is able to regulate collagen content by inhibiting TGF-βRIII expression, which would then contribute to development of fibrosis. It has previously been demonstrated [110, 111] that TGF-β1 increases miR-1 and TGF-βRIII, which in turn represses TGF-β1 signaling pathways. These findings indicate a role of TGF-βRIII in the regulation and expression of miR-21 through inhibition of the TGF-β1 pathway. The first stage of this pathway involves the interaction of TGF-β1 with a complex of transmembrane serine/threonine kinase receptors (TGFβR1/TGF-βRII), leading to phosphorylation of transcription factors Smad2/Smad3, to form a Smad4 complex [109]. The heteromeric complex (Smad4) is translocated into the nucleus where it modulates other target genes via direct DNA binding or by interacting with promoter-specific transcription factors [112]. TGF-βRIII, formally known as beta-glycan, is found to be involved in the process of tumor cell proliferation, differentiation as well as apoptosis [113]. TGF-βRIII has the potential to prevent fibrosis, as it is able to directly neutralize TGF-β1, which indirectly prevented the formation of TGF-β1/TGF-βII compounds [110]. However, the mechanism is still poorly understood.

16 Transforming Growth Factor-β (TGF-β1) in Tissue Injury and Repair

TGF-β1 has been widely implicated in tissue repair, but it is the agent whose excessive action may be responsible for tissue damage from scarring [87]. A number of events involving TGF-β1 occur during the process of tissue repair. High concentrations of TGF-β1 are present in platelets that on degranulation at the site of an injury release TGF-β1 into the surrounding tissue. TGF-β1 then initiates a complex sequence of events that lead to healing, including chemoattraction of monocytes and leukocytes, induction of angiogenesis, and control of the production of cytokines and other inflammatory markers [112]. TGF-β1 stimulates the synthesis of individual matrix components including fibronectin and its variant forms including collagens and proteoglycans. TGF-β1 concurrently inhibits matrix degradation by increasing the synthesis of proteases and also the levels of protease inhibitors [106]. It is also known to increase the expression of integrins and alter their relative proportions on the surface of cells in a way that facilitates adhesion to matrix. Each of these events is found to be beneficial in tissue repair. However, TGF-β1 also induces deposition of ECM at the sites of injury, leading to scarring and fibrosis. In addition, the ability of TGF-β to induce its own production is found to be a major factor in the development of fibrosis and scarring into chronic, progressive conditions that alter the tissue structure [85, 99].

17 Natriuretic Peptides

Atrial natriuretic peptide (ANP), first discovered by Bold et al. [114, and brain natriuretic peptide (BNP), discovered by Sudoh et al. [115], are predominantly synthesized in the myocardium and pursue a crucial role in autocrine and paracrine effects of the heart as well as acting in an endocrine manner to regulate blood pressure [28]. ANP and BNP are heart-derived peptides activated via an increase in either wall tension or stretch of the cardiac chambers. Elevated levels of ANP and BNP have been observed in patients with various heart diseases including myocardial infarction (MI), hypertrophy, DCM, and HF [116]. Although these peptides are classically known to be secreted and synthesised by cardiomyocytes, evidence shows that they can also be produced by CFs [117]. ANP and BNP are involved in stimulating cyclic guanosine monophosphate (cGMP) accumulation in CFs. These findings support the fact that the heart itself is also amongst the target organs of these peptides, suggesting an important physiological role of these peptides in the structural remodeling of the myocardium by regulating fibroblast growth [116]. Therefore, the local accumulation of these peptides in the heart may be of important pathological significance. The concentration of BNP in a healthy person has been reported to be lower than that of ANP [118]. Conversely, in chronic heart failure (CHF), the plasma BNP/ANP ratio is seen to be reversed. BNP levels are found to increase more rapidly than ANP levels in the acute phase of MI and in an obstructive form of hypertrophic cardiomyopathy. CHF involves cardiac overload with the release of both ANP and BNP that activates the particulate guanylyl cyclase (pGC) NPR-A. In turn, this results in the generation of a second messenger cyclic GMP and its effector, protein kinase G (PKG) [119].

18 Endothelial Cell Dysfunction (ECD)

HG is the major contributory factor in the development of endothelial cell dysfunction (ECD), although the mechanisms underlying this event are multifactorial [120]. ECD, defined as an imbalance of endothelium-derived vasoconstrictor and vasodilator substances, is a common denominator in the pathogenesis and progression of both microvascular (retinopathy, nephropathy, neuropathy) and macrovascular (heart disease, stroke, peripheral arterial disease) complications [120]. Many experimental trials have acknowledged HG as the key determinant that has immediate biochemical sequelae, directly altering ECD and leading to the growth of chronic diabetic complications [43, 44]. Both the increase and accumulation of intracellular glucose and its metabolites can increase the production of extracellular matrix components, including collagen and fibronectin. ECD is thus determined as the key initiation stage for the development of vascular complications including ischemic heart disease, peripheral vascular disease, and cerebrovascular diseases, which are all important contributors of morbidity and mortality in both T1DM and T2DM patients [121].

Physical and chemical stimuli that take place inside the vessels are sensed by a receptor-effector endothelial cell (EC), which in turn either modifies the shape of the vessel or releases necessary regulatory mediators such as nitric oxide (NO), endothelin-1 (ET-1), ANG-II, plasminogen activator inhibitor-1 (PAI-1), von Willebrand factor (vWF), and many others to maintain vascular homeostasis [122, 123]. Under physiological conditions, a balanced release of these products occurs, but this delicate balance is altered during CVD risk factors such as diabetes and atherosclerosis, thereby altering the capacity of the EC and leading to further progression of vascular end-organ damage [124]. Among the chief molecules synthesized by EC is the vaso-protective molecule nitric oxide (NO), which is constitutively produced by endothelial NO synthase (eNOS), an enzyme known to be the major weapon of ECs that fights vascular disease [22, 125]. NO is considered as the most important endothelium-derived mediator because of its vasodilator, antiproliferative, antiinflammatory, antiplatelet, and permeability-decreasing properties [126].

Previous reports claim that HG can affect ECD through increased oxidative stress [127, 128]. A series of cellular events are induced by HG during a process known as eNOS-uncoupling where NOS switches from the production of NO to generate the superoxide anion (O_2^-). This stage in turn increases the production of reactive oxygen species (ROS, superoxide anion) that inactivate NO, leading to the formation of peroxynitrite ($ONNO^-$) [129]. HG initiates this process by increasing the production of the superoxide anion via the mitochondrial electron-transport chain. This superoxide anion then promotes a surge of endothelial processes that employ increasing numbers of cellular elements such as PKC to manufacture oxygen-derived free radicals [130, 131]. The peroxynitrite ion oxidizes the NOS cofactor tetrahydrobiopterin (BH_4) [132, 133], which uncouples the enzyme and increases the superoxide anion formation over NO production. Hence, the consequence is an ever-increasing production of superoxide anion and inactivation of NO. However, some studies also suggest that HG is a necessary but not a sufficient cause of clinically important microangiopathy. The risk of microangiopathy is closely related to risk factors for atherothrombosis including age, hypertension, smoking, dyslipidemia, obesity, hypercholesterolemia, insulin resistance, and impaired fibrinolysis. These factors constantly cause progressive damage to vascular walls, leading to endothelial activation and thus ECD (see the flow diagram in Fig. 7 for details).

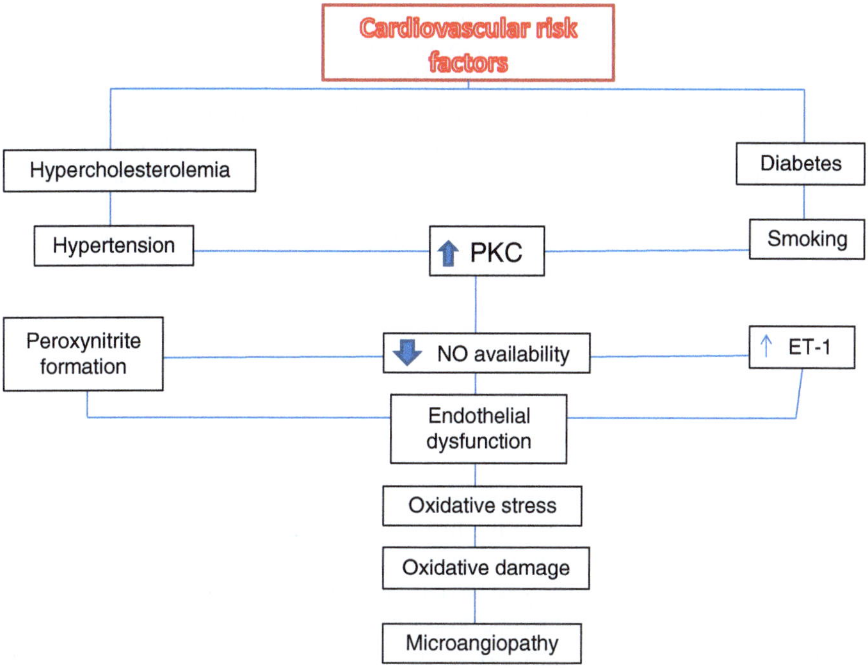

Fig. 7 Potential mechanisms in the development of endothelial dysfunction and eventually microangiopathy caused by various cardiovascular risk factors. *ET-1* endothelin-1, *PKC* protein kinase C, *NO* nitric oxide

19 HG-Induced Changes to Internal Structures of the Heart

Studies in our laboratory using electron microscopy have revealed that the structures of mitochondria are deranged in ventricular myocytes of diabetic rats. They are swollen with irregular shapes, and moreover their levels are much less in diabetic myocytes compared to age-matched controls. Similar findings have been reported by other studies, which also suggested that diminished mitochondria can cause dysfunction of the heart that may result in heart failure [134–136]. Similar to mitochondria, the intercalated discs (ICDs) of the cardiomyocytes in the diabetic rats are grossly impaired compared to controls. ICDs bind to adjacent cardiomyocytes, thus allowing fast and optimal conduction of electrical impulses from one cell to the other. In the case of the diabetic cardiomyocytes, defective ICDs can contribute to poor impulse conduction, resulting in arrhythmias and heart failure that are common in diabetic patients [137, 138]. DM is also known to be associated with a loss of cardiac myofibrils [136], which are the main contractile components for muscle contraction. Severe loss of myofibrils can lead to decreased contraction and cardiac output normally seen in diabetic patients. Previous studies have also demonstrated that the structures of many other cytoplasmic organelles including the sarcoplasmic reticulum, T-tubules and plasma membrane, Golgi apparatus, and microfilaments are deranged after the onset of long-term DM [139]. These ultrastructural derangements of the internal components of the heart include invagination of the plasma membranes and swelling and reduction in the number of cytoplasmic organelles, all of which can lead to dysfunction of the heart [139]. These results have clearly demonstrated that diabetes-induced HG is responsible for the structural changes in the myocardium. The cellular, subcellular, and molecular mechanisms involved in these processes may lie in changes in signaling molecules within the heart.

20 Conclusion

In summary, this review described the relationship between HG-induced DM, biochemical changes, and structural damage/alteration to the heart. HG seems to have both direct and indirect effects to the myocardium via a number of metabolic and signalling pathways, resulting in formation of fibrosis and subsequent remodeling of the heart. Cardiac remodeling may be broadly defined as alterations in cardiac structure resulting from altered hemodynamic load and/or cardiac injury. The process may be physiological or pathological. Cardiac remodeling is a common denominator in the etiology of several primary cardiovascular diseases, notably DM, coronary atherosclerosis, hypertension, cardiomyopathy, and myocarditis. Several neurohormonal and inflammatory pathways are activated, including the rennin–angiotensin-aldosterone system (RAAS), adrenergic system, inflammatory cytokine systems, and a host of other autocrine and paracrine mechanisms as compensatory mechanisms to maintain stroke volume at a reduced ejection fraction. All these pathways have been described in this review.

References

1. Rahman S, Rahman T, Ismail AA-S, Rashid AR (2007) Diabetes-associated macrovasculopathy: pathophysiology and pathogenesis. Diabetes Obes Metab 9:767–780
2. Jacobsen IB, Henriksen JE, Hother-Nielsen O et al (2009) Evidence-based insulin treatment in type 1 diabetes mellitus. Diabetes Res Clin Pract 86:1–10
3. Fujisawa T, Ikegami H, Kawaguchi Y (2004) Common genetic basis between type 1 and type 2 diabetes mellitus indicated by interview-based assessment of family history. Diabetes Res Clin Pract 66:S91–S95
4. Thorve VS, Kshirsagar AD, Vyawahare NS (2003) Diabetes-induced erectile dysfunction: epidemiology, pathophysiology and management. J Diabetes Complications 25:129–136
5. Kakleas K, Kandyla B, Karayianni C, Karavanaki K (2009) Psychosocial problems in adolescents with type 1 diabetes mellitus. Diabetes Metab 35:339–350
6. Goldstein BJ (2002) Insulin resistance as the core defect in type 2 diabetes mellitus. Am J Cardiol 90:3–10
7. Hyer SL, Shehata HA (2005) Gestational diabetes mellitus. Curr Obstet Gynaecol 15: 368–374

8. Choi SW, Benzie IFF, Ma SW et al (2008) Acute hyperglycemia and oxidative stress: direct cause and effect? Free Radic Biol Med 44:1217–1231
9. Boudina S, Abel ED (2007) Diabetic cardiomyopathy revisited. Circulation 115:3213–3223
10. Rolo AP, Palmeira CM (2006) Diabetes and mitochondrial function: role of hyperglycaemia and oxidative stress. Toxicol Appl Pharmacol 212:167–178
11. Diabetes UK (2007) Diabetes heartache: the hard reality of cardiovascular care for people. Report can be accessed at http://www.diabetes.org.uk/Documents/News/Heartache_report07.pdf
12. Tuttle KR (2005) Linking metabolism and immunology: diabetic nephropathy is an inflammatory disease. Am J Soc Nephrol 16:1537–1538
13. Schalkwijk CG, Stehouwer CDA (2005) Vascular complications in diabetes mellitus: the role of endothelial dysfunction. Clin Sci 109:143–159
14. Winer N, Sowers JR (2004) Epidemiology of diabetes. J Clin Pharmacol 44:397–405
15. Ceriello A, Testa R (2009) Antioxident and anti-inflammatory treatment in type 2 diabetes. Diabetes Care 32:S232–S236
16. Rana JS, Nieuwdorp M, Jukema JW, Kastelein JJP (2007) Cardiovascular metabolic syndrome—an interplay of obesity, inflammation, diabetes and coronary heart disease. Diabetes Obes Metab 9:218–232
17. Pires IF, Moreira LAF (2012) Diabetic cardiomyopathy: understanding the molecular and cellular basis to progress in diagnosis and treatment. Heart Fail Rev 17:325–344
18. Fang ZY, Prins JB, Marwick TH (2004) Diabetic cardiomyopathy: evidence, mechanisms, and therapeutic implications. Endocr Rev 25:543–567
19. Rubler S, Dlugash J, Yuceoglu YZ (1972) New type of cardiomyopathy associated with glomerulosclerosis. Am J Cardiol 30:595–602
20. Somaratne JB, Whalley GA, Poppe KK et al (2011) Screening for left ventricular hypertrophy in patients with type 2 diabetes mellitus in the community. Cardiovasc Diabetol 10:29
21. Shizukuda Y, Reyland ME, Buttrick PM (2002) Protein kinase C-modulates apoptosis induced by hyperglycemia in adult ventricular myocytes. Am J Physiol Heart Circ Physiol 282:H1625–H1634
22. Forstermann U, Munzel T (2006) Endothelial nitric oxide synthase in vascular disease: from marvel to menace. Circulation 113:1708–1714
23. Bertoni AG, Tsai A, Kasper EK, Francati F (2003) Diabetes and idiopathic cardiomyopathy: a nationwide case–control study. Diabetes Care 26:2791–2795
24. Ghosh AK, Bradham WS et al (2010) Genetic deficiency of plasminogen activator inhibitor-1 promotes cardiac fibrosis in aged mice: involvement of constitutive transforming growth factor-β signaling and endothelial-to-mesenchymal transition. Am J Physiol Heart Circ Physiol 122:1200–1209
25. Boudina S, Abel ED (2010) Diabetic cardiomyopathy, causes and effects. Rev Endocr Metab Disord 11:31–39
26. Dobrin JS, Lebeche D (2010) Diabetic cardiomyopathy: signaling defects and therapeutic approaches. Expert Rev Cardiovasc Ther 8:373–391
27. Johansen JS, Harris K, Rychly DJ, Ergul A (2005) Oxidative stress and the use of antioxidants in diabetes: linking basic science to clinical practice. Cardiovasc Diabetol 4:1–11
28. D'Souza A, Howarth FC, Yanni J et al (2011) Left ventricle structural remodelling in the prediabetic Goto-Kakizaki rat. Exp Physiol 96:875–888
29. Edwards JL, Vincent AM, Cheng HT, Feldman EL (2008) Diabetic neuropathy: mechanisms to management. Pharmacol Exp Ther 120:1–34
30. Filippo CD, Marfella R, Cuzzocrea S et al (2006) Hyperglycemia in streptozotocin-induced diabetic rat increases infarct size associated with low levels of myocardial HO-1 during ischemia/reperfusion. Diabetes 54:808–810
31. Chen S, Evans T, Mukherjee K et al (2000) Diabetes-induced myocardial structural changes: role of endothelin-1 and its receptors. J Mol Cell Cardiol 32:1621–1629
32. Ceriello A (2008) Cardiovascular effects of acute hyperglycaemia: pathophysiological underpinnings. Diab Vasc Dis Res 5:260–268

33. Doron A (2003) Cross-linking of glycated collagen in the pathogenesis of arterial and myocardial stiffening of ageing and diabetes. J Hypertens 21:3–12
34. Asbun J, Villareal FJ (2006) The pathogenesis of myocardial fibrosis in the setting of diabetic cardiomyopathy. J Am Coll Cardiol 47:693–700
35. Masoudi FA, Inzucchi SE (2007) Diabetes mellitus and heart failure: epidemiology, mechanisms, and pharmacotherapy. Am J Cardiol 99:113–132
36. Brownlee M (2001) Biochemistry and molecular cell biology of diabetic complications. Nature 414:813–820
37. Brownlee M (2005) The pathobiology of diabetic complications: a unifying mechanism. Nature 54:1615–1625
38. Giacco F, Brownlee M (2010) Oxidative stress and diabetic complications. Circ Res 107:1058–1070
39. Diabetes Control and Complications Trial Research Group (1993) The effect of intensive treatment of diabetes on the development and progression of long-term complications in insulin-dependent diabetes mellitus. N Engl J Med 329:977–986
40. Nishikawa T, Edelstein D, Du XL et al (2002) Normalizing mitochondrial superoxide production blocks three pathways of hyperglycaemic damage. Nature 404:787–790
41. Robertson RP (2004) Chronic oxidative stress as a central mechanism for glucose toxicity in pancreatic islet beta cells in diabetes. J Biol Chem 279:42351–42354
42. Yasuda H, Terada M, Maeda K et al (2003) Diabetic neuropathy and nerve regeneration. Prog Neurobiol 69:229–285
43. Brownlee M (2006) The pathobiology of diabetic complications: a unifying mechanism. Diabetes 54:1615–1625
44. Wold LE, Ceylan AF, Ren J (2005) Oxidative stress and stress signalling: menace of diabetic cardiomyopathy. Acta Pharmacol Sin 26:908–917
45. Aronson D, Rayfield EJ (2002) How hyperglycemia promotes atherosclerosis: molecular mechanisms. Cardiovasc Diabetol 1:1–10
46. Farhangkhoee H, Khan ZA, Kaur H et al (2006) Vascular endothelial dysfunction in diabetic cardiomyopathy: pathogenesis and potential treatment targets. Pharmacol Exp Ther 111:384–399
47. Bidasee KR, Nallani K, Yu Y et al (2006) Chronic diabetes increases advanced glycation end products on cardiac ryanodine receptors/calcium-release channels. Diabetes 52:1825–1836
48. Deivedi J, Sarkar D (2010) Oxidative stress with homocysteine, lipoprotein (a) and lipid profile in diabetic nephropathy. Int J ABPT 1:840–846
49. Mohora M, Greabu M, Muscurel C et al (2007) The sources and the targets of oxidative stress in the aetiology of diabetic complications. Rom J Biophys 17:63–84
50. Geraldes P, King GL (2010) Activation of protein kinase C isoforms and its impact on diabetic complications. Circ Res 106:1319–1331
51. Twigg SM, Chen MM, Joly AH et al (2001) Advanced glycosylation end products up-regulate connective tissue growth factor (insulin-like growth factor-binding protein-related protein) in human fibroblasts: a potential mechanism for expansion of extracellular matrix in diabetes mellitus. Endocrinology 142:1760–1769
52. Pugliese G, Pricci F, Leto G et al (2002) The diabetic milieu modulates the advanced glycation end product—receptor complex in the mesangium by inducing or up-regulating galectin-3 expression. Diabetes 49:1249–1257
53. Candido R, Forbes JM, Thomas MC et al (2003) A breaker of advanced glycation end products attenuates diabetes-induced myocardial structural changes. Circ Res 92:785–792
54. Aronson D (2003) Cross-linking of glycated collagen in the pathogenesis of arterial and myocardial stiffening of ageing in diabetes. J Hypertens 21:3–12
55. Soulis-Liparota T, Cooper ME, Dunlop M, Jerums J (1995) The relative roles of advanced glycation, oxidation and aldose reductase inhibition in the development of experimental diabetic nephropathy in the Sprague–Dawley rat. Diabetologia 38:387–394
56. Forbes JM, Cooper ME, Oldfield MD, Thomas ME (2003) Role of advanced glycation end products in diabetic nephropathy. J Am Soc Nephrol 14:S254–S258

57. Idris I, Gray S, Donnelly R (2001) Protein kinase C activation: isozyme specific effects on metabolism and cardiovascular complications in diabetes. Diabetologia 44:659–673
58. Idris I, Gray S, Donnelly R (2004) Protein kinase C-beta inhibition and diabetic microangiopathy: effects on endothelial permeability responses in vitro. Eur J Pharmacol 485:141–144
59. Park JYP, Ha SU, King GL (1999) The role of protein kinase C activation in the pathogenesis of diabetic vascular complications. Peri Dial Int 19(2):S222–S227
60. Ways DK, Sheetz MJ (2002) The role of protein kinase C in the development of the complications of diabetes. Vitam Horm 60:149–193
61. Eichberg J (2002) Protein kinase C changes in diabetes: Is the concept relevant to neuropathy? Neurobiol Diabet Neuropathy 50:61–82
62. Evcimen ND, King GL (2007) The role of protein kinase C activation and the vascular complications of diabetes. Pharmacol Res 55:498–510
63. Scott JA, King GL (2004) Oxidative stress and antioxidant treatment in diabetes. Ann N Y Acad Sci 1031:204–213
64. Koya D, King GL (1998) Protein kinase C activation and the development of diabetic complications. Diabetes 47:859–866
65. Zhu Q, Xu X, Xia X, Qing G (2005) Role of protein kinase C on the alteration of retinal endothelin-1 in streptozotocin-induced diabetic rats. Exp Eye Res 81:200–206
66. Sayeski PP, Craven PA, Derubertis FR (1993) Role for protein kinase C in the mediation of increased fibronectin accumulation by mesangial cells grown in high-glucose medium. Diabetes 42:118–126
67. Sayeski PP, Wang D, Su K, Han IO et al (1997) Cloning and partial characterization of the mouse glutamine: fructose-6-phosphate amidotransferase (GFAT) gene promoter. Nucleic Acids Res 25:1458–1466
68. Schleicher ED, Weigert C (2000) Role of the hexosamine biosynthetic pathway in diabetic nephropathy. Kidney Int 58:S-13–S-18
69. Evans T, Deng DX, Chen S, Chakrabarti S (2000) Endothelin receptor blockade prevents augmented extracellular matrix component mRNA expression and capillary basement membrane thickening in the retina of diabetic and galactose-fed rats. Diabetes 49:662–666
70. West IC (2000) Radicals and oxidative stress in diabetes. Diabet Med 17:171–180
71. Rosen P, Nawroth PP, King G et al (2001) The role of oxidative stress in the onset and progression of diabetes and its complications. Diabetes/Metab Res Rev 17:189–212
72. Jay D, Hitomi H, Griendling KK (2006) Oxidative stress and diabetic cardiovascular complications. Free Radic Biol Med 40:183–192
73. Du X, Edelstein D, Obici S et al (2006) Insulin resistance reduces arterial prostacyclin synthase and eNOS activities by increasing endothelial fatty acid oxidation. J Clin Invest 116:1071–1080
74. Shen X, Zheng S, Metreveli NS, Epstein PN (2006) Protection of cardiac mitochondria by overexpression of MnSOD reduces diabetic cardiomyopathy. Diabetes 55:798–805
75. Nascimento NR, Lessa LM, Kerntopf MR et al (2006) Inositols prevent and reverse endothelial dysfunction in diabetic rat and rabbit vasculature metabolically and by scavenging superoxide. Proc Natl Acad Sci USA 103:218–223
76. Stitt AW, Li YM, Gardiner TA, Bucala R (1997) Advanced glycation end products (AGEs) co-localize with AGE receptors in the retinal vasculature of diabetic and of AGE-infused rats. Am J Pathol 150:523–531
77. Yao D, Brownlee M (2010) Hyperglycemia-induced reactive oxygen species increase expression of the receptor for advanced glycation end products (RAGE) and RAGE ligands. Diabetes 59:249–255
78. Cohn JN, Ferrari R, Sharpe N (2000) Cardiac remodelling—concepts and clinical implications: a consensus paper from an international forum on cardiac remodelling. J Am Coll Cardiol 35:569–582
79. Pichler M, Rainer PP, Schauer S, Hoefler G (2012) Cardiac fibrosis in human transplanted hearts is mainly driven by cells of intracardiac origin. J Am Coll Cardiol 59:1008–1016

80. Li B, Zheng Z, Wei Y et al (2011) Therapeutic effects of neuregulin-1 in diabetic cardiomyopathy rats. Cardiovasc Diabetol 10:69–77
81. Ahmed SH, Clark LL, Pennington WR et al (2006) Matrix metalloproteinases/tissue inhibitors of metalloproteinases: relationship between changes in proteolytic determinants of matrix composition and structural, functional, and clinical manifestations of hypertensive heart disease. Circulation 113:2089–2096
82. Cai L, Wang Y, Zhou G et al (2006) Attenuation by metallothionein of early cardiac cell death via suppression of mitochondrial oxidative stress results in a prevention of diabetic cardiomyopathy. J Am Coll Cardiol 48:1688–1697
83. Aneja A, Tang WH, Bansilal S et al (2008) Diabetic cardiomyopathy: insights into pathogenesis, diagnostic challenges and therapeutic options. Am J Med 121:748–757
84. Sabbah HN, Sharov VG, Lesch M, Goldstein S (1995) Progression of heart failure: a role for interstitial fibrosis. Mol Cell Biochem 147:29–34
85. Burlew BS, Weber KT (2002) Cardiac fibrosis as a cause of diastolic dysfunction. Herz 27:92–98
86. Bosman FT, Stamenkovic I (2003) Functional structure and composition of the extracellular matrix. J Pathol 200:423–428
87. Manabe I, Shindo T, Nagai R (2002) Gene expression in fibroblasts and fibrosis: involvement in cardiac hypertrophy. Circ Res 91:1103–1113
88. De Leeuw N, Ruiter DJ, Balk AH et al (2001) Histopathologic findings in explanted heart tissue from patients with end-stage idiopathic dilated cardiomyopathy. Transpl Int 14:299–306
89. Weber KT (2004) Fibrosis in hypertensive heart disease: focus on cardiac fibroblasts. J Hypertens 22:47–50
90. Camelliti P, Borg TK, Kohl P (2005) Structural and functional characterisation of cardiac fibroblasts. Cardiovasc Res 65:40–51
91. Brown RD, Ambler SK, Mitchell MD, Long CS (2005) The cardiac fibroblast: therapeutic target in myocardial remodeling and failure. Annu Rev Pharmacol Toxicol 45:657–687
92. Pauschinger M, Knopf D, Petschauer S et al (1999) Dilated cardiomyopathy is associated with significant changes in collagen type I/III ratio. Circulation 99:2750–2756
93. Bouzegrhane F, Thibault G (2002) Is angiotensin II a proliferative factor of cardiac fibroblasts? Cardiovasc Res 53:304–312
94. Cuspidi C, Ciulla M, Zanchetti A (2006) Hypertensive myocardial fibrosis. Nephrol Dial Transplant 21:20–23
95. Sun Y, Weber KT (2000) Infarct scar: a dynamic tissue. Cardiovasc Res 46:250–256
96. Dinh DT, Frauman AG, Johnston CI, Fabiani ME (2001) Angiotensin receptors: distribution, signalling and function. Clin Sci (Lond) 100:481–492
97. Lijnen PJ, Petrov VV, Fagard RH (2001) Angiotensin II-induced stimulation of collagen secretion and production in cardiac fibroblasts is mediated via angiotensin II subtype 1 receptors. J Renin Angiot Aldost Syst 2:117–122
98. Staufenberger S, Jacobs M, Brandstatter K et al (2001) Angiotensin II type 1 receptor regulation and differential trophic effects on rat cardiac myofibroblasts after acute myocardial infarction. J Cell Physiol 187:326–335
99. Rosenkranz S (2004) TGF-β1 and angiotensin networking in cardiac remodelling. Cardiovasc Res 63:423–432
100. Jiang XY, Gao GD, Du XJ et al (2007) The signalling of AT2 and the influence on the collagen metabolism of AT2 receptor in adult rat cardiac fibroblasts. Acta Cardiol 62:429–438
101. Pan CH, Wen CH, Lin CS (2008) Interplay of angiotensin II and angiotensin (1–7) in the regulation of matrix metalloproteinases of human cardiocytes. Exp Physiol 93:599–612
102. Sato H, Watanabe A, Tanaka T et al (2003) Regulation of the human tumour necrosis factor-α promoter by angiotensin II and lipopolysaccharide in cardiac fibroblasts: different cis-acting promoter sequences and transcriptional factors. J Mol Cell Cardiol 35:1197–1205
103. Chao HH, Chen JJ, Chen CH et al (2005) Inhibition of angiotensin II induced endothelin-1 gene expression by 17-β-oestradiol in rat cardiac fibroblasts. Heart 91:664–669

104. Wang S, Wang X, Yan J et al (2007) Resveratrol inhibits proliferation of cultured rat cardiac fibroblasts: correlated with NO-cGMP signaling pathway. Eur J Pharmacol 567:26–35
105. Chintalgattu V, Nair DM, Katwa LC (2003) Cardiac myofibroblasts: a novel source of vascular endothelial growth factor (VEGF) and its receptors Flt-1 and KDR. J Mol Cell Cardiol 35:277–286
106. Border WA, Ruoslahtit E (1999) Transforming growth factor-β in disease: the dark side of tissue repair. J Clin Invest 90:1–7
107. Border WA, Noble NA (1994) Transforming growth factor β in tissue fibrosis. N Engl J Med 331:1286–1292
108. Wu L, Derynck R (2009) Essential role of TGF-β1 signalling in glucose-induced cell hypertrophy. Dev Cell 17:35–48
109. Liang H, Zhang C, Ban T (2012) A novel reciprocal loop between microRNA-21 and TGF-RIII is involved in cardiac fibrosis. Int J Biochem Cell Biol 44:2152–2160
110. Chu W, Li X, Li C et al (2011) TGFBR3, a potential negative regulator of TGF-beta signaling, protects cardiac fibroblasts from hypoxia-induced apoptosis. J Cell Physiol 226:2586–2594
111. Derynck R, Zhang YE (2003) Smad-dependent and Smad-independent pathways in TGF beta family signalling. Nature 425:577–584
112. Porter KE, Turner NA (2009) Cardiac fibroblasts: at the heart of myocardial remodelling. Pharmacol Exp Ther 123:255–278
113. Eickelberg O, Centrella M, Reiss M et al (2002) Betaglycan inhibits TGF-beta signalling by preventing type I-type II receptor complex formation. Glycosaminoglycan modifications alter beta-glycan function. J Biol Chem 277:823–829
114. Bold AJ, Borenstein HB, Veress AT, Sonnenberg H (1981) A rapid and potent natriuretic response to intravenous injection of atrial myocardial extract in rats. Life Sci 28:89–94
115. Sudoh T, Kangawa K, Minamino N, Matsuo H (1988) A new natriuretic peptide in porcine brain. Nature 332:78–81
116. Tanaka T, Hasegawa K, Fujita M et al (1998) Marked elevation of brain natriuretic peptide levels in pericardial fluid is closely associated with left ventricular dysfunction. J Am Coll Cardiol 31:399–403
117. Burnett JC, Kao PC, Hu DC et al (1986) Atrial natriuretic peptide elevation in congestive heart failure in the human. Science 231:1145–1147
118. Tsuruda T, Boerrigter G, Huntley BK et al (2002) Brain natriuretic peptide is produced in cardiac fibroblasts and induces matrix metalloproteinases. Circ Res 91:1127–1134
119. Garbers DL, Chrisman TD, Wiegn P et al (2006) Membrane guanylyl cyclase receptors: an update. Trends Endocrinol Metab 17:251–258
120. Ergul A (2011) Endothelin-1 and diabetic complications: focus on the vasculature. Pharmacol Res 63:477–482
121. Schleicher ED, Nerlich A (1996) The role of hyperglycemia in the development of diabetic complications. Horm Metab Res 28:367–373
122. Esper RJ, Vilarino JO, Machado RA, Parango A (2008) Endothelial dysfunction in normal and abnormal glucose metabolism. Adv Cardiol 45:17–43
123. Guo Z, Xia Z, Jiang J, McNeil JH (2007) Down-regulation of NADPH oxidase, antioxidant enzymes and inflammatory markers in the heart of streptozotocin-induced diabetic rats by *N* enzymes. Am J Physiol Heart Circ Physiol 292:H1728–H1736
124. Tan KC, Chow WS, Ai VH et al (2002) Effects of angiotensin II receptor antagonist on endothelial vasomotor function and urinary albumin excretion in type 2 diabetic patients with microalbuminuria. Diabetes Metab Res Rev 18:71–76
125. Creager MA, Luscher TF (2003) Diabetes and vascular disease pathophysiology, clinical consequences, and medical therapy: Part I. Circulation 108:1527–1532
126. Kawashima S (2004) The two faces of endothelial nitric oxide synthase in the pathophysiology of atherosclerosis. Endothelium 11:99–107
127. Nomura S, Shouzu A, Omoto S et al (2000) Significance of chemokines and activated platelets in patients with diabetes. Clin Exp Immunol 121:437–443

128. Williams SB, Goldfine AB, Timimi FK et al (1998) Acute hyperglycemia attenuates endothelium-dependent vasodilation in humans in vivo. Circulation 97:1695–1701
129. Beckman JA, Goldfine AB, Gordon MB et al (2001) Ascorbate restores endothelium-dependent vasodilation impaired by acute hyperglycemia in humans. Circulation 103:1618–1623
130. Hink U, Li H, Mollnau H et al (2001) Mechanisms underlying endothelial dysfunction in diabetes mellitus. Circ Res 88:E14–E22
131. Kinlay S, Libby P, Ganz P (2001) Endothelial function and coronary artery disease. Curr Opin Lipidol 12:383–389
132. Mehta JL, Rasouli N, Sinha AK, Molavi B (2006) Oxidative stress in diabetes: a mechanistic overview of its effects on atherogenesis and myocardial dysfunction. Int J Biochem Cell Biol 38:794–803
133. Laursen JB, Somers M, Kurz S et al (2001) Endothelial regulation of vasomotion in apoE-deficient mice: implications for interactions between peroxynitrite and tetrahydrobiopterin. Circulation 103:1282–1288
134. Feng Z, Hu W, Tang MS (2004) Trans-4-hydroxy-2-nonenal inhibits nucleotide excision repair in human cells: a possible mechanism for lipid peroxidation-induced carcinogenesis. Proc Natl Acad Sci USA 101:8598–8602
135. Seager MJ, Singal PK, Orchard R, Pierce GN, Dhalla NS (1984) Cardiac cell damage: a primary myocardial disease in streptozotocin-induced chronic diabetes. Br J Exp Pathol 65:613–623
136. Hsiao YC, Suzuki K, Abe H, Toyota T (1987) Ultrastructural alterations in cardiac muscle of diabetic BB Wistar rats. Virchows Arch A Pathol Anat Histopathol 411:45–51
137. Valensi PE, Johnson NB, Maison-Blanche P, Extramania F, Motte G, Coumel P (2002) Influence of cardiac autonomic neuropathy on heart rate dependence of ventricular repolarization in diabetic patients. Diabetes Care 25:918–923
138. Pfister R, Cairns R, Erdmann E, Schneider CA (2011) PROactive investigators: prognostic impact of electrocardiographic signs in patients with type 2 diabetes and cardiovascular disease: results from the PROactive study. Diabet Med 28:1206–1212
139. Adeghate E (2004) Molecular and cellular basis of the aetiology and management of diabetic cardiomyopathy: a short review. Mol Cell Biochem 261:187–191

Hyperglycemia, Oxidative Stress, and Vascular Complications: Role of Epigenetic Mechanisms

Estelle R. Simo Cheyou and Ashok K. Srivastava

Abstract Vascular disease as evidenced by aberrant endothelial and vascular smooth muscle cell physiology represents one of the major complications of diabetes. Although the metabolic disturbances such as oxidative stress, inflammation, and hyperlipidemia have been well described as main players in the process of vascular dysfunction, epigenetic modifications of gene expression also occur under the hyperglycemic state and modulate cardiovascular homeostasis. The main epigenetic mechanisms that can modify chromatin structure and gene expression include chromatin remodeling via histone modifications or DNA methylation, and gene silencing by small noncoding RNA molecules termed microRNAs. Recent studies have suggested that these epigenetic events either alone or in concert are capable of modulating the expression of multiple target genes involved in redox homeostasis, vascular cell proliferation, and migration, as well as in proinflammatory pathways associated with vascular dysfunction. This review highlights some epigenetic changes induced by hyperglycemic and oxidative states in the vascular system and discusses their potential role in the pathogenesis of diabetes-associated vascular complications.

E.R. Simo Cheyou
Laboratory of Cell Signaling, Montreal Diabetes Research Centre, Centre de recherche-Centre hospitalier de l'Université de Montréal (CRCHUM) – CRCHUM, 900 St-Denis street, Montreal, QC H2X 0A9, Canada

Department of Nutrition, University of Montreal, Montreal, QC, Canada

A.K. Srivastava (✉)
Laboratory of Cell Signaling, Montreal Diabetes Research Centre, Centre de recherche-Centre hospitalier de l'Université de Montréal (CRCHUM) – CRCHUM 900 St-Denis street, Montreal, QC H2X 0A9, Canada

Department of Medicine, University of Montreal, Montreal, QC, Canada

Department of Nutrition, University of Montreal, Montreal, QC, Canada
e-mail: ashok.srivastava@umontreal.ca

B. Turan and N.S. Dhalla (eds.), *Diabetic Cardiomyopathy: Biochemical and Molecular Mechanisms*, Advances in Biochemistry in Health and Disease 9, DOI 10.1007/978-1-4614-9317-4_6, © Springer Science+Business Media New York 2014

Keywords Hyperglycemia • Oxidative stress • Endothelial dysfunction • Gene silencing • Inflammation • MicroRNAs • Vessel remodeling • DNA methylation • Atherosclerosis • Histone deacetylases • Histone demethyltransferases

1 Introduction

The cluster of cardiovascular complications that occur during diabetes mellitus includes atherosclerosis, peripheral vascular disease, cardiomyopathy, and stroke [1]. Although they are distinct with regard to their pathophysiological features (endothelial dysfunction, cell proliferation and migration, myocyte hypertrophy, vessel narrowing, and apoptosis), these pathologies share the same underlying mechanisms that mainly involve major intracellular processes taking place under hyperglycemic conditions and for which oxidative stress plays a central role [2]. First, the increase in glucose accumulation leads to an increased activation of the polyol pathway that consists of the reduction of glucose or other carbonyl compounds into sorbitol and other corresponding sugar alcohols. This reaction requires the catalytic activity of the enzyme aldose reductase that utilizes the cofactor nicotinamide adenine dinucleotide phosphate (NAD(P)H) [3]. Second, hyperglycemia also induces protein glycosylation that contributes to the increase in the quantity of advanced glycation end product (AGE) and reactive oxygen species (ROS) [4]. AGE binding to their receptors (RAGE) on cell surfaces can activate NAD(P)H oxidase and thereby increase ROS production. Hyperglycemia also activates protein kinase C (PKC) isoforms partly by AGE receptor activation as well as from ROS-dependent increase in intracellular diacylglycerol, the upstream activator of PKC [5]. In diabetes, PKC activation induces massive ROS generation mainly via the stimulation of NAD(P)H, resulting in increased redox signaling [6–9]. Thus, oxidative stress via hyperactivation of mitogenic, inflammatory, and proliferative signaling cascades plays a critical role in vascular complications associated with diabetes [5, 10, 11]. In addition to a role of hyperglycemia and oxidative stress-mediated signaling pathways in inducing diabetic vascular complications, implication of epigenetic mechanisms in the pathophysiology of cardiovascular dysfunction has also been suggested recently [12, 13]. The purpose of this chapter is to highlight some key epigenetic signals that are activated by hyperglycemia and oxidative stress in the diabetic state and to discuss their potential role in vascular dysfunction in diabetes.

2 Basic Epigenetic Mechanisms

Epigenetic changes modify the patterns of gene expression without affecting the DNA structure. These changes occur in response to a variety of stimuli in the physical or physiological environment. Epigenetic mechanisms include reversible posttranslational modifications of histone structure within the nucleosomes in the chromatin, methylation of specific sites of the DNA strand, and gene silencing by small sequences of noncoding RNA (microRNAs) [14].

2.1 Histone Modifications

The chromatin is constructed of multiple groups consisting of four pairs of histones (H3, H4, H2A, H2B) wrapped by a 147-base-pair fragment of DNA and linked together with a fifth type of histone, H1 [14]. Each monomer of histone at the center of a nucleosome extends as a histone tail with a C-terminal domain and an N-terminal tail. With regard to gene expression, the standard requirement for an active transcription is a loosened chromatin opened for the binding of transcription factors; such conformation is called euchromatin. In contrast, a compact chromatin, called heterochromatin, disables transcriptional events. The transition from heterochromatin to euchromatin is achieved through reversible posttranscriptional modifications within the N-terminal fragment of histone tails. Such modifications mainly include acetylation and methylation, catalyzed by histone acetyltransferases (HATs) and histone methyltransferases (HMTs), respectively; these enzymes transfer acetyl or methyl groups mostly to lysine, serine, or arginine residues within the N-terminal tail of histone. These reactions are reversible because of the presence of other enzymes with opposite effects, histone deacetylases (HDACs) and histone demethyltransferases (HDMTs). Altogether, chromatin changes at the histone level form a histone code that can be read by other proteins and translated into a functional signal depending on the underlying context [14]. Implication of such a code in vascular biology has been suggested and is being further explored in the context of diabetic cardiovascular dysfunction [15].

2.2 DNA Methylation

Methylation of DNA is another form of epigenetic modification that occurs at the 5′-position of a cytosine residue separated by a phosphate group from a guanosine residue in the longitudinal DNA sequence (CpG island) [14]. Hypomethylation or absence of methylation favors gene transcription whereas hypermethylation or methylation of previously unmethylated CpG causes gene silencing. CpG methylation is catalyzed by DNA methyltransferases (DNMTs). It is the best studied epigenetic mechanism, and to date, three types of DNMTs have been identified: DNMT1 is responsible for the heritable DNA methylation state as it helps to preserve the parental methylation patterns after DNA replication whereas DNMTs 3a and 3b are de novo methyltransferases that methylate previously unmethylated fragments of DNA [14]. The loss of DNMT1 has been shown to be lethal in mice [16]. Attempts have been made to understand the correlation between DNA methylation state and diabetic nephropathy in patients with type 1 diabetes [17], as well as to identify and to replicate the alterations of DNA methylation signatures of diabetic nephropathy in renal mesangial and epithelial cells, suggesting the importance of this process in diabetes and cardiovascular homeostasis [18].

2.3 Gene Silencing by MicroRNAs

The central dogma in molecular biology requires that DNA is transcribed into RNA and that RNA is translated into proteins. However, it has been found that there are tiny RNA fragments that are expressed in the nucleus but remain untranslated. These fragments are termed microRNAs (miRNAs, named as miR-specific number) that are frequently referred to as "micromanagers of gene expression" [19]. They are broadly expressed in eukaryotic cells, and it is estimated that approximately 1,000 miRNAs are encoded by the human genome [20]. Mature miRNAs are made up of approximately 22 nucleotides. Primary miRNA genes located in the introns of either coding or noncoding genes [21] are transcribed by *RNA polymerase II* or *III* in the nucleus to form large pre-miRNA transcripts. These transcripts remain in the nucleus and are processed by an RNase III enzyme, Drosha, and by a double-stranded RNA-binding protein, *Pasha*, into approximately 70 nucleotide pre-miRNAs. Two transporters, *RanGTP* and *exportin 5*, export the pre-miRNA into the cytoplasm where it is digested by another RNase III, *Dicer,* to form a transient 18- to 24-nucleotide duplex. The duplex is incorporated as a single strand into a multiprotein RNA-induced silencing complex and forms the mature miRNA. miRNAs function through binding with complementary sites in the mRNA transcript to induce either translational repression or gene silencing by RNA degradation [22]. miRNA function has been associated with chromatin modulation as well as with genome stability [23], resulting in the regulation of multiple genes involved in the control of cardiovascular functions [24].

3 HDACs- and HATs-Mediated Responses in Diabetic Cardiovascular Complications

Depending on their domain organization or sequence identity, HDACs are divided into four classes: I, IIA, IIB, and III. They are expressed in multiple atherosclerosis-relevant cell types in the vasculature including monocytes, endothelial cells, and smooth muscle cells [25]. Inflammation, monocyte adhesion, and aberrant smooth muscle cell migration and proliferation are among the critical events that contribute to the atherogenic process. Several groups have investigated the roles of HDAC- and HAT-mediated deacetylation or acetylation of histone and non-histone proteins in the events associated with diabetic complications. For example, de Kreutzenberg et al. reported that high glucose downregulated the expression of Sirt-1, a protein that belongs to the class III HDACs, in human peripheral blood mononuclear cells [26]. In these studies, the downregulation of Sirt-1 gene and protein expression was associated with increased acetylation of p53, a cell-cycle regulator, and enhanced phosphorylation of the stress signaling molecule c-Jun N-terminal kinase (JNK) [26]. Because of a negative association of Sirt-1 expression and the carotid intima-media thickness, a potential role of Sirt-1 in regulating early atherosclerosis was suggested from these studies [26]. Downregulation of Sirt-1 was also observed in endothelial cells in response to high glucose [27]. In fact, hyperglycemia was shown to induce atherosclerotic endothelial cell aging through reduction of Sirt-1 expression [27]. Sirt-1 has also the ability to modify oxidative stress by its effects on the transcription of enzymes implicated in regulating ROS generation. Sirt-1, via the deacetylation of the transcription factor FOXO [28], positively regulates the transcription of the manganese superoxide dismutase (MnSOD), an antioxidant enzyme that attenuates the oxidative stress induced by hyperglycemia [29]. Hyperglycemia-dependent Sirt-1 downregulation, through a regulatory feedback interaction between Sirt-1 and HAT p300, stimulates the acetylation of FOXO and decreases MnSOD levels, resulting in increased oxidative stress [30]. Thus, in diabetic endothelial senescence leading to vascular dysfunction, a persistent redox state is maintained by high glucose-induced change in the acetylation state of the non-histone protein FOXO, and by modulation of the epigenetic markers Sirt-1 and p300 that exhibit HDAC and HAT properties, respectively [30]. In accordance with this, decreased glucose concentrations have been associated with the increase in Sirt-1 in cancer cells [31].

The thioredoxin interacting protein (TXNIP), which is an endogenous antagonist of the ROS scavenging protein thioredoxin, is known to be upregulated under hyperglycemic conditions [32] and contributes to hyperglycemia-induced oxidative stress as well as vascular complications [33, 34]. An elegant study reported that glucose-induced upregulation of the TXNIP gene expression is dependent on the recruitment of the HAT p300 at the TXNIP promoter and its histone H4 acetylation [35]. These studies demonstrated that H4 acetylation plays an important role in the control of TXNIP transcription and showed that a pharmacological HDAC inhibitor, trichostatin A (TSA), increased TXNIP expression in pancreatic beta cells highlighting the role of histone acetylation in TXNIP gene transcription [35].

Moreover, hyperglycemia has also been shown to induce the expression of TXNIP as well as several pro-inflammatory genes such as cyclooxygenase 2 (COX-2), vascular endothelial growth factor (VEGF), and the intercellular adhesion molecule type 1 (ICAM-1) in retinal capillary endothelial cells [36]. In these studies, hyperglycemia-induced TXNIP gene expression also contributed to the histone H3 lysine 9 acetylation of the COX-2 gene promoter via a p38-dependent signaling pathway [36]. Such hyperglycemia-induced histone acetylation has not only been described in the context of inflammation but also in other atherosclerotic events such as the expression of vasoactive factors and extracellular matrix proteins [37, 38]. For example, high glucose-induced HAT activity via p300 upregulation has been shown to be an essential upstream component for glucose-induced expression of endothelin-1, fibronectin, and VEGF in vascular smooth muscle cells (VSMCs) and endothelial cells [37]. Increased H3 acetylation was associated with these glucose-induced responses and, interestingly, overexpression of the HAT p300 potently induced glucose-like effects in these cells [37]. Moreover, because the HAT p300 is one of the primary targets of the nuclear factor kappa B (NF-κB) [39], a marker of glucose-induced inflammation [40, 41], it has been proposed to participate in the expression of the pro-inflammatory cytokine interleukin 6 (IL-6). Accordingly, p300 was shown to mediate the lysine acetylation at IL-6 as well as the tumor necrosis factor-α (TNF-α) gene promoters and enhance their transcription in monocytes [42]. Similarly, hyperglycemia-induced recruitment of HAT and lysine acetylation at key inflammatory gene promoters were also observed in monocytes from patients with either type 1 or type 2 diabetes, suggesting a contribution of chromatin remodeling in diabetes-induced inflammation [13, 43].

Oxidative stress, as observed in diabetes, is an inhibitor of HDAC2, a member of the class 1 HDACs, and this inhibition was shown to stimulate the expression of another pro-inflammatory cytokine, interleukin 8 (IL-8) [44]. Experimental production of peroxynitrite as well as treatment of human airway epithelial cells with TSA was reported to attenuate HDAC2 activity and expression, resulting in elevated production of IL-8 in response to IL-1β [44]. More importantly, a similar attenuation of HDAC activity by TSA in the context of dyslipidemia has been shown to exacerbate the atherogenic process; this was observed in hyperlipidemic mice in which treatment with TSA was shown to enhance the levels of atherogenic markers such as the scavenger receptor A, the lymphocyte CD36, TNF-α, and vascular cell adhesion molecule 1 (VCAM-1) [45], suggesting a critical role of HDAC activity for the maintenance of vascular homeostasis in pathophysiological circumstances. The inhibition of HDAC2 by hyperglycemia-induced redox signaling could also be deleterious per se because HDAC2 itself is an antagonist of the atherosclerotic plaque formation. HDAC2 acts in this sense by deacetylating the class II transactivator (CIITA), whose role is to repress the collagen promoter and to activate the major histocompatibility complex promoter in smooth muscle cells and macrophages [46]. Thus, it appears that, in these cell types, HDAC2 is a good antagonist of transcriptional activity, as well as a positive regulator of collagen formation, both mechanisms contributing to a delay in the atherogenic events [46, 47].

In view of the importance of the histone acetylation state in vascular abnormalities, epigenetic therapy by reversal or mimicking of HAT or HDAC activities, respectively, has generated much interest for the treatment of diabetic cardiovascular complications. In this regard, dietary polyphenols have attracted some attention because of their ability to modulate HDAC or HAT activity and, subsequently, atherosclerotic gene expression [48, 49]. Among these polyphenolic compounds, curcumin, a derivative of the curry spice turmeric, has been shown to exhibit cardiovascular protective, as well as antiinflammatory, antithrombotic and antioxidant properties in the context of type 2 diabetes [50–53]. Interestingly, curcumin has been reported to inhibit p300 HAT activity [54]. It has been shown to inhibit high glucose-induced relA/p65 activation, the critical regulatory subunit of NF-κB, in monocytes through suppression of HAT activity and upregulation of the HDAC activity [42]. RelA/p65 is regulated on one hand through its own acetylation and deacetylation but on the other hand through histone acetylation at the NF-κB promoter [42, 55]. Furthermore, in the context of VSMC physiology, curcumin has been demonstrated to inhibit growth of rat and human VSMCs through the induction of the antiproliferative enzyme heme oxygenase via the translocation of the nuclear transcription factor E2-related factor-2 (Nrf-2) and the subsequent activation of the antioxidant response element (ARE) within the heme oxygenase promoter [56]. This effect of curcumin on Nrf-2 translocation could also arise from its HAT inhibitory property because it has recently been reported that TSA-mediated inhibition of HDAC is a potent inducer of Nrf-2 expression, translocation, and binding to ARE on the heme oxygenase [57].

Another interesting compound with epigenetic properties is resveratrol, found in red grape skin as a polyphenolic phytoalexin, that exhibits the ability to activate Sirt-1 [58, 59]. Because of its antioxidant and antiinflammatory properties, resveratrol has been broadly demonstrated to be beneficial for cardiovascular health and disease [60–62] and also to protect against diabetic nephropathy [63] as well as diabetic atherogenic events in the vasculature [64]. In vessels isolated from alloxan-induced diabetic rabbits, resveratrol attenuated the production of ROS, improved vascular reactivity to acetylcholine, and helped in maintaining the integrity of the endothelium under diabetic conditions [65]. It has been reported that high glucose-induced mitochondrial ROS production was decreased in a resveratrol-Sirt-1-dependent fashion in human coronary arterial endothelial cells (CAECs) [66]. In these studies, siRNA- or electroporation-mediated silencing of Sirt-1 in hyperglycemic CAECs abolished the resveratrol-induced increase in the pro-oxidant molecules, MnSOD and glutathione, as well as the resveratrol-induced reduction in intracellular H_2O_2 [66]. Such positive effect on endothelial cell physiology was also recently reported from studies using bovine aortic endothelial cells (BAECs) under hyperglycemic conditions [67]. Resveratrol was also shown to attenuate glucose-induced early atherosclerosis events such as endothelial hyperpermeabilty and overexpression of caveolin as well as the glucose-induced expression of VEGF and its receptor [67]. It should be noted that VEGF induction has been previously reported to be Sirt-1 dependent in HT1080 cell line [68]. Furthermore, a positive effect of resveratrol in reversing the p53 acetylation induced by high glucose and high palmitate in human monocytes was reported and demonstrated to be a Sirt-1-dependent process, suggesting the antiapoptotic effect of resveratrol [26]. Thus, given the potential of natural HDAC/HAT modulators in regulating cardiovascular homeostasis and cardiovascular protection, further investigations are required to appropriately assign them a beneficial role in diabetic vascular dysfunction.

4 MicroRNA-Mediated Responses in Diabetic Atherosclerosis

Multiple studies have assessed the implication of microRNAs in diabetes pathobiology [12, 69, 70], suggesting a possible involvement of these molecules in diabetic cardiovascular diseases. The process of diabetic atherosclerosis is a result of endothelial dysfunction, chronic inflammation, and vessel remodeling [71].

4.1 Endothelial Dysfunction

Depending on their specificity, microRNAs can be considered as either positive or negative modulators of diabetic endothelial dysfunction. Caporali et al. have assessed the role of miR-503 on endothelial cell function in diabetic conditions and demonstrated that hyperglycemia resulted in increased levels of miR-503 in cultured endothelial cells [72]. They also showed that lentiviral overexpression of miR-503 reduced endothelial cell proliferation and migration; these functional capacities were upregulated by inhibition of miR-503 suggesting the negative involvement of this miRNA in diabetic-induced endothelial dysfunction. Their observations were further strengthened by the finding that heightened levels of miR-503 are expressed in ischemic muscular tissues from diabetic patients [72]. Similarly, high glucose was previously shown to induce an increase of miR-221 in human umbilical vein endothelial cells [73]. Hyperglycemia is known to alter the transmigratory capacity of endothelial cells that is essential for angiogenesis following vascular damage, and a lack of this capacity exacerbates endothelial dysfunction as seen in numerous cardiovascular disorders. In the latter study [73], the authors found that c-kit, a specific stem-cell receptor that plays a role in endothelial cell transmigration, was downregulated under hyperglycemic conditions. Antisense oligonucleotide-mediated inhibition of miR-221 was able to prevent this downregulation and to restore the endothelial transmigratory capacities altered by high glucose, suggesting a role of miR-221 in facilitating endothelial dysfunction. These two studies proposed that miR-221 and miR-503 inhibition in endothelial cells can reinforce their functional capacities under diabetic conditions and help in preventing atherosclerotic disease. However, some opposite actions of miRNAs have also been reported. For example, a study from Meng et al. showed that diabetes was associated with reduced levels of miR-126, miR-21, miR-27a, miR-27b, and miR-130a in endothelial progenitor cells (EPC) [74]; among these, miR-126 was the most affected. Overexpression of miR-126 resulted in enhanced EPC function and reduced EPC apoptosis through the downregulation of Spred-1 gene, a known angiogenic inhibitor in endothelial cells. These observations were in accordance with another study where miR-126 was demonstrated to be a positive regulator of vascular integrity [75]. Endothelial cell-specific deletion of miR-126 in mice resulted in the impairment of vascular function as illustrated by defective cell migration, proliferation, and angiogenesis [75]. Thus, to clarify the role of various miRNAs in vascular dysfunction, more in-depth studies on the contribution of miRNAs in the endothelial system pathobiology under diabetic conditions are needed.

4.2 Inflammation

MicroRNAs also play a role in chronic inflammation that occurs in the vessels of diabetic patients. Molecular hallmarks of inflammation include the heightened production of adhesion molecules and proinflammatory enzymes such as COX-2.

COX-2 mRNA is a target of miR-16 in endothelial cells. Past studies have demonstrated that under diabetic conditions, inflammation and atherosclerosis are in part caused by the binding of RAGE with their ligand, S100b, leading to their subsequent activation [76]. Recent work from Shanmugam et al. has proposed that this is achieved through the S100b-dependent inhibition of the binding of miR-16 with COX-2 mRNA [77]. S100b was in fact shown to displace the nuclear ribonucleoprotein K into the cytoplasm, where it interacts with COX-2 mRNA and disables miR-16 attachment [77]. These observations support an antiinflammatory role of miR-16 in diabetes. Moreover, several other microRNAs have been implicated in endothelial atherosclerotic inflammation through pro-inflammatory gene repression. Overexpression of miR-155 and miR-121/122 effectively lowered angiotensin-II induced vascular cell adhesion molecule 1 and monocyte chemoattractant protein 1 (MCP-1) in a study [78]. It was concluded that these microRNAs participate in endothelial cell inflammation and migration by targeting angiotensin II receptor type 1 or the transcription factor Ets [78]. However, in contradiction to this, miR-200b has been demonstrated to exhibit pro-inflammatory properties in diabetic VSMCs [79]. The overexpression of this miRNA enhanced COX-2 promoter activity and induced the expression of MCP-1. In the same study, monocyte binding in diabetic VSMCs was also stimulated by miR-200b, suggesting a positive regulation of inflammatory processes by this miRNA [79]. Another group has also addressed the role of miR-125b in the epigenetic regulation of inflammatory genes in VSMCs [80]. They observed that miR-125b was upregulated in diabetic mice concomitantly to a downregulation of histone methyltransferase Suv39H1. Overexpression of miR-125b reduced Suv39H1 transcription and resulted in pro-inflammatory diabetic phenotype, as illustrated by the enhanced monocyte binding [80]. This last set of studies demonstrates that, on the basis of their pro-inflammatory properties, some miRNAs contribute to the development of diabetic atherosclerosis.

4.3 Vessel Remodeling

Vessel remodeling is a common feature of cardiovascular diseases and vascular dysfunction that involves the aberrant growth and proliferation of vascular cells and rearrangement of the extracellular matrix. miRNAs are involved in VSMC proliferation and hypertrophy, leading to vessel remodeling [81, 82]. In a recent study [83], miR-208 was shown to mediate insulin-induced VSMC proliferation through the downregulation of the cell-cycle regulator protein p21, suggesting a role of this miRNA in diabetic neointima formation. In addition, a role of miR-143 and miR-145 in the maintenance of the beneficial contractile VSMC phenotype, and in turn in vascular homeostasis, has recently been suggested [82, 84–86]. It has been demonstrated that miR-143 and miR-145 control VSMC proliferation, migration, and differentiation, as well as overall vessel reactivity [87]. In fact, overexpression of miR-143/145 in VSMCs resulted in significant attenuation of DNA synthesis [87]. In addition, VSMCs isolated from miR-143/145 knockout mice exhibited a heightened migratory response to the platelet-derived growth factor (PDGF) and presented decreased levels of the VSMC differentiation markers smooth muscle myosin heavy chain and smooth muscle α-actin [87]. It has also been postulated that miR-145 could regulate the atherogenic processes in relationship to plaque formation and stability. This hypothesis was recently confirmed in an atherosclerotic animal model, the apolipoprotein E knockout mouse, where a VSMC-targeted overexpression of miR-145 decreased the risk of plaque rupture by reducing its size and by promoting the VSMC differentiation toward the contractile phenotype [88]. Similar conclusions were made from studies that focused on the role of miR-133 in vascular physiology [89]. Adenoviral transfection of miR-133 in rat carotid aorta led to a reduction of neointima formation, VSMC proliferation, and the phenotypic switch from the contractile to the synthetic state after balloon injury, all of these via the repression of the transcription factor Sp-1, a mediator of smooth muscle gene expression [89]. Therefore, given the altered expression of multiple miRNAs in diabetes, it can be postulated that miRNAs represents potential targets for the modulation of diabetic atherosclerosis as they can interfere with endothelial dysfunction, inflammatory processes, and vessel remodeling.

5 DNA Methylation in Diabetic Cardiovascular Complications

The DNA methylation process has not been very well explored in the context of diabetic cardiovascular complications; however, in addition to genome-wide studies that have attempted to characterize the DNA methylation signatures in diabetes and to establish a correlation with the risk of diabetic nephropathy in patients with type 1 diabetes [17], few studies have also associated DNA hypomethylation with atherosclerosis in human atherosclerotic smooth muscle cells as well as in obese animal models such as high-fat-fed rabbits and ApoE null mice [90–92]. Normal chow-fed mice ApoE null mice were also shown to exhibit altered DNA methylation patterns in leukocytes before the onset of atherosclerosis [93]. Furthermore, global DNA methylation in peripheral blood leukocytes from patients with chronic kidney disease was associated with inflammation and increased mortality, suggesting the involvement of DNA methylation state in the exacerbation of kidney disease and related vascular disorders [94]. An alteration in the DNA methylation of several vascular homeostasis-relevant genes such as c-fos, p53, matrix metalloproteinase, endothelial nitric oxide synthase, hypoxia-induced factor 1, and growth factors has been reported in VSMCs and endothelial cells isolated from animal models, suggesting a role of these modifications in the development of vascular dysfunction [25, 95]. Although a direct role of DNA methylation in diabetes and its complications remains unclear, studies using animal models have implicated increased methylation at key gene promoters in islet dysfunction and diabetes [96, 97]. However, no differences in DNA methylation have been noticed at the promoters of high glucose-induced genes in streptozotocin-induced type 1 diabetic rats [18, 98], suggesting further studies are required to better enlighten the role of this epigenetic process in diabetes-associated disorders.

6 Conclusions

Epigenetic modulation of gene expression focused on the effect of DNA methylation and histone modifications has been studied for a long time in the context of cancer. However, it is only recently that there has been a surge of interest in investigating the involvement of epigenetic pathways in the pathogenesis of vascular dysfunction in diabetes. As a result, evidence has accumulated to show that histone acetylation, methylation, and miRNA-induced processes modulate the expression and function of the genes that are linked to ROS generation, inflammation, cell-cycle regulation, migration, proliferation, and other proatherogenic events (Fig. 1). Nevertheless,

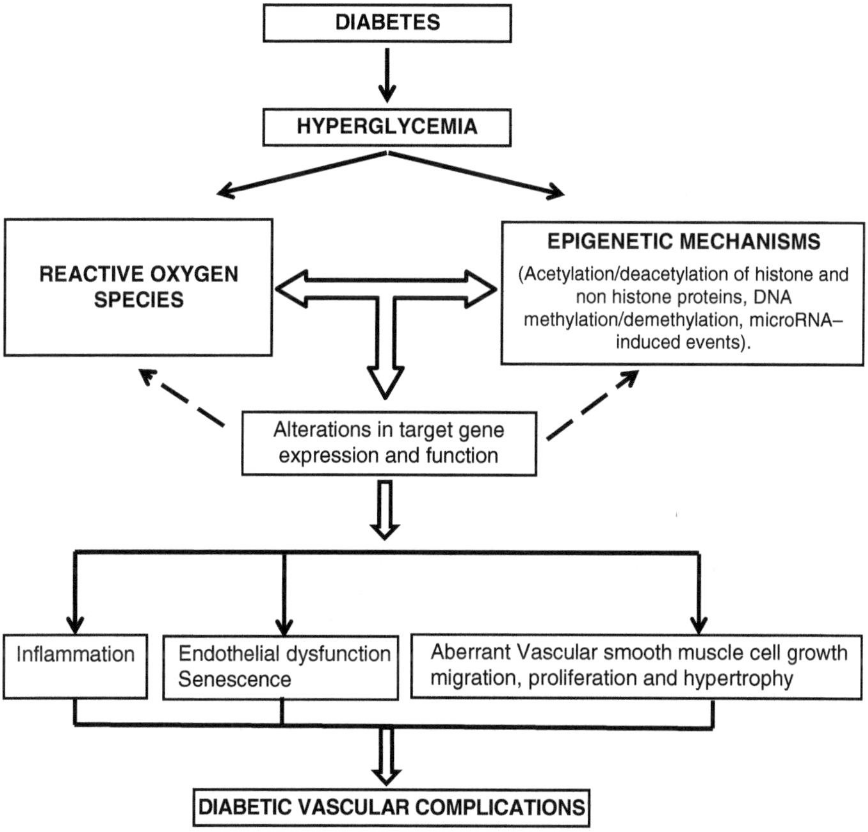

Fig. 1 *Schematic model depicting the interplay between hyperglycemia, reactive oxygen species, and epigenetic events in diabetes-associated vascular complications*. Diabetes/hyperglycemia promotes redox signaling through an increase in the generation of reactive oxygen species (*ROS*). ROS activity can either be modulated by epigenetic events that result from hyperglycemia or exacerbate them, leading to alterations in the expression and function of the genes that are linked to ROS generation, inflammation, cell-cycle regulation, migration, proliferation, and other proatherogenic events. The consequences of these events are activation of the proinflammatory cascades in the vessel wall as well as the impairment of endothelial and vascular smooth muscle physiology, both contributing to aberrant vascular functions

more in-depth studies on the epigenetic modifications in the vascular system of various experimental models of diabetes and in humans are needed to precisely establish a role of these modifications in the pathogenesis of vascular dysfunction associated with diabetes.

Acknowledgments This work was supported by funding from the Canadian Institutes of Health Research (CIHR) operating grant number 67037 to A.K.S. E.R.S.C. is a recipient of a studentship from the Faculty of Graduate and Postdoctoral Studies of the University of Montreal.

References

1. Grundy SM, Benjamin IJ, Burke GL et al (1999) Diabetes and cardiovascular disease: a statement for healthcare professionals from the American Heart Association. Circulation 100:1134–1146
2. Niedowicz DM, Daleke DL (2005) The role of oxidative stress in diabetic complications. Cell Biochem Biophys 43:289–330
3. Chung SS, Ho EC, Lam KS, Chung SK (2003) Contribution of polyol pathway to diabetes-induced oxidative stress. J Am Soc Nephrol 14:S233–S236
4. Cantero AV, Portero-Otin M, Ayala V et al (2007) Methylglyoxal induces advanced glycation end product (AGEs) formation and dysfunction of PDGF receptor-beta: implications for diabetic atherosclerosis. FASEB J 21:3096–3106
5. Giacco F, Brownlee M (2010) Oxidative stress and diabetic complications. Circ Res 107: 1058–1070
6. Inoguchi T, Li P, Umeda F et al (2000) High glucose level and free fatty acid stimulate reactive oxygen species production through protein kinase C-dependent activation of NAD(P)H oxidase in cultured vascular cells. Diabetes 49:1939–1945
7. Inoguchi T, Sonta T, Tsubouchi H et al (2003) Protein kinase C-dependent increase in reactive oxygen species (ROS) production in vascular tissues of diabetes: role of vascular NAD(P)H oxidase. J Am Soc Nephrol 14:S227–S232
8. Singh U, Jialal I (2006) Oxidative stress and atherosclerosis. Pathophysiology 13:129–142
9. Srivastava A, Anand-Srivastava M (2008) Role of hyperglycemia and redox-induced signaling in vascular complications of diabetes. In: Srivastava A, Anand-Srivastava M (eds) Signal transduction in the cardiovascular system in health and disease, 3rd edn. Springer, New York, pp 177–192
10. Forbes JM, Cooper ME (2013) Mechanisms of diabetic complications. Physiol Rev 93: 137–188
11. Srivastava AK (2002) High glucose-induced activation of protein kinase signaling pathways in vascular smooth muscle cells: a potential role in the pathogenesis of vascular dysfunction in diabetes. Int J Mol Med 9:85–89
12. Muhonen P, Holthofer H (2009) Epigenetic and microRNA-mediated regulation in diabetes. Nephrol Dial Transplant 24:1088–1096
13. Villeneuve LM, Reddy MA, Natarajan R (2011) Epigenetics: deciphering its role in diabetes and its chronic complications. Clin Exp Pharmacol Physiol 38:451–459
14. Korkmaz A, Manchester LC, Topal T, Ma S, Tan DX, Reiter RJ (2011) Epigenetic mechanisms in human physiology and diseases. J Exp Integr Med 1:139–147
15. Yla-Herttuala S, Glass CK (2011) Review focus on epigenetics and the histone code in vascular biology. Cardiovasc Res 90:402–403
16. Li E, Bestor TH, Jaenisch R (1992) Targeted mutation of the DNA methyltransferase gene results in embryonic lethality. Cell 69:915–926

17. Bell CG, Teschendorff AE, Rakyan VK, Maxwell AP, Beck S, Savage DA (2010) Genome-wide DNA methylation analysis for diabetic nephropathy in type 1 diabetes mellitus. BMC Med Genomics 3:33
18. Brennan EP, Ehrich M, O'Donovan H et al (2010) DNA methylation profiling in cell models of diabetic nephropathy. Epigenetics 5:396–401
19. Bartel DP, Chen CZ (2004) Micromanagers of gene expression: the potentially widespread influence of metazoan microRNAs. Nat Rev Genet 5:396–400
20. Berezikov E, Guryev V, van de Belt J, Wienholds E, Plasterk RH, Cuppen E (2005) Phylogenetic shadowing and computational identification of human microRNA genes. Cell 120:21–24
21. Rodriguez A, Griffiths-Jones S, Ashurst JL, Bradley A (2004) Identification of mammalian microRNA host genes and transcription units. Genome Res 14:1902–1910
22. Bartel DP (2009) MicroRNAs: target recognition and regulatory functions. Cell 136:215–233
23. van Wolfswinkel JC, Ketting RF (2010) The role of small non-coding RNAs in genome stability and chromatin organization. J Cell Sci 123:1825–1839
24. Natarajan R, Putta S, Kato M (2012) MicroRNAs and diabetic complications. J Cardiovasc Transl Res 5:413–422
25. Matouk CC, Marsden PA (2008) Epigenetic regulation of vascular endothelial gene expression. Circ Res 102:873–887
26. de Kreutzenberg SV, Ceolotto G, Papparella I et al (2010) Downregulation of the longevity-associated protein sirtuin 1 in insulin resistance and metabolic syndrome: potential biochemical mechanisms. Diabetes 59:1006–1015
27. Mortuza R, Chen S, Feng B, Sen S, Chakrabarti S (2013) High glucose induced alteration of SIRTs in endothelial cells causes rapid aging in a p300 and FOXO regulated pathway. PLoS One 8:e54514
28. Daitoku H, Sakamaki J, Fukamizu A (2011) Regulation of FoxO transcription factors by acetylation and protein–protein interactions. Biochim Biophys Acta 1813:1954–1960
29. Malik AI, Storey KB (2011) Transcriptional regulation of antioxidant enzymes by FoxO1 under dehydration stress. Gene 485:114–119
30. Han L, Zhou R, Niu J, McNutt MA, Wang P, Tong T (2010) SIRT1 is regulated by a PPAR{gamma}-SIRT1 negative feedback loop associated with senescence. Nucleic Acids Res 38:7458–7471
31. Mousa SA, Gallati C, Simone T et al (2009) Dual targeting of the antagonistic pathways mediated by Sirt1 and TXNIP as a putative approach to enhance the efficacy of anti-aging interventions. Aging (Albany NY) 1:412–424
32. Shalev A, Pise-Masison CA, Radonovich M et al (2002) Oligonucleotide microarray analysis of intact human pancreatic islets: identification of glucose-responsive genes and a highly regulated TGF-beta signaling pathway. Endocrinology 143:3695–3698
33. Schulze PC, Yoshioka J, Takahashi T, He Z, King GL, Lee RT (2004) Hyperglycemia promotes oxidative stress through inhibition of thioredoxin function by thioredoxin-interacting protein. J Biol Chem 279:30369–30374
34. Turturro F, Friday E, Welbourne T (2007) Hyperglycemia regulates thioredoxin-ROS activity through induction of thioredoxin-interacting protein (TXNIP) in metastatic breast cancer-derived cells MDA-MB-231. BMC Cancer 7:96
35. Cha-Molstad H, Saxena G, Chen J, Shalev A (2009) Glucose-stimulated expression of Txnip is mediated by carbohydrate response element-binding protein, p300, and histone H4 acetylation in pancreatic beta cells. J Biol Chem 284:16898–16905
36. Perrone L, Devi TS, Hosoya K, Terasaki T, Singh LP (2009) Thioredoxin interacting protein (TXNIP) induces inflammation through chromatin modification in retinal capillary endothelial cells under diabetic conditions. J Cell Physiol 221:262–272
37. Chen S, Feng B, George B, Chakrabarti R, Chen M, Chakrabarti S (2010) Transcriptional coactivator p300 regulates glucose-induced gene expression in endothelial cells. Am J Physiol Endocrinol Metab 298:E127–E137
38. Kaur H, Chen S, Xin X, Chiu J, Khan ZA, Chakrabarti S (2006) Diabetes-induced extracellular matrix protein expression is mediated by transcription coactivator p300. Diabetes 55:3104–3111

39. Vanden Berghe W, De BK, Boone E, Plaisance S, Haegeman G (1999) The nuclear factor-kappaB engages CBP/p300 and histone acetyltransferase activity for transcriptional activation of the interleukin-6 gene promoter. J Biol Chem 274:32091–32098
40. Shanmugam N, Reddy MA, Guha M, Natarajan R (2003) High glucose-induced expression of proinflammatory cytokine and chemokine genes in monocytic cells. Diabetes 52:1256–1264
41. Guha M, Bai W, Nadler JL, Natarajan R (2000) Molecular mechanisms of tumor necrosis factor alpha gene expression in monocytic cells via hyperglycemia-induced oxidant stress-dependent and -independent pathways. J Biol Chem 275:17728–17739
42. Yun JM, Jialal I, Devaraj S (2011) Epigenetic regulation of high glucose-induced proinflammatory cytokine production in monocytes by curcumin. J Nutr Biochem 22:450–458
43. Miao F, Gonzalo IG, Lanting L, Natarajan R (2004) In vivo chromatin remodeling events leading to inflammatory gene transcription under diabetic conditions. J Biol Chem 279:18091–18097
44. Ito K, Hanazawa T, Tomita K, Barnes PJ, Adcock IM (2004) Oxidative stress reduces histone deacetylase 2 activity and enhances IL-8 gene expression: role of tyrosine nitration. Biochem Biophys Res Commun 315:240–245
45. Choi JH, Nam KH, Kim J et al (2005) Trichostatin A exacerbates atherosclerosis in low density lipoprotein receptor-deficient mice. Arterioscler Thromb Vasc Biol 25:2404–2409
46. Kong X, Fang M, Li P, Fang F, Xu Y (2009) HDAC2 deacetylates class II transactivator and suppresses its activity in macrophages and smooth muscle cells. J Mol Cell Cardiol 46:292–299
47. Fernandez AZ, Siebel AL, El-Osta A (2010) Atherogenic factors and their epigenetic relationships. Int J Vasc Med 2010:437809
48. Rahman I, Biswas SK, Kirkham PA (2006) Regulation of inflammation and redox signaling by dietary polyphenols. Biochem Pharmacol 72:1439–1452
49. Wolfram S (2007) Effects of green tea and EGCG on cardiovascular and metabolic health. J Am Coll Nutr 26:373S–388S
50. Khajehdehi P, Pakfetrat M, Javidnia K et al (2011) Oral supplementation of turmeric attenuates proteinuria, transforming growth factor-beta and interleukin-8 levels in patients with overt type 2 diabetic nephropathy: a randomized, double-blind and placebo-controlled study. Scand J Urol Nephrol 45:365–370
51. Usharani P, Mateen AA, Naidu MU, Raju YS, Chandra N (2008) Effect of NCB-02, atorvastatin and placebo on endothelial function, oxidative stress and inflammatory markers in patients with type 2 diabetes mellitus: a randomized, parallel-group, placebo-controlled, 8-week study. Drugs R D 9:243–250
52. Wongcharoen W, Phrommintikul A (2009) The protective role of curcumin in cardiovascular diseases. Int J Cardiol 133:145–151
53. Kapakos G, Youreva V, Srivastava AK (2012) Cardiovascular protection by curcumin: molecular aspects. Indian J Biochem Biophys 49:306–315
54. Marcu MG, Jung YJ, Lee S et al (2006) Curcumin is an inhibitor of p300 histone acetylatransferase. Med Chem 2:169–174
55. Kiernan R, Bres V, Ng RW et al (2003) Post-activation turn-off of NF-kappa B-dependent transcription is regulated by acetylation of p65. J Biol Chem 278:2758–2766
56. Pae HO, Jeong GS, Jeong SO et al (2007) Roles of heme oxygenase-1 in curcumin-induced growth inhibition in rat smooth muscle cells. Exp Mol Med 39:267–277
57. Wang B, Zhu X, Kim Y et al (2012) Histone deacetylase inhibition activates transcription factor Nrf2 and protects against cerebral ischemic damage. Free Radic Biol Med 52:928–936
58. Brooks CL, Gu W (2009) How does SIRT1 affect metabolism, senescence and cancer? Nat Rev Cancer 9:123–128
59. Borra MT, Smith BC, Denu JM (2005) Mechanism of human SIRT1 activation by resveratrol. J Biol Chem 280:17187–17195
60. Petrovski G, Gurusamy N, Das DK (2011) Resveratrol in cardiovascular health and disease. Ann N Y Acad Sci 1215:22–33
61. Das M, Das DK (2010) Resveratrol and cardiovascular health. Mol Aspects Med 31:503–512
62. Markus MA, Morris BJ (2008) Resveratrol in prevention and treatment of common clinical conditions of aging. Clin Interv Aging 3:331–339

63. Sharma S, Anjaneyulu M, Kulkarni SK, Chopra K (2006) Resveratrol, a polyphenolic phytoalexin, attenuates diabetic nephropathy in rat. Pharmacology 76:69–75
64. Zang M, Xu S, Maitland-Toolan KA et al (2006) Polyphenols stimulate AMP-activated protein kinase, lower lipids, and inhibit accelerated atherosclerosis in diabetic LDL receptor-deficient mice. Diabetes 55:2180–2191
65. Akar F, Pektas MB, Tufan C et al (2011) Resveratrol shows vasoprotective effect reducing oxidative stress without affecting metabolic disturbances in insulin-dependent diabetes of rabbits. Cardiovasc Drugs Ther 25:119–131
66. Ungvari Z, Labinskyy N, Mukhopadhyay P et al (2009) Resveratrol attenuates mitochondrial oxidative stress in coronary arterial endothelial cells. Am J Physiol Heart Circ Physiol 297: H1876–H1881
67. Tian C, Zhang R, Ye X et al (2013) Resveratrol ameliorates high-glucose-induced hyperpermeability mediated by caveolae via VEGF/KDR pathway. Genes Nutr 8:231–239
68. Lim JH, Lee YM, Chun YS, Chen J, Kim JE, Park JW (2010) Sirtuin 1 modulates cellular responses to hypoxia by deacetylating hypoxia-inducible factor 1alpha. Mol Cell 38:864–878
69. Gallagher IJ, Scheele C, Keller P et al (2010) Integration of microRNA changes in vivo identifies novel molecular features of muscle insulin resistance in type 2 diabetes. Genome Med 2:9
70. Lovis P, Roggli E, Laybutt DR et al (2008) Alterations in microRNA expression contribute to fatty acid-induced pancreatic beta-cell dysfunction. Diabetes 57:2728–2736
71. Gonzalez-Chavez A, Elizondo-Argueta S, Gutierrez-Reyes G, Leon-Pedroza JI (2011) Pathophysiological implications between chronic inflammation and the development of diabetes and obesity. Cir Cir 79:209–216
72. Caporali A, Meloni M, Vollenkle C et al (2011) Deregulation of microRNA-503 contributes to diabetes mellitus-induced impairment of endothelial function and reparative angiogenesis after limb ischemia. Circulation 123:282–291
73. Li Y, Song YH, Li F, Yang T, Lu YW, Geng YJ (2009) MicroRNA-221 regulates high glucose-induced endothelial dysfunction. Biochem Biophys Res Commun 381:81–83
74. Meng S, Cao JT, Zhang B, Zhou Q, Shen CX, Wang CQ (2012) Downregulation of microRNA-126 in endothelial progenitor cells from diabetes patients, impairs their functional properties, via target gene Spred-1. J Mol Cell Cardiol 53:64–72
75. Wang S, Aurora AB, Johnson BA et al (2008) The endothelial-specific microRNA miR-126 governs vascular integrity and angiogenesis. Dev Cell 15:261–271
76. Bucciarelli LG, Wendt T, Qu W et al (2002) RAGE blockade stabilizes established atherosclerosis in diabetic apolipoprotein E-null mice. Circulation 106:2827–2835
77. Shanmugam N, Reddy MA, Natarajan R (2008) Distinct roles of heterogeneous nuclear ribonuclear protein K and microRNA-16 in cyclooxygenase-2 RNA stability induced by S100b, a ligand of the receptor for advanced glycation end products. J Biol Chem 283:36221–36233
78. Zhu N, Zhang D, Chen S et al (2011) Endothelial enriched microRNAs regulate angiotensin II-induced endothelial inflammation and migration. Atherosclerosis 215:286–293
79. Reddy MA, Jin W, Villeneuve L et al (2012) Pro-inflammatory role of microrna-200 in vascular smooth muscle cells from diabetic mice. Arterioscler Thromb Vasc Biol 32:721–729
80. Villeneuve LM, Kato M, Reddy MA, Wang M, Lanting L, Natarajan R (2010) Enhanced levels of microRNA-125b in vascular smooth muscle cells of diabetic db/db mice lead to increased inflammatory gene expression by targeting the histone methyltransferase Suv39h1. Diabetes 59:2904–2915
81. Quintavalle M, Condorelli G, Elia L (2011) Arterial remodeling and atherosclerosis: miRNAs involvement. Vascul Pharmacol 55:106–110
82. Rangrez AY, Massy ZA, Metzinger-Le Meuth V, Metzinger L (2011) miR-143 and miR-145: molecular keys to switch the phenotype of vascular smooth muscle cells. Circ Cardiovasc Genet 4:197–205
83. Zhang Y, Wang Y, Wang X et al (2011) Insulin promotes vascular smooth muscle cell proliferation via microRNA-208-mediated downregulation of p21. J Hypertens 29:1560–1568
84. Norata GD, Pinna C, Zappella F et al (2012) MicroRNA 143–145 deficiency impairs vascular function. Int J Immunopathol Pharmacol 25:467–474

85. Boettger T, Beetz N, Kostin S et al (2009) Acquisition of the contractile phenotype by murine arterial smooth muscle cells depends on the Mir143/145 gene cluster. J Clin Invest 119: 2634–2647
86. Cordes KR, Sheehy NT, White MP et al (2009) miR-145 and miR-143 regulate smooth muscle cell fate and plasticity. Nature 460:705–710
87. Elia L, Quintavalle M, Zhang J et al (2009) The knockout of miR-143 and -145 alters smooth muscle cell maintenance and vascular homeostasis in mice: correlates with human disease. Cell Death Differ 16:1590–1598
88. Lovren F, Pan Y, Quan A et al (2012) MicroRNA-145 targeted therapy reduces atherosclerosis. Circulation 126:S81–S90
89. Torella D, Iaconetti C, Catalucci D et al (2011) MicroRNA-133 controls vascular smooth muscle cell phenotypic switch in vitro and vascular remodeling in vivo. Circ Res 109: 880–893
90. Laukkanen MO, Mannermaa S, Hiltunen MO et al (1999) Local hypomethylation in atherosclerosis found in rabbit ec-sod gene. Arterioscler Thromb Vasc Biol 19:2171–2178
91. Hiltunen MO, Yla-Herttuala S (2003) DNA methylation, smooth muscle cells, and atherogenesis. Arterioscler Thromb Vasc Biol 23:1750–1753
92. Hiltunen MO, Turunen MP, Hakkinen TP et al (2002) DNA hypomethylation and methyltransferase expression in atherosclerotic lesions. Vasc Med 7:5–11
93. Lund G, Andersson L, Lauria M et al (2004) DNA methylation polymorphisms precede any histological sign of atherosclerosis in mice lacking apolipoprotein E. J Biol Chem 279: 29147–29154
94. Stenvinkel P, Karimi M, Johansson S et al (2007) Impact of inflammation on epigenetic DNA methylation—a novel risk factor for cardiovascular disease? J Intern Med 261:488–499
95. Turunen MP, Aavik E, Yla-Herttuala S (2009) Epigenetics and atherosclerosis. Biochim Biophys Acta 1790:886–891
96. Park JH, Stoffers DA, Nicholls RD, Simmons RA (2008) Development of type 2 diabetes following intrauterine growth retardation in rats is associated with progressive epigenetic silencing of Pdx1. J Clin Invest 118:2316–2324
97. Ling C, Del GS, Lupi R et al (2008) Epigenetic regulation of PPARGC1A in human type 2 diabetic islets and effect on insulin secretion. Diabetologia 51:615–622
98. Williams KT, Garrow TA, Schalinske KL (2008) Type I diabetes leads to tissue-specific DNA hypomethylation in male rats. J Nutr 138:2064–2069

A Critical Balance Between Oxidative Stress and Antioxidant Defense in Cardiovascular System Under Hyperglycemia: A Summary of Experimental Studies

Murat Ayaz and Belma Turan

Abstract Diabetes mellitus is a disorder resulting from a lost in control of blood glucose level by insufficient insulin release (type 1), impaired insulin function, or insulin resistance (type 2). The main etiology for mortality and a great percent of the morbidity in patients with diabetes is cardiovascular disease. In addition to hyperglycemia, enhanced oxidative stress plays a major role in the pathogenesis of diabetes. Although reactive oxygen species (ROS) are known to be mediators of intracellular signaling pathways under physiological conditions, excessive production of ROS can be detrimental to the cells as a result of increased oxidative stress and thereby cellular dysfunction. Hence, well-tuned, balanced, and responsive antioxidant systems are vital for proper regulation of the redox status of the cells. Studies have reported valuable effects of antioxidant agents, including trace elements, on diabetes-induced cardiovascular system dysfunctions, either directly or indirectly. Thus, several approaches have been carried out to either diminish an elevated ROS production or improve the endogenous levels of antioxidants. Indeed, reduced fatty acid oxidation and use of trace elements in treatment strategies result in promising prevention hints for diabetes-induced cardiovascular dysfunctions. Our scope here is to review the important role of antioxidants, particularly selenium, as cardioprotective agents in several types of disease states including diabetes, presenting our research results on cardiac function by using experimental animal models for diabetes. Although the paradigm that inhibiting overproduction of superoxides and peroxides would prevent cardiac dysfunction diabetes-induced damage has been difficult to verify using conventional antioxidants such as sele-

M. Ayaz (✉)
Department of Biophysics, Faculty of Medicine, Selcuk University,
Alaaddin Keykubat Kampusu, Konya 42075, Turkey
e-mail: ayaz72@yahoo.com

B. Turan
Department of Biophysics, Faculty of Medicine, University of Ankara,
Ankara 06100, Turkey

B. Turan and N.S. Dhalla (eds.), *Diabetic Cardiomyopathy: Biochemical and Molecular Mechanisms*, Advances in Biochemistry in Health and Disease 9,
DOI 10.1007/978-1-4614-9317-4_7, © Springer Science+Business Media New York 2014

nium, of special note is that its role as hyperglycemia controller, insulin sensitizer, or antioxidant therapy remains to be further explored as well as the effect of hypolipidemic therapy.

Keywords Diabetes • Oxidative stress • Antioxidants • Action potential • Ion channels • Contraction • Selenium

1 Introduction

The word *diabetes* is derived from Latin *diabētēs*, which in turn comes from Ancient Greek *diabētēs* as a combination of dia (=through) and bainein (=to go). Diabetes in that language literally means "a passer through" with the intended meaning "excessive discharge of urine" as the name for the disease. Diabetes mellitus became known as a disease of pancreatic insufficiency or failure when scientists (Minkowski in the 1880s and later Banting and Best in the 1920s) modelled this condition in dogs by removing a part of or the entire pancreas.

Diabetes Mellitus or shortly diabetes is a condition that occurs when the body cannot use glucose (a type of sugar) normally. Glucose is the main source of energy for the body cells. The levels of glucose in the blood are controlled by a hormone called insulin, which is made by the pancreas. Insulin helps glucose enter the cells. In diabetes, the pancreas does not make enough insulin or the body does not respond normally to the insulin that is made; this causes glucose levels in the blood to rise, leading to symptoms such as increased urination, extreme thirst, and unexplained weight loss. Hyperglycemia, or elevated blood sugar, is a common effect of uncontrolled diabetes and over time leads to serious damage to many systems of the body, especially the cardiovascular system.

The classification of diabetes and the tests used for its diagnosis were brought into order by the National Diabetes Data Group of the United States and the second World Health Organization Expert Committee *Diabetes Mellitus* on in 1979 and 1980. Apart from minor modifications by WHO in 1985, little has been changed since that time. There is, however, considerable new knowledge regarding the etiology of different forms of diabetes as well as more information on the predictive value of different blood glucose values for the complications of diabetes [1].

This group of metabolic disorders is characterized by hyperglycemia resulting from defects in insulin secretion or insulin action. Besides hyperglycemia, several other factors such as hyperlipidemia and enhanced oxidative stress play a major role in pathogenesis of diabetes. The disease is progressive and is linked with a high risk of complications [1]. Diabetes results from one of two mechanisms: (1) inadequate production of insulin (which is made mainly by the pancreas) and (2) inadequate sensitivity of cells to the action of insulin. Two main types of diabetes correspond to these two mechanisms, called insulin-dependent (type 1) and non-insulin-dependent (type 2) diabetes. In type 1 diabetes, there is either no insulin or not enough insulin for the cells. In type 2 diabetes, there is generally enough insulin but the cells upon which it should act are not normally sensitive to its action. The signs and symptoms of both types of diabetes include, generally, increased urine output and decreased appetite, as well as general fatigue in the entire body of the subjects.

It is well accepted that hyperglycemia increases the production of reactive oxygen species (ROS), alters the cellular redox status, and causes rapid changes in membrane function, followed by contractile dysfunction within weeks in the diabetic heart [2–6]. Significant increases in oxidants trigger a cascade of pathological events, including contractile dysfunction [5, 6]. Oxidative stress, being an imbalance between endogenous ROS and antioxidant systems in favor of the former, is involved in the etiology of diabetes-induced downregulation of heart function. Moreover, there is a close relationship between impaired insulin signaling and alteration in the heart function via the depressed endogenous antioxidant defense mechanism [5–7].

2 Signs and Symptoms of Diabetes in Humans

Both type 1 and type 2 diabetes are multifactorial diseases in which a very complex genetic background interacts with environmental factors contributing to the disease development [8]. The symptomatology of diabetes is more readily recognizable in children than in adults. Diabetes is a great imitator: influenza, gastroenteritis, and appendicitis are the conditions more often diagnosed, only to find that the disease is certainly diabetes. The sequence of chemical events can be summarized as polyuria, polydipsia, polyphagia, progressive cachexia, glucosuria, and increased specific gravity of urine. These events result in hyperglycemia and acidosis, which in turn produce the three "polys" of diabetes—polyphagia, polydipsia, and polyuria—the cardinal symptoms of the disease. In non-insulin-dependent diabetes, the insulin values are found to be increased, and there is often tiredness and frequent infections in the subjects [9].

The epidemic proportion of people with diabetes is alarming, and it has been estimated that by the year 2025, 300 million people will become affected by the disease [10]. Among the vast array of secondary problems associated with diabetes, cardiovascular complications significantly contribute to increasing rates of morbidity and mortality [10, 11]. Nearly 80% of the deaths associated with diabetes are caused by cardiac complications [12, 13]. Although previous studies have focused on coronary artery disease and autonomic neuropathy as the primary cardiac complication, during the past 40 years, diabetic cardiomyopathy has also been identified as a significant entity [14, 15].

Cardiac dysfunction in diabetic patients, named diabetic cardiomyopathy, was introduced first by Rubler et al. in 1972 [14]. Following the topic orientation of many researchers, a typical definition includes structural and functional abnormalities in the myocardium of diabetic patients without coronary artery disease or hypertension [16]. In addition, it is known that diabetic cardiomyopathy can also be developed in diabetic patients parallel to coronary artery disease and/or hypertension in the same subjects. Indeed, diabetes is also accompanied by an extensive increase in atherosclerotic pathology of the large vessels, including cardiac, cerebral, and peripheral vascular systems, named cardiovascular diseases.

Diabetes has serious and compelling effects on the human cardiovascular system by hyperglycemia-induced damage in a number of organs/tissues/cells via a number of pathways. Indeed, these processes do not develop independently; each process can accelerate or worsen the states of the others. As mentioned previously, a diabetes-specific cardiac dysfunction or diabetic cardiomyopathy leads generally to inability of the heart to circulate blood efficiently, which is defined as heart failure. Later, diabetic cardiomyopathy can culminate in both pulmonary and peripheral edema. For the majority of diabetic patients, the earliest sign of cardiac dysfunction is a mild left ventricular diastolic dysfunction with little effect on ventricular filling. They may also show subtle signs of diabetic cardiomyopathy related to decrease in left ventricular compliance, left ventricular hypertrophy, or a combination of both syndromes. Following the development of systolic dysfunction, left ventricular dilation and symptomatic heart failure generally develop in diabetic patients [14, 17]. The 60% of patients with diabetic cardiomyopathy are accompanied by a variety of electrocardiographic changes with or without structural damages in the heart tissue. Later in the sequence, a prolonged QT interval may indicate fibrosis. Although the definition of diabetic cardiomyopathy excludes concomitant atherosclerosis or hypertension, there are no changes in perfusion or in atrial natriuretic peptide levels until the very late stages of the disease [17] when hypertrophy and fibrosis become very pronounced.

Functional assessment of diabetic cardiomyopathy is determined by ventricular dilation, cardiomyocyte hypertrophy, prominent interstitial fibrosis, and decreased or preserved systolic function in the presence of a diastolic dysfunction. Even though these complications are secondary outcomes for diabetes, they are also related mainly to hyperglycemia as well as several other factors implicated in the pathogenesis of this disease. Indeed, the etiology of diabetic cardiomyopathy may include (a) microangiopathy and related endothelial dysfunction, (b) autonomic neuropathy, (c) metabolic alterations that consist of abnormal glucose use and increased fatty acid oxidation, (d) generation and accumulation of free radicals, and (e) alterations in ion homeostasis, particularly intracellular Ca^{2+} regulation.

Among many other suggestions, it is hypothesized that extracellular hyperglycemia leads to intracellular hyperglycemia in endothelial cells unable to regulate their glucose uptake. Furthermore, in contrast to myocytes, endothelial cells do not have mechanisms allowing them to internalize their glucose transporter. Together with this the enhanced intracellular glucose concentration is nearly 4-fold, all resulting from increasing concentration of glycolytic intermediates upstream of the rate-limiting glyceraldehyde-3-phosphate reaction, which is inhibited by mechanisms activated by increased free radical formation, as a common procedure in diabetes [18]. Another research in this field focused on lipid oxidation in the pathogenesis of diabetic cardiomyopathy, resulting in increased oxidized LDL cholesterol in patients, would seem to provide a rationale for antioxidant therapy [19]. Both observational and epidemiologic studies have suggested that persons with high plasma levels of antioxidants have lower rates of coronary heart disease, and an inverse relationship has been found between β-carotene and cardiovascular mortality as well [20].

3 Animal Models for Study of Diabetes Mellitus

Animal models have enormously contributed to the study of outcomes of this disease as well as giving researchers the opportunity to control in vivo the genetic and environmental factors that may influence the development of the disease and establishment of its complications and thus gain new information about its handling and treatment in humans. Although there is much debate about the true value of using animal models in the study of diabetes [21] and the ability of implementing an animal-derived therapeutic protocol in clinical use [22], it must be admitted that experimental models are essential tools for understanding the molecular basis, pathogenesis of complications, and utility of therapeutic agents in a multifactorial disease such as diabetes mellitus [23].

Animal models can mimic diabetes by either spontaneously or by using chemical, surgical, genetic, or other techniques, and depict many clinical features or related phenotypes of the disease. There are many ways to classify animal models of diabetes. Based on methodology, there are three types of experimental models, that is, chemical, surgical, and genetic (immunological) modifications [24–26]. Early studies used pancreatectomized dogs to confirm the central role of the pancreas in glucose homeostasis, culminating in the discovery and purification of insulin. Today, animal experimentation is conscientious and subject to legal and ethical restrictions that vary throughout the world. Most experiments are carried out on rodents, although some studies are still performed on larger animals. Several toxins, including streptozotocin (STZ) and alloxan (ALL), induce hyperglycemia in rats and mice. Selective inbreeding has produced several strains of animals that are considered reasonable models of type 1 diabetes, type 2 diabetes, and related phenotypes such as obesity and insulin resistance. Apart from their use in studying the pathogenesis of the disease and its complications, all new treatments for diabetes, including islet cell transplantation and preventative strategies, are initially investigated in animals. In recent years, molecular biological techniques have produced a large number of new animal models for the study of diabetes, including knock-in, generalized knockout, and tissue-specific knockout mice. The surgical method is based on the partial or complete removal of the pancreas (pancreatectomy). Genetic models of diabetes include either spontaneously developed diabetic rats or genetically engineered mice.

Among the animal models, a chemical-induced model appears to be the most popularly used procedure because less technical expertise (compared to surgical method) is required, with a low less infection risk, more animals survive, and application of the chemicals to animals can be friendly. For an example, following a single dose of STZ administration, experimental animals (rats or mice) should be housed for 5–7 days for full induction of diabetes. On the other hand, once a model is established, it can be used for studies of diabetes-induced complications in different organs, tissues, or cells, including cardiovascular diseases in the same animals.

As a summary, rodents are commonly used models for testing the new pharmacologically active substances, not only in the context of transplantation (immunosuppressive), but also with regard to therapy or prevention of human diseases. Rats and mice are commonly used in safety and effectiveness testing of new orally active compounds. Moreover, for more than 50 years, investigators have unsuccessfully tried to recreate the cardiovascular complications of diabetes in experimental animals. Particularly, accelerated atherosclerosis and dilated cardiomyopathy, the major causes of mortality in patients with diabetes, have been conspicuously absent in many murine models of the disease. Under the auspices of the NIH, the Animal Models of Diabetic Complications Consortium has worked to address this issue [27]. Finally, for maximum benefits from these animal models in the study of cardiomyopathy, atherosclerosis, and other diabetic complications, a system is needed for sharing both animal models and accumulated phenotypic data with the greater scientific community.

4 Relationship Between Oxidative Stress/Antioxidant Defense and Hyperglycemia

Oxidative stress contributes to the development of a wide range of diseases including cardiovascular diseases, the pathologies caused by diabetes [28–32]. Oxidative stress is caused generally by an inequality between the production of reactive oxygen species and the biological system's ability to readily detoxify the reactive intermediates or easily repair the resulting damage. There are many sources for generation of the reactive oxygen species (ROS) within the cardiomyocyte, such as mitochondrial electron transport chain-derived ROS, NADPH oxidase-derived ROS, uncoupled nitric oxide (NO) synthases, NOS, xanthine oxidase induced ROS, and enzyme systems of mitochondria in addition to electron transport systems such as 2-oxoglutarate, pyruvate dehydrogenase, and flavoprotein acyl-CoA dehydrogenase [33, 34]. Apart from the increased ROS production, decreased system ability can also be a cause of increased oxidative stress in cardiomyocytes. A sophisticated enzymatic and nonenzymatic antioxidant defense system including catalase (CAT), superoxide dismutase (SOD), and reduced glutathione (GSH) counteracts and regulates overall ROS levels to maintain physiological homeostasis in the cells.

Considering the high energy needs of cardiomyocytes, it is generally accepted that the mitochondria are the main source of ROS during homeostasis. In fact, mitochondria have many biochemical properties that favor the organelles as permanent producers of ROS. This source of ROS is thought to be generated mainly at complexes I and III of the electron transport chain (ETC), mitochondrial one-electron carriers, through "leakage" during respiration [34–36]. The important point here is that mitochondrial ROS levels are influenced not only by the ROS generation rate but also by ROS-scavenging systems. In this regard, there is close linkage and cross-talk between the redox couples involved in substrate oxidation and the ETC (i.e., $NADH/NAD^{+}$) and those involved in antioxidant defense through NADPH-regenerating reactions that maintain reduced pools of glutathione, glutaredoxin, and thioredoxin, that is, NADPH/NADP$^+$ [37]. Generation of NADPH as the oxidative part of the pentose phosphate pathway has been proposed as a fuel for NADPH oxidase and sustained ROS production in the heart [38]. Among many others, hyperglycemia-induced ROS production via the NADPH oxidase pathway is a well-documented situation for both cardiac and endothelial cells [39, 40].

Increased production of mitochondrial ROS by hyperglycemia is recognized as a major cause of the clinical complications associated with diabetes and obesity [41]. The metabolic abnormalities of diabetes cause mitochondrial superoxide overproduction in endothelial cells of both large and small vessels, as well as in the myocardium. This increased superoxide production causes the activation of five major pathways involved in the pathogenesis of complications: polyol pathway flux, increased formation of advanced glycation end products (AGEs), increased expression of the receptor for AGEs and its activating ligands, activation of protein kinase C isoforms, and overactivity of the hexosamine pathway. It also directly inactivates two critical anti-atherosclerotic enzymes, endothelial nitric oxide synthase and prostacyclin synthase [42].

5 Redox Status and Excitation–Contraction Coupling in Heart Under Hyperglycemia

Hyperglycemia is a major etiological component in the development of diabetic cardiomyopathy and is known to promote the production of reactive oxygen species (ROS) and reactive nitrogen species (RNS), and to deplete antioxidant mechanisms in many cell types. Increasing evidence indicates that oxidative stress contributes markedly to the alterations observed during diabetes [43]. Activities of enzymes that play important roles in antioxidant defense mechanisms of the cells, including GR, GSHPx, glutathione-S-transferase, GST, glucose-6-phosphate dehydrogenase (G6PD), and thioredoxin, TRX-P, decreased significantly in the diabetic rat heart [44]. It should be noted that some previous studies reported that both the levels and activities of the antioxidant enzymes G6PD, 6-phosphogluconate dehydrogenase (6PGD), GR, GSH-Px, and CAT, but not GST, are increased in the diabetic heart [45, 46].

Diabetes alters the activity and expression of nitric oxide synthase (NOS) [47]. Furthermore, it was recently reported that an increase in ROS activity can upregulate nitric oxide synthase, NOS expression in vitro and also in vivo. This effect appears to be, in part, mediated by limiting the availability of NO, thereby exerting a negative feedback influence on NOS expression through activation of nuclear factor kappa B, NF-κB [48]. In the past decade, evidence indicating upregulation of the cardiac renin-angiotensin system (RAS) in the diabetic heart is continuing to accumulate [49]. Ang II, via AT1 receptor stimulation, is also suggested to increase the production of free oxygen radicals [50, 51] and to activate NAD(P)H oxidase enzymes that may generate superoxide anions such as O_2^- in various tissues including the heart [50–52]. AT1 receptors also activate PKC, as was found in the diabetic heart [53]. Supportingly, Fiordaliso et al. [49] also observed elevated myocyte death associated with upregulation of RAS in the diabetic heart as well.

Activation of the RAS, and subsequent signaling through the AT1 receptors, appears to contribute to the development of diabetic cardiomyopathy [54]. Tissue concentrations of Ang II and AT1 density increase in diabetic rat myocardium [55]. Ang II is well known to increase vascular smooth muscle O_2^- production by activation of a membrane-bound NAD(P)H oxidase. Angiotensin-converting enzyme (ACE) inhibitors improve the outcome of heart failure in diabetic patients, even to a greater extent than in nondiabetic subjects [56]. AT1 blockade, as well as NADPH oxidase inhibition, protects the heart from enhanced ROS production induced by high glucose levels applied to rat cardiomyocytes. Thus, it has been suggested that diabetes-related oxidative stress attenuates K^+ currents through Ang II-generated increased superoxide ion levels [57]. AT1 blockade restores action potential duration, transient-outward K^+ current amplitude, Ca^{2+} homeostasis including Ca^{2+} transient kinetics, SR-Ca^{2+} load, spatio-temporal properties of Ca^{2+} sparks, and the basal Ca^{2+} level, as well as the β-adrenergic-mediated enhancement of glucose uptake [53, 58]. The latter report [53] further described that these effects were associated with reduction of the increased PKC level and oxidized protein thiol level in the membrane fraction of the diabetic rat heart. Interestingly, in STZ-induced diabetic rat cardiomyocytes, bosentan, an endothelin-1 receptor inhibitor, induced similar recovery effects on K^+ current and action potential duration [59] as well as did spironolactone, a mineralocorticoid antagonist in male but not in female diabetic rats, reported to be associated with a reduction in aldosterone-induced oxidative stress [60]. A recent study also demonstrated that a depressed level of Na^+ currents contributed a slowdown in depolarization and depression in maximum amplitude of action potential measured in diabetic rat cardiomyocytes [61].

6 Role of Antioxidants in Diabetic Complications

Increased oxidative stress-induced damage in cells contributes to damage in tissues following damage/dysfunction of organs that in turn induces several diseases in humans. The paradigm that interrupting the overproduction of superoxide radicals and hydrogen peroxides would normalize most alterations that contribute to cardiac dysfunction has been difficult to accomplish using conventional antioxidants. Conventional scavenging antioxidants act on a one-to-one basis whereas hyperglycemia-induced overproduction of superoxide radicals is continuous, which led to the use of catalytic antioxidants such as SOD/CAT mimetic. Various trials have examined developing a possible protective/therapeutic strategy by using different types of antioxidants as supplements to diabetic subjects [62–65]. In this sense, adherence to a Mediterranean diet (not confounded by genetic or shared environmental factors) [66] or energy-restricted diets [67] seems to have beneficial effects on oxidative stress-induced damage.

Selenium, considered as a dangerous poison for a long time, became recognized to play a role in biology other than threatening life. Selenium compounds exert their biological effects by either directly or being incorporated into enzymes and other bioactive proteins. The main two inorganic dietary forms of selenium are sodium selenite (Na_2SeO_3) and sodium selenate (Na_2SeO_4). The catalytic pro-oxidant attribute of some selenium compounds appears to account for their toxicity when such activity exceeds plant and animal methylation reactions and antioxidant defenses. This pro-oxidant activity may also account for cellular apoptosis and is used for pharmaceutical application of selenium compounds as antibacterial, antiviral, antifungal, and anticancer agents. In the body, selenium is incorporated into proteins to make important antioxidant enzymes called selenoproteins. The antioxidant properties of selenoproteins help prevent cellular damage from free radicals by scavenging them and reducing them to water and other harmless molecules.

Patients with diabetes demonstrate significantly lower levels of selenoalbumine [68]. The potential relevance of selenium deficiency to human health remained for long supported only by epidemiological studies, which suggested a protective role of selenium against cancer and cardiovascular diseases [69]. The recommended dietary allowance for selenium is 55 μg/day for healthy adults in the U.S., although to allow full expression of selenoproteins, an intake of 75 μg selenium/day as selenomethionin is suggested to be an optimal value [70]. Selenium supplementation results in a significant increase in plasma selenium and glutathione peroxidase (GSH-Px) activity and decreases lipid peroxidation (LPO) [71–73]. Some reports have suggested that selenium may be beneficial in treating diabetes [74, 75]. Other studies indicate that selenium supplementation has no overall benefit in prevention of cardiovascular diseases [76], or even that high serum selenium levels were positively associated with the prevalence of diabetes [77].

Vitamin E as a lipid-soluble antioxidant mainly scavenges oxidized products that result from damaged molecules by hydroxyl radicals and peroxinitrite. However it does not provide protection against the damage of these molecules (key enzymes, lipids, and DNA). In randomized clinical trials, vitamin E did not provide significant benefit [78, 79].

In the first of these studies, in (n-3) polyunsaturated fatty acids (PUFA), omega-3 was shown to lower cardiovascular death significantly. Increasing evidence related to increased consumption of PUFA benefiting subjects without adversely affecting glucose control has prompted the American Diabetes Association to recommend the consumption of fish. Experimentally, a regimen including omega-3 together with vitamin E given to STZ-induced diabetic rats caused almost complete normalization of the antioxidant defense enzymes. This regimen also caused significant recovery of the left ventricular developed pressure by normalizing the lengthening of relaxation in male rats, although it had little effect in females [44].

Alpha-lipoic acid is a disulfide compound that functions as a coenzyme in pyruvate dehydrogenase and α-ketoglutarate dehydrogenase mitochondrial reactions, leading to the production of cellular energy. α-lipoic and its reduced form, dihydrolipoic acid, reduce oxidative stress by scavenging a number of free radicals by chelating transition metals in biological systems, by preventing membrane lipid peroxidation and protein damage through the redox regeneration of other antioxidants such as vitamin C and E, and by increasing intracellular glutathione. Specifically, in STZ-induced diabetic female rats, α-lipoic acid normalized the reduced SOD activity in heart after 14 days of treatment [80].

Aldose reductase (AR) has been implicated in the pathogenesis of various diabetic complications. It catalyzes the reduction of aldehydes, including the adehylde form of glucose using NADPH as a cofactor, initiating the polyol pathway that elicits an increased NADH/NAD^+ ratio. At the early stage of hyperglycemia in the STZ-induced diabetic mouse model, AR activity was markedly increased. AR possesses a redox-sensitive cysteine residue that modulates the enzyme activity. Thus, *N*-acetyl-L-cysteine (NAC), a precursor of GSH, increases the GSSH/GSSG ratio along with a decrease in thiobarbituric acid-reactive substances, TBARS [81]. In fact, the levels of GSH in STZ-induced or insulin-dependent diabetic rats are elevated in the heart, suggesting that the cardiac tissue exposed to ambient high glucose may initiate a defense response by enhancing its antioxidant systems. Augmented activities of antioxidative enzymes such as SOD, CAT, and GSH-Px were also documented at an early stage of diabetes without any changes in TBARS [82]. More recently, this was also shown for the protein expression of Cu-Zn-SOD, heme oxygenase-1, and total SOD activity although manganese-superoxide dismutase (MnSOD or SOD2) activity was clearly reduced [83, 84]. NAC treatment prevented the increased expression of Cu-Zn-SOD and HO-1 whereas the reduced left ventricular developed pressure and rate of relaxation were only partially compensated.

Patients having type 1 diabetes have an elevated red blood cell malondialdehyde (MDA) level, which is a good candidate as an oxidative stress biomarker, together with an increased blood ketone level [85]. Furthermore, red blood cells incubated with ketones were found to have increased production of hydroxyl radicals, membrane lipid peroxidation, and reduced cellular GSH levels. These effects of ketones were significantly suppressed by incubation with antioxidants (such as vitamin E, NAC). De Mattia and coworkers in their clinical trials have shown that the plasma vascular cell adhesion molecule (VCAM)-1 was higher in type 1 diabetic patients compared to healthy individuals and that administration of NAC significantly prevented this increase [86]. In view of the VCAM-1 contribution to deep vein thrombosis and atherosclerosis [87], it is highly probable that antioxidants can prevent diabetes-induced vascular disease through inhibition of VCAM-1 production.

Although many antioxidants have been shown to exert positive effects in experimental cardiovascular complications, their beneficial actions in clinical trials are seen to be controversial [88]. These controversies can arise from identifying the effective dose and determining the efficacy of antioxidants without checking blood oxidant levels. It is becoming evident that for the patients having cardiovascular complications antioxidants may not be able to preserve the pathological changes. Among the applications of antioxidants either separately or with their combination treatments, literature data represent promising results with resveratrol supplementation [89–91]. This antioxidant, reducing blood cholesterol, seems to prevent and retard the development of atherosclerosis and coronary heart diseases related to diabetes. From the clinical trials, it can be recommended that antioxidants have the potential to serve as adjuncts to therapeutic approaches for attenuating the progression of diabetes-induced cardiovascular disorders.

7 Selenium Effects on the Control of Cardiac Activity in Diabetic Samples

Selenium provides a classic example of dichotomy of effects and has generated concerns at both ends of its supplement spectrum. Selenium compounds are known to play an antioxidant role. However, selenium may have opposite effects [92]. Thus, although selenium at low essential levels (about nM ranges) is required for synthesis of redox active selenoenzymes such as GSH-Px and Trx-R, at higher toxic levels (>5–10 mM) selenite can react with essential thiol groups on enzymes to form RS–Se–SR adducts with resultant inhibition of enzyme activity. Also, toxic effects could be directly or indirectly related to increased endogenous release of hydrogen peroxide [93], or to a reduced binding of NF-kB to a nuclear responsive element [94].

Sodium selenite supplementation for 4 weeks to control rats increased blood glucose and lowered plasma insulin [95]. GSH-Px but not SOD activity was increased, leading to a significant increase (70%) in GSSG level associated with a limited decrease in GSH. This treatment also caused a slight prolongation in action potential with no significant effect on spontaneous contraction parameters or intracellular Ca^{2+} transients. L-type Ca^{2+} current or transient K^+ current kinetics, but not density, showed marked alterations, resulting in ≈50% increase or decrease in total charges carried by Ca^{2+} or K^+ currents, respectively, of the selenium-supplemented rat cardiomyocytes [95].

The classical idea that selenium is toxic to the heart at levels higher than available in a balanced diet is not always supported by experimental work. In a study on rat papillary muscles, it was shown that selenite applied in the millimolar range had biphasic contractile effects [96]. The initial transient increase in force, attributable to Ca^{2+} sensitization of the myofilaments, is later on counterbalanced by a reduction of the Ca^{2+} currents and Ca^{2+} transients associated with increased diastolic Ca^{2+} level. These effects occurred mainly through oxidative alteration of protein thiols as the disulfide-reducing agent, DTT, restored control observations. These cardiomyocytes acutely exposed to selenite demonstrate significant decreases in both GSH and protein thiol levels [96]. The increase in resting tension and decrease in contractile force induced by selenite were protected by adding ATP to the bathing solution [97]. Furthermore, millimolar selenite, by reducing the Na^+ current, shortens the action potential recorded on rat isolated papillary muscles, without affecting resting membrane potential [98].

Experimental studies showed that in STZ-induced diabetic rats, daily sodium selenate treatment reduced or normalized high blood glucose level and restored left

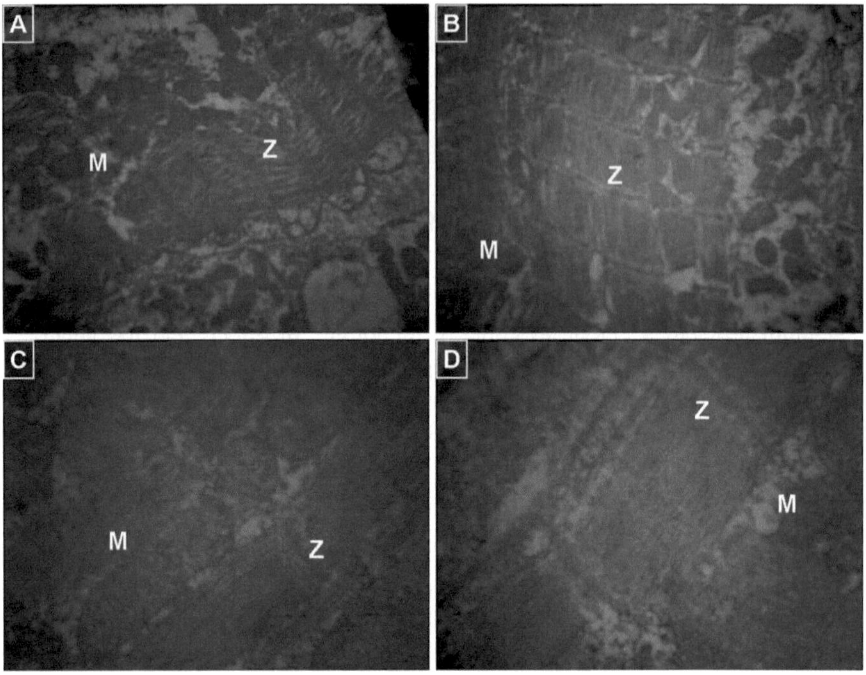

Fig. 1 *Effect of antioxidant treatment on diabetes-induced ultrastructural alterations in cardiac tissue*. *Irregular* and *thickened* Z-lines (*Z*) of myofibers, and loss of cristae and a granular matrix in mitochondria (*M*), were observed in diabetic left ventricular part of heart (**a**, **c**) whereas these alterations were found to be normal in appearance in either sodium selenate- or *N*-acetyl-L-cysteine (NAC)-treated rat heart samples (**b** and **d**, respectively). **a**, **c** ×4,646; **b, d** ×12,930

ventricular pressure parameters without any positive effect on low insulin level [99]. Either selenite or selenate, or NAC treatment could prevent the loss of myofibrils and Z-lines, the reduction of cardiomyocyte diameter and the alterations of the *discus intercalaris* seen in heart tissue of diabetic rats [100] as seen also Fig. 1.

Sodium selenite treatment reversed the prolongation in both action potential duration and twitch duration of the diabetic rats by restoring both fast transient, I_{to} and sustained, I_{ss} K^+ currents [4], whereas treatment of cardiomyocytes from diabetic rats with GSH, similar to insulin application, have been shown to upregulate I_{to} density [3]. Sodium selenite treatment of the diabetic rats caused a significant normalization of cationic homeostasis. Thus, selenite treatment restored basal $[Zn^{2+}]_i$ and $[Ca^{2+}]_i$ values and normalized Ca^{2+} transients of cardiomyocytes isolated from diabetic rats [5]. Similarly, in a later study, we showed that either sodium selenate or omega-3E treatment of diabetic rats for 4 weeks prevented the diabetes-induced depression in left ventricular developed pressure as well as the rates of changes in the developed pressure. Moreover, these treatments reduced the diabetes-induced increase in myocardial oxidized protein sulfhydryl and nitrite concentrations, besides normalization of the reduced TnI and α-actinin protein levels, as well as protected the heart against diabetes-induced depression in β-adrenergic-receptor responses [101, 102]. A schematic of preventive effects of selenium treatment on diabetes-induced cardiac dysfunction is summarized in Table 1.

The authors postulated that during ischemia–reperfusion procedure, polyol pathway activity increases the NADH/NAD^- ratio, leading to the activation of protein kinase C (PKC) and inhibition of Na^+/K^+-ATPase activity. Because the NADH/NAD^- ratio is also increased in high glucose-treated hearts because of polyol pathway activity, activation of PKC and reduction of Na^+/K^+-ATPase activity might be a potential contributory factor for abnormal Ca^{2+} signaling in hearts exposed to hyperglycemia. Presently, no data are available in cardiac tissues; however, it was reported that selenium supplementation resulted in a complete normalization of the Na^+/K^+ pump activity in diabetic aorta homogenates [103].

8 Conclusions

Diabetes mellitus is a disorder resulting in impaired control of blood glucose levels by either impaired insulin release (type 1) or impaired insulin function or insulin resistance (type 2). Selenium has been reported to exert insulin-like cellular functions both in vivo and in vitro. Addition of selenium as sodium selenate to isolated primary rat adipocytes stimulated glucose transport [104]. In STZ-induced diabetic rats, sodium selenate was shown to improve glucose homeostasis [105]. However, insulin release in response to a glucose challenge is markedly reduced in selenate-treated normal rats but not selenium-treated diabetic rats, although selenium potentially promotes an overall improvement in islet function [106]. It has been demonstrated that selenium stimulates glucose transport and anti-lipolysis by stimulating the tyrosine kinases involved in the distal signaling of the insulin signaling cascade but independent of insulin receptor activation [107]. Thus, selenium exerts both insulin-like and non-insulin-like actions in cells. It is evident that a critical balance in the ratio of oxidants to antioxidant defense system is involved in a wide array of cellular functions in almost all organisms under different pathological conditions including hyperglycemia. Antioxidants, therefore, have pivotal importance for cellular functioning ranging from maintenance of redox status to cellular signal transduction. However, as outlined in this review, it is evident that we are quickening our pace toward better ideas in using the antioxidant system and its related molecules as potential therapeutic targets for the treatment of a number of diseases such as cardiovascular diseases in diabetes.

To conclude, although the importance of selenium for heart physiology is known, its dietary supplementation has been popularized for its beneficial effects in general health. Much work is still needed to understand its multiple effects and the numerous pathways that are implicated.

Table 1 Effects of an antioxidant selenium treatment on the parameters of cardiac excitation–contraction coupling and its regulators in diabetic rats

	Parameters	DM	Selenite	Selenate	Reference no
Excitation					
Resting membrane potential	RMP	↔	↔	↔	4,101
Time to peak	TP	↔	↔	↔	4,101
25 % of action potential duration	APD_{25}	↑	↓	↓	4,101
50 % of action potential duration	APD_{50}	↑	↓	↓	4,101
75 % of action potential duration	APD_{75}	↑	↓	↓	4,101
90 % of action potential duration	APD_{90}	↑	↓	↓	4,101
Transient outward current	I_{to}	↓	↑		4
Steady-state current	I_{ss}	↓	↑		4
L-type Ca current	I_{CaL}	↔	↔		4
Contractile activity of Langendorff-perfused hearts					
Left ventricular end-diastolic pressure	LVEDP	↔		↔	101
Left ventricular developed pressure	LVDP	↓		↑	101
Rate of increase in the developed pressure	+dP/dt	↓		↑	101
Rate of decay in the developed pressure	–dP/dt	↓		↑	101
Time to peak of the developed pressure	TP	↑		↓	101
Time to half of decay of the developed pressure	DT_{50}	↑		↓	101
Matrix metalloproteinase-2	MMP-2 activity	↓		↑	101
Tissue inhibitor of matrix metalloproteinase-4	TIMP-4	↓		↑	101
Troponin I	TnI expression	↓		↑	101
Alfa actin	A-actin	↓		↑	101
Beta-adrenergic receptor density (fmol/mg)	β-ARD	↓		↑	102
Contractile activity of papillary muscle trabecules					
Peak tension	PT	↔	↔		4
Twitch duration	TD	↑	↓		4
Time required to 50 % relaxation	RT_{50}	↑	↓		4

All comparisons are given for DM group (diabetic animals) with respect to age-matched controls
No difference ↔, increased ↑, decreased ↓. Sodium compounds as elenite or selenate are used 4–5 weeks following induction of diabetes in male 3-month-old rats

References

1. World Health Organisation (1999) Definition, diagnosis and classification of diabetes mellitus and its complications. http://whqlibdoc.who.int/hq/1999/WHO_NCD_NCS_99.2.pdf
2. Pierce GN, Russel JC (1997) Regulation of intracellular Ca^{2+} in the heart during diabetes. Cardiovasc Res 34:41–47
3. Xu Z, Patel KP, Lou MF, Rozanski GJ (2002) Up-regulation of K(+) channels in diabetic rat ventricular myocytes by insulin and glutathione. Cardiovasc Res 53:80–88
4. Ayaz M, Ozdemir S, Ugur M, Vassort G, Turan B (2004) Effects of selenium on altered mechanical and electrical cardiac activities of diabetic rat. Arch Biochem Biophys 426:83–90

5. Ayaz M, Turan B (2006) Selenium prevents diabetes-induced alterations in $[Zn^{2+}]^i$ and metallothionein level of rat heart via restoration of cell redox cycle. Am J Physiol Heart Circ Physiol 290:H1071–H1080
6. Boudina S, Abel ED (2007) Diabetic cardiomyopathy revisited. Circulation 115:3213–3223
7. Bjelakovic G, Nikolova D, Gluud LL, Simonetti RG, Gluud C (2008) Antioxidant supplements for prevention of mortality in healthy participants and patients with various diseases. Cochrane Database Syst Rev 2, CD007176
8. Ikegami H, Fujisawa T, Ogihara T (2004) Mouse models of type 1 and type 2 diabetes derived from the same closed colony: genetic susceptibility shared between two types of diabetes. ILAR J 45:268–277
9. Cooke DW, Plotnick L (2008) Type 1 diabetes mellitus in pediatrics. Pediatr Rev 29: 374–384
10. Sicree R, Shaw J, Zimmet P, Tapp R (2003) The global burden of diabetes. In: Gan D (ed) Diabetes atlas 2003, 2nd edn. International Diabetes Federation, Brussels
11. Garcia MJ, McNamara PM, Gordon T, Kannel WB (1974) Morbidity and mortality in diabetics in the Framingham population: sixteen year follow-up study. Diabetes 23:105–111
12. Consensus statement/American Diabetes Association (1993) Role of cardiovascular risk factors in prevention and treatment of macrovascular disease in diabetes. Diabetes Care 16:72–78
13. Hayat SA, Patel B, Khattar RS, Malik RA (2004) Diabetic cardiomyopathy: mechanisms, diagnosis and treatment. Clin Sci (Lond) 107:539–557
14. Rubler S, Dlugash J, Yuceoglu YZ et al (1972) New type of cardiomyopathy associated with diabetic glomerulosclerosis. Am J Cardiol 30:595–602
15. Bell DS (1995) Diabetic cardiomyopathy. A unique entity or a complication of coronary artery disease? Diabetes Care 18:708–714
16. Aneja A, Tang WH, Bansilal S et al (2008) Diabetic cardiomyopathy: insights into pathogenesis, diagnostic challenges, and therapeutic options. Am J Med 121:748–757
17. Ferri C, Piccoli A, Laurenti O et al (1994) Atrial natriuretic factor in hypertensive and normotensive diabetic patients. Diabetes Care 17:195–200
18. Du XL, Edelstein D, Rossetti L et al (2000) Hyperglycemia-induced mitochondrial superoxide overproduction activates the hexosamine pathway and induces plasminogen activator inhibitor-1 expression by increasing Sp1 glycosylation. Proc Natl Acad Sci U S A 97:12222–12226
19. Witztum JL (1994) The oxidation hypothesis of atherosclerosis. Lancet 344(8925):793–795
20. Steinberg D (1992) Antioxidants in the prevention of human atherosclerosis. Summary of the proceedings of a National Heart, Lung, and Blood Institute Workshop. Circulation 85:2337–2344
21. Roep BO, Atkinson M (2004) Animal models have little to teach us about type 1 diabetes: 1. In support of this proposal. Diabetologia 47:1650–1656
22. Serreze DV, Chen YG (2005) Of mice and men: use of animal models to identify possible interventions for the prevention of autoimmune type 1 diabetes in humans. Trends Immunol 26:603–607
23. Chen D, Wang MW (2005) Development and application of rodent models for type 2 diabetes. Diabetes Obes Metab 7:307–317
24. Etuk EU (2010) Animal models for studying diabetes mellitus. Agric Biol J North Am 1:130–134
25. Chatzigeorgio A, Halapas A, Kalafatakis K, Kamper E (2009) The use of animal models in the study of diabetes mellitus. In Vivo 23:245–258
26. Le NN, Rose MB, Levinson H, Klitzman B (2011) Implant healing in experimental animal models of diabetes. J Diabetes Sci Technol 5:605–618
27. Hsueh W, Abel D, Breslow JL et al (2007) Recipes for creating animal models of diabetic cardiovascular disease. Circ Res 100:1415–1427
28. Baynes JW, Thorpe SR (1999) Role of oxidative stress in diabetic complications: a new perspective on an old paradigm. Diabetes 48:1–9

29. Van Gaal LF, Mertens IL, De Block CE (2006) Mechanisms linking obesity with cardiovascular disease. Nature 444:875–880
30. Aviram M (2000) Review of human studies on oxidative damage and antioxidant protection related to cardiovascular diseases. Free Radic Res 33:S85–S97
31. Giugliano D, Ceriello A, Paolisso G (1996) Oxidative stress and diabetic vascular complications. Diabetes Care 19:257–267
32. Turan B (2010) Role of antioxidants in redox regulation of diabetic cardiovascular complications. Curr Pharm Biotechnol 11:819–836
33. Santos CX, Anilkumar N, Zhang M et al (2011) Redox signaling in cardiac myocytes. Free Radic Biol Med 50:777–793
34. Kowaltowski AJ, de Souza-Pinto NC, Castilho RF et al (2009) Mitochondria and reactive oxygen species. Free Radic Biol Med 47:333–343
35. St.-Pierre J, Buckingham JA, Roebuck SJ et al (2002) Topology of superoxide production from different sites in the mitochondrial electron transport chain. J Biol Chem 277:44784–44790
36. Murphy MP (2009) How mitochondria produce reactive oxygen species. Biochem J 417:1–13
37. Aon MA, Cortassa S, O'Rourke B (2010) Redox-optimized ROS balance: a unifying hypothesis. Biochim Biophys Acta 1797:865–877
38. Zuurbier CJ, Eerbeek O, Goedhart PT et al (2004) Inhibition of the pentose phosphate pathway decreases ischemia–reperfusion-induced creatine kinase release in the heart. Cardiovasc Res 62:145–153
39. Balteau M, Tajeddine N, de Meester C et al (2011) NADPH oxidase activation by hyperglycaemia in cardiomyocytes is independent of glucose metabolism but requires SGLT1. Cardiovasc Res 92:237–246
40. Ceolotto G, Gallo A, Papparella I et al (2007) Rosiglitazone reduces glucose-induced oxidative stress mediated by NAD(P)H oxidase via AMPK-dependent mechanism. Arterioscler Thromb Vasc Biol 27:2627–2633
41. Brownlee M (2001) Biochemistry and molecular cell biology of diabetic complications. Nature 414:813–820
42. Giacco F, Brownlee M (2010) Oxidative stress and diabetic complications. Circ Res 107:1058–1070
43. Forstermann U (2008) Oxidative stress in vascular disease: causes, defence mechanisms and potential therapies. Nat Clin Pract Cardiovasc Med 5:338–349
44. Tuncay E, Seymen AA, Tanriverdi E, Yaras N et al (2007) Gender related differential effects of Omega-3E treatment on diabetes-induced left ventricular dysfunction. Mol Cell Biochem 304:255–263
45. Kakkar R, Kalra J, Mantha SV, Prasad K (1995) Lipid peroxidation and activity of antioxidant enzymes in diabetic rats. Mol Cell Biochem 151:113–119
46. Okutan H, Ozcelik N, Yilmaz HR, Uz E (2005) Effects of caffeic acid phenethyl ester on lipid peroxidation and antioxidant enzymes in diabetic rat heart. Clin Biochem 38(2):191–196
47. Stockklauser-Farber K, Ballhausen T, Laufer A, Rosen P (2000) Influence of diabetes on cardiac nitric oxide synthase expression and activity. Biochim Biophys Acta 1535:10–20
48. Zhen J, Lu H, Wang XQ, Vaziri ND, Zhou XJ (2008) Upregulation of endothelial and inducible nitric oxide synthase expression by reactive oxygen species. Am J Hypertens 21:28–34
49. Fiordaliso F, Li B, Latini R, Sonnenblick EH et al (2000) Myocyte death in streptozotocin-induced diabetes in rats in angiotensin II-dependent. Lab Invest 80:513–527
50. Bendall JK, Cave AC, Heymes C, Gall N, Shah AM (2002) Pivotal role of a gp91(phox)-containing NADPH oxidase in angiotensin II-induced cardiac hypertrophy in mice. Circulation 105:293–296
51. Griendling KK, Ushio-Fukai M (2000) Reactive oxygen species as mediators of angiotensin II signaling. Regul Pept 91:21–27
52. Ricci C, Pastukh V, Leonard J, Turrens J et al (2008) Mitochondrial DNA damage triggers mitochondrial-superoxide generation and apoptosis. Am J Physiol Cell Physiol 294: C413–C422

53. Yaras N, Bilginoglu A, Vassort G, Turan B (2007) Restoration of diabetes-induced abnormal local Ca^{2+} release in cardiomyocytes by angiotensin II receptor blockade. Am J Physiol Heart Circ Physiol 292:H912–H920
54. Privratsky JR, Wol LE, Sowers JR, Quinn MT, Ren J (2003) AT1 blockade prevents glucose-induced cardiac dysfunction in ventricular myocytes: role of the AT1 receptor and NADPH oxidase. Hypertension 42:206–212
55. Brown L, Wall D, Marchant C, Sernia C (1997) Tissue-specific changes in angiotensin II receptors in streptozotocin-diabetic rats. J Endocrinol 154:355–362
56. Shekelle PG, Rich MW, Morton SC, Atkinson CS et al (2003) Efficacy of angiotensin-converting enzyme inhibitors and beta-blockers in the management of left ventricular systolic dysfunction according to race, gender, and diabetic status: a meta-analysis of major clinical trials. J Am Coll Cardiol 41:1529–1538
57. Shimoni Y, Hunt D, Chuang M, Chen KY et al (2005) Modulation of potassium currents by angiotensin and oxidative stress in cardiac cells from the diabetic rat. J Physiol 567: 177–190
58. Raimondi L, De Paoli P, Mannucci E, Lonardo G et al (2004) Restoration of cardiomyocyte functional properties by angiotensin II receptor blockade in diabetic rats. Diabetes 53: 1927–1933
59. Ding Y, Zou R, Judd RL, Zhong J (2006) Endothelin-1 receptor blockade prevented the electrophysiological dysfunction in cardiac myocytes of streptozotocin-induced diabetic rats. Endocrine 30:121–127
60. Shimoni Y, Chen K, Emmett T, Kargacin G (2008) Aldosterone and the autocrine modulation of potassium currents and oxidative stress in the diabetic rat heart. Br J Pharmacol 154: 675–687
61. Bilginoglu A, Kandilci HB, Turan B (2013) Intracellular levels of Na(+) and TTX-sensitive Na(+) channel current in diabetic rat ventricular cardiomyocytes. Cardiovasc Toxicol 13(2):138–147
62. Valdecantos MP, Pérez-Matute P, Martínez JA (2009) Obesity and oxidative stress: role of antioxidant supplementation. Rev Invest Clin 61:127–139
63. Campión J, Milagro FI, Fernández D et al (2006) Differential gene expression and adiposity reduction induced by ascorbic acid supplementation in a cafeteria model of obesity. J Physiol Biochem 62:71–80
64. Prieto-Hontoria PL, Pérez-Matute P, Fernández-Galilea M et al (2009) Lipoic acid prevents body weight gain induced by a high fat diet in rats: effects on intestinal sugar transport. J Physiol Biochem 65:43–50
65. Venditti P, Bari A, Di Stefano L et al (2007) Vitamin E attenuates cold-induced rat liver oxidative damage reducing H_2O_2 mitochondrial release. Int J Biochem Cell Biol 39:1731–1742
66. Dai J, Jones DP, Goldberg J et al (2008) Association between adherence to the Mediterranean diet and oxidative stress. Am J Clin Nutr 88:1364–1370
67. Crujeiras AB, Parra MD, Rodríguez MC et al (2006) A role for fruit content in energy-restricted diets in improving antioxidant status in obese women during weight loss. Nutrition 22:593–599
68. Roman M, Lapolla A, Jitaru P, Sechi A et al (2010) Plasma selenoproteins concentrations in type 2 diabetes mellitus—a pilot study. Transl Res 156:242–250
69. Neve J (1996) Selenium as a risk factor for cardiovascular diseases. J Cardiovasc Risk 3:42–47
70. Xia Y, Hill KE, Li P, Xu J, Zhou D et al (2010) Optimization of selenoprotein P and other plasma selenium biomarkers for the assessment of the selenium nutritional requirement: a placebo-controlled, double-blind study of selenomethionine supplementation in selenium-deficient Chinese subjects. Am J Clin Nutr 92:525–531
71. Brown BG, Zhao XQ, Chait A, Fisher LD et al (2001) Simvastatin and niacin, antioxidant vitamins, or the combination for the prevention of coronary disease. N Engl J Med 345:1583–1592
72. Faure P, Ramon O, Favier A, Halimi S (2004) Selenium supplementation decreases nuclear factor-kappa B activity in peripheral blood mononuclear cells from type 2 diabetic patients. Eur J Clin Invest 34:475–481

73. Xia Y, Hill KE, Byrne DW, Xu Ji Burk RF (2005) Effectiveness of selenium supplements in a low-selenium area of China. Am J Clin Nutr 81:829–834
74. Rayman MP (2000) The importance of selenium to human health. Lancet 356:233–241
75. Faure P (2003) Protective effects of antioxidant micronutrients (vitamin E, zinc and selenium) in type 2 diabetes mellitus. Clin Chem Lab Med 41:995–998
76. Stranges S, Marshall JR, Trevisan M, Natarajan R et al (2006) Effects of selenium supplementation on cardiovascular disease incidence and mortality: secondary analyses in a randomized clinical trial. Am J Epidemiol 163:694–699
77. Bleys J, Navas-Acien A, Guallar E (2007) Serum selenium and diabetes in U.S. adults. Diabetes Care 30:829–834
78. Gruppo Italiano per lo Studio della Sopravvivenza nell'Infarto miocardico. (1999) Dietary supplementation with n-3 polyunsaturated fatty acids and vitamin E after myocardial infarction: results of the GISSI-Prevenzione trial. Lancet 54(9177):447–455
79. Lonn E, Bosch J, Yusuf S, Sheridan P, Pogue J et al (2005) Effects of long-term vitamin E supplementation on cardiovascular events and cancer: a randomized controlled trial. JAMA 293:1338–1347
80. Maritim AC, Sanders RA, Watkins JB (2003) Effects of alpha-lipoic acid on biomarkers of oxidative stress in streptozotocin-induced diabetic rats. J Nutr Biochem 14:288–294
81. Iwata K, Nishinak T, Matsuno K, Kakehi T, Katsuyama M et al (2007) The activity of aldose reductase is elevated in diabetic mouse hart. J Pharmacol Sci 103:408–416
82. Kakkar R, Mantha SV, Kalra J, Prasad K (1996) Time course study of oxidative stress in aorta and heart of diabetic rat. Clin Sci (Lond) 91:441–448
83. Guo Z, Xia Z, Jiang J, McNeill JH (2007) Downregulation of NADPH oxidase, antioxidant enzymes, and inflammatory markers in the heart of streptozotocin-induced diabetic rats by *N*-acetyl-L-cysteine. Am J Physiol Heart Circ Physiol 292:H1728–H1736
84. Xia Z, Guo Z, Nagareddy PR, Yuen V, Yeung E, McNeill JH (2006) Antioxidant *N*-acetylcysteine restores myocardial Mn-SOD activity and attenuates myocardial dysfunction in diabetic rats. Eur J Pharmacol 544:118–125
85. Jain SK, McVie R (1999) Hyperketonemia can increase lipid peroxidation and lower glutathione levels in human erythrocytes in vitro and in type 1 diabetic patients. Diabetes 49: 1850–1855
86. De Mattia G, Bravi MC, Laurenti O et al (1998) Reduction of oxidative stress by oral *N*-acetyl-L-cysteine treatment decreases plasma soluble vascular cell adhesion molecule-1 concentrations in non-obese, non-dyslipidaemic, normotensive, patients with non-insulin-dependent diabetes. Diabetologia 41:1392–1396
87. González-Ordóñez AJ, Fernández-Carreira JM, Fernández-Alvarez CR et al (2003) The concentration of soluble vascular cell adhesion molecule-1 and lipids are independently associated with venous thromboembolism. Haematologica 88:1035–1043
88. Diaz MN, Frei B, Vita JA et al (1997) Antioxidants and atherosclerotic heart disease. N Engl J Med 337:408–416
89. Voloshyna I, Hussaini SM, Reiss AB (2012) Resveratrol in cholesterol metabolism and atherosclerosis. J Med Food 15:763–773
90. Tomé-Carneiro J, Gonzálvez M, Larrosa M et al (2012) One-year consumption of a grape nutraceutical containing resveratrol improves the inflammatory and fibrinolytic status of patients in primary prevention of cardiovascular disease. Am J Cardiol 110:356–363
91. Bhatt JK, Thomas S, Nanjan MJ (2012) Resveratrol supplementation improves glycemic control in type 2 diabetes mellitus. Nutr Res 32:537–541
92. Oldfield JE (1987) The two faces of selenium. J Nutr 117:2002–2008
93. Simmons TW, Jamall IS, Lockshin RA (1987) The effect of selenium deficiency on peroxidative injury in the house fly, *Musca domestica*. A role for glutathione peroxidase. FEBS Lett 218:251–254
94. Kim IY, Stadtman TC (1997) Inhibition of NF-kappaB DNA binding and nitric oxide induction in human T cells and lung adenocarcinoma cells by selenite treatment. Proc Natl Acad Sci U S A 94:12904–12907

95. Ayaz M, Ozdemir S, Yaras N, Vassort G, Turan B (2005) Selenium-induced alterations in ionic currents of rat cardiomyocytes. Biochem Biophys Res Commun 327:163–173
96. Turan B, Desilets M, Acan LN, Hotomaroglu O, Vannier C, Vassort G (1996) Oxidative effects of selenite on rat ventricular contractility and Ca movements. Cardiovasc Res 32:351–361
97. Ugur M, Turan B (2001) Adenosine triphosphate alters the selenite-induced contracture and negative inotropic effect on cardiac muscle contractions. Biol Trace Elem Res 79:235–245
98. Ugur M, Ayaz M, Ozdemir S, Turan B (2002) Toxic concentrations of selenite shortens repolarization phase of action potential in rat papillary muscle. Biol Trace Elem Res 89:227–238
99. Battell ML, Delgatty HL, McNeill JH (1998) Sodium selenate corrects glucose tolerance and heart function in STZ diabetic rats. Mol Cell Biochem 179:27–34
100. Ayaz M, Can B, Ozdemir S, Turan B (2002) Protective effect of selenium treatment on diabetes-induced myocardial structural alterations. Biol Trace Elem Res 89:215–226
101. Aydemir-Koksoy A, Bilginoglu A, Sariahmetoglu M, Schulz R, Turan B (2010) Antioxidant treatment protects diabetic rats from cardiac dysfunction by preserving contractile protein targets of oxidative stress. J Nutr Biochem 21:827–833
102. Bilginoglu A, Seymen A, Tuncay E et al (2009) Antioxidants but not doxycycline treatments restore depressed beta-adrenergic responses of the heart in diabetic rats. Cardiovasc Toxicol 9(1):21–29
103. Aydemir-Koksoy A, Turan B (2008) Selenium inhibits proliferation signaling and restores sodium/potassium pump function of diabetic rat aorta. Biol Trace Elem Res 126:237–245
104. Ezaki O (1990) The insulin-like effects of selenate in rat adipocytes. J Biol Chem 265:1124–1128
105. Becker DJ, Reul B, Ozcelikay AT, Buchet JP, Henquin JC, Brichard SM (1996) Oral selenate improves glucose homeostasis and partly reverses abnormal expression of liver glycolytic and gluconeogenic enzymes in diabetic rats. Diabetologia 39:3–11
106. McNeill JH, Delgatty HL, Battell ML (1991) Insulin-like effects of sodium selenate in streptozocin-induced diabetic rats. Diabetes 40:1675–1678
107. Li S, Li X, Li YL, Shao CH, Bidasee KR, Rozanski GJ (2007) Insulin regulation of glutathione and contractile phenotype in diabetic rat ventricular myocytes. Am J Physiol Heart Circ Physiol 292:H1619–H1629

Aldose Reductase and Diabetic Cardiovascular Disease

Mariane Abdillahi and Ravichandran Ramasamy

Abstract Cardiovascular disease is a major cause of morbidity and mortality in patients with diabetes mellitus. Studies by others and ourselves have implicated aldose reductase (AR) as a key player in mediating diabetic cardiovascular complications. Findings by us and others demonstrate that increased flux via AR in diabetics perpetuates increased injury after myocardial infarction, accelerates atherosclerotic lesion formation, and promotes restenosis via multiple mechanisms. Taken together, these findings place AR in the center of biochemical and molecular stresses that characterize the cardiovascular complications of diabetes. Blockade of AR-dependent signaling may hold the key to interrupting cycles of cellular perturbation and tissue damage in diabetic cardiovascular disease.

Keywords Aldose reductase • Aldose reductase inhibitor • Sorbitol dehydrogenase • Reactive oxygen species • Advanced glycation end products • Protein kinase C • Diacyl glycerol • 4-Hydroxy nonenal • Mitochondrial permeability transition pore • Ischemia/reperfusion • Malondialdehyde • Receptor for advanced glycation end products

1 Introduction

Diabetes mellitus (DM) is a chronic disease affecting a large population of the world; it is estimated that 300 million people will be diagnosed with DM by the year 2050. Patients with diabetes have an increased risk of mortality as a result of heart

M. Abdillahi • R. Ramasamy, Ph.D. (✉)
Diabetes Research Program, Department of Medicine, New York University
Langone Medical Center, Smilow Research Building Room 901, 550 First Avenue,
New York, NY 10016, USA
e-mail: ramasr02@nyumc.org

B. Turan and N.S. Dhalla (eds.), *Diabetic Cardiomyopathy: Biochemical and Molecular Mechanisms*, Advances in Biochemistry in Health and Disease 9,
DOI 10.1007/978-1-4614-9317-4_8,

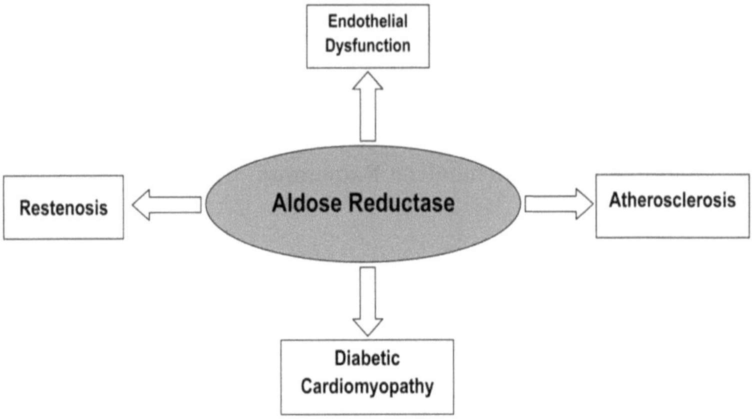

Fig. 1 Consortium of diabetic complications that occur as a result of increased polyol/aldose reductase (AR) pathway

failure in the absence of coronary heart disease (CAD). Cardiac complications that arise from macrovascular and microvascular diseases are the major causes of morbidity and mortality in patients with diabetes [1]. A large body of literature from both experimental and clinical data illustrates the cardiac structural and functional changes that occur as a result of DM in the absence of CAD and hypertension. Several pathophysiological mechanisms have been proposed to contribute to diabetic complications, including protein kinase C (PKC) activation, increased flux via the polyol/aldose reductase pathway, formation of advanced glycation end products (AGEs), disruption of the mitochondrial respiratory chain, and increased reactive oxygen species (ROS) production [2]. However, glucose metabolism via the polyol/aldose reductase (AR) pathway has been proposed as a major contributor to the consortium of diabetic complications that occur (Fig. 1).

The primary focus of this chapter is on AR, the first and rate-limiting enzyme of the polyol pathway that channels the entry of excess glucose into the cytosol of cardiovascular cells. We present evidence that glucose flux through AR plays a central role in mediating diabetic cardiovascular complications, with reference to ischemia–reperfusion injury, atherosclerosis, and restenosis. Finally we discuss the clinical application of various AR inhibitors and their beneficial role in reducing these cardiovascular complications in diabetic subjects.

2 Metabolism Under Normal and Pathological Conditions

Understanding myocardial metabolism under normal conditions is critical before gaining insight into metabolism under the conditions of diabetes and ischemic stress. The heart can oxidize various energy substrates including fatty acids (FA), glucose, lactate, and ketone bodies to meet its energy demands. Under normal aerobic conditions, healthy cardiac tissue preferentially oxidizes FA under beta oxidation, a process that yields 70 % of adenosine triphosphate (ATP) production [3]. The remaining 30 % of ATP is generated from oxidation of other substrates including glucose, lactate, and pyruvate. The process of beta oxidation is energetically less favorable when compared to glucose oxidation because 11–12 % more oxygen is consumed per given amount of ATP produced [4]. Under fasting conditions, FA are readily available in the blood stream and undergo beta oxidation. After consumption of a meal, insulin levels increase immediately following the increase in blood glucose levels. As a result of increased insulin levels, FA oxidation is suppressed and the heart switches from utilizing FA to glucose.

A substantial amount of evidence exists describing the decreased utilization of FA and an increased utilization of pyruvate generated via glycolysis in the ischemic heart [5–7]. In a study using radiolabeled tracers, patients with dilated cardiomyopathy were found to have FA oxidation reduced by 70 % when compared to healthy control subjects [6]. Several hypotheses aim to explain the switch from FA oxidation under ischemic conditions, and it has been hypothesized that the switch in energy substrates occurs to protect the failing heart. The breakdown of myocardial glycogen stores that increases blood glucose concentrations as well as rapid translocation of GLUT4 transporters to the plasma membrane results in rapid glucose uptake by the heart during ischemia. Increased glucose utilization via glycolysis during ischemia significantly reduces oxygen consumption while allowing for continued ATP production. Despite the reduced production of ATP via glycolysis during ischemia, the metabolic switch from FA to glucose oxidation preserves myocardial contractile function and is critical to rescue damaged tissue. The beneficial effect of glucose and glycolysis to energy production depends on the severity and duration of ischemia. If ischemia is severe and prolonged, the positive effects of glycolysis will eventually be inhibited by the accumulation of glycolytic by-products.

The question of whether the metabolic shift in energy substrate during ischemia is adaptive or maladaptive is evolving [8]. Evidence in a study utilizing heart-specific lipoprotein lipase knockout (hLpL0) mice demonstrated that after abdominal aortic constriction, glucose oxidation alone was insufficient to protect the myocardium against chronic injury. Under normal conditions, increased pyruvate dehydrogenase (PDH) activity resulting in increased glucose oxidation and uptake and decreased FA oxidation have been observed in hLpL0 mice. The evidence demonstrating the association between negative outcomes and increased FA oxidation during ischemia–reperfusion (I/R) have highlighted potential therapeutic strategies that can increase glucose oxidation and decrease FA beta oxidation. Carnitine palmitoyltransferase-I (CPT-1) is a key enzyme in the process of beta oxidation as it mediates the transfer of FA to the mitochondria. Glucose oxidation and PDH activity are also influenced by L-carnitine, which also is important in the delivery of FA to the mitochondria. Several pharmacological agents, including ranolazine and trimetazidine, have been developed that have been shown in some studies to reduce FA beta oxidation and improve glucose oxidation. To date, trimetazidine and perhexiline remain the most effective pharmacological agents that inhibit FA oxidation and improve myocardial function [3]. The effects of trimetazidine have been evaluated extensively, and studies have shown that it is well tolerated in patients with heart failure without affecting blood pressure or heart rate [9–12]. Other pharmacological agents including dichloroacetate (DCA) and acipimox, which decrease FA beta oxidation, have been tested in clinical trials. DCA, a PDH kinase inhibitor has been shown to improve glucose oxidation and myocardial contractility when administered to patients with heart failure [13, 14]. Use of acipimox in studies has been shown to improve glucose oxidation while decreasing FA oxidation [15]. Pharmacological agents that alter energy metabolism in the heart offer exciting new methods to treat cardiovascular disorders such as ischemic heart disease.

Under normal physiological conditions in which insulin production is sufficient, lipolysis is inhibited. Under diabetic conditions, insulin production is low, thereby relieving lipolytic inhibition; this in turn increases circulating free fatty acids that inhibit glucose uptake. Additionally, GLUT 4 transporters that mediate glucose uptake are decreased, which cumulatively results in impaired glucose metabolism. Our studies in the diabetic rat heart have shown that increased AR activity led to increases in the NADH/NAD$^+$ ratio [16–18]. Inhibition of AR in this model improved cardiac function by lowering the NADH/NAD$^+$ ratio and thus improving glucose metabolism under basal and ischemia–reperfusion conditions [16].

3 Aldose Reductase (AR)

The aldose reductase enzyme, first described in 1956 by Hers, is a member of the aldo-keto reductase superfamily [19]. It is a nicotinamide adenine dinucleotide (NAD$^+$) phosphate (NADPH)-dependent enzyme that catalyzes the conversion of glucose to sorbitol in the first, rate-limiting step of the polyol pathway (Fig. 2). Sorbitol dehydrogenase (SDH) utilizes NAD$^+$ to convert sorbitol to glucose in the second part of the polyol pathway (Fig. 2). AR is widely expressed in several tissues including heart, lens, brain, and kidney [20, 21]. Inagaki et al. [22] and Grimshaw [23] investigated the glucose anomer specificity of AR and calculated that AR acts on the aldehyde form of D-glucose with a K_m of 0.66 μmol/l; that is, it is a higher affinity substrate than many others. Although AR has a greater preference for glucose as a substrate (binds with a higher affinity than any other substrate), it has the ability to efficiently catalyze other substrates including retinoids, methylglyoxyl, 4-hydroxynonenal (4-HNE), and 2-methylpentanal [24–26].

The activity of AR increases under elevated glucose conditions as well as in ischemic conditions [27–30]. Under normal glucose conditions, the majority of glucose is shunted toward the glycolytic pathway where glucose is phosphorylated by hexokinase, forming glucose-6-phosphate, whereas less than 5 % is fluxed through the AR pathway. Under hyperglycemic and ischemic conditions, glucose flux via the polyol pathway increases significantly, accounting for more than 15–30 % of glucose use.

Increased glucose flux via the AR pathway limits glycolytic flux by the increased demand for NAD$^+$ in both pathways. In glycolysis, GAPDH requires NAD$^+$ for

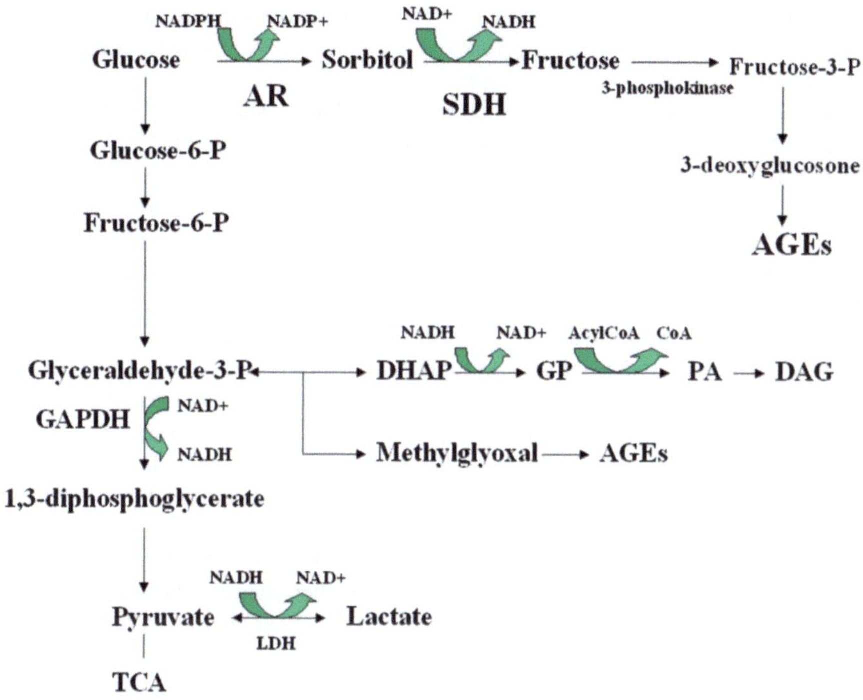

Fig. 2 Comprehensive overview of the interplay between glucose flux via aldose reductase, glycolysis, and advanced glycation end product (AGEs) generation

converting glyceraldehyde-3-phosphate to 1,3-diphosphoglycerate. Similarly, SDH requires NAD^+ to catalyze sorbitol into fructose, and the increased demand for limited cytosolic NAD^+ creates a competition between the two pathways. A comprehensive overview of both the AR and glycolytic pathways is shown in Fig. 2. Altered metabolism caused by increased AR flux has been proposed to mediate diabetic complications from increases in reactive oxygen species (ROS) oxidative stress, advanced glycation end product generation, decreased ATP levels, and imbalances in Ca^+-ATPase and Na^+/K^+-ATPase, resulting in increased calcium accumulation (Fig. 2).

4 AR and Myocardial Ischemia–Reperfusion Injury: Mechanisms

Ischemia–reperfusion (I/R) injury in the myocardium is one of the leading causes of morbidity and mortality in patients with DM. As mentioned earlier, increased AR activity is observed under I/R conditions. Increased polyol flux has been shown to contribute to increased injury after I/R in the myocardium of both diabetics and non-diabetics [16, 17, 31–33]. Studies that have utilized the isolated perfused heart and transient occlusion of the left anterior descending (LAD) coronary artery have demonstrated enhanced glucose flux via AR [31, 32, 34, 35]. Pharmacological interventions via use of AR inhibitors reduced ischemic injury, attenuated ROS generation, improved glycolysis, increased ATP levels, and maintained the ionic balance (sodium and calcium homeostasis) in the heart [31, 32, 34, 35].

Proof-of-concept data demonstrating AR as a mediator of ischemia–reperfusion injury comes from transgenic mice studies. Inherently, mice display much lower levels of AR compared to human subjects. Hence, to develop a more relevant means to test the role of AR in murine models, a transgenic mouse line in which human AR (hAR) was expressed via a histocompatibility gene promoter [36] was used to study the effect of diabetes on cardiovascular stress. This transgene had AR activity comparable to that of humans. Studies utilizing transgenic mice that broadly overexpress human-relevant levels of AR showed increased cardiac injury during ischemia–reperfusion [32]. A more direct relationship of glucose flux via AR to oxidative stress was demonstrated in cardiac injury. AR influenced the opening of the mitochondrial permeability transition pore (MPTP) and was linked to generation of hydrogen peroxide and diminished antioxidant status as measured by glutathione (GSH). Antioxidants or ARIs significantly reduced generation of ROS and inhibited MPTP opening in AR transgenic mitochondria after I/R [35]. Taken together, these studies implicate AR and ROS generated through the AR pathway as a key player in mediating I/R injury in the heart. In rabbit hearts subjected to I/R injury, inhibition of AR was protective [37], although it has also been reported to abolish the cardioprotective effects of ischemic preconditioning [30]. Others [38] have shown increases in AR activity during ischemia consistent with our earlier publication [31]; in contrast, they were unable to demonstrate cardioprotection with ARIs in a glucose-perfused isolated rat heart ischemia–reperfusion (I/R) model. Although reasons for these contrasting findings are not clear, it may be speculated that model-dependent variations and substrate availability during ischemia may underlie the apparent differences.

4-Hydroxynonenal (4-HNE) accumulates during I/R. AR has been proposed to detoxify these aldehydes that accumulate during I/R. Although studies by Chen et al. show reduction of 4-HNE by activation of ALDH-2 and protection of hearts from ischemic damage [39], other reports [40], including studies from our group, have demonstrated increased injury, poor functional recovery, and increased oxidative stress after myocardial I/R in mice hearts overexpressing AR. Further, mice expressing human AR displayed greater injury and higher malonyldialdehyde (MDA) content [32]. Studies in rodent hearts subjected to I/R showed increases in the polyol pathway activity associated with oxidative damage. In contrast, AR null mice were reported to have reduced oxidative stress and protection against ischemic injury [41]. In rat hearts subjected to I/R, increases in polyol pathway activity exacerbated oxidative damage [42]. Furthermore, AR inhibition in animals did not cause increases in lipid peroxidation products such as MDA [42–46]. However, further studies on comprehensive measurements of 4-HNE and the role of ALDH-2 will enable us to understand the role of AR as a potential detoxifying enzyme in ischemic hearts.

Reductions in ATP, increases in calcium, and ROS favor the opening of the MPTP, which, when opened, results in apoptotic and necrotic cell death mechanisms Studies that have utilized the ex vivo Langendorff perfusion system and the in vivo (LAD) models have demonstrated that when hAR transgenic mice are subjected to I/R, their hearts displayed greater injury. After I/R, hAR mice hearts when compared to wild-type (WT) mice hearts displayed decreased ATP production, increased generation of ROS, and increased MPTP opening, all resulting in decreased contractile function.

Because AR has been shown to be a key component of I/R injury, it has emerged as a potential therapeutic target to prevent against the negative pathologies that occur through increased AR flux. The protective effects of AR inhibition have been demonstrated to reduce ischemic injury and improve myocardial function after I/R. Furthermore, inhibition of AR increases cytosolic availability of NAD^+, thus improving glycolytic flux and maintenance of ATP levels. Mice in which the AR gene is knocked out (AR null mice) have been shown to be protected from injury that is associated with I/R. These observances have illustrated the benefits of AR inhibitors as potential therapeutic agents given in adjunct with current therapies for treating patients with myocardial infarction. It is clear from the studies of Ramasamy et al. and others that poor functional recovery, ROS generation, and increased MPTP opening are all associated with increased expression and activity of AR.

There has been increased interest in recent years on the role of janus kinase (JAK) and signal transducer and activator of transcription (STAT) signaling pathways in various cardiac pathologies including I/R [47, 48]. Several studies have demonstrated increased activation of JAK-STAT signaling pathways in rat and mice models as a result of I/R. Specifically, Hwang et al. have shown that in hearts isolated from rats and transgenic hAR mice subjected to global ischemia, the AR pathway mediates myocardial ischemic injury via JAK2 activation followed by STAT5 activation [49]. Changes in the $NADH/NAD^+$ levels proved to be a critical mediator of JAK2/STAT5 activation. An increased flux via AR reduces the cytosolic availability of NAD^+ impacting the $NADH/NAD^+$ ratio. Data from this study also indicate that lowering of cytosolic $NADH/NAD^+$ by the use of niacin and an aldose reductase inhibitor is an important step by which inhibition of AR prevents JAK2/STAT5 activation [49]. Increased $NADH/NAD^+$ further influenced diacylglycerol (DAG) levels and impacts PKC α/β activation. Hearts perfused with chelyrethrine, inhibitor of most PKC isoforms, as well as Gö 6976 (inhibitor of PKC α/β-isoforms), establish that AR-mediated JAK-STAT signaling involves the α- and β-isoforms of PKC. Furthermore, use of the aldose reductase inhibitor zopolrestat inhibited phosphorylation of JAK-2 and STAT5 proteins [49]. These studies are consistent with others that have demonstrated activation of JAK/STAT signaling in vascular smooth muscle cells through involvement of AR and PKC [50].

In addition to JAK2/STAT5 activation, increased AR flux has also been demonstrated to mediate injury via decreased phosphorylation and inhibition of glycogen synthase kinase 3β (p-GSK3β) [51]. GSK3β, a serine/threonine kinase, has been identified by several studies as a key signaling protein that mediates cardioprotection and reduces cell death [52]. Phosphorylation of GSK3β is a key determinant of MPTP opening. Studies that utilized hAR transgenic mice as well as an AR null mouse model (ARKO) demonstrated that AR mediates I/R injury, in part, via decreased phosphorylation of GSK3β (Ser^9), resulting in impaired mitochondrial

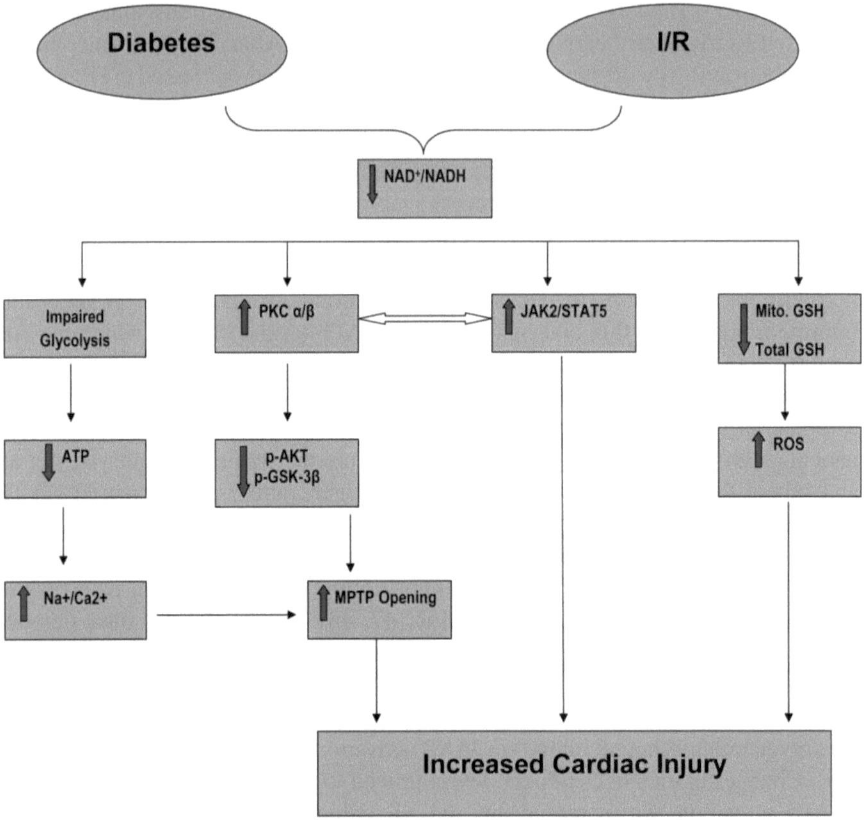

Fig. 3 Established signaling mechanisms by which AR mediates ischemia–reperfusion injury in diabetic and nondiabetic hearts

function as well as impaired functional recovery of the heart [51]. The observed decreased levels of p-GSK3β correlated with increased apoptosis in ARTg hearts. Furthermore, data from this study illustrated that increases in AR flux lead to PKC-α/β activation followed by decreases in Akt and GSK3β phosphorylation and was linked to I/R injury. In summary, these studies demonstrate that AR mediates I/R injury, in part, via modulation of GSK3β phosphorylation. Second, increased flux of AR impairs myocardial functional recovery after I/R because of changes in the cytosolic redox state, which in turn mediate activation of ROS, JAK2/STAT5, and PKC signaling pathways (Fig. 3).

5 AR in Failing Hearts

Studies have demonstrated increases in AR expression in failing hearts of diabetic subjects. In our studies we demonstrated increased glucose flux via polyol pathway in type 2 diabetic rat hearts and that this increased flux is a key factor in mediating increased vulnerability to reperfusion injury and contractile dysfunction after ischemic stress [53]. Studies in dogs revealed attenuation in pacing induced heart failure caused by AR expression [54]. Clinical studies have shown AR expression increases in patients with ischemic cardiomyopathy and diabetic cardiomyopathy [55]. These emerging studies underscore the importance of AR in the etiology of heart failure.

6 AR and Diabetic Atherosclerosis

Emerging studies demonstrate that integrated actions of AR are key to understanding mechanisms driving diabetic cardiovascular complications. Upon crossing onto the atherogenic LDL receptor null background, the hAR transgene had an effect on acceleration of atherosclerosis, particularly in streptozotocin-induced diabetic mice [56]. In contrast, no significant effect of the hAR transgene was observed in nondiabetic mice. In parallel with increased atherosclerosis, the diabetic mice overexpressing hAR in the LDL receptor null background showed a decrease in antioxidant defenses, as observed by alteration of GSH. In contrast, expression of hAR in mice fed a high-fat diet with mild insulin resistance without hyperglycemia had no effect on vascular lesions, at least over the time-course studied [57]. Thus, plausible evidence substantiated the fact that hyperglycemia is needed and might need to be sufficiently increased to provide a substrate for AR in the acceleration of atherosclerosis.

In contrast, another study demonstrated increased lesion area in diabetic and nondiabetic AR knockout mice in an apoE null background [58]. The lesions showed increased presence of 4-HNE, which correlated with the lesion size, and this was attributed to defective clearance of toxic phospholipid aldehydes because of the AR-deficient genotype. The lesions were shown to be highly stable, as demonstrated by more collagen [58]. These contrasting reports lead to a paradox on the precise role of AR in cardiovascular disease. A plethora of questions arose regarding the ambiguity; wherein it was assumed either association of compensatory regulation in the AR null mice leads to alteration in vascular function or genetic overexpression simulates the disease pathophysiology. Emerging studies on the impact of aldehyde dehydrogenase 2 (ALDH-2) in detoxifying 4-HNE [39] add further complexity to the interpretation of lipotoxoic aldehyde levels in AR-overexpressing and AR null mice. Thus, further investigation of the interplay between AR and ALDH-2 in detoxifying 4-HNE and other lipotoxic aldehydes in hAR and AR null mice is critical to resolving these contrasting findings.

Our recent study has elucidated that overexpression of hAR in apoE null mice made diabetic by streptozotocin is proatherogenic and that expression specifically in endothelial cells leads to increased pathology [59]. However, the highlight of the study was that use of a competitive inhibitor that reduces AR activity was found to reduce the lesion size significantly [59]. Overexpression of hAR led to endothelial dysfunction and increased expression of VCAM-1 and MMP-2. Aortic rings of diabetic hAR-overexpressing mice demonstrated decreased acetylcholine-mediated endothelial vasorelaxation compared to the wild-type aorta. Endothelial cell-specific overexpression of hAR imparted a similar effect [59]. In support of this study, other pharmacological studies of AR inhibition have shown improvement in acetylcholine-induced relaxation of diabetic aorta in various murine models. These findings suggest that increased activity of the AR pathway in hyperglycemia is partly responsible for the abnormal endothelium-dependent relaxation in the diabetic blood vessel.

AR expression has been reported in CD68+ cells (monocytes/macrophages) in human atherosclerotic plaque macrophages [60]. Monocyte-derived macrophages isolated from human blood when incubated with oxidized LDL (oxLDL) demonstrated increased AR gene expression and activity along with increased ROS. Inhibition of AR in these oxLDL-stimulated cells attenuated ROS generation. Similarly, endothelial function was improved and VCAM-1 and MMP-2 expression reduced by both pharmacological inhibition and targeted silencing of AR in ECs exposed to high oxLDL [59].

An additional mechanism proposed for AR-mediated accelerated atherosclerosis is through advanced glycation end (AGE)-linked RAGE activation. In support of this, studies in smooth muscle cells incubated with AGE-bovine serum albumin (AGE-BSA) resulted in greater increments of ICAM-1 and monocyte chemoattractant protein 1 (MCP-1), migration, and monocyte adhesion in AR transgenic versus wild-type cells [61]. These AGE adduct-mediated increases were suppressed by either pharmacological inhibition of AR using ARI (zopolrestat) or molecular intervention using AR antisense oligonucleotides [62].

7 AR and Vascular Injury

AR is implicated in proliferation of smooth muscle cell (SMC) growth in a model of vascular repair. Intervention of AR prevents SMC growth in culture and in situ in balloon-injured carotid arteries [63–68]. Increased glucose flux via the AR pathway was mediated through high glucose-induced DAG accumulation and PKC activation in SMCs [64]. Inhibition of AR prevented high glucose-induced stimulation of the extracellular signal-related kinase/mitogen-activated protein kinase and phosphatidylinositol-3-kinase [65], activated nuclear factor-κB [66], and decreased SMC chemotaxis, vascular inflammation, and adhesion. These studies provide fundamental evidence for further evaluation of ARIs in diabetic patients undergoing angioplasty.

8 ARIs and Translational Applications in Humans Relevant to Cardiovascular Studies

Genetic variability in the human AR genes, ALD2 or AKR1B, have been studied extensively and have been found to be associated with rapid progression of diabetic complications including nephropathy, neuropathy, and retinopathy [69].

Additionally, diabetic microangiopathy has been linked to genetic polymorphisms of AR in Japanese type 2 diabetic patients [70]. Finally, the onset of cardiorenal complications has been associated with genetic polymorphisms of AR in a population of diabetic Chinese patients [71].

Gene polymorphism and animal model studies clearly demonstrate the benefit of targeting aldose reductase. Although novel classes of ARIs are in various stages of development, the ARIs that belong to the carboxylic acid and the hydantoins class have been already tested for diabetic neuropathy with mixed results [72–74]. Diabetic subjects (with neuropathy) treated with zopolrestat for 1 year displayed increased left ventricular ejection fraction (LVEF), cardiac output, left ventricle stroke volume, and exercise LVEF [75]. In contrast, placebo-treated subjects demonstrated decreased exercise cardiac output, stroke volume, and end-diastolic volume [75]. In another clinical study, ARI treatment was associated with improved autonomic variability in diabetic patients with autonomic neuropathy [76]. These relatively small but key studies in human subjects with established diabetic complications underscore the promising potential of inhibiting AR in the heart in long-term diabetes.

Increased platelet aggregation as a result of high blood glucose is another factor contributing to atherothrombosis and cardiovascular complications observed in diabetic populations [77, 78]. Platelets from patients with diabetes (chronic hyperglycemic conditions) display an increased responsiveness to collagen [79]. Streptozotocin-induced diabetic rats display platelet hyperaggregation [80]. Inhibition of AR in these models has been shown to reverse these outcomes [2]. Treatment with Sorbinil in diabetic patients normalized increased platelet responsiveness to collagen, whereas in rats AR inhibition reversed platelet hyperaggregation [79–81]. Schulz et al. found AR to be highly expressed in human platelets and platelet aggregation to be reduced by AR inhibitor epalrestat [82]. Taken together, these studies allude to the central role of AR in platelet aggregation under high glucose conditions.

Clinical studies in type 1 and type 2 diabetic patients with nephropathy treated with ARIs have been reported. Administration of tolrestat for 6 months to type 1 diabetic subjects reduced urinary albumin excretion [83]. Similarly, type 2 diabetic subjects treated with epalrestat for 5 years had a similar clinical outcome [84]. In other studies, zopolrestat administered to normotensive type 1 diabetic subjects for 1 year resulted in a progressive reduction in urinary albumin excretion that was not correlated with changes in glycosylated hemoglobin or blood pressure [85].

These data strongly suggest that AR promotes diabetic cardiovascular and renal complications and provides key evidence in human subjects for the benefits of ARIs.

9 Conclusions

It is evident from accumulating data in the literature that AR plays a central role in the pathophysiology of diabetic cardiovascular complications. Animal studies have offered insight into the pathophysiological mechanisms by which increased

AR activity mediates cardiovascular dysfunction and complications via oxidative stress and mitochondrial damage. Animal studies utilizing AR inhibitors or genetically modified AR null mice have shown promising effects in reducing injury associated with increased AR activity, and have highlighted AR as a promising target for therapeutic interventions. Additional animal studies are required to understand various downstream targets of the AR pathway, specifically addressing ROS generation, source of the ROS at the subcellular level, and interrelationships with other pathways such as the AGE-RAGE pathway in relevance to cardiovascular diseases. Although there are ongoing studies of ARIs in humans, the need of the hour is a well-designed, large randomized multicenter human trial using an ARI, relatively free from side effects, that will help establish its therapeutic potential in cardiovascular diseases associated with diabetes and myocardial injury.

Acknowledgments Supported by grants from U.S. National Institutes of Health (HL61783, HL68954, HL60901, AG026467).

References

1. Duckworth W, Abraira C, Moritz T et al (2009) Glucose control and vascular complications in veterans with type 2 diabetes. N Engl J Med 360:129–139
2. Tang WH, Martin KA, Hwa J (2012) Aldose reductase, oxidative stress, and diabetic mellitus. Front Pharmacol 3:87
3. Lionetti V, Stanley WC, Recchia FA (2011) Modulating fatty acid oxidation in heart failure. Cardiovasc Res 90:202–209
4. Stanley WC, Recchia FA, Lopaschuk GD (2005) Myocardial substrate metabolism in the normal and failing heart. Physiol Rev 85:1093–1129
5. Yazaki Y, Isobe M, Takahashi W et al (1999) Assessment of myocardial fatty acid metabolic abnormalities in patients with idiopathic dilated cardiomyopathy using ^{123}I BMIPP SPECT: correlation with clinicopathological findings and clinical course. Heart 81:153–159
6. Neglia D, De Caterina A, Marraccini P et al (2007) Impaired myocardial metabolic reserve and substrate selection flexibility during stress in patients with idiopathic dilated cardiomyopathy. Am J Physiol Heart Circ Physiol 293:H3270–H3278
7. Davila-Roman VG, Vedala G, Herrero P et al (2002) Altered myocardial fatty acid and glucose metabolism in idiopathic dilated cardiomyopathy. J Am Coll Cardiol 40:271–277
8. Taegtmeyer H, Stanley WC (2011) Too much or not enough of a good thing? Cardiac glucolipotoxicity versus lipoprotection. J Mol Cell Cardiol 50:2–5
9. Rosano GM, Vitale C, Sposato B et al (2003) Trimetazidine improves left ventricular function in diabetic patients with coronary artery disease: a double-blind placebo-controlled study. Cardiovasc Diabetol 2:16
10. McClellan KJ, Plosker GL (1999) Trimetazidine. A review of its use in stable angina pectoris and other coronary conditions. Drugs 58:143–157
11. Dalla-Volta S, Maraglino G, Della-Valentina P et al (1990) Comparison of trimetazidine with nifedipine in effort angina: a double-blind, crossover study. Cardiovasc Drugs Ther 4(suppl 4): 853–859
12. Lu C, Dabrowski P, Fragasso G et al (1998) Effects of trimetazidine on ischemic left ventricular dysfunction in patients with coronary artery disease. Am J Cardiol 82:898–901
13. Bersin RM, Stacpoole PW (1997) Dichloroacetate as metabolic therapy for myocardial ischemia and failure. Am Heart J 134:841–855

14. Bersin RM, Wolfe C, Kwasman M et al (1994) Improved hemodynamic function and mechanical efficiency in congestive heart failure with sodium dichloroacetate. J Am Coll Cardiol 23: 1617–1624
15. Layden JD, Malkova D, Nimmo MA (2004) Fat oxidation after acipimox-induced reduction in plasma nonesterified fatty acids during exercise at 0 degrees C and 20 degrees C. Metabolism 53:1131–1135
16. Ramasamy R, Oates PJ, Schaefer S (1997) Aldose reductase inhibition protects diabetic and nondiabetic rat hearts from ischemic injury. Diabetes 46:292–300
17. Trueblood N, Ramasamy R (1998) Aldose reductase inhibition improves altered glucose metabolism of isolated diabetic rat hearts. Am J Physiol 275:H75–H83
18. Vedantham S, Ananthakrishnan R, Schmidt AM et al (2012) Aldose reductase, oxidative stress and diabetic cardiovascular complications. Cardiovasc Hematol Agents Med Chem 10: 234–240
19. Hers HG (1956) The mechanism of the transformation of glucose in fructose in the seminal vesicles. Biochim Biophys Acta 22:202–203
20. Srivastava SK, Ansari NH, Hair GA et al (1984) Aldose and aldehyde reductases in human tissues. Biochim Biophys Acta 800:220–227
21. Wermuth B, Burgisser H, Bohren K et al (1982) Purification and characterization of human-brain aldose reductase. Eur J Biochem 127:279–284
22. Inagaki K, Miwa I, Okuda J (1982) Affinity purification and glucose specificity of aldose reductase from bovine lens. Arch Biochem Biophys 216:337–344
23. Grimshaw CE (1986) Direct measurement of the rate of ring opening of D-glucose by enzyme-catalyzed reduction. Carbohydr Res 148:345–348
24. Crosas B, Hyndman DJ, Gallego O et al (2003) Human aldose reductase and human small intestine aldose reductase are efficient retinal reductases: consequences for retinoid metabolism. Biochem J 373:973–979
25. Vander Jagt DL, Kolb NS, Vander Jagt TJ et al (1995) Substrate specificity of human aldose reductase: identification of 4-hydroxynonenal as an endogenous substrate. Biochim Biophys Acta 1249:117–126
26. Wetzelberger K, Baba SP, Thirunavukkarasu M et al (2010) Postischemic deactivation of cardiac aldose reductase: role of glutathione S-transferase P and glutaredoxin in regeneration of reduced thiols from sulfenic acids. J Biol Chem 285:26135–26148
27. Kaneko M, Bucciarelli L, Hwang YC et al (2005) Aldose reductase and AGE-RAGE pathways: key players in myocardial ischemic injury. Ann N Y Acad Sci 1043:702–709
28. Ramasamy R, Goldberg IJ (2010) Aldose reductase and cardiovascular diseases, creating human-like diabetic complications in an experimental model. Circ Res 106:1449–1458
29. Yabe-Nishimura C (1998) Aldose reductase in glucose toxicity: a potential target for the prevention of diabetic complications. Pharmacol Rev 50:21–33
30. Shinmura K, Bolli R, Liu SQ et al (2002) Aldose reductase is an obligatory mediator of the late phase of ischemic preconditioning. Circ Res 91:240–246
31. Hwang YC, Sato S, Tsai JY et al (2002) Aldose reductase activation is a key component of myocardial response to ischemia. FASEB J 16:243–245
32. Hwang YC, Kaneko M, Bakr S et al (2004) Central role for aldose reductase pathway in myocardial ischemic injury. FASEB J 18:1192–1199
33. Ramasamy R, Trueblood N, Schaefer S (1998) Metabolic effects of aldose reductase inhibition during low-flow ischemia and reperfusion. Am J Physiol 275:H195–H203
34. Ramasamy R, Liu H, Oates PJ et al (1999) Attenuation of ischemia induced increases in sodium and calcium by the aldose reductase inhibitor zopolrestat. Cardiovasc Res 42: 130–139
35. Ananthakrishnan R, Kaneko M, Hwang YC et al (2009) Aldose reductase mediates myocardial ischemia-reperfusion injury in part by opening mitochondrial permeability transition pore. Am J Physiol Heart Circ Physiol 296:H333–H341
36. Yamaoka T, Nishimura C, Yamashita K et al (1995) Acute onset of diabetic pathological changes in transgenic mice with human aldose reductase cDNA. Diabetologia 38:255–261

37. Tracey WR, Magee WP, Ellery CA et al (2000) Aldose reductase inhibition alone or combined with an adenosine A(3) agonist reduces ischemic myocardial injury. Am J Physiol Heart Circ Physiol 279:H1447–H1452
38. Kaiserova K, Tang XL, Srivastava S et al (2008) Role of nitric oxide in regulating aldose reductase activation in the ischemic heart. J Biol Chem 283:9101–9112
39. Chen CH, Budas GR, Churchill EN et al (2008) Activation of aldehyde dehydrogenase-2 reduces ischemic damage to the heart. Science 321:1493–1495
40. Iwata K, Matsuno K, Nishinaka T et al (2006) Aldose reductase inhibitors improve myocardial reperfusion injury in mice by a dual mechanism. J Pharmacol Sci 102:37–46
41. Lo AC, Cheung AK, Hung VK et al (2007) Deletion of aldose reductase leads to protection against cerebral ischemic injury. J Cereb Blood Flow Metab 27:1496–1509
42. Tang WH, Wu S, Wong TM et al (2008) Polyol pathway mediates iron-induced oxidative injury in ischemic-reperfused rat heart. Free Radic Biol Med 45:602–610
43. Lee AY, Chung SS (1999) Contributions of polyol pathway to oxidative stress in diabetic cataract. FASEB J 13:23–30
44. Chung SS, Ho EC, Lam KS et al (2003) Contribution of polyol pathway to diabetes-induced oxidative stress. J Am Soc Nephrol 14:S233–S236
45. Ho EC, Lam KS, Chen YS et al (2006) Aldose reductase-deficient mice are protected from delayed motor nerve conduction velocity, increased c-Jun NH_2-terminal kinase activation, depletion of reduced glutathione, increased superoxide accumulation, and DNA damage. Diabetes 55:1946–1953
46. Obrosova IG, Minchenko AG, Vasupuram R et al (2003) Aldose reductase inhibitor fidarestat prevents retinal oxidative stress and vascular endothelial growth factor overexpression in streptozotocin-diabetic rats. Diabetes 52:864–871
47. Changelian PS, Flanagan ME, Ball DJ et al (2003) Prevention of organ allograft rejection by a specific Janus kinase 3 inhibitor. Science 302:875–878
48. Podewski EK, Hilfiker-Kleiner D, Hilfiker A et al (2003) Alterations in Janus kinase (JAK)-signal transducers and activators of transcription (STAT) signaling in patients with end-stage dilated cardiomyopathy. Circulation 107:798–802
49. Hwang YC, Shaw S, Kaneko M et al (2005) Aldose reductase pathway mediates JAK-STAT signaling: a novel axis in myocardial ischemic injury. FASEB J 19:795–797
50. Shaw S, Wang X, Redd H et al (2003) High glucose augments the angiotensin II-induced activation of JAK2 in vascular smooth muscle cells via the polyol pathway. J Biol Chem 278:30634–30641
51. Abdillahi M, Ananthakrishnan R, Vedantham S et al (2012) Aldose reductase modulates cardiac glycogen synthase kinase-3beta phosphorylation during ischemia-reperfusion. Am J Physiol Heart Circ Physiol 303:H297–H308
52. Murphy E, Steenbergen C (2008) Does inhibition of glycogen synthase kinase protect in mice? Circ Res 103:226–228
53. Li Q, Hwang YC, Ananthakrishnan R et al (2008) Polyol pathway and modulation of ischemia-reperfusion injury in type 2 diabetic BBZ rat hearts. Cardiovasc Diabetol 7:33
54. Srivastava S, Chandrasekar B, Bhatnagar A et al (2002) Lipid peroxidation-derived aldehydes and oxidative stress in the failing heart: role of aldose reductase. Am J Physiol Heart Circ Physiol 283:H2612–H2619
55. Yang J, Moravec CS, Sussman MA et al (2000) Decreased SLIM1 expression and increased gelsolin expression in failing human hearts measured by high-density oligonucleotide arrays. Circulation 102:3046–3052
56. Vikramadithyan RK, Hu Y, Noh HL et al (2005) Human aldose reductase expression accelerates diabetic atherosclerosis in transgenic mice. J Clin Invest 115:2434–2443
57. Wu L, Vikramadithyan R, Yu S et al (2006) Addition of dietary fat to cholesterol in the diets of LDL receptor knockout mice: effects on plasma insulin, lipoproteins, and atherosclerosis. J Lipid Res 47:2215–2222
58. Srivastava S, Vladykovskaya E, Barski OA et al (2009) Aldose reductase protects against early atherosclerotic lesion formation in apolipoprotein E-null mice. Circ Res 105:793–802

59. Vedantham S, Noh H, Ananthakrishnan R et al (2011) Human aldose reductase expression accelerates atherosclerosis in diabetic apolipoprotein E–/– mice. Arterioscler Thromb Vasc Biol 31:1805–1813
60. Gleissner CA, Sanders JM, Nadler J et al (2008) Upregulation of aldose reductase during foam cell formation as possible link among diabetes, hyperlipidemia, and atherosclerosis. Arterioscler Thromb Vasc Biol 28:1137–1143
61. Harja E, Bu DX, Hudson BI et al (2008) Vascular and inflammatory stresses mediate atherosclerosis via RAGE and its ligands in apoE–/– mice. J Clin Invest 118:183–194
62. Dan Q, Wong R, Chung SK et al (2004) Interaction between the polyol pathway and non-enzymatic glycation on aortic smooth muscle cell migration and monocyte adhesion. Life Sci 76:445–459
63. Srivastava S, Ramana KV, Tammali R et al (2006) Contribution of aldose reductase to diabetic hyperproliferation of vascular smooth muscle cells. Diabetes 55:901–910
64. Campbell M, Trimble ER (2005) Modification of PI3K- and MAPK-dependent chemotaxis in aortic vascular smooth muscle cells by protein kinase C-beta II. Circ Res 96:197–206
65. Ramana KV, Friedrich B, Srivastava S et al (2004) Activation of nuclear factor-kappaB by hyperglycemia in vascular smooth muscle cells is regulated by aldose reductase. Diabetes 53:2910–2920
66. Ramana KV, Chandra D, Srivastava S et al (2002) Aldose reductase mediates mitogenic signaling in vascular smooth muscle cells. J Biol Chem 277:32063–32070
67. Ruef J, Liu SQ, Bode C et al (2000) Involvement of aldose reductase in vascular smooth muscle cell growth and lesion formation after arterial injury. Arterioscler Thromb Vasc Biol 20:1745–1752
68. Ramana KV, Tammali R, Reddy AB et al (2007) Aldose reductase-regulated tumor necrosis factor-alpha production is essential for high glucose-induced vascular smooth muscle cell growth. Endocrinology 148:4371–4384
69. Oates PJ (2008) Aldose reductase, still a compelling target for diabetic neuropathy. Curr Drug Targets 9:14–36
70. Watarai A, Nakashima E, Hamada Y et al (2006) Aldose reductase gene is associated with diabetic macroangiopathy in Japanese type 2 diabetic patients. Diabet Med 23:894–899
71. So WY, Wang Y, Ng MC et al (2008) Aldose reductase genotypes and cardiorenal complications: an 8-year prospective analysis of 1,074 type 2 diabetic patients. Diabetes Care 31:2148–2153
72. Mylari BL, Beyer TA, Siegel TW (1991) A highly specific aldose reductase inhibitor, ethyl 1-benzyl-3-hydroxy-2(5H)-oxopyrrole-4-carboxylate, and its congeners. J Med Chem 34:1011–1018
73. Sarges R, Oates PJ (1993) Aldose reductase inhibitors: recent developments. Prog Drug Res 40:99–161
74. Alexiou P, Pegklidou K, Chatzopoulou M et al (2009) Aldose reductase enzyme and its implication to major health problems of the 21st century. Curr Med Chem 16:734–752
75. Johnson BF, Nesto RW, Pfeifer MA et al (2004) Cardiac abnormalities in diabetic patients with neuropathy: effects of aldose reductase inhibitor administration. Diabetes Care 27:448–454
76. Didangelos TP, Athyros VG, Karamitsos DT et al (1998) Effect of aldose reductase inhibition on heart rate variability in patients with severe or moderate diabetic autonomic neuropathy. Clin Drug Investig 15:111–121
77. Kessler L, Wiesel ML, Attali P et al (1998) Von Willebrand factor in diabetic angiopathy. Diabetes Metab 24:327–336
78. Boden G, Rao AK (2007) Effects of hyperglycemia and hyperinsulinemia on the tissue factor pathway of blood coagulation. Curr Diab Rep 7:223–227
79. Jennings PE, Nightingale S, Le Guen C et al (1990) Prolonged aldose reductase inhibition in chronic peripheral diabetic neuropathy: effects on microangiopathy. Diabet Med 7:63–68
80. Hara T, Nakamura J, Koh N et al (1995) An aldose reductase inhibitor, TAT, reduces ADP-induced platelet hyperaggregation in streptozotocin-induced diabetic rats with neuropathy. J Lab Clin Med 126:541–547
81. Tang WH, Stitham J, Gleim S et al (2011) Glucose and collagen regulate human platelet activity through aldose reductase induction of thromboxane. J Clin Invest 121:4462–4476

82. Schulz C, Leuschen NV, Frohlich T et al (2010) Identification of novel downstream targets of platelet glycoprotein VI activation by differential proteome analysis: implications for thrombus formation. Blood 115:4102–4110
83. Passariello N, Sepe J, Marrazzo G et al (1993) Effect of aldose reductase inhibitor (tolrestat) on urinary albumin excretion rate and glomerular filtration rate in IDDM subjects with nephropathy. Diabetes Care 16:789–795
84. Iso K, Tada H, Kuboki K et al (2001) Long-term effect of epalrestat, an aldose reductase inhibitor, on the development of incipient diabetic nephropathy in type 2 diabetic patients. J Diabetes Complications 15:241–244
85. Oates J, Klioze S, Schwartz P et al (2008) Aldose reductase inhibitor zopolrestat reduces elevated urinary albumin excretion rate in T1DM subjects with incipient diabetic nephropathy. J Am Soc Nephrol 19:abst 642A

Sex Differences and Diabetes Mellitus in Cardiovascular Function

Semir Ozdemir, Nazmi Yaras, and Belma Turan

Abstract Diabetes is an increasingly widespread epidemic disease, with type 2 diabetes accounting for most of the cases. Cardiovascular complications are the most common cause of morbidity and mortality in diabetic patients. Some studies have also reported that gender difference has a profound impact in the pathogenesis, development, and severity of cardiovascular diseases in diabetic patients, although this assertion was not documented and reviewed in detail. Indeed, cardiovascular diseases are the leading cause of death among women in developed countries. Similarly, results of human and animal studies have shown that sex differences cannot be ruled out in diabetes-induced cardiovascular abnormalities. The proposed underlying mechanisms of gender-related differences in response to different stimuli in healthy and diabetic subjects are the distinction in regulation of cytosolic Ca^{2+} levels and the varied rate of oxidative damage. The female rat myocardium is more resistant to diabetes-induced cardiac dysfunction than that of male rats, but this female advantage is canceled in postmenopausal individuals. Therefore, it is possible to suggest that estrogen can exert protective effects against diabetes through modulation of altered Ca^{2+} dynamics and reduction of oxidative damage in the heart. Although the current findings provide convincing evidence about the sex-related differences in diabetes-induced cardiovascular pathology, further studies are needed to clarify the underlying mechanisms of this distinction.

Keywords Diabetes • Gender differences • Action potential • Ion channels • Contraction • Ca homeostasis • Oxidative stress

S. Ozdemir (✉) • N. Yaras
Faculty of Medicine, Department of Biophysics, Akdeniz University, Antalya, Turkey
e-mail: osemir@akdeniz.edu.tr

B. Turan
Faculty of Medicine, Department of Biophysics, University of Ankara, Ankara 06100, Turkey

B. Turan and N.S. Dhalla (eds.), *Diabetic Cardiomyopathy: Biochemical and Molecular Mechanisms*, Advances in Biochemistry in Health and Disease 9,
DOI 10.1007/978-1-4614-9317-4_9,

1 Introduction

Diabetes Mellitus, or in short diabetes, is an increasingly prevalent pathology, with cardiovascular disease recognized as a major long-term and life-threatening complication, although pharmacological treatment continues to prolong the lifespan of diabetic patients. Sex-related differences in the development of cardiovascular diseases are not yet well defined between the genders of diabetic patients, although roles of gender differences in the pathogenesis of cardiovascular diseases became relatively well known in the past century [1]. Indeed, cardiovascular diseases are the leading cause of death among women in developed countries. Importantly, results of animal studies have shown that sex differences in any organ function cannot be ruled out in pathological conditions, and particularly in cardiovascular responses to certain neurohumoral stimuli [1–4]. Studies in humans and animals have consistently demonstrated intrinsic sex differences in cardiac function as well as a relationship between sex and cardiovascular disease [5, 6]. Therefore, caution is needed during investigation of the mechanisms and the importance of sex difference in the etiology of cardiac dysfunction that arises in response to several types of systemic disorders. As an example, diabetes is associated with cardiovascular disease; with women showing a greater increase in the risk of sex-specific changes in Ca^{2+} homeostasis that contribute directly to diabetes-induced heart dysfunction [1, 7].

In the diabetic condition, the energetic processes of the heart become particularly dependent on the metabolism of fatty acids. Such an increase in the myocardial uptake of fatty acids results in an abnormal increase in fatty acid oxidation and reduced glucose oxidation, which leads to decreased ATP production and overproduction of reactive oxygen species (ROS) because of increased mitochondrial uncoupling [8, 9]. Thus, increased oxidation of fatty acids would result in increased ROS production. The role of increased oxidative stress and depressed antioxidant defense system under hyperglycemia is discussed in breadth in the review article by Turan and Vassort [10]. Furthermore, hyperglycemia can not only stimulate excessive ROS production and reactive nitrogen species (RNS) release [11, 12] but also reduces activities of antioxidant enzymes that act as the primary defense against nitro-oxidative damage and modify other nonstructural proteins [8, 10, 13, 14]. Accordingly, recent studies have suggested that an imbalance in redox status of cellular pathways may be involved in the pathogenesis of cardiovascular diseases, most likely because of upregulated generation of ROS [13, 14].

In the past decade, the inherent influence of gender on myocardial contractile performance of both normal persons and diabetics has been examined widely by several research groups [2, 3, 15, 16]. Among general findings, sex-dependent alterations in regulation of diastolic and systolic Ca^{2+} levels are proposed as one of the predominant mechanisms underlying the sex-based distinction between responses to different stimuli in both healthy and diseased individuals. Particularly, under different experimental conditions, the increase in intracellular concentration of free Ca^{2+} level ($[Ca^{2+}]_i$) has been shown to be lower in cardiomyocytes from normal female rats compared to that of male rats [3]. In parallel to this finding, Schwertz et al. have also reported a significant sex-related difference in myocardial Ca^{2+} regulation compared to that of age-matched males [15]. This difference was suggested to relate to the female rat myocardium being more resistant against diabetes-induced cardiac dysfunction than that of the male rat [2]. Given the important role of $[Ca^{2+}]_i$ in the regulation of cardiac function, it is tempting to suggest a pivotal role for differential Ca^{2+} regulation in cardiomyocytes from male and female individuals under pathological conditions. However, since these studies provide very limited information so far, a substantial effort is needed to elicit explicitly the role of sex differences in the development and severity of diabetes-induced cardiac dysfunction.

2 Impact of Diabetes on Longevity

Chronic diseases including diabetes can induce destructive abnormalities in individual lives. The impact of diabetes on lifespan is a pivotal jeopardy, and thereby substantially accentuated. In general, a diabetic patient has shorter life expectancy than that of age-matched non-diabetics. It is most likely that there is an inverse correlation between life expectancy and diabetes. Moreover, it can be clearly correlated that diabetes-induced complications in humans will be more deleterious if they are suffered by other pathological conditions (i.e., aging) together with diabetes. As an example, besides differences in the development of coronary disease, sex difference has been demonstrated to influence the propensity for various types of cardiac arrhythmias [17] and sudden cardiac death in males [5].

Diabetes is one of the major public health problems and ultimately leads to several acute and chronic complications. Cardiac dysfunction occurs in both types of diabetes as a result of parameters including glucotoxicity, lipotoxicity, fibrosis, and mitochondrial uncoupling. However, heart disease is becoming a main cause of death among diabetics as a result of improved treatment of diabetic complications other than cardiovascular disorders. Diabetes is a major risk factor for coronary heart disease, stroke, and other vascular complications. On the other hand, hyperglycemia leads directly to heart damage and diabetic cardiomyopathy, which is suggested to contribute to severe cardiac abnormalities independent of any major vascular disease [18, 19].

Although the incidence of cardiovascular disease in females is much less than that of age-matched males, this sex advantage disappears with the onset of diabetes [16]. Furthermore, a significant increase in mortality has been observed in diabetic women after menopause compared to age-matched diabetic men [16]. This diabetes-associated increase in mortality has been attributed largely to the development of diabetic cardiovascular complications. Although for many years estradiol has been considered one of the most important hormones involved in female physiology, later studies showed that it plays important roles also in pathological conditions including glucose homeostasis and insulin resistance. Therefore, hormones related to the female gender are important indicators for the role of sex differences in diabetic complications [20].

Some sex differences are seen in acute and chronic complications of this disease. Almost twice as many women were diagnosed with nonketotic hyperosmolar coma than men, according to the United States National Hospital Survey, with almost 1.5 times more women diagnosed with hypoglycemia compared to men [21]. In another population-based study, the rate of diabetic acidosis in females was 1.5 times that of males [22]. No notable sex differences exist with respect to chronic complications except for a significantly higher frequency of blindness in women than men [23]. Speculations include that women have a longer duration of diabetes, registration of blind men may be less complete, and diabetic men may be more likely than women to die before they develop severe retinopathy.

Perreault and coworkers performed a survey for the Diabetes Prevention Program (DPP) on the roles of sex differences in diabetes risk and the effect of intensive lifestyle modification [24]. Their survey demonstrated that in participants of the DPP randomized to intensive lifestyle modification, meeting intensive lifestyle modification goals strongly correlated with prevention of diabetes in the group as a whole. Men met significantly more intensive lifestyle modification goals than women but had a similar incidence of diabetes. Therefore, they explored sex differences in risk factors for diabetes and the effect of intensive lifestyle modification on these factors.

The effect of diabetes on life expectancy can also be relevant to the quality of care a patient receives during the siege. A person who receives a regular salary and thus can obtain high-quality care is expected to live longer than a person who has not had preventive treatment or received inadequate treatment. In addition, it is clear that the effect of diabetes on life expectancy is closely associated with obesity. Because, obesity increases the risk of other complications that may lower the life expectancy of a diabetic person. Although recent medical advances have greatly improved the quality of life of diabetic patients, recent studies have indicated that people with diabetes still have a much shorter lifespan than those of the non-diabetics [25].

3 General View of Diabetes and Cardiovascular Disorders

Clinical studies demonstrate a marked increase in the incidence of diabetic complications, particularly cardiovascular disorders, despite improved treatment approaches and techniques. Further, these studies emphasized the importance of sex disparities in heart dysfunction under different pathological conditions. Yet, there are no clear data about these disparities in either incidence or mortality of any type of disease resulting from sex differences. Similarly, various clinical outcomes showed gender differences in heart function under both physiological and pathophysiological conditions. In addition to the noted sex differences in cardiac physiology, studies show that women have higher mortality rates after myocardial infarction and exhibit greater diastolic heart failure compared to men [26].

Clinical investigations using different imaging techniques have clearly established the presence of early diastolic dysfunction, which is a nonsymptomatic stage, as a prelude to subsequent progression of systolic dysfunction. In the experimental settings, it is possible to dissect components of systolic and diastolic dysfunction more accurately by using ex vivo and in vivo approaches. However it is interesting to note that with only a few notable exceptions, these experiments involved males only.

3.1 Importance of Sex Differences in Type 1 Diabetic Individuals

Diastolic dysfunction is widely demonstrated in rodent models of type 1 diabetes, mostly those treated with streptozotocin (STZ). Accordingly, this is characterized as reduced end-diastolic volume, slowed relaxation associated with a significant increase in the left ventricle end-diastolic pressure, in part caused by increased rigidity of the left ventricle [27–30]. In vivo evaluation of animal models showed generally a significant reduction in the heart rate of type 1 diabetes model [31, 32] with some exceptional situations [33]. Progressive systolic dysfunction was also reported in many studies of type 1 diabetes, characterized by a reduced ejection fraction as assessed by echocardiography [31, 34]. However, in animal studies with diabetes, contractile diminution signs and resultant systolic dysfunction usually appeared in perfused hearts ex vivo. Reduction in systolic pressure and rate of pressure development has been reported in hearts of STZ-induced diabetic animals [35, 36]. On the other hand, in the ex vivo settings, although relaxation parameters can be measured very precisely, it is difficult to establish the presence of diastolic dysfunction accurately in basic conditions.

Surprisingly, although clinical studies demonstrate sex disparities in the heart, potential sex specificity in cardiac function in healthy and diabetic individuals has received very little attention. There is little information about this topic: most of these are experimental studies. In a study with STZ-diabetic rats, females exhibited less impaired left ventricular pressure development rate compared to those of age-matched males [37]. Other studies of adult rats injected with STZ have suggested that development of left ventricular pressure is less affected in females, despite a similar level of glucose in both sexes. Furthermore, reduced contractility, prolonged duration, and slowed rate of contraction and relaxation have been measured in muscle preparations and ventricular myocytes from diabetic hearts [19, 36, 38, 39]. These findings are mostly associated with prolonged action potentials, altered ionic currents, depressed Ca^{2+} reuptake by sarcoplasmic reticulum (SR), reduced SR Ca^{2+} content, isozyme switch from α- to β-myosin heavy chain, and alterations in cardiac troponin T expression and troponin I phosphorylation [7, 39–42].

In our STZ-diabetic female and male rats, light and electron microscopic morphometry of left ventricular samples revealed typical diabetic changes such as altered volume fraction of myofibrils, vacuolation, loss of myofibrils and Z-lines, alterations in sarcoplasmic reticulum and t-tubules, and reduced cardiomyocyte dimension, increase in collagen content, interstitial volume, mitochondrial size,

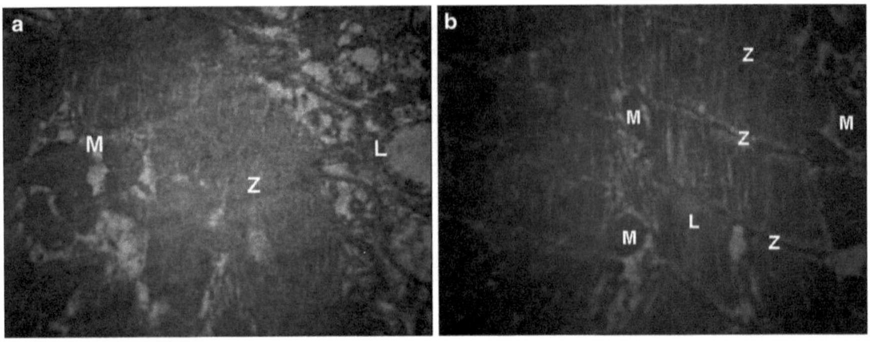

Fig. 1 Marked morphological changes, including destruction, loss, and degeneration in the morphology of myofilaments and Z-lines (*Z*) of myofibers over sarcomere lengths, loss of cisterna and granular matrix in the mitochondria (*M*), and increased number of lipid droplets (*L*) inside the cytosol when compared with those of the controls using electron microscopy. Number and size of lipid droplets in the male heart (**a**) are significantly higher (~20 %) than in the female heart (**b**). ×6,000. (For methods see Ayaz et al. [43])

volume fraction, and lipid droplet size and numbers [43, 44]. Ultrastructure of heart samples from diabetic rats showed marked morphological changes including destruction, loss, and degeneration in morphology of myofilaments and Z-lines of myofibers over the length of sarcomeres when compared with controls [44]. Loss of cisterna and granular matrix in the mitochondria and an increased number of lipid droplets inside the cytosol of the cardiomyocytes were also observed.

Comparison of electron microscopy findings among female and male diabetic hearts showed basically that the number and size of the lipid droplets in the female heart are significantly lower (~20 %) than in the male heart (Fig. 1). This basic finding simply demonstrates that cardiac dysfunction, a major complication of diabetes, is profoundly related with oxidative stress degree which is much higher in males compared to those of females.

In the Framingham Heart Study, diabetes increased the prevalence of congestive heart failure in men with a greater difference in younger patients than in those more than 65 years old [45]. Reduced antioxidant activity and increased oxidative stress occur early after the diagnosis of type 1 diabetes, specifically in women, accounting for an increased susceptibility of diabetic women to cardiovascular complications [46]. The greater effect of diabetes in women compared to men as a risk factor for congestive heart failure is in agreement with a greater contribution of diabetes to arteriosclerosis in women [47].

3.2 *Role of Sex Differences in Type 2 Diabetic Individuals*

Clinical investigations characterized type 2 diabetes with systolic dysfunction in patients. In experimental animal model studies with type 2 diabetes induced by high fat/low dose of STZ, significant reduction in fractional shortening and ejection fraction in the heart has been reported [48]. In addition, a decrease in left ventricular developed pressure plus decreased fractional shortening has been shown in diabetic db/db mice [49, 50], and reduced ejection fraction and fractional shortening were basic alterations in non-obese diabetic Torii rats [51]. Nevertheless, in Otsuka Long-Evans Tokushima fatty (OLETF) rats fractional shortening was not reduced, although diastolic dysfunction was prominent [52]. Isolated heart experiments have demonstrated reduced left ventricular pressure, contractility, and lusitropy [53] as well as decreased heart rate, diminished cardiac output, and slowed left ventricular pressure development in type 2 diabetic mice models [54]. Similarly, less contractile work associated with reduced heart rate was found in isolated Zucker rat heart [55] and sucrose-fed rat heart [56].

As can be seen from the literature, specific gender-related complications of type 2 diabetes models are not significant compared with models for type 1 diabetes. The differences in functional effects of diabetes in the experimental parameters were not apparent early in the development of the disease, but it seems to become a more important factor with aging. Despite these findings, documents related with the role of sex differences in cardiac dysfunction of type 2 diabetic subjects are very limited. Thus, current studies do not allow definitive conclusions, and apparently much more information is needed to suggest a link between sex and type 2 diabetes-associated differences observed in clinical and experimental models. Therefore, it requires further verification and convincing results to prove gender-dependent differences in diabetes-induced cardiac abnormalities unequivocally.

4 Impact of Sex Differences on Cardiac Function

Cardiac function differences between healthy men and women have been compared over years. Merri et al. [57] found longer QT intervals in the heart of healthy women than in men. Despite the confounding factors such as age, heart size, and physiological status that may lead to conflicting conclusions, gender differences have been clearly demonstrated in the incidence and manifestation of heart disease [26, 58]. The results of studies in human subjects showed that incidence of heart disease and failure are lower in premenopausal women than in age-matched men and postmenopausal women as well [26, 45]. However, in the later phases of life, the prevalence of heart failure is also higher in women than men, and older women are more likely to have heart failure with diastolic dysfunction, whereas men with failure are more susceptible to systolic dysfunction [59].

Animal studies have also shown gender-based differences in cardiac performance and cardiac pathology [1, 60, 61]. Studies in animal models have shown gender-dependent differences in myocardial function, but the nature of these differences vary between studies. Schaible and Scheuer [60] showed moderately higher cardiac function in men than in women. However, echocardiographic data of aged rats revealed better systolic function in females compared to males [62]. Experimental findings indicated that there are significant differences in atrial function between female and male rats and that atria from female rats have faster maximal rate of force development (+dF/dt), rate of relaxation (−dF/dt), and time to 50 % relaxation. Furthermore, several lines of evidence obtained from animal studies indicated that it is more likely to induce cardiac hypertrophy and failure in males than females. Accordingly the heart failure progresses faster and the likelihood of impaired relaxation is greater in males than females [63]. Furthermore, females with heart failure have been claimed to have a survival advantage over males [62]. Therefore, female sex seems to exhibit a cardioprotective effect in animals, similar to human counterparts.

4.1 Role of Estrogen in Diabetic Cardiovascular Dysfunction

In recent years estrogen has been recognized to exert significant cardiovascular protection. This proposal comes from the epidemiologic evidence for sex-based differences in the incidence of cardiovascular disease, which was found to be significantly lower in premenopausal women than in age-matched men [47, 64]. After menopause, the risk of cardiovascular disease increases among women comparably with men, which can be controlled following estrogen replacement therapy [64]. This prominent reduction in risk is most likely related to the sex hormone estrogen and its actions on the heart itself [65]. Although the exact nature of the possible direct effects of estrogen on the heart is still equivocal, estrogen may exert its positive effects through modulation of altered Ca^{2+} dynamics in the heart [66, 67].

Johnson et al. [68] studied the effects of estrogen on the heart and found an increased expression of L-type Ca^{2+} channel in hearts of estrogen receptor knockout mice (ERKO). This Ca^{2+} channel is increased by 49 % in cardiac myocytes of ERKO mice compared to the control group, suggesting that the cardiac L-type Ca^{2+} channel is regulated by the estrogen receptor. This scheme infers a modulation of Ca^{2+} regulation by estrogen to limit the amount of Ca^{2+} in the ventricular myocytes. Estrogen has also been shown to exert direct effect on the heart and is thus suggested to be involved at fast regulation of the transcription. Also, acute application of 17β-estradiol has a negative inotropic effect [69] and inhibits the L-type Ca^{2+} current [66, 70] in heart muscle preparations and isolated ventricular cells.

4.2 Role of Sex Differences in Development of Diabetic Cardiac Dysfunction at the Cellular Level

Previous studies designed to investigate sex-related differences in the contractile function of the heart have displayed contradictory results. For example, contractile performance of papillary muscles isolated from female rats was found to be greater than that of male rats [1]. However, another study group reported smaller cardiac function in female rats than male rats by using perfused hearts [60] although more recent reports showed no difference in intrinsic contractile performance of male and female rats' heart [2, 71]. Decreased papillary muscle contraction was also documented in females compared to males [67].

Curl et al. [3] attempted to provide direct evidence for gender-specific differences in Ca^{2+} handling of ventricular myocytes from both male and female rat hearts resulting from the central role of Ca^{2+}in regulation of the contractile function of the heart. The male cardiomyocytes showed a higher $[Ca^{2+}]_i$ than those of females under different experimental conditions [3]. Furthermore, data on reduced activation of Ca^{2+} influx and smaller Ca^{2+} transients in the female ventricular cells are reported, mostly caused by inhibitory effects of 17β-estradiol on the L-type Ca^{2+} current [66, 69, 70]. To understand whether sex-dependent differences in Ca^{2+} transients under electrical stimulation could be responsible for their varied contractile performances; Curl et al. [3] further evaluated contractility of male and female cardiomyocytes via detection of change in length. Both the degree and the rate of shortening were reported to be significantly lower in the female cells than in the male cells. These results are consistent with the data of Ca^{2+} and correlate with lesser Ca^{2+} influx and mobilization of intracellular Ca^{2+} in the female cell. Reduced SR Ca^{2+} content and a resultant decrement in Ca^{2+} release could be another possible reason for the smaller amplitude of the Ca^{2+} transients and peak $[Ca^{2+}]_i$ in female cardiac myocytes. Accordingly, several lines of evidence have suggested the possibility of sex-related differences in SR function as the main culprit for this dimorphism [1, 53, 67]. It was also reported that transient Ca^{2+} changes in 10-month-old female rats were reduced significantly and rate of decay slowed remarkably when compared with those of age-matched male rats [67]. The most obvious reason for this outcome would be a reduction in SR Ca^{2+}-ATPase (SERCA2) activity [53]. Despite the abundance of studies that ascribe the differences in the rate of decay of Ca^{2+} transients to diversity in the SR Ca^{2+}-ATPase activity between the two sexes, the differences in other Ca^{2+} extrusion mechanisms such as Na^+/Ca^{2+} exchange cannot be fully excluded.

The role of sex differences in β-adrenergic receptor stimulation of diabetic heart samples was also assessed. The responses of male and female cardiomyocytes to the β-adrenergic agonist isoproterenol were reported to be different among the sexes. In male ventricular myocytes, isoprenaline induced greater increase in baseline $[Ca^{2+}]_i$, peak $[Ca^{2+}]_i$, and the amplitude of Ca^{2+} transients compared to those of the females, although it was demonstrated to stimulate Ca^{2+} uptake to a similar extent in cardiomyocytes from male and female rats [3].

These functional sex-dependent changes among diabetic individuals were associated with both the activity and expression levels of Ca^{2+}-handling proteins such as beta-adrenergic receptors (β-AR), L-type Ca^{2+} channels (Cav1.2), ryanodine receptors (RyR2), SERCA2, phospholamban (PLB), and Na^+/Ca^{2+} exchanger (NCX) [38, 72]. Indeed, it has been demonstrated that the ventricular abundance of Cav1.2, RyR2, and NCX differs significantly between sexes and is found to be higher in females than males [73], although the Ca^{2+} current values cannot reflect this difference [66, 68–70]. In addition, it is obviously hypothesized that a combination of more Ca^{2+} channels and more RyR2 protein in female ventricles would translate into greater developed force in the female heart, which also contradicts the experimental findings. However, female myocardia seem to possess lower myofilament Ca^{2+} sensitivity than those from males, making the female myocardia more predisposed to diabetes-induced reduction in myofilament Ca^{2+} sensitivity. In accordance, left ventricular myofilament Ca^{2+} sensitivity has been shown to increase in ovariectomized rats [74]. However, atrial preparations from female rats developed greater force than atria from male rats, and skinned fibers were reported to be more responsive to Ca^{2+} in the female rat, which implicated increased responsiveness of female myofibrils to Ca^{2+} [61]. These differences could be tissue specific and need further investigation to be elucidated.

Although sex-related differences have not been studied as much in diabetic conditions, Ren and Ceylan-Isik [16] reported that gender-specific difference in myocardial function was "canceled off" by diabetes. Female myocytes exhibited weaker peak shortening and maximal velocity of shortening/relengthening associated with longer contraction duration of the ventricle [16, 75]. Consistent with this finding, a sex-related difference in cardiomyocyte function has been demonstrated in an insulin-resistant rat model. Males exhibited more dramatic mechanical impairment and developed cardiac dysfunction at an earlier stage than females, and manipulations of intracellular Ca^{2+} (i.e., increasing $[Ca^{2+}]_o$ and stimulus frequency) were suggested to be more effective in restoring normal function in females than in males. Thus, cardiomyocyte dysfunction was indicated to appear earlier and be more severe in males in a sucrose-fed diabetic model, suggesting that female rats have some cardioprotection during the course of metabolic changes [76].

Altered Ca^{2+} signaling is a distinguishing characteristic of diabetic cardiomyopathy and well studied. Yaras et al. [38] attempted to investigate the relevance of gender differences to intracellular Ca^{2+} regulating mechanisms in STZ-induced diabetic rats. They demonstrated defective intracellular Ca^{2+} signaling with depressed amplitude and slowed kinetics of the Ca^{2+} transients with decreased SR Ca^{2+} loads in cardiomyocytes isolated from STZ-diabetic male and female rats (Fig. 2). Furthermore, spatiotemporal properties of the Ca^{2+} sparks were found to be altered with depressed amplitude and prolonged time to peak in males but not females, which was ascribed to defective RyR2 and resultant abnormal SR Ca^{2+} release, although the protein levels of RyR2 and FKBP12.6 were higher in male ventricular cardiomyocytes compared to females. In addition, these important differences were maintained under diabetic conditions. However, the apparent hyperphosphorylation of RyR with less thiol oxidation in diabetic females can account for a lower risk of cardiac abnormalities in female rats with respect to male rats [38]. Overall, although our data and previous reports have demonstrated gender-based differences in both electrical and mechanical cardiac activities in either healthy or diabetic conditions, there are some contradictory findings. The main controversy may arise from either animal and sample differences or experimental protocols.

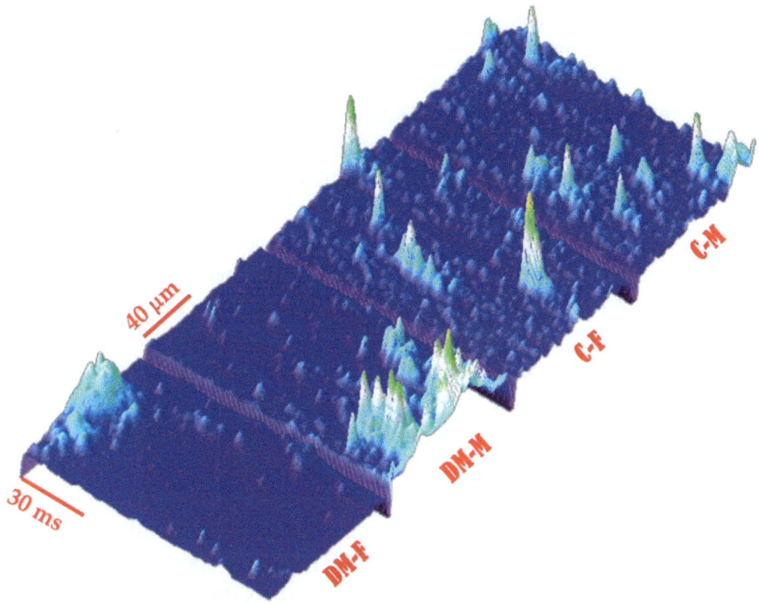

Fig. 2 Line scan images with spontaneously arising Ca^{2+} sparks from control (C-M and C-F) and diabetic (DM-M and DM-F) male and female rat heart cardiomyocytes. (For methods see Yaras et al. [44])

5 Role of Oxidative Stress in Sex-Dependent Cardiac Dysfunction in Diabetic Animals

Despite improved treatment over the years, cardiovascular disorders continue to develop and are presently the leading cause of diabetes-induced mortality. Changes in cardiac energy metabolism, abnormal glucose uptake, and increased fatty acid oxidation in relationship to mitochondrial uncoupling and altered Ca^{2+} homeostasis, leading to diminished contractile activity, are likely [9]. Several mechanisms have been proposed to explain how all the pathologies involved in the progression of diabetic cardiomyopathy might result from hyperglycemia. Diabetes has been shown to induce activation of both systemic and cardiac renin-angiotensin systems (RAS) [77, 78]. It is known that angiotensin II (Ang II) has direct effects on cardiomyocytes through the Ang II type 1 (AT_1) receptor [77, 79]. Thus, stimulation of the AT_1 receptor generates oxygen-derived free radicals, which have detrimental effects on the cardiovascular system [80, 81]. The AT_1 receptor has also been shown to be coupled to several postreceptor signaling pathways, including NADPH oxidase [82]. Therefore, upregulation of cardiac RAS and increased PKC activity in diabetic animals suggests the importance of these pathways in the development of cardiac complications.

In supporting work, Gassanov et al. [83] described marked recovery in the Ang II-induced depressed parameters of Ca^{2+} sparks with candesartan. We and others have shown that chronic administration of Ang II-receptor blockers could protect the heart from the development of cellular alterations typically related to diabetes [39, 84]. Our study showed that candesartan, an AT_1 blocker, exerts protective action against anomalous Ca^{2+} homeostasis that is attributable to reduced RyR2 phosphorylation level and recovery of protein levels of both RyR2 and FKBP12.6 in STZ-induced diabetic cardiomyocytes. Furthermore, candesartan downregulates PKC translocation to the membranes. Consistent with these findings, it was demonstrated that activation of AT_1 receptor plays a major role in inducing functional changes of heart under diabetic conditions via a defective interaction of FKBP12.6 with RyR2 that is triggered by the hyperphosphorylation of RyR2 via modulation with PKA as well as upregulation of the PKC and increase in ROS generation [7, 39, 81].

Several mechanisms in the development of cardiomyopathy have been postulated, including alterations in intracellular ion homeostasis and glucose metabolism and enhanced oxidative stress. Although alteration of Ca^{2+} signaling via changes in critical processes that regulate intracellular Ca^{2+} has become a hallmark of this type of cardiomyopathy, controversies are currently related to specific alterations in Ca^{2+} signaling pathways contributing to the cardiac defects in diabetes [85]. It is widely accepted that metabolic shifts brought about by diabetes increase the production of ROS and RNS. Nowadays, great interest has risen on the possible role of increased oxidative stress in the development of diabetes complications [8, 18]. In patients with diabetes, oxidative stress induced by either increased production of ROS or reduction of antioxidant defense mechanisms appears to result from the consequent increase in lipid peroxidation [10, 18, 86]. Although differences in ovarian sex hormones, especially the potent antioxidant capacity associated with estrogen, have been speculated to play a major role in this "sex bias" of myocardial contractile function [16, 87], the precise mechanism of action behind sex differences in basal cardiac function and diabetic cardiomyopathy remains unclear.

Ceylan-Isik et al. [16, 75] demonstrated varied contractile function caused by gender difference in diabetic cardiomyocytes and suggested that the relatively higher glutathione antioxidant capacity in female hearts may contribute to the sex differences in cardiomyocyte contractile function. This proposal was supported by the observation of increased glutathione capacity and thus elevated antioxidant capacity in males with cardiac metallothionein overexpression. Metallothionein overexpresion nullified the sex difference in glutathione capacity and canceled the sex difference in cardiomyocyte contractile function of STZ-induced diabetic mice. Other findings [16, 88] depicted elevated cardiac protein carbonyl formation in diabetes in both sexes although there was no sex differences in basal cardiac protein carbonyl levels. These results preferably suggest a significant role of cardiac protein damage in the onset of diabetic heart dysfunction but not a gender-based difference in myocyte contractility.

Moreover, Shimoni and Liu [4] proposed that diabetic cardiomyocytes of female rats show lesser thiol oxidation compared to those of males, accounting for less variation in K^+ currents. The authors also reported that the activation of autocrine/paracrine mechanisms is absent or less pronounced in cardiac cells from type 1 diabetic females because of the protective action of estrogen [4, 89]. Therefore, some of the cytoprotective effects of estrogen can be attributed to its antioxidative properties [90, 91]. Shimoni and Liu [89] recently demonstrated that induction of diabetic cardiomyopathy in females leads to a lower level of oxidative stress expressed as lower superoxide ion generation. Several studies, including those of our group, showed that oxidative stress is involved in the etiology of diabetes-induced down-regulation of transient outward K^+ currents [4, 38, 89]. Thus, different levels of intra-cellular oxidants could contribute to the sex-related differences in cardiac function under control and diabetic conditions. Total and free SH levels measured in isolated cardiomyocytes of diabetic rats were significantly less than levels in controls, with less reduction in females [38]. These data suggest that sex-dependent differences in diabetes-induced electrophysiological changes of the hearts could in part be caused by sex-dependent generation of oxidative stress in diabetes.

Some authors have observed increased oxidative stress in the early stages of type 1 diabetes in children and adolescents [92]. Recently, it has been demonstrated in patients with well-controlled type 1 diabetes that total plasma antioxidant capacity was significantly lower and that two different markers of lipid peroxidation were significantly augmented [46, 93]. Oxidative stress occurs early in the course of the disease in type 1 diabetes patients, with a difference between men and women. Thus, diabetes rapidly initiates destruction of plasma antioxidant capacity and increases oxidative stress, even in the early stages of the disease. Therefore, type 1 diabetic patients show antioxidant capacity imbalance even at the early phase of hyperglycemia, and impaired metabolic control leads to increased lipid peroxidation, even in the absence of complications. Instead, the uric acid levels contribute to this phenomenon, seen particularly in diabetic women. Independent of other factors women with diabetes display a greater decrease in antioxidant capacity and thus upregulated lipid peroxidation compared to diabetic men, which implies a canceled advantage against cardiovascular risk in women during diabetes [46].

6 Conclusions

In the Framingham Heart Study, diabetes increased the prevalence of congestive heart failure in men more than in women, although this difference significantly diminished with advanced ages [94]. Reduced antioxidant activity and increased oxidative stress occur early after the diagnosis of type 1 diabetes, specifically in elderly women, accounting for an increased susceptibility of diabetic women to cardiovascular complications [46]. The greater effect of diabetes in women compared to men as a risk factor for congestive heart failure is in agreement with higher contribution of diabetes to arteriosclerosis in women [95]. However, at the cardiomyocyte level, gender differences indicate an increased Ang II level only in male rats with a consequent more marked reduction in K^+ currents [89]. Also, diabetes-induced alterations in RyR2 phosphorylation and FKBP12 unbinding, total and free sulfydryl group reduction, and PKC increase were less marked in females than in male rats [38]. It should be noted that antioxidant treatment with omega-3 induces significant recovery in depressed left ventricular function and its rates of changes in males, whereas it further lengthens the diabetes-induced increase in time to peak of the developed pressure in females, although restoration in the altered antioxidant enzyme activities occured without significant gender differences [36]. It is most likely that the presentation of overall cardiac pathologies differs between sexes and deserves intensive further investigation to elucidate the underlying mechanisms unequivocally.

References

1. Capasso JM, Remily RM, Smith RH et al (1983) Sex differences in myocardial contractility in the rat. Basic Res Cardiol 78(2):156–171
2. Brown RA, Filipovich P, Walsh MF et al (1996) Influence of sex, diabetes and ethanol on intrinsic contractile performance of isolated rat myocardium. Basic Res Cardiol 91(5):353–360
3. Curl CL, Wendt IR, Kotsanas G (2001) Effects of gender on intracellular [Ca^{2+}]i in rat cardiac myocytes. Pflugers Arch 441(5):709–716
4. Shimoni Y, Liu XF (2003) Sex differences in the modulation of K^+ currents in diabetic rat cardiac myocytes. J Physiol 550:401–412
5. Dahlberg ST (1990) Gender difference in the risk factors for sudden cardiac death. Cardiology 77(Suppl 2):31–40
6. Gadrin J, Wagenknecht L, Anton-Culver H et al (1995) Relationship of cardiovascular risk factors to echocardiographic left ventricular mass in healthy young black and white adult men and women. Circulation 92:380–387
7. Yaras N, Ugur M, Ozdemir S et al (2005) Effects of diabetes on ryanodine receptor Ca release channel (RyR2) and Ca^{2+} homeostasis in rat heart. Diabetes 54:3082–3088
8. Baynes JW, Thorpe SR (1999) Role of oxidative stress in diabetic complications: a new perspective on an old paradigm. Diabetes 48:1–9
9. Turan B (2010) Role of antioxidants in redox regulation of diabetic cardiovascular complications. Curr Pharm Biotechnol 11:819–836
10. Vassort G, Turan B (2010) Protective role of antioxidants in diabetes-induced cardiac dysfunction. Cardiovasc Toxicol 10:73–86
11. Bidasee KR, Nallani K, Besch HR et al (2003) Streptozotocin-induced diabetes increases disulfide bond formation on cardiac ryanodine receptor (RyR2). J Pharmacol Exp Ther 305:989–998
12. Dhalla NS, Liu X, Panagia V et al (1998) Subcellular remodeling and heart dysfunction in chronic diabetes. Cardiovasc Res 40:239–247
13. Aydemir M, Ozturk N, Dogan S et al (2012) Sodium tungstate administration ameliorated diabetes-induced electrical and contractile remodeling of rat heart without normalization of hyperglycemia. Biol Trace Elem Res 148:216–223
14. Yamawaki H, Berk BC (2005) Thioredoxin: a multifunctional antioxidant enzyme in kidney, heart and vessels. Curr Opin Nephrol Hypertens 14:149–153
15. Schwertz DW, Beck JM, Kowalski JM et al (2004) Sex differences in the response of rat heart ventricle to calcium. Biol Res Nurs 5:286–298
16. Ren J, Ceylan-Isik AF (2004) Diabetic cardiomyopathy: do women differ from men? Endocrine 25:73–83
17. Pham TV, Rosen MR (2002) Sex, hormones, and repolarization. Cardiovasc Res 53:740–751

18. Cai L, Kang Y (2001) Oxidative stress and diabetic cardiomyopathy: a brief review. Cardiovasc Toxicol 1:181–193
19. Fein FS, Sonnenblick EH (1994) Diabetic cardiomyopathy. Cardiovasc Drugs Ther 8:65–73
20. Barros RPA, Machado UF, Gustafsson JA (2006) Estrogen receptors: new players in diabetes mellitus. Trends Mol Med 12:425–431
21. Fishbein H, Palumbo PJ (1995) Acute metabolic complications in diabetes. In: National Diabetes Data group (ed) Diabetes in America, 2nd edn. DIANE Publishing Co., Darby, PA, pp 283–291
22. Faich GA, Fishbein HA, Ellis SE (1983) The epidemiology of diabetic acidosis: a population-based study. Am J Epidemiol 117:551–558
23. Icks A, Trautner C, Haastert B et al (1997) Blindness due to diabetes: population-based age- and sex-specific incidence rates. Diabet Med 14:571–575
24. Perreault L, Ma Y, Dagogo-Jack S et al (2008) Sex differences in diabetes risk and the effect of intensive lifestyle modification in the Diabetes Prevention Program. Diabetes Care 31:1416–1421
25. Hayes AJ, Leal J, Kelman CW et al (2011) Risk equations to predict life expectancy of people with type 2 diabetes mellitus following major complications: a study from Western Australia. Diabet Med 28:428–435
26. Johnson BE, Johnson CA (2001) Cardiovascular disease and differences between the sexes. Am Fam Physician 63:1290–1292
27. Huynh K, McMullen JR, Julius TL et al (2010) Cardiac-specific IGF-1 receptor transgenic expression protects against cardiac fibrosis and diastolic dysfunction in a mouse model of diabetic cardiomyopathy. Diabetes 59:1512–1520
28. Khong FL, Zhang Y, Edgley AJ et al (2011) 3′,4′-Dihydroxyflavonol antioxidant attenuates diastolic dysfunction and cardiac remodeling in streptozotocin-induced diabetic m(Ren2)27 rats. PLoS One 6:e22777
29. Tian C, Shao CH, Moore CJ et al (2011) Gain of function of cardiac ryanodine receptor in a rat model of type 1 diabetes. Cardiovasc Res 91:300–309
30. Li Y, Ma J, Zhu H et al (2011) Targeted inhibition of calpain reduces myocardial hypertrophy and fibrosis in mouse models of type 1 diabetes. Diabetes 60:2985–2994
31. Hoit BD, Castro C, Bultron G et al (1999) Noninvasive evaluation of cardiac dysfunction by echocardiography in streptozotocin-induced diabetic rats. J Card Fail 5:324–333
32. Radovits T, Korkmaz S, Loganathan S et al (2009) Comparative investigation of the left ventricular pressure-volume relationship in rat models of type 1 and type 2 diabetes mellitus. Am J Physiol Heart Circ Physiol 297:H125–H133
33. Hadour G, Ferrera R, Sebbag L et al (1998) Improved myocardial tolerance to ischaemia in the diabetic rabbit. J Mol Cell Cardiol 30:1869–1875
34. Sharma V, Sharma A, Saran V et al (2011) β-receptor antagonist treatment prevents activation of cell death signaling in the diabetic heart independent of its metabolic actions. Eur J Pharmacol 657:117–125
35. Shimabukuro M, Higa S, Shinzato T et al (1996) Cardioprotective effects of troglitazone in streptozotocin-induced diabetic rats. Metabolism 45:1168–1173
36. Tuncay E, Seymen AA, Tanriverdi E et al (2007) Gender related differential effects of Omega-3E treatment on diabetes-induced left ventricular dysfunction. Mol Cell Biochem 304:255–263
37. Rodrigues B, McNeill JH (1987) Comparison of cardiac function in male and female diabetic rats. Gen Pharmacol 18:421–423
38. Yaras N, Tuncay E, Purali N et al (2007) Sex-related effects on diabetes-induced alterations in calcium release in the rat heart. Am J Physiol Heart Circ Physiol 293:H3584–H3592
39. Ozdemir S, Ugur M, Gürdal H et al (2005) Treatment with AT(1) receptor blocker restores diabetes-induced alterations in intracellular Ca(2+) transients and contractile function of rat myocardium. Arch Biochem Biophys 435:166–174
40. Dillmann WH (1980) Diabetes mellitus induces changes in cardiac myosin of the rat. Diabetes 29:579–582
41. Liu X, Takeda N, Dhalla NS (1996) Troponin I phosphorylation in heart homogenate from diabetic rat. Biochim Biophys Acta 1316:78–84

42. Ozturk N, Yaras N, Ozmen A et al (2013) Long-term administration of rosuvastatin prevents contractile and electrical remodelling of diabetic rat heart. J Bioenerg Biomembr 45:343–352
43. Ayaz M, Can B, Ozdemir S et al (2002) Protective effect of selenium treatment on diabetes-induced myocardial structural alterations. Biol Trace Elem Res 89:215–226
44. Yaras N, Sariahmetoglu M, Bilginoglu A et al (2008) Protective action of doxycycline against diabetic cardiomyopathy in rats. Br J Pharmacol 155:1174–1184
45. Ho KK, Pinsky JL, Kannel WB et al (1993) The epidemiology of heart failure: the Framingham Study. J Am Coll Cardiol 22:6A–13A
46. Marra G, Cotroneo P, Pitocco D et al (2002) Early increase of oxidative stress and reduced antioxidant defenses in patients with uncomplicated type 1 diabetes: a case for gender difference. Diabetes Care 25:370–375
47. Barrett-Connor E, Bush TL (1991) Estrogen and coronary heart disease in women. JAMA 265:1861–1867
48. Zhou H, Li YJ, Wang M et al (2011) Involvement of RhoA/ROCK in myocardial fibrosis in a rat model of type 2 diabetes. Acta Pharmacol Sin 32:999–1008
49. Khalid AM, Hafstad AD, Larsen TS et al (2011) Cardioprotective effect of the PPAR ligand tetradecylthioacetic acid in type 2 diabetic mice. Am J Physiol Heart Circ Physiol 300:H2116–H2122
50. Semeniuk LM, Kryski AJ, Severson DL (2002) Echocardiographic assessment of cardiac function in diabetic db/db and transgenic db/db-hGLUT4 mice. Am J Physiol Heart Circ Physiol 283:H976–H982
51. Lakshmanan AP, Harima M, Sukumaran V et al (2012) Modulation of AT-1R/AMPK-MAPK cascade plays crucial role for the pathogenesis of diabetic cardiomyopathy in transgenic type 2 diabetic (Spontaneous Diabetic Torii) rats. Biochem Pharmacol 83:653–660
52. Kim SK, Zhao ZS, Lee YJ et al (2003) Left-ventricular diastolic dysfunction may be prevented by chronic treatment with PPAR-alpha or -gamma agonists in a type 2 diabetic animal model. Diabetes Metab Res Rev 19:487–493
53. Belke DD, Swanson EA, Dillmann WH (2004) Decreased sarcoplasmic reticulum activity and contractility in diabetic db/db mouse heart. Diabetes 53:3201–3208
54. Hafstad AD, Solevåg GH, Severson DL et al (2006) Perfused hearts from Type 2 diabetic (db/db) mice show metabolic responsiveness to insulin. Am J Physiol Heart Circ Physiol 290:H1763–H1769
55. Wang P, Chatham JC (2004) Onset of diabetes in Zucker diabetic fatty (ZDF) rats leads to improved recovery of function after ischemia in the isolated perfused heart. Am J Physiol Endocrinol Metab 286:E725–E736
56. Cárdenas G, Carlos Torres J, Zamora J et al (2006) Isolated heart function after ischemia and reperfusion in sucrose-fed rats: influence of gender and treatment. Clin Exp Hypertens 28:85–107
57. Merri M, Benhorin J, Alberti M et al (1989) Electrocardiographic quantitation of ventricular repolarization. Circulation 80:1301–1308
58. Carroll JD, Carroll EP, Feldman T et al (1992) Sex-associated differences in left ventricular function in aortic stenosis of the elderly. Circulation 86:1099–1107
59. Hayward CS, Kalnins WV, Kelly RP (2001) Gender-related differences in left ventricular chamber function. Cardiovasc Res 49:340–350
60. Schaible TF, Scheuer J (1984) Comparison of heart function in male and female rats. Basic Res Cardiol 79:402–412
61. Schwertz DW, Vizgirda V, Solaro RJ et al (1999) Sexual dimorphism in rat left atrial function and response to adrenergic stimulation. Mol Cell Biochem 200:143–153
62. Forman DE, Cittadini A, Azhar G et al (1997) Cardiac morphology and function in senescent rats: gender-related differences. J Am Coll Cardiol 30:1872–1877
63. Gardner JD, Brower GL, Janicki JS (2002) Gender differences in cardiac remodeling secondary to chronic volume overload. J Card Fail 8:101–107
64. Grady D, Rubin SM, Petitti DB et al (1992) Hormone therapy to prevent disease and prolong life in postmenopausal women. Ann Intern Med 117:1016–1037

65. Grohé C, Kahlert S, Löbbert K et al (1997) Cardiac myocytes and fibroblasts contain functional estrogen receptors. FEBS Lett 416:107–112
66. Jiang C, Poole-Wilson PA, Sarrel PM et al (1992) Effect of 17 beta-oestradiol on contraction, Ca^{2+} current and intracellular free Ca^{2+} in guinea-pig isolated cardiac myocytes. Br J Pharmacol 106:739–745
67. Leblanc N, Chartier D, Gosselin H et al (1998) Age and gender differences in excitation-contraction coupling of the rat ventricle. J Physiol 511:533–548
68. Johnson BD, Zheng W, Korach KS et al (1997) Increased expression of the cardiac L-type calcium channel in estrogen receptor-deficient mice. J Gen Physiol 110:135–140
69. Sitzler G, Lenz O, Kilter H et al (1996) Investigation of the negative inotropic effects of 17 beta-oestradiol in human isolated myocardial tissues. Br J Pharmacol 119:43–48
70. Meyer R, Linz KW, Surges R et al (1998) Rapid modulation of L-type calcium current by acutely applied oestrogens in isolated cardiac myocytes from human, guinea-pig and rat. Exp Physiol 83:305–321
71. Tuncay E, Seymen AA, Sam P et al (2009) Effects of beta-adrenergic receptor blockers on cardiac function: a comparative study in male versus female rats. Can J Physiol Pharmacol 87:310–317
72. Chu SH, Sutherland K, Beck J et al (2005) Sex differences in expression of calcium-handling proteins and beta-adrenergic receptors in rat heart ventricle. Life Sci 76:2735–2749
73. Chen J, Petranka J, Yamamura K et al (2003) Gender differences in sarcoplasmic reticulum calcium loading after isoproterenol. Am J Physiol Heart Circ Physiol 285:H2657–H2662
74. Wattanapermpool J, Riabroy T, Preawnim S (2000) Estrogen supplement prevents the calcium hypersensitivity of cardiac myofilaments in ovariectomized rats. Life Sci 66:533–543
75. Ceylan-Isik AF, LaCour KH, Ren J (2006) Gender disparity of streptozotocin-induced intrinsic contractile dysfunction in murine ventricular myocytes: role of chronic activation of Akt. Clin Exp Pharmacol Physiol 33:102–108
76. Schwanke ML, Dutta K, Podolin DA et al (2006) Cardiomyocyte dysfunction in insulin-resistant rats: a female advantage. Diabetologia 49:1097–1105
77. Fiordaliso F, Li B, Latini R et al (2000) Myocyte death in streptozotocin-induced diabetes in rats in angiotensin II-dependent. Lab Invest 80:513–527
78. Sechi LA, Griffin CA, Schambelan M (1994) The cardiac renin-angiotensin system in STZ-induced diabetes. Diabetes 43:1180–1184
79. Dostal DE (2000) The cardiac renin-angiotensin system: novel signaling mechanisms related to cardiac growth and function. Regul Pept 91:1–11
80. Nickenig G, Harrison DG (2002) The AT(1)-type angiotensin receptor in oxidative stress and atherogenesis. Part II: AT(1) receptor regulation. Circulation 105:530–536
81. Ozdemir S, Tandogan B, Ulusu NN et al (2009) Angiotensin II receptor blockage prevents diabetes-induced oxidative damage in rat heart. Folia Biol (Praha) 55:11–16
82. Privratsky JR, Wold LE, Sowers JR et al (2003) AT1 blockade prevents glucose-induced cardiac dysfunction in ventricular myocytes: role of the AT1 receptor and NADPH oxidase. Hypertension 42:206–212
83. Gassanov N, Brandt MC, Michels G et al (2006) Angiotensin II-induced changes of calcium sparks and ionic currents in human atrial myocytes: potential role for early remodeling in atrial fibrillation. Cell Calcium 39:175–186
84. Hanatani A, Yoshiyama M, Takeuchi K et al (1998) Angiotensin II type 1-receptor antagonist candesartan cilexitil prevents left ventricular dysfunction in myocardial infarcted rats. Jpn J Pharmacol 78:45–54
85. Ishikawa T, Kajiwara H, Kurihara S (1999) Alterations in contractile properties and Ca^{2+} handling in streptozotocin-induced diabetic rat myocardium. Am J Physiol 277:H2185–H2194
86. Nourooz-Zadeh J, Rahimi A, Tajaddini-Sarmadi J et al (1997) Relationships between plasma measures of oxidative stress and metabolic control in NIDDM. Diabetologia 40:647–653
87. Brown RA, Walsh MF, Ren J (2001) Influence of gender and diabetes on vascular and myocardial contractile function. Endocr Res 27:399–408
88. Ceylan-Isik AF, LaCour KH, Ren J (2006) Sex difference in cardiomyocyte function in normal and metallothionein transgenic mice: the effect of diabetes mellitus. J Appl Physiol 100:1638–1646

89. Shimoni Y, Liu XF (2004) Gender differences in ANG II levels and action on multiple K^+ current modulation pathways in diabetic rats. Am J Physiol Heart Circ Physiol 287:H311–H319
90. Ceriello A (2003) New insights on oxidative stress and diabetic complications may lead to a "causal" antioxidant therapy. Diabetes Care 26:1589–1596
91. Strehlow K, Rotter S, Wassmann S et al (2003) Modulation of antioxidant enzyme expression and function by estrogen. Circ Res 93:170–177
92. Dominguez C, Ruiz E, Gussinye M et al (1998) Oxidative stress at onset and in early stages of type 1 diabetes in children and adolescents. Diabetes Care 21:1736–1742
93. Santini SA, Marra G, Giardina B et al (1997) Defective plasma antioxidant defenses and enhanced susceptibility to lipid peroxidation in uncomplicated IDDM. Diabetes 46:1853–1858
94. Kannel WB, Hjortland M, Castelli WP (1974) Role of diabetes in congestive heart failure: the Framingham Study. Am J Cardiol 34:29–34
95. Regitz-Zagrosek V, Lehmkuhl E, Mahmoodzadeh S (2007) Gender aspects of the role of the metabolic syndrome as a risk factor for cardiovascular disease. Gend Med 4:S162–S177

Part II
Cellular Mechanisms of Diabetic Cardiomyopathy

MicroRNomics of Diabetic Cardiomyopathy

Paras K. Mishra and Suresh C. Tyagi

Abstract MicroRNAs (miRNAs) are a novel class of noncoding, conserved, tiny (19–24 nt) RNAs that regulate gene expression either by RNA interference (RNAi), where they target 3′-UTR and degrade mRNA or repress translation, or by RNA activation (RNAa), where they target promoter elements at 5′-UTR and induce gene transcription. They have emerged as a therapeutic target for diabetes and cardiovascular diseases because each miRNA has several targets that allows it to make a layer of regulatory network. Diabetes is recognized as a multifactorial metabolic disease that increases the chances of heart failure and exacerbates mortality. MiRNAs regulates insulin production, beta-cell differentiation, cardiac hypertrophy, fibrosis, and rhythm, and thereby plays a crucial role in cardiac remodeling in diabetes. Differential expression of circulatory miRNAs has potential as a biomarker for heart failure in diabetes. It is documented that miRNAs regulate inflammation, epigenetic modifications, and autophagy and are altered by matrix metalloproteinase 9, homocysteine, and exercise, which are associated with diabetic cardiomyopathy. This chapter embodies the differentially expressed miRNAs in diabetic hearts, their plausible causes of deregulation, and the therapeutic potential of miRNAs in ameliorating diabetic cardiomyopathy.

Keywords miRNA • Diabetes • Heart failure • Therapy • MMP • Homocysteine • Exercise • Epigenetics

P.K. Mishra (✉)
Department of Cellular and Integrative Physiology, University of Nebraska Medical Center, 985850 Nebraska Medical Center, Omaha, NE 68198, USA
e-mail: paraskumar.mishra@unmc.edu

S.C. Tyagi
Department of Physiology & Biophysics, University of Louisville, Louisville, KY 40202, USA

B. Turan and N.S. Dhalla (eds.), *Diabetic Cardiomyopathy: Biochemical and Molecular Mechanisms*, Advances in Biochemistry in Health and Disease 9, DOI 10.1007/978-1-4614-9317-4_10, © Springer Science+Business Media New York 2014

1 Introduction

MicroRNAs (miRNAs) are conserved, noncoding RNAs of 19–24 nucleotides that maintain the precise levels of genes by either mRNA degradation (if the seed sequence of miRNA matches perfectly with the mRNA of the target gene) or translational repression (if the seed sequence matches imperfectly with the mRNA) [1, 2]. Canonically, miRNAs are recognized to inhibit gene by acting on their 3′-UTR in the cytoplasm. However, recent studies revealed that mature miRNA can reinter the nucleus [3, 4] and bind to the promoter to modulate gene transcription [5–7]. MiRNAs can also target the 5′-UTR [8] and gene termini [9] to regulate gene expression. More than 1,000 miRNAs have been discovered in humans, and they target nearly one third of the total genes, thereby regulating almost all biological processes [10, 11]. miRNAs are associated with cardiovascular diseases and diabetes [2, 12–17] and have emerged as a therapeutic target for diabetic cardiomyopathy [2, 12, 18, 19]. Diabetes is caused by either the deficiency or absence of insulin [destruction of pancreatic beta cells, type 1 (T1D)] or insulin intolerance (type 2, T2D). It is documented that insulin production and its transcription are regulated by miR-15a [20] and miR-30d [21], respectively. There are several miRNAs differentially regulated in the diabetic heart [22] and contribute to cardiac remodeling in diabetes [22–28]. However, the causes of alterations in the levels of miRNAs in the diabetic heart are not clearly understood. Recent studies have revealed that miRNAs play a crucial regulatory role in stem-cell differentiation, apoptosis, and epigenetic modifications [13, 29–32] and have heterogonous targets that regulate different functions [33]. This novel class of small noncoding RNAs opened a new subdiscipline of genomics called microRNomics that is concerned with biogenesis, structure, expression, and regulation and finding the target genes of miRNAs to understand their biology and function [33, 34]. The synergy of miRNA and stem cells is emerging as a novel future therapeutic approach for diabetic cardiomyopathy [2]. Understanding the specific role of differentially expressed miRNAs in the diabetic heart will provide a candidate to develop intervention tools to ameliorate diabetic cardiomyopathy. This chapter describes key miRNAs involved in regulation of cardiac remodeling, stem cell differentiation, and epigenetic modifications, and the future perspective of miRNAs as a therapeutic target for diabetic cardiomyopathy.

2 Biogenesis and Functions of miRNAs

miRNAs are transcribed from introns of protein-coding genes or intergenic regions (individually) or polycistronic transcripts (in cluster) [18, 35]. If the miRNA is transcribed from a very short intron, it is called a mirtron [36]. The mirtrons and intronic miRNAs are transcribed by RNA polymerase II, but intergenic miRNAs are transcribed by either RNA polymerase II or III [37]. The primary transcript of miRNA, called primary miRNA (pri-miRNA), is ~200 kb; it has a 5′- and a 3′-polyA tail. During processing for maturation, at first the 5′-cap is cleaved by a microprocessor complex that contains RNase III endonuclease Drosha and DiGeorge syndrome critical region gene 8 (DGCR8) that reduces the size to ~70 nucleotides, which is called precursor miRNA (pre-miRNA) [38–43]. For maturation, pre-miRNA is translocated from the nucleus to cytoplasm with the help of Exportin5 and Ran GTP, which also protect pre-miRNA from cytoplasmic degradation [44–48]. Mirtron is also transported in the same manner in *Drosophila*; however, in mammals the method of transport is unclear [10, 49]. The pre-miRNA has two strands. The strand that becomes mature miRNA is called the guide strand and is denoted by miRNA*. The other strand is called passenger miRNA and is degraded during maturation [50]. The maturation of pre-miRNA is processed by dicer (an RNase III endonuclease) that degrades the passenger strand and releases the single-stranded mature miRNA, which is immediately loaded into an RNA-induced silencing complex (the detailed mechanism is elaborated in [2]). The mature miRNA inhibits the gene at three levels: (1) if the seed sequence of the miRNA complements perfectly with the 3′-UTR of mRNA sequence, the mRNA will be degraded; (2) if the seed sequence pairs imperfectly with the 3′-UTR of mRNA, the translational machinery will be obstructed by pairing of miRNA to mRNA; and (3) miRNA can be imported into the nucleus with the help of importin8, where it targets promoter and binds to either DNA sequence or nascent cognate transcript derived from the promoter. It can either induce transcription (a process called RNA activation, RNAa) or suppress transcription by transcriptional gene silencing (TGS). The activation (RNAa) or silencing (TGS) depends on recruitment of chromatin-modifying proteins (CMPs) to methylate either H3K4 (RNAa) or H3K9/K27 (TGS) [4].

3 Diabetic Cardiomyopathy

Diabetes, the most rapidly increasing disease in the world [51–53], increases the chances of heart failure [54, 55]. Diabetic cardiomyopathy is defined as diabetes-mediated pathological cardiac remodeling in the absence of coronary artery disease, hypertension, or valvular disease [56]. Diabetes is a multifactorial disease. Therefore, the mechanism of diabetic cardiomyopathy is complex. However, understanding the mechanism of diabetic cardiomyopathy is important for mitigating heart failure caused by diabetes or reversing the pathological cardiac remodeling [57–60]. There are three major stages of diabetic cardiomyopathy: (i) The early stage is asymptomatic wherein changes occur at cellular and metabolic levels but are not recognized by systolic dysfunction; (ii) the middle stage is recognized by decrease in ejection fraction (EF <50 %) accompanied by a slight increase in left ventricular size and diastolic dysfunction. At this stage apoptosis, insulin resistance, and fibrosis are induced. (iii) The late stage is characterized by both systolic and diastolic dysfunction, coronary artery diseases, and cardiovascular autonomic neuropathy [13]. One of the mechanisms of diabetic cardiomyopathy is metabolic perturbation in which the heart is unable to utilize carbohydrates at increased workloads or starvation and utilizes free fatty acid, which decreases glucose transporter 1 and -4 [61–63]. Other mechanisms include (1) insulin resistance [58, 64] and activation of peroxisome proliferator-activated receptor-α [65] by elevated levels of free fatty acids; (2) induction of reactive oxygen species [66–68] that induce matrix metalloproteinase 9 [69], which contributes to cardiac fibrosis [69, 70] and contractile dysfunction [71], (3) up regulation of inflammatory cytokines [22, 72], (4) attenuation of beta-1 and -2-adrenergic receptors [73–75], and (5) changes in the myocardial autonomic neurotransmitters [76]. A high-fat diet and obesity also contributes to insulin resistance, which along with beta-cell destruction causes structural, functional, and regulatory remodeling in the heart that leads to diabetic cardiomyopathy [13].

4 miRNAs in the Diabetic Heart

miRNAs are the key regulator of diabetic cardiomyopathy as they regulate the processes from beta-cell differentiation to insulin production and cardiac remodeling to heart failure, suggesting that diabetes is a miRNA-related disease [2, 20, 21, 77–80]. In the diabetic heart, miR-133 is attenuated [81]. MiR-133 regulates cardiac hypertrophy [82], and fibrosis [83, 84] and arterial remodeling [85]. Similarly, miR-1, -29, -30, and -150 are downregulated in the diabetic heart and contribute to cardiac hypertrophy [86, 87]. On the other hand, upregulation of miR-23, -24,-125,-129, -195, -199, and -208 are associated with hypertrophy [87]. Cardiac fibrosis is upregulated by inhibition of miR-29 [88] and induction of miR-21 [89]. The upregulation of miR-503 in diabetes causes endothelial dysfunction, which is mitigated by inhibition of miR-503 [24]. miR-221 is induced in the diabetic condition and contributes to endothelial dysfunction and attenuation of c-kit [25]. Glucose transporter 4 is attenuated in diabetics and is regulated by miR-223, which induces glucose transporter 4 [26]. The role of different miRNAs in diabetic heart failure is also elaborated in several excellent reviews [2, 12, 13, 18, 78]. miRNAs are also associated with epigenetic modifications. Recently, miR-133 has been shown to regulate DNA methyl transferase 1, -3a, and 3b. These methyl transferases are altered in diabetic hearts and induce DNA methylation [29].

miRNAs are also present in the blood, and they circulate through the blood in the body. These miRNAs are called circulating miRNAs. The differential expression of circulating miRNAs correlates with tumor progression and has emerged as a potential biomarker for cancer [90–94]. Similarly, circulating miRNAs are documented as biomarkers for heart failure [14, 95–99]. The differential expression of circulating miRNAs can be assessed in the different stages of diabetic cardiomyopathy and can be used as a biomarker for diabetic cardiomyopathy.

5 miRNAs as Therapeutic Targets for Diabetic Cardiomyopathy

Diabetic cardiomyopathy is a progressive disease with three different stages, starting from the asymptomatic (early) stage to diastolic and systolic dysfunction (late) stage. The heart is a sophisticated and vulnerable organ and is tightly regulated by miRNAs [18]. In diabetics, the glucose-mediated alterations in the metabolism, inflammation, reactive oxygen species-triggered apoptosis, and extracellular matrix degradation changes the microenvironment of the heart and induces pathological cardiac remodeling [13, 18]. These alterations also deregulate those miRNAs that maintain the homeostasis of the heart. Therefore, investigating those miRNAs that are altered in the diabetic heart is important. For example, abrogation of only miR-133 results in cardiac hypertrophy [82], and it is attenuated in the diabetic heart [81]. Also, the transgenic expression of miR-133 protects the heart from pressure overload-induced cardiac fibrosis [83]. Therefore, miR-133 is an important miRNA in diabetic cardiomyopathy. To protect the diabetic heart from pathological remodeling, overexpression of miR-133 could be an important intervention tool. The levels of different circulating miRNAs can be assessed in the different stages of diabetic cardiomyopathy and compared with the control to diagnose the specific miRNAs differentially expressed in the three stages of diabetic cardiomyopathy. Similarly, the unique miRNAs in the different stages of diabetic cardiomyopathy can be analyzed and used as biomarkers for specific stages of cardiomyopathy. Therefore, miRNA has great promise as a biomarker and therapeutic target for diabetic cardiomyopathy.

6 Conclusions

miRNAs play pivotal roles in diabetic cardiomyopathy. MiRNA profiling and targeting specific miRNA(s) is a promising approach for the therapy of diabetic cardiomyopathy.

Acknowledgments This work is supported in part with by National Institute of Heart HL-113281 and HL-116205 to Paras K. Mishra and HL-108621 and HL-74185 to Suresh C Tyagi.

References

1. Bartel DP (2009) MicroRNAs: target recognition and regulatory functions. Cell 136:215–233
2. Tyagi AC, Sen U, Mishra PK (2011) Synergy of microRNA and stem cell: a novel therapeutic approach for diabetes mellitus and cardiovascular diseases. Curr Diabetes Rev 7:367–376
3. Huang V, Li LC (2012) miRNA goes nuclear. RNA Biol 9:269–273
4. Liao JY, Ma LM, Guo YH, Zhang YC et al (2010) Deep sequencing of human nuclear and cytoplasmic small RNAs reveals an unexpectedly complex subcellular distribution of miRNAs and tRNA 3′ trailers. PLoS One 5:e10563

5. Huang V, Place RF, Portnoy V, Wang J et al (2012) Upregulation of cyclin B1 by miRNA and its implications in cancer. Nucleic Acids Res 40:1695–1707
6. Place RF, Li LC, Pookot D, Noonan EJ et al (2008) MicroRNA-373 induces expression of genes with complementary promoter sequences. Proc Natl Acad Sci USA 105:1608–1613
7. Younger ST, Corey DR (2011) Transcriptional gene silencing in mammalian cells by miRNA mimics that target gene promoters. Nucleic Acids Res 39:5682–5691
8. Lytle JR, Yario TA, Steitz JA (2007) Target mRNAs are repressed as efficiently by microRNA-binding sites in the 5′ UTR as in the 3′ UTR. Proc Natl Acad Sci USA 104:9667–9672
9. Younger ST, Corey DR (2011) Transcriptional regulation by miRNA mimics that target sequences downstream of gene termini. Mol Biosyst 7:2383–2388
10. Berezikov E, Chung WJ, Willis J, Cuppen E et al (2007) Mammalian mirtron genes. Mol Cell 28:328–336
11. Lewis BP, Burge CB, Bartel DP (2005) Conserved seed pairing, often flanked by adenosines, indicates that thousands of human genes are microRNA targets. Cell 120:15–20
12. Asrih M, Steffens S (2013) Emerging role of epigenetics and miRNA in diabetic cardiomyopathy. Cardiovasc Pathol 22:117–125
13. Chavali V, Tyagi SC, Mishra PK (2013) Predictors and prevention of diabetic cardiomyopathy. Diabetes Metab Syndr Obes 6:151–160
14. Tijsen AJ, Pinto YM, Creemers EE (2012) Circulating microRNAs as diagnostic biomarkers for cardiovascular diseases. Am J Physiol Heart Circ Physiol 303:H1085–H1095
15. Latronico MV, Condorelli G (2011) MicroRNAs in hypertrophy and heart failure. Exp Biol Med (Maywood) 236:125–131
16. Port JD, Sucharov C (2010) Role of microRNAs in cardiovascular disease: therapeutic challenges and potentials. J Cardiovasc Pharmacol 56:444–453
17. Topkara VK, Mann DL (2011) Role of microRNAs in cardiac remodeling and heart failure. Cardiovasc Drugs Ther 25:171–182
18. Mishra PK, Tyagi N, Kumar M, Tyagi SC (2009) MicroRNAs as a therapeutic target for cardiovascular diseases. J Cell Mol Med 13:778–789
19. Montgomery RL, Hullinger TG, Semus HM, Dickinson BA et al (2011) Therapeutic inhibition of miR-208a improves cardiac function and survival during heart failure. Circulation 124: 1537–1547
20. Sun LL, Jiang BG, Li WT, Zou JJ et al (2011) MicroRNA-15a positively regulates insulin synthesis by inhibiting uncoupling protein-2 expression. Diabetes Res Clin Pract 91:94–100
21. Tang X, Muniappan L, Tang G, Ozcan S (2009) Identification of glucose-regulated miRNAs from pancreatic beta cells reveals a role for miR-30d in insulin transcription. RNA 15:287–293
22. Chavali V, Tyagi SC, Mishra PK (2013) Differential expression of dicer, miRNAs, and inflammatory markers in diabetic Ins2$^{+/-}$ Akita hearts. Cell Biochem Biophys. June 15, Epub ahead of print
23. He A, Zhu L, Gupta N, Chang Y et al (2007) Overexpression of micro ribonucleic acid 29, highly up-regulated in diabetic rats, leads to insulin resistance in 3T3-L1 adipocytes. Mol Endocrinol 21:2785–2794
24. Caporali A, Meloni M, Vollenkle C, Bonci D et al (2011) Deregulation of microRNA-503 contributes to diabetes mellitus-induced impairment of endothelial function and reparative angiogenesis after limb ischemia. Circulation 123:282–291
25. Li Y, Song YH, Li F, Yang T et al (2009) MicroRNA-221 regulates high glucose-induced endothelial dysfunction. Biochem Biophys Res Commun 381:81–83
26. Lu H, Buchan RJ, Cook SA (2010) MicroRNA-223 regulates Glut4 expression and cardiomyocyte glucose metabolism. Cardiovasc Res 86:410–420
27. Shan ZX, Lin QX, Deng CY, Zhu JN et al (2010) miR-1/miR-206 regulate Hsp60 expression contributing to glucose-mediated apoptosis in cardiomyocytes. FEBS Lett 584:3592–3600
28. Wang XH, Qian RZ, Zhang W, Chen SF (2009) MicroRNA-320 expression in myocardial microvascular endothelial cells and its relationship with insulin-like growth factor-1 in type 2 diabetic rats. Clin Exp Pharmacol Physiol 36:181–188

29. Chavali V, Tyagi SC, Mishra PK (2012) MicroRNA-133a regulates DNA methylation in diabetic cardiomyocytes. Biochem Biophys Res Commun 425:668–672
30. Glass C, Singla DK (2011) MicroRNA-1 transfected embryonic stem cells enhance cardiac myocyte differentiation and inhibit apoptosis by modulating the PTEN/Akt pathway in the infarcted heart. Am J Physiol Heart Circ Physiol 301:H2038–H2049
31. Liu C, Teng ZQ, McQuate AL, Jobe EM et al (2013) An epigenetic feedback regulatory loop involving microRNA-195 and MBD1 governs neural stem cell differentiation. PLoS One 8:e51436
32. Vinas JL, Ventayol M, Brune B, Jung M et al (2013) miRNA let-7e modulates the Wnt pathway and early nephrogenic markers in mouse embryonic stem cell differentiation. PLoS One 8:e60937
33. Wang Y, Du L, Li X, Zhang S et al (2011) Functional homogeneity in microRNA target heterogeneity: a new sight into human microRNomics. OMICS 15:25–35
34. Zhang C (2008) MicroRNomics: a newly emerging approach for disease biology. Physiol Genomics 33:139–147
35. van Rooij E, Olson EN (2007) MicroRNAs: powerful new regulators of heart disease and provocative therapeutic targets. J Clin Invest 117:2369–2376
36. Krol J, Loedige I, Filipowicz W (2010) The widespread regulation of microRNA biogenesis, function and decay. Nat Rev Genet 11:597–610
37. Macfarlane LA, Murphy PR (2010) MicroRNA: biogenesis, function and role in cancer. Curr Genomics 11:537–561
38. Denli AM, Tops BB, Plasterk RH, Ketting RF et al (2004) Processing of primary microRNAs by the microprocessor complex. Nature 432:231–235
39. Gregory RI, Yan KP, Amuthan G, Chendrimada T et al (2004) The microprocessor complex mediates the genesis of microRNAs. Nature 432:235–240
40. Han J, Lee Y, Yeom KH, Kim YK et al (2004) The Drosha-DGCR8 complex in primary microRNA processing. Genes Dev 18:3016–3027
41. Han J, Lee Y, Yeom KH, Nam JW et al (2006) Molecular basis for the recognition of primary microRNAs by the Drosha-DGCR8 complex. Cell 125:887–901
42. Lee Y, Jeon K, Lee JT, Kim S et al (2002) MicroRNA maturation: stepwise processing and subcellular localization. EMBO J 21:4663–4670
43. Lee Y, Ahn C, Han J, Choi H et al (2003) The nuclear RNase III Drosha initiates microRNA processing. Nature 425:415–419
44. Bohnsack MT, Czaplinski K, Gorlich D (2004) Exportin 5 is a RanGTP-dependent dsRNA-binding protein that mediates nuclear export of pre-miRNAs. RNA 10:185–191
45. Lund E, Dahlberg JE (2006) Substrate selectivity of exportin 5 and Dicer in the biogenesis of microRNAs. Cold Spring Harb Symp Quant Biol 71:59–66
46. Okada C, Yamashita E, Lee SJ, Shibata S (2009) A high-resolution structure of the pre-microRNA nuclear export machinery. Science 326:1275–1279
47. Yi R, Qin Y, Macara IG, Cullen BR (2003) Exportin-5 mediates the nuclear export of pre-microRNAs and short hairpin RNAs. Genes Dev 17:3011–3016
48. Zeng Y, Cullen BR (2004) Structural requirements for pre-microRNA binding and nuclear export by Exportin 5. Nucleic Acids Res 32:4776–4785
49. Okamura K, Hagen JW, Duan H, Tyler DM (2007) The mirtron pathway generates microRNA-class regulatory RNAs in *Drosophila*. Cell 130:89–100
50. Du T, Zamore PD (2005) microPrimer: the biogenesis and function of microRNA. Development 132:4645–4652
51. King H, Aubert RE, Herman WH (1998) Global burden of diabetes, 1995–2025: prevalence, numerical estimates, and projections. Diabetes Care 21:1414–1431
52. Tabak AG, Herder C, Rathmann W, Brunner EJ et al (2012) Prediabetes: a high-risk state for diabetes development. Lancet 379:2279–2290
53. Wild S, Roglic G, Green A, Sicree R et al (2004) Global prevalence of diabetes: estimates for the year 2000 and projections for 2030. Diabetes Care 27:1047–1053

54. Mathew V, Gersh BJ, Williams BA, Laskey WK et al (2004) Outcomes in patients with diabetes mellitus undergoing percutaneous coronary intervention in the current era: a report from the Prevention of REStenosis with Tranilast and its Outcomes (PRESTO) trial. Circulation 109:476–480
55. Pignone M, Alberts MJ, Colwell JA, Cushman M et al (2010) Aspirin for primary prevention of cardiovascular events in people with diabetes: a position statement of the American Diabetes Association, a scientific statement of the American Heart Association, and an expert consensus document of the American College of Cardiology Foundation. Diabetes Care 33:1395–1402
56. Rubler S, Dlugash J, Yuceoglu YZ, Kumral T et al (1972) New type of cardiomyopathy associated with diabetic glomerulosclerosis. Am J Cardiol 30:595–602
57. Aneja A, Tang WH, Bansilal S, Garcia MJ, Farkouh ME (2008) Diabetic cardiomyopathy: insights into pathogenesis, diagnostic challenges, and therapeutic options. Am J Med 121:748–757
58. Maisch B, Alter P, Pankuweit S (2011) Diabetic cardiomyopathy—fact or fiction? Herz 36: 102–115
59. Miki T, Yuda S, Kouzu H, Miura T (2013) Diabetic cardiomyopathy: pathophysiology and clinical features. Heart Fail Rev 18:149–166
60. Sharma V, McNeill JH (2006) Diabetic cardiomyopathy: where are we 40 years later? Can J Cardiol 22:305–308
61. Depre C, Shipley GL, Chen W, Han Q (1998) Unloaded heart in vivo replicates fetal gene expression of cardiac hypertrophy. Nat Med 4:1269–1275
62. Goodwin GW, Taylor CS, Taegtmeyer H (1998) Regulation of energy metabolism of the heart during acute increase in heart work. J Biol Chem 273:29530–29539
63. Rodrigues B, Cam MC, McNeill JH (1998) Metabolic disturbances in diabetic cardiomyopathy. Mol Cell Biochem 180:53–57
64. Poornima IG, Parikh P, Shannon RP (2006) Diabetic cardiomyopathy: the search for a unifying hypothesis. Circ Res 98:596–605
65. Boudina S, Abel ED (2006) Mitochondrial uncoupling: a key contributor to reduced cardiac efficiency in diabetes. Physiology (Bethesda) 21:250–258
66. Du X, Matsumura T, Edelstein D, Rossetti L et al (2003) Inhibition of GAPDH activity by poly(ADP-ribose) polymerase activates three major pathways of hyperglycemic damage in endothelial cells. J Clin Invest 112:1049–1057
67. Nishikawa T, Edelstein D, Brownlee M (2000) The missing link: a single unifying mechanism for diabetic complications. Kidney Int Suppl 77:S26–S30
68. Nishikawa T, Edelstein D, Du XL, Yamagishi S et al (2000) Normalizing mitochondrial superoxide production blocks three pathways of hyperglycaemic damage. Nature 404:787–790
69. Mishra PK, Tyagi N, Sen U, Joshua IG et al (2010) Synergism in hyperhomocysteinemia and diabetes: role of PPAR-gamma and tempol. Cardiovasc Diabetol 9:49
70. Mishra PK, Chavali V, Metreveli N, Tyagi SC (2012) Ablation of MMP9 induces survival and differentiation of cardiac stem cells into cardiomyocytes in the heart of diabetics: a role of extracellular matrix. Can J Physiol Pharmacol 90:353–360
71. Mishra PK, Metreveli N, Tyagi SC (2010) MMP-9 gene ablation and TIMP-4 mitigate PAR-1-mediated cardiomyocyte dysfunction: a plausible role of dicer and miRNA. Cell Biochem Biophys 57:67–76
72. Wakasaki H, Koya D, Schoen FJ, Jirousek MR et al (1997) Targeted overexpression of protein kinase C beta2 isoform in myocardium causes cardiomyopathy. Proc Natl Acad Sci USA 94:9320–9325
73. Mishra PK, Givvimani S, Metreveli N, Tyagi SC (2010) Attenuation of beta2-adrenergic receptors and homocysteine metabolic enzymes cause diabetic cardiomyopathy. Biochem Biophys Res Commun 401:175–181
74. Mishra PK, Awe O, Metreveli N, Qipshidze N et al (2011) Exercise mitigates homocysteine—beta2-adrenergic receptor interactions to ameliorate contractile dysfunction in diabetes. Int J Physiol Pathophysiol Pharmacol 3:97–106

75. Wang G, Zhu X, Xie W, Han P et al (2010) Rad as a novel regulator of excitation-contraction coupling and beta-adrenergic signaling in heart. Circ Res 106:317–327
76. Tarquini R, Lazzeri C, Pala L, Rotella CM, Gensini GF (2011) The diabetic cardiomyopathy. Acta Diabetol 48:173–181
77. Guay C, Roggli E, Nesca V, Jacovetti C et al (2011) Diabetes mellitus, a microRNA-related disease? Transl Res 157:253–264
78. Hennessy E, O'Driscoll L (2008) Molecular medicine of microRNAs: structure, function and implications for diabetes. Expert Rev Mol Med 10:e24
79. Poy MN, Eliasson L, Krutzfeldt J, Kuwajima S et al (2004) A pancreatic islet-specific microRNA regulates insulin secretion. Nature 432:226–230
80. Zampetaki A, Kiechl S, Drozdov I, Willeit P et al (2010) Plasma microRNA profiling reveals loss of endothelial miR-126 and other microRNAs in type 2 diabetes. Circ Res 107:810–817
81. Feng B, Chen S, George B, Feng Q et al (2010) miR133a regulates cardiomyocyte hypertrophy in diabetes. Diabetes Metab Res Rev 26:40–49
82. Care A, Catalucci D, Felicetti F, Bonci D et al (2007) MicroRNA-133 controls cardiac hypertrophy. Nat Med 13:613–618
83. Matkovich SJ, Wang W, Tu Y, Eschenbacher WH et al (2010) MicroRNA-133a protects against myocardial fibrosis and modulates electrical repolarization without affecting hypertrophy in pressure-overloaded adult hearts. Circ Res 106:166–175
84. Duisters RF, Tijsen AJ, Schroen B, Leenders JJ et al (2009) miR-133 and miR-30 regulate connective tissue growth factor: implications for a role of microRNAs in myocardial matrix remodeling. Circ Res 104:170–178, 6p
85. Shan H, Zhang Y, Lu Y, Zhang Y et al (2009) Downregulation of miR-133 and miR-590 contributes to nicotine-induced atrial remodelling in canines. Cardiovasc Res 83:465–472
86. Chen JF, Murchison EP, Tang R, Callis TE et al (2008) Targeted deletion of Dicer in the heart leads to dilated cardiomyopathy and heart failure. Proc Natl Acad Sci USA 105:2111–2116
87. Latronico MV, Elia L, Condorelli G, Catalucci D (2008) Heart failure: targeting transcriptional and post-transcriptional control mechanisms of hypertrophy for treatment. Int J Biochem Cell Biol 40:1643–1648
88. van Rooij E, Sutherland LB, Thatcher JE, DiMaio JM et al (2008) Dysregulation of microRNAs after myocardial infarction reveals a role of miR-29 in cardiac fibrosis. Proc Natl Acad Sci USA 105:13027–13032
89. Thum T, Gross C, Fiedler J, Fischer T et al (2008) MicroRNA-21 contributes to myocardial disease by stimulating MAP kinase signalling in fibroblasts. Nature 456:980–984
90. Brase JC, Wuttig D, Kuner R, Sultmann H (2010) Serum microRNAs as non-invasive biomarkers for cancer. Mol Cancer 9:306
91. Brase JC, Johannes M, Schlomm T, Falth M et al (2011) Circulating miRNAs are correlated with tumor progression in prostate cancer. Int J Cancer 128:608–616
92. Cortez MA, Bueso-Ramos C, Ferdin J, Lopez-Berestein G et al (2011) MicroRNAs in body fluids—the mix of hormones and biomarkers. Nat Rev Clin Oncol 8:467–477
93. Kosaka N, Iguchi H, Ochiya T (2010) Circulating microRNA in body fluid: a new potential biomarker for cancer diagnosis and prognosis. Cancer Sci 101:2087–2092
94. Tie Y, Liu B, Fu H, Zheng X (2009) Circulating miRNA and cancer diagnosis. Sci China C Life Sci 52:1117–1122
95. Ajit SK (2012) Circulating microRNAs as biomarkers, therapeutic targets, and signaling molecules. Sensors (Basel) 12:3359–3369
96. Dimmeler S, Zeiher AM (2010) Circulating microRNAs: novel biomarkers for cardiovascular diseases? Eur Heart J 31:2705–2707
97. Fichtlscherer S, Zeiher AM, Dimmeler S (2011) Circulating microRNAs: biomarkers or mediators of cardiovascular diseases? Arterioscler Thromb Vasc Biol 31:2383–2390
98. Li C, Pei F, Zhu X, Duan DD et al (2012) Circulating microRNAs as novel and sensitive biomarkers of acute myocardial infarction. Clin Biochem 45:727–732
99. Scholer N, Langer C, Kuchenbauer F (2011) Circulating microRNAs as biomarkers—True Blood? Genome Med 3:72

Cellular Mechanism Underlying the Misfunction of Cardiac Ionic Channels in Diabetes

Mónica Gallego and Oscar Casis

Abstract The alterations observed in the ECG of diabetic patients correspond to alterations in the repolarizing currents of the cardiac action potential. Diabetes affects the amplitude and kinetics of most of the potassium currents by modifying the biophysical behavior and even the expression levels of the potassium channel-forming proteins. We review here the effects of diabetes mellitus on individual currents, as well as the intracellular mechanisms that mediate them. Then, we discuss how diabetes modifies the regulation of calcium and potassium channels, with particular emphasis on the transient outward potassium current. Impaired sympathetic activity, loss of insulin-mediated trophic effect, and impaired metabolic status have major effects on the normal functioning of potassium channels in the diabetic heart.

Keywords AMPK • Ion channels • Diabetes • Repolarizing • Expression • Insulin • Sympathetic • MAPK • Metabolic status • Adrenoceptors

1 Introduction

The functioning of the heart consists of a mechanical process of blood pumping, which takes place in response to a specific electrical order: the cardiac action potential (AP). In humans, the cardiac action potential is the result of depolarizing sodium

M. Gallego • O. Casis (✉)
Lascaray Research Center, University of the Basque Country (UPV/EHU), Av. Miguel de Unamuno 3, 01006 Vitoria, Spain

Departamento de Fisiología, Facultad de Farmacia, University of the Basque Country (UPV/EHU), Paseo de la Universidad 7, 01006 Vitoria, Spain
e-mail: oscar.casis@ehu.es

B. Turan and N.S. Dhalla (eds.), *Diabetic Cardiomyopathy: Biochemical and Molecular Mechanisms*, Advances in Biochemistry in Health and Disease 9,
DOI 10.1007/978-1-4614-9317-4_11,

and calcium currents and several repolarizing potassium currents, each of them driven by a particular combination of membrane-located ionic channels. Alterations in these proteins often affect the shape or duration of the action potential, causing arrhythmias than can ultimately alter the mechanical function of the heart.

Among the pathologies that severely affect the electrical properties of the cardiac tissue, diabetes mellitus appears [1]. The fact that diabetic patients may die suddenly and without any obvious cause has long intrigued scientists. In this sense, the electrocardiogram of diabetic patients shows prolonged QTc (QT interval corrected for heart rate) and increased QT dispersion (QTd), alterations typically associated with a greater predisposition to suffer arrhythmias and sudden death [2–5].

Diabetic laboratory animals have been used extensively in the study of the function of cardiac ionic channels. Diabetes is typically induced by the administration of the pancreatic beta-cell toxins streptozotocin or alloxan [6]. Thus, the lengthening of repolarization observed in the electrocardiograms of diabetic patients is reproduced in diabetic animals [7].

Although one single sodium current, I_{Na}, causes the rapid initial depolarization phase of the cardiac action potential, the complex repolarization takes place by the balance between inward calcium and outward potassium currents. The results from the ECG and the action potential recordings indicate that the repolarization is the major phase affected by diabetes. The prolonged QTc and the increase in QTd reflect a lengthened repolarization in diabetic cardiac myocytes that is corroborated by prolonged action potential duration, especially at the endocardium [8, 9].

2 Diabetes Alters the Cardiac Ionic Currents

2.1 *Alterations in Na^+ and Ca^{2+} Inward Currents*

The absence of changes in both cardiac conduction in the ECG and the depolarization phase of the action potential indicates that there is no significant effect of diabetes on cardiac sodium currents. The only alteration reported over the I_{Na} comes from Feuvray's laboratory, who found that diabetes decreases a slowly inactivating component of the sodium current [10].

Because the plateau phase of the cardiac AP is the result of a fine balance between inward calcium and outward potassium currents, any alterations of that balance might cause the increased action potential duration observed in diabetic cardiomyocytes. In adult ventricular myocytes from most species, the L type is the only calcium channel found in the sarcoplasmic membrane, and the increase in the amplitude of L-type calcium current (I_{Ca-L}), as well as alteration of the current kinetics, could lengthen AP.

The first research work on the characteristics of ionic currents in the diabetic heart showed no alteration in the L-type calcium current [11, 12]. More recent work has confirmed that I_{Ca-L} amplitude, inactivation kinetics, and voltage dependence remain unchanged in diabetic ventricular myocytes, and that the expression of the pore-forming subunit of the channel is not significantly altered by diabetes [7, 13]. However, it is necessary to point out that these experiments were performed in conditions of intracellular calcium buffering; the relevance of this is discussed next.

2.2 *Alterations in K^+ Outward Currents*

Sodium and calcium channels consist of one α-subunit that forms the pore, senses the voltage, and provides ion selectivity, but in the case of potassium channels four α-subunits must combine to form a functional pore. The great variety of α-subunits that exists for K^+ channels makes this group the most diverse among the voltage-dependent ionic channels. Further, α-subunits form either homo- or heterotetramers, usually associated with auxiliary subunits such as beta subunits (Kvβ), K^+ channel-interacting proteins (KChIPs), and dipeptidylpeptidases (DPPX), thereby increasing even more the diversity of functional K^+ channels [14].

Although important variations between species exist, the main K^+ repolarizing currents in the human heart are the fast and the slow transient outward currents $I_{to,f}$ and $I_{to,s}$; the ultrarapid, the rapid, and the slow delayed rectifiers I_{Kur}, I_{Kr} and I_{Ks}; and the inward rectifier I_{K1} [15–19]. The transient outward potassium current determines the rapid repolarization or phase 1 of the cardiac action potential and affects the plateau phase. The delayed rectifiers maintain and finish the plateau and eventually repolarize the cell. Last, the inward rectifier keeps the resting membrane potential.

In contrast with the unaltered calcium current, reduced potassium permeability in the membranes of diabetic cells has been consistently reported, from initial work performed over total K^+ current up to the recent biophysical studies of individual channels. The picture that emerges is that diabetes alters most of the potassium currents, particularly $I_{to,fast}$, $I_{to,slow}$, I_{Kur}, and I_{Ks}, whereas I_{K1} and I_{Kr} remain unchanged and are the exception [7–9, 13, 20].

The early studies in potassium currents were performed over total K^+ current, which was subsequently analyzed by fragments. In the rat heart, this indirect method suggested that diabetes mellitus reduced the amplitude of the transient outward and the delayed rectifiers, but not the inward rectifier I_{K1} [8, 11, 21]. However, it is important to remember that in the normal heart the AP duration is not uniform throughout the ventricles. In rats, it is shorter in the right ventricle than in the left, and also shorter in the epicardium compared to the endocardium [22, 23]. So, when the regional variability was included in the study, the results showed that diabetes mellitus reduces the amplitude of the transient outward and the delayed rectifier K^+ currents, but in a different proportion. Thus, the decrease of I_{to} and I_{ss} currents amplitude is marked and homogeneous. Conversely, the decrease in I_K current varies between 10 % in the right ventricle and the 50 % in the endocardium of the left ventricle [9]. Therefore, this uneven reduction in I_K amplitude is the responsible for the regional differences in APD in the diabetic rat heart.

Later, new electrophysiological and pharmacological protocols were designed enabling the recording, and therefore the direct measurement, of individual currents without overlapping or contamination [24, 25]. Then, the easy method for inducing diabetes in rodents, a single injection of streptozotocin, together with its large amplitude, has made the rat cardiac I_{to} by far the most extensively studied repolarizing current in diabetes. The amplitude of this fast transient outward K^+ current is reduced in cardiac myocytes isolated from diabetic rats, mice, and dogs. At the same time there is a reduction in the expression of the Kv4.2 and Kv4.3 proteins, the pore-forming α-subunits responsible for this current in the different species [8, 9, 13, 20].

When the decrease in the cardiac I_{to} current amplitude was first reported in diabetes [8], the delayed rectifier potassium currents I_{Kr} and I_{Ks} were just described [26]. After some attempts with the non-human-like rat I_K [9, 21], and some nonreproducible work in diabetic rabbits [27, 28], the first reliable studies on the effects of diabetes on the delayed rectifiers I_{Kr} and I_{Ks} came at last, showing that the two currents are not affected by diabetes in a similar manner.

Two consecutive studies in diabetic rabbits and then in diabetic dogs reported no effect of diabetes in the rapid delayed rectifier I_{Kr} [7, 13]. Thus, the amplitude and the biophysical behavior of the current, including voltage dependence and activation/deactivation kinetics, were similar in healthy and diabetic ventricular myocytes. Moreover, the expression of the ERG protein, the pore-forming α-subunit of I_{Kr}, is also unaffected by diabetes in the dog heart. Similar results have been found very recently in diabetic rabbits [20].

In contrast, I_{Ks} current amplitude is reduced in cardiomyocytes isolated from diabetic rabbits and dogs. Other parameters such as current voltage dependence or kinetics are not altered [7, 13]. Interestingly, the decreased I_{Ks} amplitude is not caused by reduced synthesis of the channel-forming protein Kv7.1, but rather the accessory subunit minK [13]. However, as Varro's group elegantly postulates, the relevance of this finding lies on the possibility that diabetes mellitus diminish the cardiac repolarization reserve. Thus, if cardiac demand increases, in a condition where several potassium currents are decreased they may not be able to compensate for each other to ensure a normal repolarization [7, 29].

It is important to point out that in the electrical behavior of ionic currents, the amplitude is not the only biophysical parameter that can be affected. A good example could be the transient outward potassium current, because diabetes slows down the capability of the I_{to} channels to recover from inactivation [21, 30], and the inactivation of I_{to} current is accelerated in cardiomyocytes isolated from diabetic rats [8, 9]. The combined effect of slower recovery and faster inactivation is that in the diabetic heart there are fewer I_{to} channels capable of driving potassium.

3 Diabetes Alters the Regulation of Ionic Channels

3.1 Alterations on the Regulation of $I_{Ca\text{-}L}$

Diabetic cardiomyopathy affects neither the expression of the pore-forming subunit of the L-type calcium channel nor current amplitude, inactivation kinetics, or voltage dependence [7, 11–13]. However, as mentioned previously, it is important to remind that those experiments were performed under intracellular calcium-buffering conditions: this is a key methodological aspect that can determine the results. In fact, experiments where intracellular medium is not artificially regulated yielded opposite results: diabetes reduces the amplitude and slows down the inactivation kinetics of $I_{Ca\text{-}L}$ in myocytes [31, 32].

Intracellular calcium handling is severely compromised in diabetic myocytes, and these alterations include reduced calcium content, release and reuptake by the sarcoplasmic reticulum, and reduced activity of the Na–Ca exchanger in the sarcoplasmic membrane [32–39]. Because the I_{Ca}-L current is strongly modulated by intracellular calcium [40], the alterations found by Wang et al. and Chattou et al. [31, 32] might be secondary to the impaired channel modulation by intracellular calcium ions. Thus, the conclusion of these apparently contradictory works might be that diabetes does not affect the abundance and behavior of the cardiac L-type calcium channel, but rather its regulation by intracellular calcium.

3.2 Alterations in the Regulation of K^+ Currents

In a given current, alterations in the amplitude can be caused by changes in the conductivity or changes in the amount of channel proteins. This second hypothesis is the more usual and can be achieved by modifications in the synthesis of the proteins that form the channel. In this regard, there are positive regulators such as insulin or norepinephrine that are required for the expression of the channel at physiological levels and therefore act as trophic factors [30, 41]. On the other hand, molecules such as AMP kinase and tumor necrosis factor (TNF) behave as negative regulators, reducing the channel expression [42, 43].

3.2.1 Acute Regulation by the Sympathetic Nervous System

One of the characteristics of diabetes mellitus is the development of cardiac autonomic neuropathy, defined by abnormalities in the parasympathetic and sympathetic nervous system, that is in fact a significant cause of morbidity and mortality among diabetic patients [44].

In the normal heart, norepinephrine reduces transient outward potassium current amplitude trough a pathway that requires the stimulation of α1-adrenoceptors [45]. However, diabetes mellitus reduces the responsiveness of several tissues such as the vas deferens, atria, or ventricles toward sympathetic stimulation [46–49]. Thus, in ventricular cardiomyocytes isolated from healthy rats norepinephrine exposure decreases I_{to} current amplitude in a concentration-dependent manner. Diabetes shifts the norepinephrine concentration–response curve to the right and reduces the maximum effect of the neurotransmitter [48]. In diabetic cells norepinephrine causes a decrease in current amplitude, but this reduction is smaller than in healthy cardiomyocytes. This resistance of diabetic myocardium toward sympathetic regulation is one of the multiple manifestations of the progressive development of diabetic cardiomyopathy, the ventricular dysfunction that occurs independently of coronary artery disease and hypertension [1, 50].

In the case of I_{to} regulation, this resistance might be caused by the reduction in cardiac α1-adrenoceptor levels that occurs in streptozotocin-diabetic animals [51]. There are no data regarding the sympathetic regulation of other K^+ currents in diabetic animals. However, because I_{Kr} is also modulated by α1-adrenoceptors [52], one can hypothesize that the response of I_{Kr} toward sympathetic stimulation would also be reduced.

3.2.2 Long-Term Regulation by the Sympathetic Nervous System

Although norepinephrine release from the sympathetic terminals acutely reduces the amplitude of K^+ currents such as I_{to} or I_{Kr} through the activation of α1-adrenoceptors [48, 52], long-term sympathetic activity, now acting through β-adrenoceptors, serves as a trophic factor ensuring the maintenance of ionic currents [41]. Moreover, the sympathetic nervous system plays a critical role in the development and/or maintenance of sodium, calcium, and potassium currents [53–56]. Hence, pathology affecting the sympathetic innervation of the heart often reduces the amplitude of ionic currents. This change is very evident in the case of the transient outward potassium current, which is diminished in the acute state of Chagas' disease when degeneration of sympathetic nerve terminals is present [57, 58]. Similarly, the reduction in I_{to} current amplitude observed in diabetic rats can be caused by an insufficient cardiac sympathetic supply as the catecholamine content in the stellate ganglion that innervates the heart is significantly smaller than in healthy animals [41, 59].

3.2.3 MAP Kinases Regulate I_{to} Channels Expression

Interestingly, both in Chagasic and in diabetic cardiomyocytes, incubation with norepinephrine for 24 h restores I_{to} current amplitude, but not current kinetics [41, 60]. Similarly, incubation of diabetic myocytes with insulin or the β-adrenoceptor agonist isoproterenol restores I_{to} current amplitude to control values [30, 41]. In all these experiments, the restoration takes place after a long incubation, suggesting that these agonists act as trophic factors that activate the synthesis of new channel proteins.

In fact, insulin affects the gene expression of a number of proteins including potassium channels [61]. Moreover, when protein synthesis is inhibited with cycloheximide, insulin fails to restore I_{to} current amplitude in diabetic cells [30]. Ultimately, the intracellular mechanism responsible for the fabrication of new I_{to} channels in response to insulin requires the activation of the MAP kinase cascade [62].

Very recently the intracellular signaling pathway connecting the activation of β-adrenoceptors with the restoration of I_{to} current amplitude in diabetic cardiomyocytes has been elucidated. As expected, it involves the MAP kinase pathway and an increased amount of channel proteins in the plasma membrane. In diabetic cardiomyocytes, β-AR stimulation by isoproterenol leads to the activation of Gαs and Gαi proteins and the dual phosphorylation of the receptor by PKA and β-ARK1. The β-AR then binds arrestin, and internalizes and recruits cSrc, which subsequently activates Ras, MEK1/2, and ERK1/2. Finally, this MAPK cascade increases the Kv4.2 and Kv4.3 protein expression and thereby restores I_{to} current amplitude [63].

3.2.4 AMP Kinase Mediates K^+-Channel Reduction in Diabetic Heart

In physiological conditions, where the insulin release is not compromised, the neurohormonal supply would ensure the correct expression of I_{to}. Thus, the natural hypotheses to explain why in cardiac muscle potassium currents are reduced in type 1 diabetes mellitus would be the reduction of protein synthesis secondary to the loss of the trophic effect of insulin.

This explanation would be satisfactory except for the fact that a 6-h incubation with metabolism improvers also restores I_{to} current amplitude in diabetic cardiomyocytes [64, 65]. Similar to insulin, metabolic improvers such as dichloroacetate, L-carnitine, or pyruvate require protein synthesis to restore the K^+ currents [20, 30, 64, 65]. Because metabolic improvers have no direct effect on protein synthesis, these studies claim the flawed glucose metabolism as the potential cause of the reduced cardiac potassium currents in diabetes mellitus.

Interestingly, insulin fulfils both functions: it directly stimulates protein synthesis and also improves the metabolic state of the cell. In a combined hypothesis, reduction of K^+ currents in diabetes could be caused by a defect in channel protein synthesis secondary to a deterioration of the metabolic status of the cells.

In this sense, the AMP-activated protein kinase acts as a cellular energy sensor in the heart [66], which in conditions of energy depletion switches on catabolic pathways to generate ATP and at the same time switches off energy-consuming anabolic pathways, including protein synthesis [67]. In heterologous expression systems, the reduction of protein synthesis by AMP kinase activity includes the downregulation of some voltage-dependent ionic channels such as Kv1.5 or Kv7.1/mink [43, 68]. In murine cardiomyocytes, AMPK activation in healthy cells reduces K^+-current amplitudes in the same way that diabetic status does, whereas metabolic improvement is unable to restore K^+-current densities in diabetic cells when AMPK is kept in an active state [20]. Therefore, AMPK might be the intracellular link between metabolic status and K^+-channel protein synthesis in the heart.

3.2.5 I_{to} Current Kinetics in Diabetic Cardiomyocytes

Most of the alterations discussed so far regard the current amplitude; however, diabetes also affects current kinetics. In diabetic myocytes, the transient outward potassium current displays slower recovery kinetics than in healthy myocytes [21, 30]. In this regard, diabetes reduces the expression of the channel-forming Kv4.x proteins [69, 70] but increases the expression of Kv1.4 [7, 70]. Although quite similar in structure, Kv4.2/Kv4.3 channels recover from inactivation more than tenfold faster than Kv1.4. Therefore, altered recovery kinetics of I_{to} observed in diabetic cells could result from a switch from the fast-recovering Kv4.x to the slow-recovering Kv1.4 isoform [71].

Moreover, diabetes accelerates the inactivation rate of rat I_{to} current [8, 9]. Experiments performed in heterologous expression systems and in rat and human cardiac myocytes indicate that inactivation of I_{to} depends on the phosphorylation state of the channel. In particular, phosphorylation of Kv4.2/Kv4.3 channels by the Ca^{2+}- and calmodulin-dependent kinase II (CaMKII) slows down the inactivation rate [72–74]. In diabetes, insulin deficiency causes a reduction in calmodulin expression, which subsequently impairs the activity of the CaMKII. As a result, phosphorylation of I_{to} channels is defective and the inactivation of the current becomes abnormally fast [75].

4 Conclusions

The electrocardiogram of diabetic patients often shows alterations that affect cardiac repolarization, especially a prolonged QTc associated with a higher risk of ventricular arrhythmias. Because the ECG is the result of the summation of the electric activity of each individual cardiac cell, the abnormalities observed in the electrocardiogram ultimately reflect alterations in either the activity or the amount of ionic channels in the membrane of the cardiomyocytes. Diabetes modifies inward and outward ionic currents in different ways. It is particularly relevant that the reduction in protein expression affects the channels responsible for the I_{to} and I_{Ks} currents. This reduction can be caused by a defective supply of trophic factors such as insulin or noradrenalin, as well as a reduction in channel protein synthesis secondary to the impaired metabolic status. Finally, diabetes also alters the currents by reducing the levels of regulatory proteins and second messengers, as is the case of the L-type Ca^{2+} current or the K^+ current I_{to}, whose intracellular regulation by Ca^{2+} and kinases is impaired.

Acknowledgments This work has been supported by Universidad del País Vasco UPV/EHU grants (PPM12/12) to O. Casis and (EHUA12/12) to M. Gallego.

References

1. Casis O, Echevarría E (2004) Diabetic cardiomyopathy: electromechanical cellular alterations. Curr Vasc Pharmacol 2:237–248
2. Kahn JK, Sisson JC, Vinik AI (1987) QT interval prolongation and sudden cardiac death in diabetic autonomic neuropathy. J Clin Endocrinol Metab 64:751–754
3. Ewing DJ, Boland O, Neilson JMM et al (1991) Autonomic neuropathy, QT interval lengthening, and unexpected deaths in male diabetic patients. Diabetologia 34:182–185
4. Cardoso C, Salles G, Bloch K et al (2001) Clinical determinants of increased QT dispersion in patients with diabetes mellitus. Int J Cardiol 79:253–262

5. Stettler C, Bearth A, Allemann S et al (2007) QTc interval and resting heart rate as long-term predictors of mortality in type 1 and type 2 diabetes mellitus: a 23-year follow-up. Diabetologia 50:186–194
6. Rees DA, Alcolado JC (2005) Animal models of diabetes mellitus. Diabet Med 22:359–370
7. Lengyel C, Virág L, Bíro T et al (2007) Diabetes mellitus attenuates the repolarization reserve in mammalian heart. Cardiovasc Res 73:512–520
8. Magyar J, Rusznák Z, Szentesi P et al (1992) Action potentials and potassium currents in rat ventricular muscle during experimental diabetes. J Mol Cell Cardiol 24:841–853
9. Casis O, Gallego M, Iriarte MM, Sánchez-Chapula JA (2000) Effects of diabetic cardiomyopathy on regional electrophysiologic characteristics of rat ventricle. Diabetologia 43:101–109
10. Chattou S, Coulombe A, Diacono J et al (2000) Slowly inactivating component of sodium current in ventricular myocytes is decreased by diabetes and partially inhibited by known Na^+-H^+ exchange blockers. J Mol Cell Cardiol 32:1181–1192
11. Jourdon P, Feuvray D (1993) Calcium and potassium currents in ventricular myocytes isolated from diabetic rats. J Physiol (Lond) 470:411–429
12. Tsuchida K, Watajima H, Otomo S (1994) Calcium current in diabetic rat ventricular myocytes. Am J Physiol Heart Circ Physiol 267:H2280–H2289
13. Lengyel C, Virág L, Kovács PP et al (2008) Role of slow delayed rectifier K^+ current in QT prolongation in the alloxan-induced diabetic rabbit heart. Acta Physiol (Oxf) 192:356–368
14. Nerbonne JM, Kass RS (2005) Molecular physiology of cardiac repolarization. Physiol Rev 85:1205–1253
15. Beuckelmann DJ, Näbauer M, Erdmann E (1993) Alterations of K^+ currents in isolated human ventricular myocytes from patients with terminal heart failure. Circ Res 73:379–385
16. Spector PS, Curran ME, Zou A et al (1996) Fast inactivation causes rectification of the IKr channel. J Gen Physiol 107:611–619
17. Li GR, Feng J, Yue L et al (1996) Evidence for two components of delayed rectifier K^+ current in human ventricular myocytes. Circ Res 78:689–696
18. Pond AL, Scheve BK, Benedict AT et al (2000) Expression of distinct ERG proteins in rat, mouse, and human heart. Relation to functional I(Kr) channels. J Biol Chem 275:5997–6006
19. Van Wagoner DR, Nerbonne JM (2000) Molecular basis of electrical remodeling in atrial fibrillation. J Mol Cell Cardiol 32:1101–1117
20. Torres-Jacome J, Gallego M, Rodríguez-Robledo JM et al (2013) Improvement of the metabolic status recovers cardiac potassium channel synthesis in experimental diabetes. Acta Physiol (Oxf) 207:447–459
21. Shimoni Y, Firek L, Severson D, Giles WR (1994) Short-term diabetes alters K^+ currents in rat ventricular myocytes. Circ Res 74:620–628
22. Watanabe T, Delbridge LM, Bustamante JO, McDonald TF (1983) Heterogeneity of the action potential in isolated rat ventricular myocytes and tissue. Circ Res 52:280–290
23. Clark RB, Bouchard RA, Salinas-Stefanon E et al (1993) Heterogeneity of action potential waveforms and potassium currents in rat ventricle. Cardiovasc Res 27:1795–1799
24. Apkon M, Nerbonne JM (1991) Characterization of two distinct depolarization-activated K^+ currents in isolated adult rat ventricular myocytes. J Gen Physiol 97:973–1011
25. Scamps F (1996) Characterization of a β-adrenergically inhibited K^+ current in rat cardiac ventricular cells. J Physiol (Lond) 491:81–97
26. Sanguinetti MC, Jurkiewicz NK (1990) Two components of cardiac delayed rectifier K^+ current. Differential sensitivity to block by class III antiarrhythmic agents. J Gen Physiol 96: 195–215
27. Zhang Y, Xiao J, Wang H et al (2006) Restoring depressed HERG K^+ channel function as a mechanism for insulin treatment of abnormal QT prolongation and associated arrhythmias in diabetic rabbits. Am J Physiol Heart Circ Physiol 291:H1446–H1455
28. Zhang Y, Xiao J, Lin H et al (2007) Ionic mechanisms underlying abnormal QT prolongation and the associated arrhythmias in diabetic rabbits: a role of rapid delayed rectifier K^+ current. Cell Physiol Biochem 19:225–238
29. Roden DM (1998) Taking the idio out of idiosyncratic-predicting torsades de pointes. Pacing Clin Electrophysiol 21:1029–1034

30. Shimoni Y, Ewat HS, Severson D (1998) Type I and type II models of diabetes reduce different modifications of K^+ currents in rat heart: role of insulin. J Physiol (Lond) 507:485–496
31. Wang DW, Kiyosue T, Shigematsu S, Arita M (1995) Abnormalities of K^+ and Ca^{2+} currents in ventricular myocytes from rats with chronic diabetes. Am J Physiol Heart Circ Physiol 269: H1288–H1296
32. Chattou S, Diacono J, Feuvray D (1999) Decrease in sodium-calcium exchange and calcium currents in diabetic rat ventricular myocytes. Acta Physiol Scand 166:137–144
33. Horackova M, Murphy MG (1988) Effects of chronic diabetes mellitus on the electrical and contractile activities, $^{45}Ca^{2+}$ transport, fatty acid profiles and ultrastructure of isolated rat ventricular myocytes. Pfügers Arch-Eur J Physiol 441:564–572
34. Penpargkul S, Fein FS, Sonnenblick SH, Scheuer J (1981) Depressed cardiac sarcoplasmic reticulum function from diabetic rats. J Mol Cell Cardiol 13:303–309
35. Allo SN, Lincoln TM, Wilson GL, Green FJ et al (1991) Non-insulin-dependent diabetes-induced defects in cardiac cellular calcium regulation. Am J Physiol Cell Physiol 260: C1165–C1171
36. Golfman LS, Takeda N, Dhalla NS (1996) Cardiac membrane Ca^{2+}-transport in alloxan-induced diabetic rats. Diabetes Res Clin Pract 31:S73–S77
37. Teshima Y, Takahashi N, Sikawa T et al (2000) Diminished expression of sarcoplasmic reticulum Ca^{2+}ATPase and Ryanodine sensitive Ca^{2+} channel mRNA in streptozotocin-induced diabetic rat heart. J Mol Cell Cardiol 32:655–664
38. Netticadam T, Temsah RM, Kent A et al (2001) Depressed levels of Ca^{2+}-cycling proteins may underlie sarcoplasmic reticulum dysfunction in the diabetic heart. Diabetes 50:2133–2138
39. Zong Y, Ahmed S, Grupp IL, Matlib MA (2001) Altered SR protein expression associated with contractile dysfunction in diabetic hearts. Am J Physiol Heart Circ Physiol 281:H1137–H1147
40. Tuckwell HC (2012) Quantitative aspects of L-type Ca^{2+} currents. Prog Neurobiol 96:1–31
41. Gallego M, Casis E, Izquierdo MJ, Casis O (2000) Restoration of cardiac transient outward potassium current by norepinephrine in diabetic rats. Pflügers Arch-Eur J Physiol 441:102–107
42. Fernández-Velasco M, Ruiz-Hurtado G, Hurtado O et al (2007) TNF-alpha downregulates transient outward potassium current in rat ventricular myocytes through iNOS overexpression and oxidant species generation. Am J Physiol Heart Circ Physiol 293:H238–H245
43. Mia S, Munoz C, Pakladok T et al (2012) Downregulation of Kv1.5 K^+ channels by the AMP-activated protein kinase. Cell Physiol Biochem 30:1039–1050
44. Vinik AI, Maser RE, Ziegler D (2011) Autonomic imbalance: prophet of doom or scope for hope? Diabet Med 28:643–651
45. Gallego M, Setién R, Puebla L et al (2005) α1-adrenoceptors stimulate a Gαs protein and reduce the transient outward K^+ current via a cAMP/PKA-mediated pathway in the rat heart. Am J Physiol Cell Physiol 288:C577–C585
46. Sato N, Hashimoyo H, Takiguchi Y, Nakashima M (1989) Altered responsiveness to sympathetic nerve stimulation and agonists of isolated left atria of diabetic rats: no evidence for involvement of hypothyroidism. J Pharmacol Exp Ther 248:367–371
47. Kamafa K, Kirisawa H (1998) Changes in electrophysiological properties and noradrenaline response in vas deferens of diabetic rats. Eur J Pharmacol 350:237–241
48. Gallego M, Casis O (2001) Regulation of cardiac transient outward potassium current by norepinephrine in normal and diabetic rats. Diabetes Metab Res Rev 17:304–309
49. Gallego M, Espiña L, Casis O (2002) Blood pressure responsiveness to sympathetic agonists in anaesthetised diabetic rats. J Physiol Biochem 58:87–93
50. Boudina S, Abel ED (2007) Diabetic cardiomyopathy revisited. Circulation 115:3213–3223
51. Bitar MS, Koulu M, Rapoport SI, Linnoila M (1987) Adrenal catecholamine metabolism and myocardial adrenergic receptors in streptozotocin diabetic rats. Biochem Pharmacol 36: 1011–1016
52. Thomas D, Wu K, Wimmer AB, Zitron E et al (2004) Activation of cardiac human ether-a-go-go related gene potassium currents is regulated by alpha(1A)-adrenoceptors. J Mol Med 82: 826–837
53. Yang T, Roden DM (1996) Regulation of sodium current development in cultured atrial tumor myocytes (AT-1) cells. Am J Physiol Heart Circ Physiol 271:H541–H547

54. Zhang JF, Robinson RB, Siegelbaum SA (1992) Sympathetic neurons mediate developmental change in cardiac sodium channel gating through long-term neurotransmitter action. Neuron 9:97–103
55. Ogawa S, Barnett JV, Sen L et al (1992) Direct contact between sympathetic neurons and rat cardiac myocytes in vitro increases expression of functional calcium channels. J Clin Invest 89:1085–1093
56. Liu QY, Rosen MR, McKinnon D, Robinson RB (1998) Sympathetic innervation modulates repolarizing K^+ currents in rat epicardial myocytes. Am J Physiol Heart Circ Physiol 274: H915–H922
57. Pacioretty LM, Barr SC, Han WP, Gilmour RF Jr (1995) Reduction of the transient outward potassium current in a canine model of Chagas' disease. Am J Physiol Heart Circ Physiol 268:H1258–H1267
58. Machado CR, De Oliveira DA, Magalhanes MJ et al (1994) Trypanosoma cruzi infection in rats induced by early lesion of the heart noradrenergic nerve terminals by a complement-independent mechanism. J Neural Transm Gen Sect 97:149–159
59. Gallego M, Setién R, Izquierdo MJ et al (2003) Diabetes-induced biochemical changes in central and peripheral catecholaminergic systems. Physiol Res 52:735–741
60. Han WP, Barr SC, Pacioretty LM, Gilmour RF Jr (1997) Restoration of the transient outward potassium current by noradrenaline in chagasic canine epicardium. J Physiol (Lond) 500: 75–83
61. O'Brien RM, Granner DK (1996) Regulation of gene expression by insulin. Physiol Rev 76:1109–1161
62. Shimoni Y, Ewart HS, Severson D (1999) Insulin stimulation of rat ventricular K^+ currents depends on the integrity of the cytoskeleton. J Physiol (Lond) 514:735–745
63. Setién R, Alday A, Diaz-Asensio C et al (2013) Mechanisms responsible for the trophic effect of beta-adrenoceptors on the Ito current density in type 1 diabetic rat cardiomyocytes. Cell Physiol Biochem 31:25–36
64. Xu Z, Patel KP, Rozanski GJ (1996) Metabolic basis of decreased transient outward K^+ current in ventricular myocytes from diabetic rats. Am J Physiol Heart Circ Physiol 271:H2190–H2196
65. Xu Z, Patel KP, Lou MF, Rozanski GJ (2002) Up-regulation of K^+ channels in diabetic rat ventricular myocytes by insulin and glutathione. Cardiovasc Res 53:80–88
66. Heidrich F, Schotola H, Popov AF et al (2010) AMP-activated protein kinase and its role in energy metabolism of the heart. Curr Cardiol Rev 6:337–342
67. Hardie DG (2012) AMP-activated protein kinase, an energy sensor that regulates all aspects of cell function. Genes Dev 25:1895–1908
68. Alesutan I, Föller M, Sopjani M et al (2011) Inhibition of the heterotetrameric K^+ channel KCNQ1/KCNE1 by the AMP-activated protein kinase. Mol Membr Biol 28:79–89
69. Qin D, Huang B, Deng L et al (2001) Downregulation of K(+) channel genes expression in type I diabetic cardiomyopathy. Biochem Biophys Res Commun 283:549–553
70. Nishiyama A, Ishii DN, Backx PH et al (2001) Altered K^+ channel gene expression in diabetic rat ventricle: isoform switching between Kv4.2 and Kv1.4. Am J Physiol Heart Circ Physiol 281:H1800–H1807
71. Gallego M, Alday A, Urrutia J, Casis O (2009) Transient outward potassium channel regulation in healthy and diabetic hearts. Can J Physiol Pharmacol 87:77–83
72. Tessier S, Karczewski P, Krause EG et al (1999) Regulation of the transient outward K^+ current by Ca^{+2}/calmodulin-dependent protein kinase II in human atrial myocytes. Circ Res 85: 810–819
73. Sergeant GP, Ohya S, Reihill JA et al (2005) Regulation of Kv4.3 currents by Ca^{2+}/calmodulin-dependent protein kinase II. Am J Physiol Cell Physiol 288:C304–C313
74. Colinas O, Gallego M, Setien R et al (2006) Differential modulation of kv4.2 and kv4.3 channels by calmodulin-dependent protein kinase II in rat cardiac myocytes. Am J Physiol Heart Circ Physiol 291:H1978–H1987
75. Gallego M, Fernández D, Ahyayauch H et al (2008) Reduced calmodulin expression accelerates transient outward potassium current inactivation in diabetic rat heart. Cell Physiol Biochem 22:625–634

Role of PPAR-δ in Diabetic Cardiomyopathy

Ying-Xiao Li, Kai-Chun Cheng, and Juei-Tang Cheng

Abstract Although diabetics are at increased risk of structural heart disease from vascular complications, the concept of diabetic cardiomyopathy suggests a direct cellular insult to the myocardium. Several investigations, mainly echocardiographic population-based studies, documented a uniform association between diabetic cardiomyopathy and the presence of cardiac hypertrophy and myocardial stiffness. In the following review, we attempt to provide a comprehensive insight to discuss the possible underlying mechanisms, especially the role of PPARs in diabetic cardiomyopathy.

Keywords Diabetic cardiomyopathy • Peroxisome proliferator-activated receptors • PPAR-δ • Fatty acid oxidation

1 Introduction

Diabetes mellitus (DM) is a common metabolic disorder with a chronic course. The prevalence of this disease is on the increase in several regions of the world [1, 2]. Diabetes is described as "a coronary heart disease equivalent," and diabetes is essentially a vascular rather than a carbohydrate malfunction [3, 4]. Individuals with diabetes are at a significantly greater risk of developing both micro- and macrovascular

Y.-X. Li • K.-C. Cheng
Department of Psychosomatic Internal Medicine, Kagoshima University Graduate School of Medical and Dental Sciences, Kagoshima 890-8520, Japan

J.-T. Cheng (✉)
Institute of Medical Sciences, Chang Jung Christian University, Guei-Ren, Tainan City 71101, Taiwan
e-mail: jtcheng5503@yahoo.com.tw; m980103@mail.chimei.org.tw

B. Turan and N.S. Dhalla (eds.), *Diabetic Cardiomyopathy: Biochemical and Molecular Mechanisms*, Advances in Biochemistry in Health and Disease 9,
DOI 10.1007/978-1-4614-9317-4_12,

disease, and have a cardiac mortality equivalent to that in nondiabetic patients with confirmed heart disease [5]. Cardiovascular complications including angina, myocardial infarction, and diabetic cardiomyopathy are major causes of morbidity and mortality in diabetic patients [6].

The presence of a diabetic cardiomyopathy was first recognized by Rubler et al. [7], who described four diabetic patients with congestive heart failure and normal coronary arteries. In one of the largest epidemiologic studies, involving more than 800,000 patients, diabetes was found to be independently associated with the occurrence of congestive heart failure after adjusting for cardiac hypertrophy, hypertension, coronary artery disease, and atrial fibrillation [8]. Diabetic cardiomyopathy has been defined as "a distinct entity characterized by the presence of abnormal myocardial performance or structure in the absence of epicardial coronary artery disease, hypertension, and significant valvular disease" [9, 10]. Although diabetics are at increased risk of structural heart disease because of vascular complications, the concept of diabetic cardiomyopathy suggests a direct cellular insult to the myocardium. Several investigations, mainly echocardiographic population-based studies, documented a uniform association between diabetic cardiomyopathy and the presence of cardiac hypertrophy and myocardial stiffness [11]. In the following review, we attempt to provide a comprehensive insight to discuss the possible underlying mechanisms, especially the role of peroxisome proliferator-activated receptors (PPARs) in diabetic cardiomyopathy.

2 Diabetic Cardiomyopathy in Human and Animal Models

In patients, cardiac ejection time is often reduced, and the length of the pre-ejection period and the ratio of the pre-ejection period to cardiac ejection time are often increased. Diastolic abnormalities have been suggested as an earliest functional effect of diabetic cardiomyopathy. In the study of normotensive asymptomatic type 2 diabetic patients with good glycemic control, 47 % were found to have diastolic dysfunction [12]. More sensitive techniques for systolic assessment, such as strain, strain rate, and myocardial tissue Doppler velocity, may detect preclinical systolic abnormalities in diabetic patients. One study using several detecting methods found that as many as 75 % of diabetic patients demonstrate abnormalities of diastolic dysfunction [13]. Fibrosis of the myocardium is usually present and can be confirmed with biopsy; myocytic hypertrophy is usually found, and increase in contractile protein glycosylation [14], all of which contribute to reduced diastolic compliance and ventricular hypertrophy in diabetic patients. In diabetes, deposition of advanced glycation end products and deposition of collagen are important determinants of the increased heart stiffness in patients having heart failure with reduced cardiac output [15].

In an animal model, the streptozotocin (STZ)-treated rat has been associated with myocardial atrophy as opposed to hypertrophy and, in particular, with loss of contractile proteins. In the isolated heart, abnormalities in diastolic function (increased ventricular end-diastolic pressure and operating chamber stiffness) and progressive reduction in heart systolic function over 1 year were reported [11]. These experimental models suggested that hyperglycemia alone may account for the functional changes observed in diabetic cardiomyopathy. Most studies of type 2 DM have been performed in genetic models of obesity and insulin resistance such as the Zucker fatty rats or db/db mice, both of which have mutations that impair leptin receptor signaling, or ob/ob mice, which lack leptin. In these experiments, cardiac manifestations depend on the onset and severity of these metabolic alterations. Hyperinsulinemic Zucker fatty (ZF) rats showing obesity and hyperinsulinemia develop early onset, and exhibit myocardial hypertrophy and variable degrees of diastolic abnormalities, whereas the diabetic ZF rats (similar to metabolic features of type 2 DM) have less increase in cardiac mass and less impairment in cardiac systolic function in isolated hearts but not in intact models [16, 17]. These findings support that hyperinsulinemia is able to induce cardiac hypertrophy in contrast to the influence of hyperglycemia, which may mitigate hypertrophy and affect systolic dysfunction to a greater extent. In other words, myocardial fibrosis is likely to be related to hyperglycemia, whereas left ventricular hypertrophy is most likely to be related to the insulin resistance syndrome [18].

3 Pathogenesis of Diabetic Cardiomyopathy

The mechanisms of this disorder are still not clear. Hyperglycemia, insulin resistance, increased fatty acid metabolism, microcirculatory changes, sympathetic dysfunction, and fibrosis are considered to collectively contribute to its pathology. Hyperglycemia may mediate its damaging effects through a series of secondary transducers, especially reactive oxygen species (ROS) and advanced glycation end products (AGEs). ROS encompass a range of highly reactive oxygen base molecules, which consist of both free radicals (superoxide) and chemicals capable of generating free radicals (hydrogen peroxide). Oxidative stress exists when the production of ROS outweighs their degradation by antioxidant defenses, and the resultant elevation of ROS has numerous deleterious effects on the cardiovascular system via cellular damage by oxidation, disruption of vascular homeostasis through interference with nitric oxide (NO), and by modulation of detrimental intracellular signaling pathways, the so-called redox signaling. Although under physiological states most of the ROS generated within cells arises from mitochondria, in diseased conditions they are produced by a range of other sources [19]. Overexpression of metallothionin [20, 21], catalase [22], and manganese superoxide dismutase [23] in the heart reversed diabetic cardiomyopathy in animal models of both type 1 and type 2 diabetes and increased glucose oxidation and mitochondrial generation of superoxide, which in turn leads to DNA damage and activation of poly (ADP ribose) polymerase (PARP) as a reparative enzyme. As a result, glucose is diverted from its glycolytic pathways and into alternative biochemical pathways that are considered the mediators of hyperglycemia-induced cellular injury [24], including increases in AGEs, increased hexosamine and polyol flux, and activation of classical isoforms of protein kinase C, as well as altered expression and function of both the ryanodine receptor (RyR) and sarco(endo)plasmic reticulum Ca^{2+}-ATPase (SERCA2). The latter are thought to be responsible for decreased systolic and diastolic function [25]. Long-lived extracellular proteins, such as collagen and elastin, are particularly vulnerable to the accumulation of AGEs [26]: this can impair the ability of collagen to be degraded, leading to collagen accumulation or fibrosis. Crosslinking of collagen and elastin and the resulting fibrosis also cause increased myocardial stiffness and impaired cardiac relaxation. Aminoguanidine (an inhibitor of AGE formation) has been shown to ameliorate changes in left ventricular structure and function [27].

Moreover, in diabetes, an increase of free fatty acid (FFA) use and oxidation by the heart could increase susceptibility to ischemia and can lead to lipid accumulation, energy deprivation, worsening insulin resistance, and ultimately cardiomyopathy [28]. Cardiac myocytes respond to increased FFA by upregulating the expression of the enzymes necessary for their disposal through mitochondrial β-oxidation. These enzymes are under transcriptional control of the nuclear transcriptional factor named peroxisome proliferator-activated receptor (PPAR). High FFA levels activate PPAR, leading to increased FFA use through myocardial fatty acid oxidation and myocardial fatty acid utilization [29]. In addition, FFAs inhibit pyruvate dehydrogenase, which impairs myocardial energy production and leads to the accumulation of glycolytic intermediates and intracellular lipids [30, 31]. It has been mentioned that an increase of ceramide levels in cardiac myocytes is associated with increased oxidative stress, apoptosis, and decreased contractile function [32].

4 The Effect of PPARs on Diabetic Cardiomyocytes

Peroxisome proliferator-activated receptors (PPARs), ligand-activated transcription factors, belong to the nuclear hormone receptor superfamily regulating expression of genes involved in different aspects of lipid metabolism, inflammation, and cardiac energy production [33]. Three isoforms of PPARs, specifically PPAR-α, PPAR-β/δ, and PPAR-γ, have distinct tissue distributions and functions. PPARs form heterodimers with retinoid X receptors, bind to specific hexanucleotide PPAR-response elements, and regulate the transcription of target genes. The PPAR isoforms are differentially expressed in tissues, and their distribution pattern differs depending on the period of organogenesis. Although PPAR-α and PPAR-γ are predominantly expressed in liver and adipose tissue, respectively, PPAR-β/δ is ubiquitously expressed [34]. The differences of PPARs in tissue expression, ligands, or activators and the main biological effects of all isoforms are shown in Table 1.

PPAR-α plays an important role in lipid metabolism by regulation of the expression of genes that encode proteins involved in the cellular free fatty acid uptake, its β-oxidation, and cellular cholesterol trafficking [35]. Parallel to this effect, the antiinflammatory activity of these factors was demonstrated [36, 37].

Table 1 Peroxisome proliferator-activated receptors (PPARs) family and their main functions

PPARs	Distributed tissue	Ligands or activators	Main functions
PPAR-δ	Heart, liver, muscle	Fenofibrate, gemfibrozil	Lipid oxidation
PPAR-δ	Adipose tissues	Thiazolidinediones	Adipogenesis and others
PPAR-β/δ	Many tissues	GW0742	Cardiac protection and others

PPAR-α is activated by natural ligands, including polyunsaturated fatty acids, such as docosahexanoic acid (DHA) and eicosapentaenoic acid (EPA), oxidized phospholipids, lipoprotein lipolytic products, and by synthetic ligands, including fibrates such as fenofibrate and gemfibrozil [38]. In diabetic cardiac hypertrophy, disordered glucose transport to the cells impairs fatty acid (FA) participation in beta-oxidation, which could inhibit ATP-dependent transporters, paralleled by insulin resistance-dependent inhibition of glycogen synthesis. These events cause the autoactivation of cell overloading with fatty acids, which are stored as lipid droplets in the cytoplasm of cardiomyocytes [39]. Impaired glucose transport, accumulation of lipid droplets in heart muscle, the additive induction of intracellular free fatty acid transport (the induction of membrane transporter of FA: FABP, CD36), and activation of enzymes taking part in β-oxidation by PPAR-α activators [40]. One experiment used transgenic mice with selective, cardiac-restricted overexpression of PPAR-α (MHC-PPAR-α) to investigate the role of PPAR-α in diabetic cardiomyocytes. All the mice were induced by injection of STZ. After 4 weeks, comparing changes in heart muscle in transgenic mice with normal PPAR-α expression, STZ diabetic mice showed a significant increase of left ventricular mass index in MHC-PPAR-$\alpha^{+/+}$ mice as compared with a control group [41]. With the addition of a high-fat diet, MHC-PPAR-$\alpha^{+/+}$ mice developed much stronger and significant cardiac hypertrophy as compared with the control group. In addition to cardiac dysfunction, wall motion abnormalities were observed only in the MHC-PPAR-$\alpha^{+/+}$ group. The possible mechanism might be that constant stimulation of the PPAR-α pathway, accompanied by the increased concentration of fatty acids in the blood, leads to increased fat acid (FA) uptake by the heart muscle (expression of FA transporters) and increased accumulation of long-chain fatty acids (LCFA) droplets in cardiomyocytes, which impairs compensatory increased glucose metabolism and glycogen synthesis. Downregulation of ATP synthesis paralleled with activated FA beta-oxidation under low energy conditions may lead to overproduction of free radicals and damage to the myocardium [40].

PPAR-γ is expressed in endothelial and vascular smooth muscle cells [42]. PPAR-γ controls adipocyte differentiation and lipid storage [43] and accordingly is heavily expressed in adipose tissue. Through its effects on adipose tissue and skeletal muscle, PPAR-γ regulates the action of insulin [44]. Natural ligands for PPAR-γ are the prostaglandin D2 derivative 15-deoxy-Δ12,14-prostaglandin J2 and forms of oxidized linoleic acid, 9- and 13(S)-HODE. Synthetic ligands for PPAR-γ include the antidiabetic agents named thiazolidinediones (TZDs), such as troglitazone, pioglitazone, and rosiglitazone [45]. These insulin-sensitizing agents decrease peripheral insulin resistance and thereby reduce blood glucose levels in patients with type

2 DM [45, 46]. The nuclear transcription factor named nuclear factor-kappa B (NF-κB) is known to promote ventricular hypertrophy, which has been widely implicated in proinflammatory and pro-growth action in isolated cardiac myocytes [45, 47] and has been shown to be activated by endothelin, catecholamines [45, 48], and predominantly angiotensin II [49, 50]. Some in vitro studies have demonstrated that PPAR-γ agonists inhibit pressure overload-induced structural changes and inhibit cardiac hypertrophy in rat models [51–53] through inhibition of the trophic effects of NF-κB signaling pathways in myocardial tissue. However, one study has described the development of cardiac hypertrophy without any effect on cardiac systolic function in rodents treated with over-therapeutic doses of rosiglitazone [54].

PPAR-β/δ (also termed PPAR-δ) is expressed in many tissues [55, 56] and plays a role in lipid metabolism by stimulating fatty acid oxidation in heart and skeletal muscle cells. PPAR-δ agonists inhibit cardiomyocyte hypertrophy [57], normalize lipid status in obese mice (db/db) and obese monkeys, and have been proposed as a putative pharmacological target for the management of obesity, insulin resistance, and dyslipidemia [58, 59]. Recent studies have shown that PPAR-δ is the predominant subtype in the heart [60], where it plays an important role in regulating the expression of genes involved in fatty acid and glucose metabolism [61, 62]. Moreover, cardiomyocyte-restricted deletion of the PPAR-δ gene results in cardiac dysfunction, cardiac hypertrophy, and progressive myocardial lipid accumulation [61]. Additionally, a PPAR-δ-specific agonist inhibited phenylephrine-induced cell hypertrophy and hydrogen peroxide-induced cell apoptosis in cultured cardiomyocytes [58, 62]. Thus, cardiac PPAR-δ has been proposed as a therapeutic target for treatment of heart failure [63]. However, little is known about the role of PPAR-δ in the development of diabetic cardiomyopathy.

5 PPAR-δ in Diabetic Cardiomyopathy

PPAR-δ expression was reduced in cardiomyocytes in response to increased glucose concentration in a dose-related manner. Moreover, PPAR-δ expression, cell size, and protein synthesis were not influenced by high glucose level in the presence of the antioxidant tiron [20]. Thus, the mechanism of ROS-mediated cell hypertrophy in high glucose-treated cells may involve a reduction of PPAR-δ expression. In the animal model, after administration of STZ, rats exhibit cardiac dysfunction and hypertrophy, whereas cardiac PPAR-δ expression is significantly reduced in diabetic hearts at both mRNA and protein levels. Reduced PPAR-δ expression may be involved in hyperglycemia-induced cardiomyopathy in diabetic rats [64]. Fatty acid oxidation is the primary source of energy in the postnatal heart [65]. Impaired fatty acid oxidation and a shift to reliance on glucose metabolism are hallmarks of myocardial diseases such as cardiac hypertrophy and congestive heart failure [65]. As in skeletal muscle, PPAR-δ is a critical regulator of fatty acid oxidation in cardiac tissue. It has been showed that cardiac-specific deletion of *PPAR-δ* suppresses the expression of oxidative genes [62]; this leads to impaired fatty acid oxidation and a reciprocal increase in glucose oxidation, along with fat accumulation in cardiomyocytes [62]. Moreover, PPAR-δ-selective agonists increase fatty acid oxidation via the induction of oxidative genes in isolated neonatal as well as adult rat cardiomyocytes [58]. The PPAR-δ-dependent maintenance of basal fatty acid oxidation is crucial for normal cardiac mechanics. *PPAR*-δ-null hearts are characterized by decreased rates of contraction and relaxation, increased left ventricular end-diastolic pressure, and decreased cardiac output, factors associated with the onset of cardiac failure [62]. Indeed, mice with cardiac-specific deletion of PPAR-δ develop age-dependent cardiac lipotoxicity, cardiac hypertrophy, end-stage dilated cardiomyopathy, and decreased survival [62]. The protective role of PPAR-δ in the heart has been confirmed by in vitro studies showing that PPAR-δ agonists attenuate phenylephrine-induced cardiac hypertrophy. Although phenylephrine suppresses fatty acid oxidation in cardiomyocytes, concomitant activation of PPARδ reverses these effects [61]. Although PPAR-δ may directly increase the transcription of fatty acid oxidative genes, at least one study suggests that effects could also be indirect because PPAR-δ interacts with and blocks the NF-κB-mediated suppression of fatty acid oxidation in cardiomyocytes [66]. PPAR-δ-dependent antagonism of NF-κB could be particularly important during sepsis, when endotoxins decrease cardiac fatty acid oxidation and initiate cardiac failure [66, 67].

6 Other Molecules in Diabetic Cardiomyopathy

Many fatty acid (FA) oxidation enzymes and mitochondrial respiratory uncoupling genes are regulated by PPAR-δ in cardiomyocytes, such as pyruvate dehydrogenase kinase and acyl-CoA oxidase [68, 69]. Deletion of cardiac PPAR-δ, which is accompanied by decreased contraction, increased left ventricular end-diastolic pressure, and lowered cardiac output, leads to decreased contraction and increased incidence of cardiac failure [62]. The effect of digoxin on contractility in patients with heart failure has been established; some studies indicated that impaired relaxation is the prominent cardiac abnormality, and it is related to depressed troponin function in the hearts of STZ diabetic rats [70, 71]. Digoxin improved cardiac contraction in STZ diabetic rats is associated with a marked increase in cardiac PPAR-δ expression without altering other PPARs, such as PPAR-α and PPAR-γ [65, 72]. In neonatal cardiomyocytes, higher expression of PPAR-δ by digoxin was observed in cells incubated under hyperglycemia and this was blocked by pretreatment with BAPTA. The increased expressions of FA oxidation genes induced by digoxin were suppressed by treatment with siRNA specific to PPARδ in hyperglycemia-incubated cells. Taken together, it can be considered that activation of PPAR-δ by digoxin is involved in cardiac FA oxidation of heart [73]. The possible mechanism was, in cardiac muscles, digoxin-induced PPAR-δ expression through a Ca^{2+}-triggered signal pathway [74–76] to regulate FA oxidation. Also, digoxin has the ability in

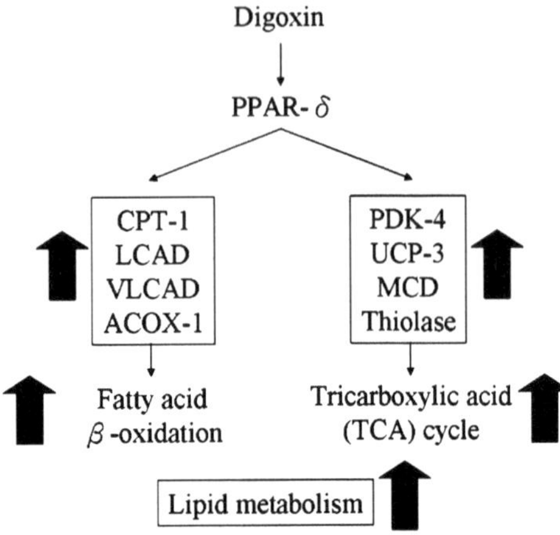

Fig. 1 Effect of agents such as digoxin that can activate peroxisome proliferator-activated receptor (PPAR)-β/δ (PPAR-δ) on lipid metabolism in the heart. Increase of two paths in lipid metabolism: one is the β-oxidation of fatty acids and another is the TCA cycle. Both pathways are increased by PPAR-δ in the heart

regulation of FA metabolism in heart of STZ diabetic rats via an increase in CPT-1, LCAD, VLCAD, ACOX-1 and PDK-4, UCP-3, MCD, thiolase, which leads to the enhancement of β-oxidation and the TCA cycle (Fig. 1). Thus, digoxin has the ability to modify the lipid metabolism in heart via the activation of PPAR-δ.

7 Conclusion

Diabetic cardiomyopathy has evolved from a nebulous concept to concrete reality over the years. Diabetes was found to be independently associated with the occurrence of congestive heart failure after adjusting for left ventricular hypertrophy, hypertension, coronary artery disease, and atrial fibrillation. The higher prevalence of biventricular cardiomyopathy in diabetes patients [27] is also suggestive of diabetes as an independent cause of cardiomyopathy. The confounding factors that could lead independently to cardiomyopathy in diabetics are still unclear, but hyperglycemia, insulin resistance, increased fatty acid metabolism, redox reaction, and fibrosis are considered to collectively contribute to its pathology. PPAR-δ activation may participate in the pathological processes such as lipotoxicity in patients with insulin resistance or diabetes mellitus. Overstimulation of this transcription factor may lead to increased fatty acid accumulation and β-oxidation in cardiomyocytes. Also, cytokines may increase the signal pathway related to Smad (TGF-β/Smad 3) to inhibit the function of PPAR-δ [77]. The role of cytokines in type 2 DM has been established. Thus, PPAR-δ is also related to type 2 diabetic disorders, although more studies are needed.

References

1. Zimmet P, Alberti KG, Shaw J (2001) Global and societal implications of the diabetes epidemic. Nature 414:782–787
2. Adeghate E, Schattner P, Dunn E (2006) An update on the etiology and epidemiology of diabetes mellitus. Ann N Y Acad Sci 1084:1–29
3. National Cholesterol Education Program (NCEP), Expert Panel on Detection, Evaluation, and Treatment of High Blood Cholesterol in Adults (Adult Treatment Panel III) (2002) Third report of the National Cholesterol Education Program (NCEP) expert panel on detection, evaluation, and treatment of high blood cholesterol in adults (Adult Treatment Panel III) final report. Circulation 106:3143–3421
4. Plutzky J (2003) Peroxisome proliferator-activated receptors as therapeutic targets in inflammation. J Am Coll Cardiol 42:1764–1766
5. Haffner SM, Lehto S, Rönnemaa T et al (1998) Mortality from coronary heart disease in subjects with type 2 diabetes and in nondiabetic subjects with and without prior myocardial infarction. N Engl J Med 339:229–234
6. Anonymous (1990) Diabetes mellitus: a major risk factor for cardiovascular disease. A joint editorial statement by the American Diabetes Association; The National Heart, Lung, and Blood Institute; The Juvenile Diabetes Foundation International; The National Institute of Diabetes and Digestive and Kidney Diseases; and The American Heart Association. Circulation 100:1132–1133
7. Rubler S, Dlugash J, Yuceoglu YZ et al (1972) New type of cardiomyopathy associated with diabetic glomerulosclerosis. Am J Cardiol 30:595–602
8. Movahed MR, Hashemzadeh M, Jamal MM (2005) Diabetes mellitus is a strong, independent risk for atrial fibrillation and flutter in addition to other cardiovascular disease. Int J Cardiol 105:315–318
9. Aneja A, Tang WH, Bansilal S et al (2008) Diabetic cardiomyopathy: insights into pathogenesis, diagnostic challenges, and therapeutic options. Am J Med 121:748–757
10. Fang ZY, Prins JB, Marwick TH (2004) Diabetic cardiomyopathy: evidence, mechanisms, and therapeutic implications. Endocr Rev 25:543–567
11. Tarquini R, Lazzeri C, Pala L et al (2011) The diabetic cardiomyopathy. Acta Diabetol 48:173–181
12. Syrovy I, Hodny Z (1992) Nonenzymatic glycosylation of myosin: effects of diabetes and ageing. Gen Physiol Biophys 11:301–307
13. Van Heerebeek L, Hamdani N, Handoko ML (2008) Diastolic stiffness of the failing diabetic heart: importance of fibrosis, advanced glycation end products, and myocyte resting tension. Circulation 117:43–45
14. Zabalgoitia M, Ismaeil MF, Anderson L et al (2001) Prevalence of diastolic dysfunction in normotensive, asymptomatic patients with well-controlled type 2 diabetes mellitus. Am J Cardiol 87:320–323
15. Boyer JK, Thanigaraj S, Schechtman KB et al (2004) Prevalence of ventricular diastolic dysfunction in asymptomatic, normotensive patients with diabetes mellitus. Am J Cardiol 93: 870–875
16. Chatham JC, Seymour AM (2002) Cardiac carbohydrate metabolism in Zucker diabetic fatty rats. Cardiovasc Res 55:104–112
17. Huisamen B, van Zyl M, Keyser A et al (2001) The effects of insulin and beta-adrenergic stimulation on glucose transport, glut 4 and PKB activation in the myocardium of lean and obese noninsulin dependent diabetes mellitus rats. Mol Cell Biochem 223:15–25
18. Bell DSH (2003) Diabetic cardiomyopathy. Diabetes Care 26:2949–2951
19. Murarka S, Movahed MR (2010) Diabetic cardiomyopathy. J Card Fail 16:971–979
20. Cai L, Wang Y, Zhou G et al (2006) Attenuation by metallothionein of early cardiac cell death via suppression of mitochondrial oxidative stress results in a prevention of diabetic cardiomyopathy. J Am Coll Cardiol 48:1688–1697

21. Liang Q, Carlson EC, Donthi RV et al (2002) Overexpression of metallothionein reduces diabetic cardiomyopathy. Diabetes 51:174–181
22. Matsushima S, Kinugawa S, Ide T et al (2006) Overexpression of glutathione peroxidase attenuates myocardial remodeling and preserves diastolic function in diabetic heart. Am J Physiol Heart Circ Physiol 291:2237–2245
23. Shen X, Zheng S, Metreveli NS et al (2006) Protection of cardiac mitochondria by overexpression of MnSOD reduces diabetic cardiomyopathy. Diabetes 55:798–805
24. Du X, Matsumura T, Edelstein D et al (2003) Inhibition of GAPDH activity by poly (ADP-ribose) polymerase activates three major pathways of hyperglycemic damage in endothelial cells. J Clin Invest 112:1049–1057
25. Poornima IG, Parikh P, Shannon RP (2006) Diabetic cardiomyopathy: the search for a unifying hypothesis. Circ Res 98:596–605
26. Susic D, Varagic J, Ahn J et al (2004) Collagen cross-link breakers: a beginning of a new era in the treatment of cardiovascular changes associated with aging, diabetes, and hypertension. Curr Drug Targets Cardiovasc Haematol Disord 4:97–101
27. Corman B, Duriez M, Poitevin P et al (1998) Aminoguanidine prevents age-related arterial stiffening and cardiac hypertrophy. Proc Natl Acad Sci U S A 95:1301–1306
28. An D, Rodrigues B (2006) Role of changes in cardiac metabolism in development of diabetic cardiomyopathy. Am J Physiol Heart Circ Physiol 291:1489–1506
29. Herrero P, Peterson LR, McGill JB et al (2006) Increased myocardial fatty acid metabolism in patients with Type 1 diabetes mellitus. J Am Coll Cardiol 47:598–604
30. Eckel J, Reinauer H (1990) Insulin action on glucose transport in isolated cardiac myocytes: signalling pathways and diabetes-induced alterations. Biochem Soc Trans 18:1125–1127
31. Liedtke AJ, DeMaison L, Eggleston AM et al (1988) Changes in substrate metabolism and effects of excess fatty acids in reperfused myocardium. Circ Res 62:535–542
32. Young ME, McNulty P, Taegtmeyer H (2002) Adaptation and maladaptation of the heart in diabetes: part II. Potential mechanisms. Circulation 109:121–130
33. Issemann I, Green S (1990) Activation of a member of the steroid hormone receptor superfamily by peroxisome proliferators. Nature 347:645–650
34. Escher P, Braissant O, Basu-Modak S et al (2001) Rat PPARs: quantitative analysis in adult rat tissues and regulation in fasting and refeeding. Endocrinology 142:4195–4202
35. Li L, Beauchamp MC, Renier G (2002) Peroxisome proliferator-activated receptor alpha and gamma agonists upregulate human macrophage lipoprotein lipase expression. Atherosclerosis 165:101–110
36. Youssef J, Badr M (2004) Role of peroxisome proliferator-activated receptors in inflammation control. J Biomed Biotechnol 3:156–166
37. Duval C, Fruchart JC, Staels B (2004) PPAR alpha, fibrates, lipid metabolism and inflammation. Arch Mal Coeur Vaiss 97:665–672
38. Desvergne B, Wahli W (1999) Peroxisome proliferator-activated receptors: nuclear control of metabolism. Endocr Rev 20:649–688
39. Tengholm A, Meyer T (2002) A PI3-kinase signaling code for insulin-triggered insertion of glucose transporters into the plasma membrane. Curr Biol 12:1871–1876
40. Finck BN, Lehman JJ, Leone TC et al (2002) The cardiac phenotype induced by PPAR-alpha overexpression mimics that caused by diabetes mellitus. J Clin Invest 109:121–130
41. Burkart EM, Sambandam N, Han X et al (2007) Nuclear receptors PPAR beta/delta and PPAR alpha direct distinct metabolic regulatory programs in the mouse heart. J Clin Invest 117:3930–3939
42. Law RE, Goetze S, Xi XP et al (2000) Expression and function of PPAR-γ in rat and human vascular smooth muscle cells. Circulation 101:1311–1318
43. Finck BN, Kelly DP (2002) Peroxisome proliferator-activated receptor alpha (PPAR-alpha) signaling in the gene regulatory control of energy metabolism in the normal and diseased heart. J Mol Cell Cardiol 34:1249–1257
44. Tontonoz P, Hu E, Spiegelman BM (1994) Stimulation of adipogenesis in fibroblasts by PPARγ2, a lipid-activated transcription factor. Cell 79:1147–1156

45. Olefsky JM, Saltiel AR (2000) PPAR gamma and the treatment of insulin resistance. Trends Endocrinol Metab 11:362–368
46. Saltiel AR, Olefsky JM (1996) Thiazolidinediones in the treatment of insulin resistance and type II diabetes. Diabetes 45:1661–1669
47. Gupta S, Purcell NH, Lin A et al (2002) Activation of nuclear factor-kappaB is necessary for myotrophin-induced cardiac hypertrophy. J Cell Biol 159:1019–1028
48. Purcell NH, Tang G, Yu C et al (2001) Activation of NF-kappaB is required for hypertrophic growth in primary rat neonatal ventricular myocytes. Proc Natl Acad Sci U S A 98:6668–6673
49. Takemoto Y, Yoshiyama M, Takeuchi K et al (1999) Increased JNK, AP-1 and NF-kappaB DNA binding activities in isoproterenol-induced cardiac remodeling. J Mol Cell Cardiol 31:2017–2030
50. Muller DN, Mervaala EM, Dechend R et al (2000) Angiotensin II (AT1) receptor blockade reduces vascular tissue factor in angiotensin II-induced cardiac vasculopathy. Am J Pathol 157:111–122
51. Muller DN, Dechend R, Mervaala EM et al (2000) NF-kappa B inhibition ameliorates angiotensin-II induced inflammatory damage in rats. Hypertension 35:193–201
52. Bishop-Bailey D (2000) Peroxisome proliferator-activated receptors in the cardiovascular system. Br J Pharmacol 129:823–824
53. Asakawa M, Takano H, Nagai T et al (2002) PPAR-gamma plays a critical role in inhibition of cardiac hypertrophy in vitro and in vivo. Circulation 105:1240–1246
54. Yamamoto K, Okhi R, Lee RT et al (2001) PPAR gamma activators inhibit cardiac hypertrophy in cardiac myocytes. Circulation 104:1670–1675
55. Balfour JA, Plosker GL (1999) Rosiglitazone. Drugs 57:921–930
56. Kliewer SA, Forman BM, Blumberg B et al (1994) Differential expression and activation of a family of murine peroxisome proliferator-activated receptors. Proc Natl Acad Sci U S A 91: 7355–7359
57. Braissant O, Foufelle F, Scotto C et al (1996) Differential expression of peroxisome proliferator-activated receptors (PPARs): tissue distribution of PPAR-alpha, -beta, and -gamma in the adult rat. Endocrinology 137:354–366
58. Planavila A, Rodríguez-Calvo R, Jové M (2005) Peroxisome proliferator-activated receptor beta/delta activation inhibits hypertrophy in neonatal rat cardiomyocytes. Cardiovasc Res 65:832–841
59. Blaschke F, Takata Y, Caglayan E et al (2006) Obesity, peroxisome proliferator-activated receptor, and atherosclerosis in type 2 diabetes. Arterioscler Thromb Vasc Biol 26:28–40
60. Muscat GE, Dressel U (2005) Cardiovascular disease and PPAR delta: targeting the risk factors. Curr Opin Invest Drugs 6:887–894
61. Cheng L, Ding G, Qin Q et al (2004) Peroxisome proliferator-activated receptor delta activates fatty acid oxidation in cultured neonatal and adult cardiomyocytes. Biochem Biophys Res Commun 313:277–286
62. Cheng L, Ding G, Qin Q et al (2004) Cardiomyocyte restricted peroxisome proliferator-activated receptor-delta deletion perturbs myocardial fatty acid oxidation and leads to cardiomyopathy. Nat Med 10:1245–1250
63. Pesant M, Sueur S, Dutartre P et al (2006) Peroxisome proliferator-activated receptor delta (PPAR delta) activation protects H9c2 cardiomyoblasts from oxidative stress-induced apoptosis. Cardiovasc Res 69:440–449
64. Barish GD, Narkar VA, Evans RM (2006) PPAR delta: a dagger in the heart of the metabolic syndrome. J Clin Invest 116:590–597
65. Yu BC, Chang CK, Ou HY et al (2008) Decrease of peroxisome proliferator-activated receptor delta expression in cardiomyopathy of streptozotocin-induced diabetic rats. Cardiovasc Res 80:78–87
66. Planavila A, Laguna JC, Vazquez-Carrera M (2005) Nuclear factor-kappaB activation leads to down-regulation of fatty acid oxidation during cardiac hypertrophy. J Biol Chem 280: 17464–17471

67. Feingold K, Kim MS, Shigenaga J et al (2004) Altered expression of nuclear hormone receptors and coactivators in mouse heart during the acute-phase response. Am J Physiol Endocrinol Metab 286:201–207
68. Finck BN, Kelly DP (2007) Peroxisome proliferator-activated receptor gamma coactivator-1 (PGC-1) regulatory cascade in cardiac physiology and disease. Circulation 115:2540–2548
69. Rimbaud S, Garnier A, Ventura-Clapier R (2009) Mitochondrial biogenesis in cardiac pathophysiology. Pharmacol Rep 61:131–138
70. Fein FS, Kornstein LB, Strobeck JE et al (1980) Altered myocardial mechanics in diabetic rats. Circ Res 47:922–933
71. Fein FS, Strobeck JE, Malhotra A et al (1981) Reversibility of diabetic cardiomyopathy with insulin in rats. Circ Res 49:1251–1261
72. Chen ZC, Yu BC, Chen LJ et al (2011) Characterization of the mechanisms of the increase in PPAR delta expression induced by digoxin in the heart using the H9c2 cell line. Br J Pharmacol 163:390–398
73. Chen ZC, Yu BC, Chen LJ et al (2012) Increase of peroxisome proliferator-activated receptor δ (PPARδ) by digoxin to improve lipid metabolism in the heart of diabetic rats. Horm Metab Res
74. Shkryl VM, Shirokova N (2006) Transfer and tunneling of Ca^{2+} from sarcoplasmic reticulum to mitochondria in skeletal muscle. J Biol Chem 281:1547–1554
75. Schaeffer PJ, Wende AR, Magee CJ et al (2004) Calcineurin and calcium/calmodulin-dependent protein kinase activate distinct metabolic gene regulatory programs in cardiac muscle. J Biol Chem 279:39593–39603
76. Tate CA, Hyek MF, Taffet GE (1991) The role of calcium in the energetics of contracting skeletal muscle. Sports Med 12:208–217
77. Tan CK, Chong HC, Tan EH et al (2012) Getting 'Smad' about obesity and diabetes. Nutr Diabetes 2:e29

The Role of Inflammation in Type 2 Diabetes-Driven Atherosclerosis

Jennifer E. Enns, Carla G. Taylor, and Peter Zahradka

Abstract Cardiovascular disease is the most frequent and costly complication of type 2 diabetes (T2D). Inflammation is a shared feature of T2D and cardiovascular disease. Vascular and obesity-mediated inflammation may be key processes responsible for the accelerated development of atherosclerosis in individuals with T2D. Atherosclerosis begins with an insult to the endothelium and progresses through several stages, including the development of endothelial dysfunction, the accumulation of lipids and immune cells in the vessel intima, and phenotypic changes to the vascular cells, all contributing to the formation of vascular lesions. Inflammation plays a central role in many of these phases. The metabolic imbalances characteristic of T2D, such as insulin resistance, hyperglycemia, and hyperlipidemia, exacerbate

J.E. Enns
St. Boniface Hospital Research Centre, Winnipeg, MB, Canada

Canadian Centre for Agri-Food Research in Health and Medicine, Department of Physiology, University of Manitoba, Winnipeg, MB, Canada

C.G. Taylor
St. Boniface Hospital Research Centre, Winnipeg, MB, Canada

Canadian Centre for Agri-Food Research in Health and Medicine, Department of Physiology, University of Manitoba, Winnipeg, MB, Canada

Department of Human Nutritional Sciences, University of Manitoba, Winnipeg, MB, Canada

P. Zahradka (✉)
St. Boniface Hospital Research Centre, Winnipeg, MB, Canada

Canadian Centre for Agri-Food Research in Health and Medicine, Department of Physiology, University of Manitoba, Winnipeg, MB, Canada

Department of Human Nutritional Sciences, University of Manitoba, Winnipeg, MB, Canada

Molecular Physiology Laboratory, Canadian Centre for Agri-Food Research in Health and Medicine, St. Boniface Hospital Research Centre, 351 Tache Ave, Winnipeg, MB R2H 2A6, Canada
e-mail: peterz@sbrc.ca

B. Turan and N.S. Dhalla (eds.), *Diabetic Cardiomyopathy: Biochemical and Molecular Mechanisms*, Advances in Biochemistry in Health and Disease 9,
DOI 10.1007/978-1-4614-9317-4_13,

vascular dysfunction and inflammation, accelerating the progression to advanced atherosclerosis. The expression and production of circulating cytokines and adipokines are altered in T2D, contributing to the proinflammatory state in blood vessels and adipose tissue. Epigenetic mechanisms are emerging as a missing link in diabetes etiology and may contribute to accelerated atherosclerosis. Two other forms of vascular disease, restenosis and graft vascular disease, share features with atherosclerosis. The presence of T2D similarly worsens these conditions, putting these individuals at risk for major adverse cardiac events. Further study of the inflammatory mechanisms and epigenetics will improve our understanding of the role of inflammation in T2D and accelerated cardiovascular disease, and may provide new personalized therapies to treat these conditions.

Keywords Type 2 diabetes • Inflammation • Macrophage • Atherosclerosis • Obesity • Adipokines • Endothelium • Nitric oxide • Insulin resistance

1 Introduction

Type 2 diabetes (T2D) affects more than 200 million people worldwide [1], and this number is increasing because of population growth, aging, urbanization, and increasing prevalence of obesity and lack of physical activity [2]. T2D patients have a twofold to threefold higher risk of developing coronary artery disease (CAD), a greater extent of coronary ischemia, and twice the risk for myocardial infarction compared to their non-T2D counterparts [3–5]; these are the primary reasons for the higher mortality associated with T2D. The underlying condition leading to the elevated incidence of cardiovascular disease in these individuals is an accelerated development of atherosclerosis, which also contributes to many of the complications of T2D.

Atherosclerosis has been described as an inflammatory disease [6–8]. It is thought to begin with injury to the endothelium, triggering a chronic inflammatory response mediated by monocyte-derived macrophages. During this process, macrophages and other immune cells accumulate in the intima–medial space of the vascular wall, along with cholesterol, calcium, and cellular debris. Cytokine release by invading immune cells initiates crosstalk among immune cells, endothelial cells, and vascular smooth muscle cells (VSMCs), and these undergo morphological changes resulting in VSMC proliferation and migration, cellular necrosis, and formation of an atherosclerotic lesion. Vascular remodeling may result in chronic luminal obstruction of blood flow and diminished oxygen supply to target organs, or acute cardiac events, should the lesion rupture and cause thrombosis in the vessels supplying the heart.

There are close ties between T2D, obesity, and inflammation. Obesity has long been considered an important risk factor for T2D and often occurs concomitantly. A large body of evidence describes a low-grade chronic inflammatory state in obesity [9]. The adipose tissue of obese subjects is host to large numbers of immune cells that are an important source of inflammatory mediators with autocrine or paracrine properties, driving whole-body insulin resistance and setting the stage for the pathophysiological events of T2D.

There is strong evidence linking inflammation to the diabetic state [10, 11] and to the development of atherosclerosis [12, 13]. The primary hypothesized mechanisms leading to insulin resistance and pancreatic β-cell dysfunction include, among others, oxidative stress, endoplasmic reticulum stress, and elevated levels of glucose and lipids in the blood. Risk factors for the development of cardiovascular disease similarly include metabolic imbalances such as dyslipidemia and obesity. It is notable that each of these stresses is thought to be exacerbated by, initiated by, or contribute to an inflammatory response [14, 15]. Given the widespread operation of inflammatory pathways in the pathophysiology of T2D and its complications, inflammation may represent a common foundation for advanced pathologies such as T2D and atherosclerosis. Thus, the purpose of this review is to examine the mechanisms of vascular and obesity-mediated inflammation in T2D and to consider their role in the development of accelerated cardiovascular disease.

2 Pathogenesis of Atherosclerosis

Atherosclerosis is a subtype of arteriosclerosis, referring specifically to the thickening and loss of elasticity in the walls of medium- to large-sized arteries [16, 17]. The disease manifests itself as lesions or "plaques" most commonly located in the coronary vessels, arteries, and abdominal aorta. Although the clinical manifestations of atherosclerosis commence primarily in middle age, it is considered a chronic disease and is believed to initiate as early as childhood [18].

The etiology of atherosclerosis is multifactorial and complex. The classical risk factors for atherosclerosis include nonmodifiable factors, such as gender, age, and genetics, as well as modifiable factors, such as smoking, dyslipidemia, hypertension, T2D, obesity, and physical activity [19]. It is these latter risk factors that are potential targets for therapeutic intervention, via either pharmaceutical agents or lifestyle modification.

The development of an atherosclerotic lesion begins with formation of a "fatty streak" in the vessel wall caused by the accumulation of lipid deposits and inflammatory cells [20]. A raised fibrous plaque may develop, consisting of VSMCs that have migrated from the medial layer to the subendothelial space to surround the lipid-filled center of the lesion, overlaid by a cap of extracellular matrix (ECM) proteins and collagen. The lipid core of the lesion contains foam cells, derived from macrophages and vascular cells that have taken up oxidized low density lipoprotein (oxLDL) through scavenger receptors [21, 22]. Further advancement of the disease leads to the formation of "complex lesions," susceptible to alteration by hemorrhage, calcification, and neovascularization. The lesion core often becomes necrotic, leading to erosion and thrombosis [23], thus destabilizing the plaque and creating a risk of rupture. Lesion rupture may result in acute thrombosis, blockage of blood flow, and consequent cardiac events [24, 25].

3 Mechanisms of Endothelial Dysfunction

The endothelium is central in the development of atherosclerosis because it forms the interface between circulating blood and the rest of the vessel wall. The endothelium is a single layer of cells lining the interior surface of blood vessels. It is a dynamic structure capable of altering its permeability and regulating the passage of small and large molecules, and its functional integrity is critical in maintaining vessel wall function. Endothelial cells synthesize important bioactive molecules and influence smooth muscle cells, platelets and peripheral leukocytes. They are also important in maintaining homeostatic control of vascular tone and blood flow, thrombosis, thrombolysis, and platelet adherence.

As a key regulator of vascular tone and blood flow, the endothelium produces vasodilator substances, most importantly nitric oxide (NO), but also prostacyclin and endothelium-derived hyperpolarizing factor. Vasoconstrictors, including endothelin 1 (ET-1) and angiotensin II [26], are also synthesized by endothelial cells. In healthy vessels, homeostatic mechanisms balance the potent vasodilator effects of NO and other vasodilators against the actions of vasoconstrictor molecules, thus helping to maintain normal arterial compliance and tension.

Atherosclerotic plaque development begins with an injury to the endothelium. Endothelial cells are subject to both mechanical and chemical stress:

- Mechanical Stress: Endothelial cells are exposed to hemodynamic forces exerted by blood flowing through the vessels. In healthy vessels, shear forces exerted on the endothelium by laminar blood flow trigger production of NO. Arterial branch points, however, experience disrupted flow (e.g., pulsatile and turbulent flow), which dysregulates vascular tone and endothelial cell function, priming these regions for lesion formation [27]. Arterial branch points are thus highly susceptible to plaque development [28].
- Chemical Stress: Blood-borne chemical signals, such as hormones, nutrients, and the noxious by-products of cigarette smoking, can damage endothelial cells by promoting the production of reactive oxygen species (ROS). In endothelial cells, NADPH oxidase produces the highly reactive and biologically toxic superoxide anion, which damages cellular components and compromises membrane integrity. Superoxide also inactivates NO and transforms it into peroxynitrite, thus uncoupling endothelial nitric oxide synthase (eNOS) and further impairing endothelial function [29].

These stress factors initiate endothelial activation, which is characterized by a change in the endothelial cell gene expression profile. Activated endothelial cells express chemotactic molecules, such as monocyte chemoattractant protein (MCP-1), which recruit leukocytes to the site of injury. They also exhibit an increase in the number of cell adhesion molecules (CAMs) on their surfaces. This increase in CAMs is believed to facilitate monocyte entry into the vascular wall in several distinct steps: rolling, attachment, and transendothelial migration [30]. This process is accelerated by the presence of activated platelets, which form bridges between monocytes and endothelial cells [31, 32]. Once the monocytes have entered the intima, the synthesis and secretion of inflammatory mediators creates a local proinflammatory microenvironment and aids in the progression of the lesion [33].

Injury to the endothelium and the subsequent interactions with the immune system ultimately lead to endothelial dysfunction, which is accompanied by further changes in gene expression and cell phenotype. Several factors influence phenotypic change: (1) disturbed blood flow sensed via mechanosensitive endothelial cell-surface receptors, which induces prothrombotic and adhesive cell surface properties [34, 35]; (2) nutrient-dependent transcriptional activation of endothelial cell proliferation genes [36]; and (3) other factors not yet identified, as demonstrated by a comparison of endothelial cell phenotype and gene expression across different vascular regions, which cannot be explained by hemodynamic forces alone [37].

Although all these factors contribute to altered endothelial function, the hallmark of endothelial dysfunction is impaired NO bioavailability. Loss of the vasorelaxative properties of NO leads to impaired vasodilation of blood vessels [38] and breakdown of the homeostatic mechanisms that govern the oxidative balance. A vicious cycle of increasing oxidative stress ensues, with uncoupling of eNOS and further reduction in NO production. To synthesize NO, eNOS function requires the cofactor tetrahydrobiopterin (BH4), which is generated from GTP. An insufficiency of BH4 results in the production of superoxide anion by eNOS [39]. BH4 itself is susceptible to oxidation by superoxide, which may further promote oxidative stress. Thus, endothelial dysfunction is the result of a number of mechanisms that reduce the ability of the endothelium to maintain its regulatory functions and drives the vascular system toward atherogenesis.

4 Metabolic Abnormalities of Type 2 Diabetes Advance Endothelial Dysfunction

The altered metabolic state that diabetes confers on patients affects both structure and function of the vessels, resulting in atherogenic changes. Several key metabolic abnormalities characteristic of T2D contribute to and accelerate the development of cardiovascular complications. Metabolic and immune pathways share many hormones, signaling proteins, transcription factors, and bioactive lipids. Thus, altered metabolism in T2D provides a link between vascular function and inflammatory status.

4.1 Insulin Resistance

Insulin is an important activator of eNOS. It acts by binding to its receptor on the surface of insulin-responsive cells. The receptor phosphorylates itself and several other substrates, including the insulin receptor substrate (IRS) family, initiating downstream signaling events [40]. Subsequent phosphorylation and activation of eNOS occurs via PI3 kinase/Akt [41], leading to stimulation of NO production [42]. NO influences several important steps of the atherogenic process: in particular, it promotes endothelial regeneration and angiogenesis, and reduces VSMC constriction, migration, and proliferation [43]. In the setting of insulin resistance, insulin-dependent NO biosynthesis is impaired, vasoconstrictor effects are unopposed, and arterial tone is increased [44]. In addition, diabetes is directly responsible for increases in endothelium-derived vasoconstrictor substances such as ET-1, angiotensin II, and prostanoids [45, 46]. Meanwhile, the effects of insulin on endothelial and VSMC proliferation and expression of ET-1 and intercellular adhesion molecule (ICAM)-1 via the MAPK pathway [47] are unhindered, leading to increased vasoconstrictive, proinflammatory, and prothrombotic activity. The inhibition of insulin signaling is an important mechanism through which inflammatory pathways lead to insulin resistance and vascular dysfunction

4.2 Hyperglycemia

Elevated levels of glucose in T2D patients induce an increase in the formation of advanced glycation end products (AGEs). AGEs arise as metabolites of glucose metabolism when the intracellular glucose concentration is excessive. They have been shown to cause crosslinking of structural proteins and thus have been implicated in the loss of arterial elasticity [48]. Recent studies have demonstrated a significant benefit to vascular function when AGE crosslinks are broken [49]. AGEs can also react with their respective receptors (RAGEs) to induce metabolic responses that contribute to vascular disease. Under normal conditions, RAGEs are expressed minimally in endothelial cells and VSMCs, but their expression is enhanced by risk factors such as hyperlipidemia during diabetes [50, 51]. Because AGE production is also accelerated in diabetes, T2D promotes atherogenesis via RAGE-dependent signaling mechanisms: (i) activation of oxidative pathways via NAPDH oxidase, leading to increased ROS production; and (ii) stimulation of nuclear factor (NF)-κB-dependent gene expression, resulting in production and secretion of proinflammatory cytokines, such as tumor necrosis factor (TNF)-α, interleukin (IL)-1β, IL-6, and IL-8 [50, 52]. These cytokines contribute to the pro-inflammatory environment in atherosclerotic lesions, promoting activation of endothelial cells, VSMCs, and immune cells, as discussed in the following section.

4.3 Hyperlipidemia

In T2D, levels of cholesterol-carrying low density lipoprotein (LDL), the oxidized form of LDL (oxLDL), and free fatty acids are elevated in the blood. Triglycerides are not considered to be directly atherogenic, but several species of triglyceride-rich lipoproteins, such as very low density lipoprotein (VLDL), VLDL remnants, and chylomicron remnants, appear to promote atherogenesis independently of LDL [53]. Cholesterol, on the other hand, is a major component of the lipid core of atherosclerotic lesions. ROS generate oxLDL by oxidizing the protein component of the lipoprotein; however, in a detrimental cycle of cell injury and atherogenesis, oxLDL also promotes ROS synthesis, which can damage endothelial cells and induce the expression of adhesion molecules [44]. Free fatty acids, elevated in the blood as a result of insulin resistance in liver and adipose tissue, interfere with NO production by inhibiting IRS-1, Akt, and eNOS phosphorylation, resulting in translocation of PKC and activation of the classic proinflammatory NF-κB pathway in the endothelium [54].

It is clear that the pathogenesis of T2D has important implications in the progression of atherosclerotic disease. The metabolic abnormalities that characterize diabetes, such as impaired insulin signaling, hyperglycemia, and hyperlipidemia, all contribute to the worsening of endothelial dysfunction by promoting oxidative stress and dysregulating arterial function. The T2D state also influences the interactions of multiple cell types involved in the development of the atherosclerotic lesion, as described next.

5 Changes in Vascular Cells, Immune Cells, and Adipose Tissue Contribute to the Progression of Type 2 Diabetes-Driven Atherosclerosis

5.1 Endothelial Cells

As endothelial dysfunction becomes more pronounced, the endothelial cells at the site of injury undergo functional and structural changes. Dysfunction is accompanied by increased apoptosis [55] and high turnover rates of endothelial cells [56]. Impaired NO production and increased oxidative stress alter the ability of the endothelium to regulate vascular tone and maintain other homeostatic mechanisms. Endothelial cells activated by elevated lipids and glucose in the blood express high levels of adhesion molecules, such as vascular cellular adhesion molecules (VCAM) and ICAM-1, on their surfaces. These molecules recruit mononuclear leukocytes (monocytes and T cells) to the vessel wall, where they transmigrate through the endothelium and enter the intima [57]. Endothelial cell production of inflammatory cytokines and chemotactic factors (e.g., MCP-1) also contributes to the creation of a proinflammatory milieu in the vessel wall.

5.2 Vascular Smooth Muscle Cells

The main function of VSMCs is regulating vessel tone, which mediates blood pressure and blood flow distribution [58]; however, they also have a prominent role in the repair process. This role requires VSMCs to maintain a degree of plasticity, which allows them to re-enter the cell cycle and proliferate [59, 60], a characteristic that has implicated them in the development of vascular disease and postsurgery repair. VSMCs exhibit a range of phenotypes, from the "contractile" type primarily found in adult vessels to the "synthetic" type, characterized by increased cell size and extracellular matrix production as well as increased proliferation and migration [61, 62]. In the early stages of atherosclerosis, VSMCs are involved in lesion formation via their proliferative and migratory responses, whereby they envelop the lipid core of the developing plaque and cause the plaque to bulge into the lumen of the blood vessel. In late-stage disease, VSMCs may play a beneficial role by synthesizing collagen, which helps to stabilize the fibrous cap and prevent lesion rupture. VSMC apoptosis has been shown to occur in advanced atherosclerotic lesions and is thought to be a prominent pathway leading to plaque destabilization [63].

5.3 Monocytes/Macrophages

Monocytes are recruited to sites on the vascular surface by chemotactic factors released from endothelial cells that have been activated by injury. The monocytes subsequently infiltrate the vascular wall by binding to adhesion molecules present on the surface of the activated endothelium and migrate across the endothelial barrier to the subendothelial space. Once in the intima, monocytes differentiate into macrophages, a process dependent on the cytokines present. Interactions with the extracellular matrix, macrophage colony-stimulating factor, and members of the tumor necrosis factor (TNF) family are known to drive monocyte differentiation [64]. Changes in gene expression occur concomitantly with differentiation, including an increase in scavenger receptor expression. Using these receptors, macrophages take up modified lipids, such as oxLDL and oxysterols, to such an extent that they become engorged with lipids in characteristic foamy deposits. Foam cells are major contributors to lesion formation via ROS production and increased secretion of proinflammatory TNFs, interleukins and interferons, matrix metalloproteinases and several other detrimental signaling molecules. Through this degenerative cycle, macrophages have been implicated in every stage of atherosclerosis, from the early stages of fatty streak formation to late-stage necrotic events and plaque rupture [65].

The accelerated development of atherosclerosis in T2D patients is driven, at least in part, by the altered properties and function of monocytes and macrophages. The classification of macrophages as one of two types, M1 (classically activated) or M2 (alternatively activated), is perhaps simplistic given the rapidly expanding M2 category, which encompasses cells with dramatically different biochemistry and physiology [66]. However, the M1 classification describes an aggressively proinflammatory cell type that expresses high levels of IL-1, IL-2, IL-6, IL-12, and TNF-α, all of which contribute to activation of vascular cells and further development of the atherosclerotic plaque. The ongoing accumulation of M1 macrophages and other activated immune cells, such as T cells, B cells, and dendritic cells, create a milieu of cytokines that act on the endothelial cells, VSMCs, and each other, driving the inflammatory process.

The diabetic state enhances the pathological processes of atherosclerosis through several mechanisms: (i) suppression of NO formation, which dysregulates arterial tone and initiates changes in vascular cell phenotypes; (ii) increased oxidative stress, which exacerbates the loss of NO bioavailability and drives immune cell recruitment; and (iii) stimulation of inflammatory processes via RAGE-dependent activation of NF-κB and activator protein 1. These transcription factors regulate the expression of several genes encoding mediators of atherogenesis, such as adhesion molecules, chemoattractants for monocytes and lymphocytes, and proinflammatory mediators found in the atheroma, including interleukins and TNF-α [67]. The hyperglycemia characteristic of T2D patients additionally activates protein kinase C (PKC), RAGEs, and NF-κB in VSMCs and endothelial cells, which further augment production of ROS and inflammatory factors [61]. These mechanisms contribute to the structural and functional changes in vascular cells and immune cells and may explain, at least in part, why T2D patients are at higher risk for cardiovascular disease and coronary events than their nondiabetic counterparts.

5.4 *Macrophages in Obesity*

Adipose tissue actively participates in inflammation and immunity through the production and release of proinflammatory molecules (e.g., leptin, TNF-α, IFN-γ, IL-6, IL-18) and antiinflammatory molecules (e.g., adiponectin, IL-10). Both the resident cells of adipose tissue (adipocytes, endothelial cells, fibroblasts, etc.) and infiltrating immune cells (T cells, macrophages) are sources of these bioactive substances. Although T cells play a prominent role in adipose tissue inflammation and may stimulate pre-adipocytes to recruit macrophages via release of chemotactic factors such as MCP-1, the number of macrophages present in adipose tissue has been positively associated with body weight, body mass index (BMI), and total body fat. This association is significant, because macrophages have been identified as a major source of proinflammatory IL-6 and TNF-α [68, 69], which contribute to the chronic systemic inflammation and insulin resistance observed in obesity. Macrophages may also modulate the phenotype of adipocytes, thus altering their secretory functions [68, 70], and further contributing to the dysregulation of this tissue. Because obesity and T2D often occur concomitantly, the proinflammatory activity of adipose tissue likely exacerbates the risk of vascular disease in these conditions [71–73]. In addition to the classical inflammatory markers produced in adipose tissue, many other bioactive compounds are secreted by adipocytes and resident immune cells. Known collectively as adipokines, these are described in more detail in the following section. Adipose tissue also contains endothelial cells that line the blood vessels supplying the tissue. These endothelial cells are susceptible to activation by adipokines and inflammatory mediators produced in their environment, as well as to circulating nutrients such as fatty acids [74], and can contribute to endothelial dysfunction and its sequelae.

6 Contribution of Circulating Inflammatory Cytokines and Adipokines

Cytokines and their receptors, described as the "currency" of inflammation, play a central role in propagating inflammatory responses between cells of the immune system and the vasculature [75]. These low molecular weight proteins have autocrine and paracrine functions [76]. Some of the downstream effectors of receptor-mediated intracellular cascades initiated by proinflammatory cytokine signaling include mitogen-activated protein kinases (MAPKs), PKC, and the signal transducer and activator of transcription (STAT) family [77]. These effectors most often activate the transcriptional "master switch" NF-κB, which controls the expression of many genes affecting immune responses [78, 79].

6.1 Interleukins

Interleukins are often classified in accordance with their effects on lymphocyte function or maturation as proinflammatory T_h1 or antiinflammatory T_h2 cytokines [80]. The T_h classifications are derived from the $CD4^+$ T helper cells, which synthesize and release interleukins and other signaling molecules, helping to direct the immune response. T_h1 cells are characterized by high expression of interferon (IFN)-γ, IL-1, IL-2, IL-12, TNF-α, and other proinflammatory cytokines, but T_h2 cells produce antiinflammatory cytokines (IL-4 and IL-10) that dampen the expression of T_h1 genes. T-helper cells make up about 10 % of the cellular content of the human atherosclerotic plaque; however, the proinflammatory T_h1 type is much more prevalent in human lesions than the T_h2 type [80, 81].

Elevated IL-1β levels are observed in obesity and T2D, and expression of IL-1β and its receptor is increased in adipose tissue of obese individuals [82]. Under normal conditions, IL-1β is derived primarily from macrophages in response to infection or injury, but in insulin-resistant states, the activated macrophages resident in insulin-sensitive tissues (such as adipose tissue) are an important source of IL-1β. This cytokine reduces insulin-stimulated glucose uptake and lipogenesis [83] by inducing suppressor of cytokine signaling (SOCS), which leads to degradation of the IRS proteins [84]. In this way, IL-1β exacerbates insulin resistance in adipose, muscle, and liver tissues. IL-1β is also an inducer of adhesion molecule expression in endothelial cells, which contributes to atherogenesis, and impairs pancreatic β-cell insulin secretion and function [85]. The actions of IL-1β and TNF-α are closely related. Both are produced locally at the site of inflammation, and they share similar signal transduction mechanisms. They also mutually enhance each other's production and act synergistically [86].

IL-18 belongs to the IL-1 superfamily. IL-18 levels are increased in obesity, insulin resistance, and T2D [87]. This cytokine has proatherogenic activity mediated by binding to the IL-18 receptor and inducing IFN-γ expression, which potentiates the inflammatory process and may interfere with formation of the fibrous cap [88, 89].

IL-18 is expressed in macrophages, adipocytes, endothelial cells, and VSMCs [90]. Recent research has suggested that the induction of IL-18 and IL-1β in metabolic diseases such as obesity, T2D, and atherosclerosis originates from the activation of the NLRP3 inflammasome, an IL-1 family cytokine-activating protein complex [91]. Inflammasome activation in atherosclerosis occurs in two steps: (1) modified LDL binds Toll-like receptor 4/6 on macrophages to stimulate inflammasome priming via NF-κB-induced upregulation of pro-IL-1β, pro-IL-18, and NLRP3; and (2) phagocytosed cholesterol crystals cause lysosomal rupture and release of proteases (cathepsins). In combination with ROS, cathepsins mediate the activation of the NLRP3 inflammasome, resulting in caspase 1 cleavage and production of mature IL-1 family members.

IL-6 plays a central role in the chronic development of atherosclerosis, promoting the creation of complex plaques. IL-6 levels are elevated in T2D, and the cytokine is considered an important determinant of T2D risk [92, 93]. It is produced at the site of inflammation, where it promotes production of proinflammatory C-reactive protein (CRP) [94]. It is secreted by a variety of tissues, including activated leukocytes, adipocytes, and endothelial cells and is involved in the acute phase of inflammation. The adipose tissue is thought to be a major source of IL-6 [93], where it promotes inflammation and adipocyte dysfunction. IL-6 is more highly expressed in adipocytes than in macrophages [68]; therefore, it is considered an important mediator of adipose tissue inflammation and adipocyte dysfunction. IL-6 has been implicated in TNF-α activation; the joint activity of these inflammatory mediators inhibits insulin signaling in adipocytes and hepatocytes via induction of SOCS family members [95, 96]. Circulating IL-6 also contributes to insulin resistance in T2D by inducing gluconeogenesis and subsequent hyperglycemia, lipolysis, and hepatic free fatty acid secretion [97]. Its damaging effects on metabolic control and insulin signaling in insulin-responsive tissues contribute to the chronic low-grade inflammatory status and atherogenesis. IL-6 is also directly involved in atherosclerotic plaque development, increasing adhesion molecule expression and contributing to clot formation by stimulating coagulation.

6.2 *TNF-α*

TNF-α acts as an amplifier of many cellular signaling events occurring during inflammation [98]. It is produced and secreted primarily by macrophages, and is elevated in obesity, where it is one of the main adipose-derived cytokines [99]. Endothelial cells and VSMCs also express TNF-α. TNF-α interferes with insulin signaling through receptor-mediated pathways in muscle cells and hepatocytes by stimulating inhibitory phosphorylation of IRS-1 [100, 101]. TNF-α influences the synthesis, secretion, and activity of other cytokines. It has been shown to decrease eNOS expression [102], increase expression of adhesion molecules VCAM-1 and ICAM-1 in endothelial cells, stimulate expression of various other proinflammatory cytokines through activation of NF-κB, stimulate VSMC proliferation, and enhance expression of matrix metalloproteinases that contribute to plaque destabilization [103, 104].

6.3 MCP-1

Chemoattractant factors, such as MCP-1 and macrophage stimulatory factor, recruit monocytes to vascular walls and promote synthesis and secretion of proinflammatory cytokines that enhance their attachment, enabling migration into the vessel wall. MCP-1 is also a key chemokine in the activation of resident macrophages in adipose tissue, stimulating proinflammatory adipokine production and supporting recruitment of more macrophages to the tissue via NF-κB regulation [105]. MCP-1 levels are increased in patients with T2D, and large-scale population-based studies have demonstrated an association between circulating MCP-1 and other risk factors for atherosclerosis, such as CRP [106]. Hyperglycemia increases the expression of MCP-1 and CAMs in endothelial cells [107], as do oxidatively modified lipoproteins isolated from the plasma of T2D patients [108]. Therapies directed at limiting MCP-1 production are now under development for this population [106].

6.4 CRP

CRP, an acute-phase protein secreted by hepatocytes in response to IL-6 signaling, is considered an independent predictor of cardiovascular complications such as atherosclerosis [109], and its association with risk factors for T2D development has been studied extensively [110]. CRP binds a large number of ligands, including native and modified plasma lipoproteins, phospholipids, and apoptotic cells, which are present in the atherosclerotic lesion. Binding of CRP to ligands activates the classical complement pathway. In endothelial cells, CRP stimulates production of ET-1, downregulates eNOS and decreases NO biosynthesis, increases ROS production, and stimulates expression of VCAM-1, ICAM-1, E-selectin, and MCP-1 [111].

6.5 Cell Adhesion Molecules

The CAM family consists of proteins located on the cell surface that bind other cells or the ECM to bring about cell adhesion. Several CAMs, including VCAM-1, ICAM-1, P-selectin, platelet/endothelial cell adhesion molecule-1, and E-selectin, have been implicated in the development of atherosclerosis by facilitating vascular infiltration [112, 113].

Although few studies have reported a specific role for CAMs in T2D, emerging research demonstrates a correlation between circulating soluble adhesion molecules (SAMs) and T2D [114]. Circulating SAMs are thought to reflect increased endothelial cell-surface expression of CAMs [115], and high serum levels of SAMs are considered markers of endothelial dysfunction, the trigger for cardiovascular disease in T2D [116].

6.6 *Adipokines*

Cytokine levels have been found to be consistently elevated in T2D [117], even when adjusted for obesity, implying an additive but not exclusive role for obesity in inflammation in diabetic individuals. The adipose tissue is not only a storage depot for energy but is also a dynamic endocrine organ that produces metabolically important signaling molecules termed adipokines. Many adipokines have immunological activity, but not all stimulate the immunological process.

6.7 *Adiponectin, Omentin, and Vaspin*

Adiponectin, omentin, and vaspin are considered antiinflammatory, vasorelaxive, and cardioprotective. Adiponectin is an adipose-specific protein that has been well documented as an antiinflammatory adipokine [118]. Circulating adiponectin levels are decreased in subjects with obesity-related insulin resistance, T2D, and CAD. Adiponectin expression is downregulated in obese/diabetic states [119, 120]. Adiponectin inhibits liver gluconeogenesis, promotes fatty acid oxidation in skeletal muscle, and improves insulin sensitivity [121]. It reduces expression of adhesion molecules in endothelial cells and decreases cytokine production from macrophages by inhibiting NF-κB signaling. It further counteracts the proinflammatory effects of TNF-α on the arterial wall and suppresses the transformation of macrophages into foam cells [122–124]. Omentin has been shown to block the TNF-α-mediated inflammatory pathway and suppress E-selectin expression [125]. Adiponectin and omentin have also been shown to induce NO-mediated endothelium-dependent vasodilation in isolated blood vessels [126–128]. Vaspin administration inhibits expression of the proinflammatory adipokines leptin, resistin, and TNF-α [129].

On the other hand, resistin specifically inhibits insulin-mediated endothelium-dependent relaxation, contributing to increased vascular tone and resistance to blood flow [130]. Visfatin is also considered a proinflammatory, proatherogenic adipokine as it mediates vascular inflammation by inducing expression of VCAM-1 and ICAM-1 via NF-κB activation [131], and stimulates endothelial [132] and VSMC [133] proliferation. Leptin is a cytokine-like hormone produced mainly by adipocytes in direct proportion to the amount of adipose tissue present in the body. It acts in the hypothalamus to regulate appetite and energy consumption, signaling in a negative-feedback manner to reduce food intake. Many obese individuals, however, are leptin resistant, and it follows that because obesity and T2D often occur concomitantly, many individuals with T2D are likewise leptin resistant [134]. In obesity, the response to leptin by the hypothalamus is dampened, and so the expanded adipose tissue produces leptin in an unregulated fashion without enacting an anorexic effect. Hyperleptinemia has been linked to cardiovascular morbidity and mortality [135, 136], and human plasma leptin concentrations are independently associated with structural and functional vascular abnormalities [137, 138].

Leptin is also considered an important immune modulator, inducing proinflammatory cytokine production in monocytes and macrophages [139, 140] and favoring T_h1 T cell differentiation [141], an important determinant of atherosclerotic lesion development.

7 Epigenetic Mechanisms in T2D

Although it is well known that key gene mutations and genetic factors are involved in the pathogenesis of diabetes, increasing evidence suggests that complex interactions between genes and the environment may play a role in metabolic disease such as diabetes and its complications [142, 143]. Epigenetics is the study of heritable alterations to DNA expression that are not caused by changes in the underlying DNA sequence. DNA is wrapped around a histone protein octamer to make up the nucleosome, the basic unit of chromatin [144]. Apart from the binding of transcription factors, transcriptional activation and repression are linked to the recruitment of proteins that alter chromatin structure via chromatin remodeling, leading to changes in accessibility for the gene expression machinery. Such remodeling includes post-translational modifications to histone proteins (e.g., methylation, acetylation, and phosphorylation) as well as DNA methylation at CpG islands [145].

Environmental factors such as diet and exercise have been shown to influence epigenetic patterns and gene expression and thus may have far-reaching effects in disease risk, development, and progression. Despite the use of medication and lifestyle changes to control glucose levels in T2D, individuals with T2D continue to suffer numerous life-threatening complications even after glycemic control has been achieved. This sequel suggests a "metabolic memory" of prior hyperglycemic exposure [145]. Recent studies have suggested that epigenetic changes in target cells may explain the missing link in diabetes etiology; exploration of gene–environment interactions may help identify new therapeutic targets.

7.1 Histone Modifications

Histone lysine acetylation and deacetylation are mediated by histone acetyltransferases (HATs) and histone deacetylases (HDACs), respectively. HATs and HDACs regulate the expression of several key genes linked to diabetes, as reviewed by Gray and De Meyts [146]. The sirtuin (SIRT) family of deacetylases has likewise been implicated in regulating several factors involved in metabolism, including insulin secretion and adipogenesis [147]. HATs and HDACs have been shown to affect NF-κB transcriptional activity, resulting in changes in inflammatory cytokine expression levels [148, 149]. High glucose levels and oxidized lipids have both been shown to influence histone lysine acetylation with the downstream effect of increasing expression of inflammatory genes [150, 151], extracellular matrix components, and vasoactive factors in endothelial cells [152].

Histone methylation is thought to be more stable than acetylation and thus may have more impact on diabetes and its complications. Monocytes from diabetic patients and monocytes exposed to hyperglycemic conditions were both shown to have altered histone methylation patterns [153]. Knockdown of histone methylases in monocytes attenuated TNF-α-dependent induction of key inflammatory gene expression via NF-κB [154], and a role for methylation was also observed during induction of inflammation in response to high glucose in endothelial cells [155, 156].

7.2 *DNA Methylation*

DNA methylation at promoter CpG islands has been associated with gene repression in the context of tumour suppressor genes and cancer. However, little is known about the role of DNA methylation in diabetes. Preliminary studies in mouse embryonic cells implicate this epigenetic mechanism in regulating insulin expression [157]. In the agouti mouse, DNA methylation of the agouti gene can affect the development of obesity in diabetes [158]. Other studies indicate a role for DNA methylation in β-cell differentiation [159], homocysteine expression (with links to atherosclerosis and T2D-related chronic kidney disease) [160, 161], and the cell cycle (affecting endothelial cell proliferation) [162].

Although T2D has long been considered to have a genetic component, recent research has improved our understanding of epigenetic mechanisms in this disease state. These studies have demonstrated the involvement of hyperglycemia in regulating gene expression via epigenetic changes, but the complexity of T2D with its multiple risk factors and comorbidities suggests that much remains to be discovered. Epigenetics will continue to contribute toward our understanding of the interactions between T2D and atherosclerosis, but further research is necessary to fully elucidate the impact of epigenetic mechanisms on complications of T2D: this may lead to the development of new therapeutic strategies for T2D and cardiovascular disease.

8 Restenosis and Graft Vascular Disease in T2D

Restenosis and graft vascular disease (GVD) are complications of vascular disease that occur following a surgical procedure. They are similar to atherosclerosis in that they are chronic, progressive processes initiated and driven by local inflammation of medium- and large-sized arteries, and their prevalence is similarly exacerbated by the presence of T2D. It should not be surprising that as the incidence of T2D escalates, the number of T2D patients requiring revascularization procedures for advanced CAD is also rising.

Restenosis, the reoccurrence of stenosis (the narrowing of a blood vessel), is a response to vessel injury after a procedure to treat the initial narrowing, such as balloon angioplasty or stenting. Many studies demonstrate that T2D is a strong risk factor for restenosis, with T2D patients experiencing poorer outcomes and more frequent revascularization procedures [163–165]. The same factors that accelerate the development of endothelial dysfunction and the other sequelae of lesion formation in T2D, hyperinsulinemia and hyperglycemia, are believed to drive restenosis [166, 167]. The thiazolidinediones, an insulin-sensitizing and antiinflammatory class of drugs, have been shown to reduce restenosis after coronary stenting [168, 169], independent of the glucose-lowering effect of these agents.

Graft vascular disease, also known as transplant vasculopathy or graft arterial disease, is a response to injury that occurs in allograft vessels following coronary bypass surgery or organ transplantation. Although immunosuppression regimens do an adequate job of preventing acute allograft rejection, GVD remains a major cause of death in patients surviving more than 1 year after transplantation [170, 171]. In fact, up to 75 % of all transplant patients have evidence of GVD after just 1 year [172]. The pathogenesis of GVD is thought to occur as follows: injured endothelial and perivascular cells induce a low-level immune response, which stimulates the secretion of growth factors to activate VSMCs, inducing proliferation and ECM synthesis [173, 174]. VSMCs and ECM components accumulate in the intima, forming diffuse concentric intimal lesions, but the internal elastic lamina is characteristically spared, a distinguishing feature from native atherosclerosis [175]. Calcification and necrosis are not typically observed unless they occur on a preexisting atherosclerotic plaque [176]. The developing lesions progressively compromise vascular flow and cause ischemia, leading to vascular graft failure. Although a strong correlation between diabetes and GVD is yet to be established, diabetes is considered a risk factor [177]. The metabolic abnormalities exhibited by type 2 diabetic individuals (dyslipidemia, insulin resistance, and glucose intolerance) likely predispose them to developing GVD by the same mechanisms through which diabetes contributes to CAD [177–179].

9 Conclusions

It is clear that, despite the enormously complex etiology of vascular disease, inflammation is a central mediator of many of the pathologies involved therein. The phenotypic changes triggered by inflammation in vascular and adipose tissues and immune cells are key drivers of the metabolic and vascular dysfunction present in T2D. Metabolic changes wrought by T2D further advance the damaging effects of vascular dysfunction, accelerating the progression of atherosclerosis and enhancing the risk of acute cardiac events. Moreover, obesity-mediated inflammation stemming from adipocyte dysfunction worsens glycemic control and contributes to systemic metabolic imbalance. The interplay of inflammatory mediators with features of atherosclerosis in T2D is summarized schematically in Fig. 1. Thus, inflammation emerges as an attractive target for new therapies committed to correcting the metabolic abnormalities in T2D. Insights derived from further studies will help to identify multi-approach interventions to curb the worldwide epidemic of debilitating cardiovascular disease and its complications.

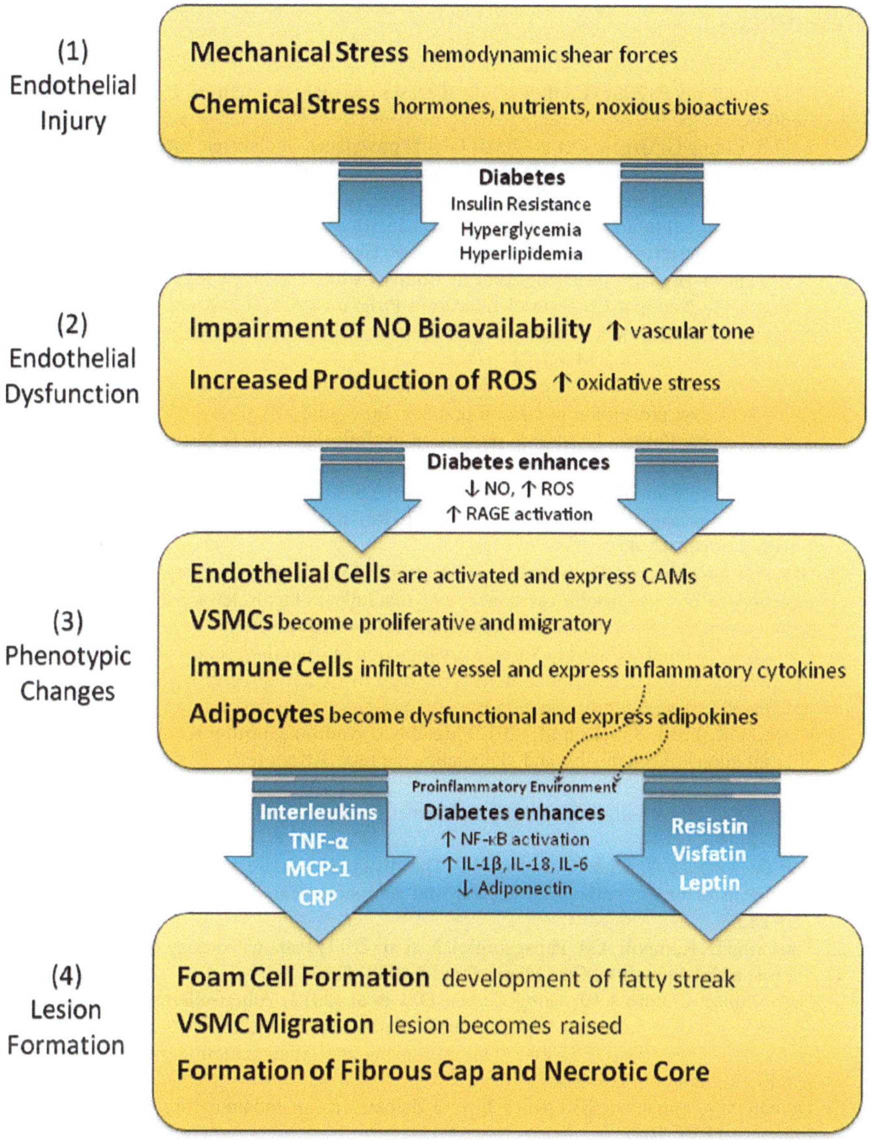

Fig. 1 The role of inflammation in the accelerated progression of atherosclerosis in type 2 diabetes (T2D). Injury to the intact endothelium initiates the process (1). The metabolic imbalances of T2D facilitate progression to endothelial dysfunction (2), where endothelial cells exhibit impaired NO production and increased generation of ROS. Features of T2D enhance these dysfunctional properties and drive phenotypic changes to endothelial, vascular smooth muscle, immune cells and adipocytes (3). The inflammatory mediators produced by these activated and dysfunctional cells create a proinflammatory environment, which aids in the accelerated formation of atherosclerotic lesions (4). *CAM* cellular adhesion molecules, *NO* nitric oxide, *ROS* reactive oxygen species, *VSMC* vascular smooth muscle cell

References

1. Bandyopadhyay P (2006) Cardiovascular diseases and diabetes mellitus. Drug News Perspect 19:369–375
2. Wild S, Roglic G, Green A et al (2004) Global prevalence of diabetes: estimates for the year 2000 and projections for 2030. Diabetes Care 27:1047–1053
3. Tavani A, Bertuzzi M, Gallus S et al (2002) Diabetes mellitus as a contributor to the risk of acute myocardial infarction. J Clin Epidemiol 55:1082–1087
4. National Cholesterol Education Program (NCEP) Expert Panel on Detection, Evaluation, and Treatment of High Blood Cholesterol in Adults (Adult Treatment Panel III) (2002) Third report of the National Cholesterol Education Program (NCEP) expert panel on detection, evaluation, and treatment of high blood cholesterol in adults (Adult Treatment Panel III) final report. Circulation 106:3143–3421
5. De Backer G, Ambrosioni E, Borch-Johnsen K et al (2003) European guidelines on cardiovascular disease prevention in clinical practice: third joint task force of European and other societies on cardiovascular disease prevention in clinical practice (constituted by representatives of eight societies and by invited experts). Eur J Cardiovasc Prev Rehabil 10:S1–S10
6. Festa A, D'Agostino R Jr, Howard G et al (2000) Chronic subclinical inflammation as part of the insulin resistance syndrome: the Insulin Resistance Atherosclerosis Study (IRAS). Circulation 102:42–47
7. Hak AE, Stehouwer CD, Bots ML et al (1999) Associations of C-reactive protein with measures of obesity, insulin resistance, and subclinical atherosclerosis in healthy, middle-aged women. Arterioscler Thromb Vasc Biol 19:1986–1991
8. Perrault R, Zahradka P (2011) Vascular dysfunction in heart disease. In: Dhalla NS, Nagano M, Ostadal B (eds) Molecular defects in cardiovascular disease. Springer, New York, pp 283–303
9. Romeo GR, Lee J, Shoelson SE (2012) Metabolic syndrome, insulin resistance, and roles of inflammation—mechanisms and therapeutic targets. Arterioscler Thromb Vasc Biol 32: 1771–1776
10. Wellen KE, Hotamisligil GS (2005) Inflammation, stress, and diabetes. J Clin Invest 115: 1111–1119
11. Puri R, Kataoka Y, Uno K et al (2012) The distinctive nature of atherosclerotic vascular disease in diabetes: pathophysiological and morphological insights. Curr Diab Rep 12: 280–285
12. Tousoulis D, Kampoli AM, Papageorgiou N et al (2011) Pathophysiology of atherosclerosis: the role of inflammation. Curr Pharm Des 17:4089–4110
13. Goldschmidt-Clermont PJ, Dong C, Seo DM et al (2012) Atherosclerosis, inflammation, genetics, and stem cells: 2012 update. Curr Atheroscler Rep 14:201–210
14. Hotamisligil GS, Erbay E (2008) Nutrient sensing and inflammation in metabolic diseases. Nat Rev Immunol 8:923–934
15. Donath MY, Shoelson SE (2011) Type 2 diabetes as an inflammatory disease. Nat Rev Immunol 11:98–107
16. Ross R (1999) Atherosclerosis—an inflammatory disease. N Engl J Med 340:115–126
17. Hahn C, Schwartz MA (2009) Mechanotransduction in vascular physiology and atherogenesis. Nat Rev Mol Cell Biol 10:53–62
18. Santos MG, Pegoraro M, Sandrini F et al (2008) Risk factors for the development of atherosclerosis in childhood and adolescence. Arq Bras Cardiol 90:276–283
19. American Heart Association (2013) Risk factors and coronary heart disease. http://www.americanheart.org/presenter.jhtml?identifier=4726. Accessed 24 Jan 2013
20. Katsiki N, Tziomalos K, Chatzizisis Y et al (2010) Effect of HMG-CoA reductase inhibitors on vascular cell apoptosis: beneficial or detrimental? Atherosclerosis 211:9–14
21. Woollard KJ, Geissmann F (2010) Monocytes in atherosclerosis: subsets and functions. Nat Rev Cardiol 7:77–86

22. Yu J, Li Y, Li M et al (2010) Oxidized low density lipoprotein-induced transdifferentiation of bone marrow-derived smooth muscle-like cells into foam-like cells in vitro. Int J Exp Pathol 91:24–33
23. Ross R, Glomset JA (1973) Atherosclerosis and the arterial smooth muscle cell: proliferation of smooth muscle is a key event in the genesis of the lesions of atherosclerosis. Science 180:1332–1339
24. Shah PK (2003) Pathophysiology of plaque rupture and the concept of plaque stabilization. Cardiol Clin 21:303–314
25. Davies MJ, Thomas A (1984) Thrombosis and acute coronary-artery lesions in sudden cardiac ischemic death. N Engl J Med 310:1137–1140
26. Tabit CE, Chung WB, Hamburg NM et al (2010) Endothelial dysfunction in diabetes mellitus: molecular mechanisms and clinical implications. Rev Endocr Metab Disord 11:61–74
27. Ku DN, Giddens DP, Zarins CK et al (1985) Pulsatile flow and atherosclerosis in the human carotid bifurcation. Positive correlation between plaque location and low oscillating shear stress. Arteriosclerosis 5:293–302
28. Chien S (2008) Role of shear stress direction in endothelial mechanotransduction. Mol Cell Biomech 5:1–8
29. Wolin MS (2009) Reactive oxygen species and the control of vascular function. Am J Physiol Heart Circ Physiol 296:H539–H549
30. Ley K, Miller YI, Hedrick CC (2011) Monocyte and macrophage dynamics during atherogenesis. Arterioscler Thromb Vasc Biol 31:1506–1516
31. Kuckleburg CJ, Yates CM, Kalia N et al (2011) Endothelial cell-borne platelet bridges selectively recruit monocytes in human and mouse models of vascular inflammation. Cardiovasc Res 91:134–141
32. Furman MI, Benoit SE, Barnard MR et al (1998) Increased platelet reactivity and circulating monocyte-platelet aggregates in patients with stable coronary artery disease. J Am Coll Cardiol 31:352–358
33. Doran AC, Meller N, McNamara CA (2008) Role of smooth muscle cells in the initiation and early progression of atherosclerosis. Arterioscler Thromb Vasc Biol 28:812–819
34. Garin G, Berk BC (2006) Flow-mediated signaling modulates endothelial cell phenotype. Endothelium 13:375–384
35. Chiu JJ, Chien S (2011) Effects of disturbed flow on vascular endothelium: pathophysiological basis and clinical perspectives. Physiol Rev 91:327–387
36. Pierce AD, Anglin IE, Vitolo MI et al (2012) Glucose-activated RUNX2 phosphorylation promotes endothelial cell proliferation and an angiogenic phenotype. J Cell Biochem 113:282–292
37. Simmons GH, Padilla J, Laughlin MH (2012) Heterogeneity of endothelial cell phenotype within and amongst conduit vessels of the swine vasculature. Exp Physiol 97:1074–1082
38. Versari D, Daghini E, Virdis A et al (2009) Endothelial dysfunction as a target for prevention of cardiovascular disease. Diabetes Care 32(Suppl 2):S314–S321
39. Vasquez-Vivar J, Kalyanaraman B, Martasek P et al (1998) Superoxide generation by endothelial nitric oxide synthase: the influence of cofactors. Proc Natl Acad Sci USA 95: 9220–9225
40. White MF (1997) The insulin signalling system and the IRS proteins. Diabetologia 40(Suppl 2):S2–S17
41. Muniyappa R, Montagnani M, Koh KK et al (2007) Cardiovascular actions of insulin. Endocr Rev 28:463–491
42. Kuboki K, Jiang ZY, Takahara N et al (2000) Regulation of endothelial constitutive nitric oxide synthase gene expression in endothelial cells and in vivo: a specific vascular action of insulin. Circulation 101:676–681
43. Moncada S, Higgs EA (2006) Nitric oxide and the vascular endothelium. Handb Exp Pharmacol 176:213–254
44. Hsueh WA, Quinones MJ (2003) Role of endothelial dysfunction in insulin resistance. Am J Cardiol 92:10J–17J

45. Christlieb AR, Janka HU, Kraus B et al (1976) Vascular reactivity to angiotensin II and to norepinephrine in diabetic subjects. Diabetes 25:268–274
46. Tesfamariam B, Brown ML, Deykin D et al (1990) Elevated glucose promotes generation of endothelium-derived vasoconstrictor prostanoids in rabbit aorta. J Clin Invest 85:929–932
47. Wheatcroft SB, Williams IL, Shah AM et al (2003) Pathophysiological implications of insulin resistance on vascular endothelial function. Diabet Med 20:255–268
48. Diez J (2007) Arterial stiffness and extracellular matrix. Adv Cardiol 44:76–95
49. Fleenor BS, Sindler AL, Eng JS et al (2012) Sodium nitrite de-stiffening of large elastic arteries with aging: role of normalization of advanced glycation end-products. Exp Gerontol 47: 588–594
50. Farmer DG, Kennedy S (2009) RAGE, vascular tone and vascular disease. Pharmacol Ther 124:185–194
51. Burke AP, Kolodgie FD, Zieske A et al (2004) Morphologic findings of coronary atherosclerotic plaques in diabetics: a postmortem study. Arterioscler Thromb Vasc Biol 24: 1266–1271
52. Nishikawa T, Edelstein D, Brownlee M (2000) The missing link: a single unifying mechanism for diabetic complications. Kidney Int Suppl 77:S26–S30
53. Talayero BG, Sacks FM (2011) The role of triglycerides in atherosclerosis. Curr Cardiol Rep 13:544–552
54. Li H, Li H, Bao Y et al (2011) Free fatty acids induce endothelial dysfunction and activate protein kinase C and nuclear factor-kappaB pathway in rat aorta. Int J Cardiol 152:218–224
55. Dimmeler S, Hermann C, Zeiher AM (1998) Apoptosis of endothelial cells. Contribution to the pathophysiology of atherosclerosis? Eur Cytokine Netw 9:697–698
56. Foteinos G, Hu Y, Xiao Q et al (2008) Rapid endothelial turnover in atherosclerosis-prone areas coincides with stem cell repair in apolipoprotein E-deficient mice. Circulation 117:1856–1863
57. van der Giessen AG, Wentzel JJ, Meijboom WB et al (2009) Plaque and shear stress distribution in human coronary bifurcations: a multislice computed tomography study. EuroIntervention 4:654–661
58. Hilgers RH, Webb RC (2005) Molecular aspects of arterial smooth muscle contraction: focus on Rho. Exp Biol Med (Maywood) 230:829–835
59. Mallika V, Goswami B, Rajappa M (2007) Atherosclerosis pathophysiology and the role of novel risk factors: a clinicobiochemical perspective. Angiology 58:513–522
60. Muto A, Fitzgerald TN, Pimiento JM et al (2007) Smooth muscle cell signal transduction: implications of vascular biology for vascular surgeons. J Vasc Surg 45(Suppl A):A15–A24
61. Rudd JHF, Weissberg PL (2002) Atherosclerosis. In: Hunt BJ, Poston L, Schachter M et al (eds) An introduction to vascular biology, 2nd edn. Cambridge University Press, Cambridge, p 302
62. Owens GK (1996) Chapter 23: Role of alterations in the differentiated state of vascular smooth muscle cells in atherogenesis. In: Fuster V, Ross R, Topol EJ (eds) Atherosclerosis and coronary artery disease. Lippincott-Raven, Philadelphia, p 401
63. Bennett MR (1999) Apoptosis of vascular smooth muscle cells in vascular remodelling and atherosclerotic plaque rupture. Cardiovasc Res 41:361–368
64. Choudhury RP, Lee JM, Greaves DR (2005) Mechanisms of disease: macrophage-derived foam cells emerging as therapeutic targets in atherosclerosis. Nat Clin Pract Cardiovasc Med 2:309–315
65. Davies MJ, Richardson PD, Woolf N et al (1993) Risk of thrombosis in human atherosclerotic plaques: role of extracellular lipid, macrophage, and smooth muscle cell content. Br Heart J 69:377–381
66. Mosser DM, Edwards JP (2008) Exploring the full spectrum of macrophage activation. Nat Rev Immunol 8:958–969
67. Schmidt AM, Stern D (2000) Atherosclerosis and diabetes: the RAGE connection. Curr Atheroscler Rep 2:430–436

68. Weisberg SP, McCann D, Desai M et al (2003) Obesity is associated with macrophage accumulation in adipose tissue. J Clin Invest 112:1796–1808
69. Voll RE, Urbonaviciute V (2008) High mobility group box 1 in the pathogenesis of inflammatory and autoimmune diseases. Isr Med Assoc J 10:26–28
70. Xu H, Barnes GT, Yang Q et al (2003) Chronic inflammation in fat plays a crucial role in the development of obesity-related insulin resistance. J Clin Invest 112:1821–1830
71. DeClercq V, Taylor C, Zahradka P (2008) Adipose tissue: the link between obesity and cardiovascular disease. Cardiovasc Hematol Disord Drug Targets 8:228–237
72. Declercq V, Enns J, Yeganeh A et al (2013) Modulation of cardiovascular function by adipokines. Cardiovasc Hematol Disord Drug Targets 13:59–72
73. Northcott JM, Yeganeh A, Taylor CG et al (2012) Adipokines and the cardiovascular system: mechanisms mediating health and disease. Can J Physiol Pharmacol 90:1029–1059
74. Sengenes C, Miranville A, Lolmede K et al (2007) The role of endothelial cells in inflamed adipose tissue. J Intern Med 262:415–421
75. England RN, Autieri MV (2012) Anti-inflammatory effects of interleukin-19 in vascular disease. Int J Inflam 2012:253583
76. Coppack SW (2001) Pro-inflammatory cytokines and adipose tissue. Proc Nutr Soc 60: 349–356
77. Paukku K, Silvennoinen O (2004) STATs as critical mediators of signal transduction and transcription: lessons learned from STAT5. Cytokine Growth Factor Rev 15:435–455
78. Brand K, Page S, Walli AK et al (1997) Role of nuclear factor-kappa B in atherogenesis. Exp Physiol 82:297–304
79. Hansson GK, Libby P (2006) The immune response in atherosclerosis: a double-edged sword. Nat Rev Immunol 6:508–519
80. Frostegard J, Ulfgren AK, Nyberg P et al (1999) Cytokine expression in advanced human atherosclerotic plaques: dominance of pro-inflammatory (Th1) and macrophage-stimulating cytokines. Atherosclerosis 145:33–43
81. von der Thusen JH, Kuiper J, van Berkel TJ et al (2003) Interleukins in atherosclerosis: molecular pathways and therapeutic potential. Pharmacol Rev 55:133–166
82. Juge-Aubry CE, Somm E, Chicheportiche R et al (2004) Regulatory effects of interleukin (IL)-1, interferon-beta, and IL-4 on the production of IL-1 receptor antagonist by human adipose tissue. J Clin Endocrinol Metab 89:2652–2658
83. Jager J, Gremeaux T, Cormont M et al (2007) Interleukin-1beta-induced insulin resistance in adipocytes through down-regulation of insulin receptor substrate-1 expression. Endocrinology 148:241–251
84. Rui L, Yuan M, Frantz D et al (2002) SOCS-1 and SOCS-3 block insulin signaling by ubiquitin-mediated degradation of IRS1 and IRS2. J Biol Chem 277:42394–42398
85. Eizirik DL, Mandrup-Poulsen T (2001) A choice of death: the signal-transduction of immune-mediated beta-cell apoptosis. Diabetologia 44:2115–2133
86. Dinarello CA (2000) Proinflammatory cytokines. Chest 118:503–508
87. Zilverschoon GR, Tack CJ, Joosten LA et al (2008) Interleukin-18 resistance in patients with obesity and type 2 diabetes mellitus. Int J Obes (Lond) 32:1407–1414
88. Robertson AK, Hansson GK (2006) T cells in atherogenesis: for better or for worse? Arterioscler Thromb Vasc Biol 26:2421–2432
89. Leon ML, Zuckerman SH (2005) Gamma interferon: a central mediator in atherosclerosis. Inflamm Res 54:395–411
90. Gerdes N, Sukhova GK, Libby P et al (2002) Expression of interleukin (IL)-18 and functional IL-18 receptor on human vascular endothelial cells, smooth muscle cells, and macrophages: implications for atherogenesis. J Exp Med 195:245–257
91. De Nardo D, Latz E (2011) NLRP3 inflammasomes link inflammation and metabolic disease. Trends Immunol 32:373–379
92. Spranger J, Kroke A, Mohlig M et al (2003) Inflammatory cytokines and the risk to develop type 2 diabetes: results of the prospective population-based European Prospective Investigation into Cancer and Nutrition (EPIC)-Potsdam Study. Diabetes 52:812–817

93. Pradhan AD, Manson JE, Rifai N et al (2001) C-reactive protein, interleukin 6, and risk of developing type 2 diabetes mellitus. JAMA 286:327–334
94. Tzoulaki I, Murray GD, Lee AJ et al (2005) C-reactive protein, interleukin-6, and soluble adhesion molecules as predictors of progressive peripheral atherosclerosis in the general population: Edinburgh Artery Study. Circulation 112:976–983
95. Senn JJ, Klover PJ, Nowak IA et al (2003) Suppressor of cytokine signaling-3 (SOCS-3), a potential mediator of interleukin-6-dependent insulin resistance in hepatocytes. J Biol Chem 278:13740–13746
96. Mooney RA, Senn J, Cameron S et al (2001) Suppressors of cytokine signaling-1 and -6 associate with and inhibit the insulin receptor. A potential mechanism for cytokine-mediated insulin resistance. J Biol Chem 276:25889–25893
97. Nonogaki K, Fuller GM, Fuentes NL et al (1995) Interleukin-6 stimulates hepatic triglyceride secretion in rats. Endocrinology 136:2143–2149
98. Johnson DR, O'Connor JC, Satpathy A et al (2006) Cytokines in type 2 diabetes. Vitam Horm 74:405–441
99. Dandona P, Aljada A, Chaudhuri A et al (2005) Metabolic syndrome: a comprehensive perspective based on interactions between obesity, diabetes, and inflammation. Circulation 111: 1448–1454
100. Hotamisligil GS, Peraldi P, Budavari A et al (1996) IRS-1-mediated inhibition of insulin receptor tyrosine kinase activity in TNF-alpha- and obesity-induced insulin resistance. Science 271:665–668
101. Aguirre V, Uchida T, Yenush L et al (2000) The c-Jun NH(2)-terminal kinase promotes insulin resistance during association with insulin receptor substrate-1 and phosphorylation of Ser(307). J Biol Chem 275:9047–9054
102. Fard A, Tuck CH, Donis JA et al (2000) Acute elevations of plasma asymmetric dimethylarginine and impaired endothelial function in response to a high-fat meal in patients with type 2 diabetes. Arterioscler Thromb Vasc Biol 20:2039–2044
103. Szekanecz Z (2008) Pro-inflammatory cytokines in atherosclerosis. Isr Med Assoc J 10: 529–530
104. Popa C, Netea MG, van Riel PL et al (2007) The role of TNF-alpha in chronic inflammatory conditions, intermediary metabolism, and cardiovascular risk. J Lipid Res 48:751–762
105. Zernecke A, Weber C (2005) Inflammatory mediators in atherosclerotic vascular disease. Basic Res Cardiol 100:93–101
106. Panee J (2012) Monocyte chemoattractant protein 1 (MCP-1) in obesity and diabetes. Cytokine 60:1–12
107. Haubner F, Lehle K, Munzel D et al (2007) Hyperglycemia increases the levels of vascular cellular adhesion molecule-1 and monocyte-chemoattractant-protein-1 in the diabetic endothelial cell. Biochem Biophys Res Commun 360:560–565
108. Takahara N, Kashiwagi A, Nishio Y et al (1997) Oxidized lipoproteins found in patients with NIDDM stimulate radical-induced monocyte chemoattractant protein-1 mRNA expression in cultured human endothelial cells. Diabetologia 40:662–670
109. Blake GJ, Ridker PM (2002) Inflammatory bio-markers and cardiovascular risk prediction. J Intern Med 252:283–294
110. Chen H (2006) Cellular inflammatory responses: novel insights for obesity and insulin resistance. Pharmacol Res 53:469–477
111. Paffen E, DeMaat MP (2006) C-reactive protein in atherosclerosis: a causal factor? Cardiovasc Res 71:30–39
112. Krieglstein CF, Granger DN (2001) Adhesion molecules and their role in vascular disease. Am J Hypertens 14:44S–54S
113. Chia MC (1998) The role of adhesion molecules in atherosclerosis. Crit Rev Clin Lab Sci 35:573–602
114. Rubio-Guerra AF, Vargas-Robles H, Serrano AM et al (2009) Correlation between the levels of circulating adhesion molecules and atherosclerosis in type-2 diabetic normotensive patients: circulating adhesion molecules and atherosclerosis. Cell Adh Migr 3:369–372

115. Galen FX (2002) Cell adhesion molecules in hypertension: endothelial markers of vascular injury and predictors of target organ damage? J Hypertens 20:813–816
116. Rubio-Guerra AF, Vargas-Robles H, Ayala GV et al (2007) Correlation between circulating adhesion molecule levels and albuminuria in type 2 diabetic normotensive patients. Med Sci Monit 13:CR349–CR352
117. Dandona P, Aljada A, Chaudhuri A et al (2003) The potential influence of inflammation and insulin resistance on the pathogenesis and treatment of atherosclerosis-related complications in type 2 diabetes. J Clin Endocrinol Metab 88:2422–2429
118. Scherer PE, Williams S, Fogliano M et al (1995) A novel serum protein similar to C1q, produced exclusively in adipocytes. J Biol Chem 270:26746–26749
119. Hotta K, Funahashi T, Arita Y et al (2000) Plasma concentrations of a novel, adipose-specific protein, adiponectin, in type 2 diabetic patients. Arterioscler Thromb Vasc Biol 20: 1595–1599
120. Ukkola O, Santaniemi M (2002) Adiponectin: a link between excess adiposity and associated comorbidities? J Mol Med 80:696–702
121. Berg AH, Combs TP, Du X et al (2001) The adipocyte-secreted protein Acrp30 enhances hepatic insulin action. Nat Med 7:947–953
122. Ouchi N, Kihara S, Arita Y et al (1999) Novel modulator for endothelial adhesion molecules: adipocyte-derived plasma protein adiponectin. Circulation 100:2473–2476
123. Ouchi N, Kihara S, Arita Y et al (2000) Adiponectin, an adipocyte-derived plasma protein, inhibits endothelial NF-kappaB signaling through a cAMP-dependent pathway. Circulation 102:1296–1301
124. Bastard JP, Maachi M, Lagathu C et al (2006) Recent advances in the relationship between obesity, inflammation, and insulin resistance. Eur Cytokine Netw 17:4–12
125. Yamawaki H (2011) Vascular effects of novel adipocytokines: focus on vascular contractility and inflammatory responses. Biol Pharm Bull 34:307–310
126. Xi W, Satoh H, Kase H et al (2005) Stimulated HSP90 binding to eNOS and activation of the PI3-Akt pathway contribute to globular adiponectin-induced NO production: vasorelaxation in response to globular adiponectin. Biochem Biophys Res Commun 332:200–205
127. Cheng KK, Lam KS, Wang Y et al (2007) Adiponectin-induced endothelial nitric oxide synthase activation and nitric oxide production are mediated by APPL1 in endothelial cells. Diabetes 56:1387–1394
128. Yamawaki H, Tsubaki N, Mukohda M et al (2010) Omentin, a novel adipokine, induces vasodilation in rat isolated blood vessels. Biochem Biophys Res Commun 393:668–672
129. Hida K, Wada J, Eguchi J et al (2005) Visceral adipose tissue-derived serine protease inhibitor: a unique insulin-sensitizing adipocytokine in obesity. Proc Natl Acad Sci USA 102: 10610–10615
130. Gentile MT, Vecchione C, Marino G et al (2008) Resistin impairs insulin-evoked vasodilation. Diabetes 57:577–583
131. Kim SR, Bae YH, Bae SK et al (2008) Visfatin enhances ICAM-1 and VCAM-1 expression through ROS-dependent NF-kappaB activation in endothelial cells. Biochim Biophys Acta 1783:886–895
132. Adya R, Tan BK, Punn A et al (2008) Visfatin induces human endothelial VEGF and MMP-2/9 production via MAPK and PI3K/Akt signalling pathways: novel insights into visfatin-induced angiogenesis. Cardiovasc Res 78:356–365
133. Wang Z, Nakayama T (2010) Inflammation, a link between obesity and cardiovascular disease. Mediators Inflamm 2010:535918
134. Coppari R, Bjorbaek C (2012) Leptin revisited: its mechanism of action and potential for treating diabetes. Nat Rev Drug Discov 11:692–708
135. Soderberg S, Ahren B, Jansson JH et al (1999) Leptin is associated with increased risk of myocardial infarction. J Intern Med 246:409–418
136. Wallace AM, McMahon AD, Packard CJ et al (2001) Plasma leptin and the risk of cardiovascular disease in the West of Scotland Coronary Prevention Study (WOSCOPS). Circulation 104:3052–3056

137. Singhal A, Farooqi IS, Cole TJ et al (2002) Influence of leptin on arterial distensibility: a novel link between obesity and cardiovascular disease? Circulation 106:1919–1924
138. Ciccone M, Vettor R, Pannacciulli N et al (2001) Plasma leptin is independently associated with the intima-media thickness of the common carotid artery. Int J Obes Relat Metab Disord 25:805–810
139. Zarkesh-Esfahani H, Pockley G, Metcalfe RA et al (2001) High-dose leptin activates human leukocytes via receptor expression on monocytes. J Immunol 167:4593–4599
140. Mancuso P, Canetti C, Gottschalk A et al (2004) Leptin augments alveolar macrophage leukotriene synthesis by increasing phospholipase activity and enhancing group IVC iPLA2 (cPLA2gamma) protein expression. Am J Physiol Lung Cell Mol Physiol 287:L497–L502
141. Lord GM, Matarese G, Howard JK et al (1998) Leptin modulates the T-cell immune response and reverses starvation-induced immunosuppression. Nature 394:897–901
142. Ling C, Groop L (2009) Epigenetics: a molecular link between environmental factors and type 2 diabetes. Diabetes 58:2718–2725
143. Litherland SA (2008) Immunopathogenic interaction of environmental triggers and genetic susceptibility in diabetes: is epigenetics the missing link? Diabetes 57:3184–3186
144. Luger K, Mader AW, Richmond RK et al (1997) Crystal structure of the nucleosome core particle at 2.8 Å resolution. Nature 389:251–260
145. Villeneuve LM, Natarajan R (2010) The role of epigenetics in the pathology of diabetic complications. Am J Physiol Renal Physiol 299:F14–F25
146. Gray SG, De Meyts P (2005) Role of histone and transcription factor acetylation in diabetes pathogenesis. Diabetes Metab Res Rev 21:416–433
147. Liang F, Kume S, Koya D (2009) SIRT1 and insulin resistance. Nat Rev Endocrinol 5:367–373
148. Ito K, Hanazawa T, Tomita K et al (2004) Oxidative stress reduces histone deacetylase 2 activity and enhances IL-8 gene expression: role of tyrosine nitration. Biochem Biophys Res Commun 315:240–245
149. Vanden Berghe W, De Bosscher K, Boone E et al (1999) The nuclear factor-kappaB engages CBP/p300 and histone acetyltransferase activity for transcriptional activation of the interleukin-6 gene promoter. J Biol Chem 274:32091–32098
150. Reddy MA, Sahar S, Villeneuve LM et al (2009) Role of Src tyrosine kinase in the atherogenic effects of the 12/15-lipoxygenase pathway in vascular smooth muscle cells. Arterioscler Thromb Vasc Biol 29:387–393
151. Miao F, Gonzalo IG, Lanting L et al (2004) In vivo chromatin remodeling events leading to inflammatory gene transcription under diabetic conditions. J Biol Chem 279:18091–18097
152. Chen S, Feng B, George B et al (2010) Transcriptional coactivator p300 regulates glucose-induced gene expression in endothelial cells. Am J Physiol Endocrinol Metab 298:E127–E137
153. Miao F, Wu X, Zhang L et al (2007) Genome-wide analysis of histone lysine methylation variations caused by diabetic conditions in human monocytes. J Biol Chem 282:13854–13863
154. Li Y, Reddy MA, Miao F et al (2008) Role of the histone H3 lysine 4 methyltransferase, SET7/9, in the regulation of NF-kappaB-dependent inflammatory genes. Relevance to diabetes and inflammation. J Biol Chem 283:26771–26781
155. El-Osta A, Brasacchio D, Yao D et al (2008) Transient high glucose causes persistent epigenetic changes and altered gene expression during subsequent normoglycemia. J Exp Med 205:2409–2417
156. Brasacchio D, Okabe J, Tikellis C et al (2009) Hyperglycemia induces a dynamic cooperativity of histone methylase and demethylase enzymes associated with gene-activating epigenetic marks that coexist on the lysine tail. Diabetes 58:1229–1236
157. Kuroda A, Rauch TA, Todorov I et al (2009) Insulin gene expression is regulated by DNA methylation. PLoS One 4:e6953
158. Morgan HD, Sutherland HG, Martin DI et al (1999) Epigenetic inheritance at the agouti locus in the mouse. Nat Genet 23:314–318
159. Park JH, Stoffers DA, Nicholls RD et al (2008) Development of type 2 diabetes follwing intrauterine growth retardation in rats is associated with progressive epigenetic silencing of Pdx1. J Clin Invest 51(2):85–91

160. Ingrosso D, Perna AF (2009) Epigenetics in hyperhomocysteinemic states. A special focus on uremia. Biochim Biophys Acta 1790:892–899
161. Ekstrom TJ, Stenvinkel P (2009) The epigenetic conductor: a genomic orchestrator in chronic kidney disease complications? J Nephrol 22:442–449
162. Jamaluddin MS, Yang X, Wang H (2007) Hyperhomocysteinemia, DNA methylation and vascular disease. Clin Chem Lab Med 45:1660–1666
163. Luscher TF, Creager MA, Beckman JA et al (2003) Diabetes and vascular disease: pathophysiology, clinical consequences, and medical therapy: part II. Circulation 108:1655–1661
164. Berry C, Tardif JC, Bourassa MG (2007) Coronary heart disease in patients with diabetes: part I: recent advances in prevention and noninvasive management. J Am Coll Cardiol 49:631–642
165. Haffner SM, Lehto S, Ronnemaa T et al (1998) Mortality from coronary heart disease in subjects with type 2 diabetes and in nondiabetic subjects with and without prior myocardial infarction. N Engl J Med 339:229–234
166. Jonas M, Edelman ER, Groothuis A et al (2005) Vascular neointimal formation and signaling pathway activation in response to stent injury in insulin-resistant and diabetic animals. Circ Res 97:725–733
167. Morss AS, Edelman ER (2007) Glucose modulates basement membrane fibroblast growth factor-2 via alterations in endothelial cell permeability. J Biol Chem 282:14635–14644
168. Piatti P, Di Mario C, Monti LD et al (2003) Association of insulin resistance, hyperleptinemia, and impaired nitric oxide release with in-stent restenosis in patients undergoing coronary stenting. Circulation 108:2074–2081
169. Takagi T, Akasaka T, Yamamuro A et al (2000) Troglitazone reduces neointimal tissue proliferation after coronary stent implantation in patients with non-insulin dependent diabetes mellitus: a serial intravascular ultrasound study. J Am Coll Cardiol 36:1529–1535
170. Costanzo MR, Naftel DC, Pritzker MR et al (1998) Heart transplant coronary artery disease detected by coronary angiography: a multiinstitutional study of preoperative donor and recipient risk factors. Cardiac Transplant Research Database. J Heart Lung Transplant 17:744–753
171. Taylor DO, Edwards LB, Boucek MM et al (2006) Registry of the International Society for Heart and Lung Transplantation: twenty-third official adult heart transplantation report—2006. J Heart Lung Transplant 25:869–879
172. Yeung AC, Davis SF, Hauptman PJ et al (1995) Incidence and progression of transplant coronary artery disease over 1 year: results of a multicenter trial with use of intravascular ultrasound. Multicenter Intravascular Ultrasound Transplant Study Group. J Heart Lung Transplant 14:S215–S220
173. Libby P, Pober JS (2001) Chronic rejection. Immunity 14:387–397
174. Salomon RN, Hughes CC, Schoen FJ et al (1991) Human coronary transplantation-associated arteriosclerosis. Evidence for a chronic immune reaction to activated graft endothelial cells. Am J Pathol 138:791–798
175. Billingham ME (1992) Histopathology of graft coronary disease. J Heart Lung Transplant 11:S38–S44
176. Mitchell RN (2009) Graft vascular disease: immune response meets the vessel wall. Annu Rev Pathol 4:19–47
177. Kass M, Haddad H (2006) Cardiac allograft vasculopathy: pathology, prevention and treatment. Curr Opin Cardiol 21:132–137
178. Valantine H, Rickenbacker P, Kemna M et al (2001) Metabolic abnormalities characteristic of dysmetabolic syndrome predict the development of transplant coronary artery disease: a prospective study. Circulation 103:2144–2152
179. Kemna MS, Valantine HA, Hunt SA et al (1994) Metabolic risk factors for atherosclerosis in heart transplant recipients. Am Heart J 128:68–72

Cardiovascular Autonomic Neuropathy in Diabetes

Takahide Arai, Masaki Ieda, and Keiichi Fukuda

Abstract The heart is extensively innervated, and its electrical and mechanical performance is controlled by the autonomic nervous system. The cardiac nervous system comprises the sympathetic, parasympathetic, and sensory nervous systems that together regulate heart function on demand. The density of cardiac innervation varies in diseased hearts, leading to unbalanced neural activation and lethal arrhythmia. Diabetic sensory neuropathy causes silent myocardial ischemia, which is characterized by loss of pain perception during myocardial ischemia and is a major cause of sudden cardiac death in diabetes mellitus (DM). Despite its clinical importance, the mechanisms underlying the control and regulation of cardiac innervation remain poorly understood. Nerve growth factor (NGF), a potent chemoattractant, is highly expressed in cardiomyocytes during development. In contrast, Sema3a, a neural chemorepellent, is highly expressed in the subendocardium of early-stage embryos, but is suppressed during development. The balance between NGF and Sema3a expression leads to epicardial to endocardial transmural sympathetic innervation patterning. Downregulation of NGF leads to diabetic neuropathy, whereas NGF supplementation rescues silent myocardial ischemia in DM. In this review, we summarize the molecular mechanisms underlying cardiac autonomic innervation, with a particular focus on DM and the clinical implications of cardiac autonomic neuropathy.

T. Arai • M. Ieda
Division of Cardiology, Department of Internal Medicine,
Keio University School of Medicine, Tokyo, Japan

K. Fukuda (✉)
Division of Cardiology, Department of Internal Medicine,
Keio University School of Medicine, Tokyo, Japan

Department of Cardiology, Keio University School of Medicine,
35 Shinanomachi, Shinjuku-ku, Tokyo 160-8582, Japan
e-mail: kfukuda@a2.keio.jp

B. Turan and N.S. Dhalla (eds.), *Diabetic Cardiomyopathy: Biochemical and Molecular Mechanisms*, Advances in Biochemistry in Health and Disease 9,
DOI 10.1007/978-1-4614-9317-4_14,

Keywords Cardiac nervous system • Diabetes mellitus • Autonomic neuropathy • Nerve growth factor • Sema3a

1 Introduction

Heart tissue is extensively innervated by the autonomic nervous system, which comprises sympathetic, parasympathetic, and sensory nerves. Cardiac autonomic diabetic neuropathy (CADN) is one of the most common complications in patients with diabetes mellitus (DM). One of the most serious consequences of CADN is the risk of death. Mortality from CADN results from the high risk of cardiac arrhythmia, silent myocardial ischemia, and sudden cardiac death. Abnormal sympathetic innervation may trigger lethal cardiac arrhythmia. Silent myocardial ischemia, characterized by loss of pain perception during ischemia, is a major complication of DM that can lead to sudden cardiac death. Despite the severity of this complication, changes in cardiac autonomic innervation and the molecular mechanisms underlying autonomic neuropathy in diabetic hearts are poorly understood. In this chapter, we review the molecular mechanisms underlying cardiac autonomic innervation, with particular focus on DM and the clinical implications of cardiac autonomic neuropathy.

2 Anatomy of the Cardiac Autonomic Nervous System

The autonomic nervous system plays a key role in the regulation of cardiovascular function. The heart is innervated by sympathetic, parasympathetic, and sensory nerves. The cardiac sympathetic nerves originate in the sympathetic neurons of the stellate ganglia, which are located bilateral to the vertebrae. Sympathetic nerve fibers are located predominantly in the subepicardium of the ventricle [1, 2]. The central conduction system, which includes the sinoatrial node, the atrioventricular node, and the bundle of His, is abundantly innervated compared with the rest of the working myocardium [3, 4]. There are reports that regional differences in cardiac sympathetic innervation are highly conserved among species [5, 6].

In contrast with the sympathetic nerves, the parasympathetic nerves extend from the parasympathetic neurons in the cardiac ganglia, which are located in the base of both atria [7, 8]. Recently, Ulphani et al. demonstrated that parasympathetic nerves innervate both the atria and ventricles, with a high density on the ventricular endocardium but a greater nerve thickness on the epicardium. In addition, the right ventricle (RV) is more densely innervated than the left ventricle (LV), whereas the LV endocardium is more densely innervated than the RV endocardium [8, 9].

The cardiac nervous system also includes afferent nerves. The sensory signals generated in the heart are conducted through cardiac afferent nerves, primarily thinly myelinated Aδ fibers and nonmyelinated C fibers that project to the upper thoracic dorsal horn (DH) via the dorsal root ganglia (DRG) [10–13].

3 Cardiac Autonomic Diabetic Neuropathy

Autonomic diabetic neuropathy is a frequent and serious complication among patients with diabetes. CADN is defined as impaired autonomic control of the cardiovascular system in the setting of diabetes. The most important feature of CADN is a significant cause of morbidity and mortality, which are reported to be caused by cardiac arrhythmia or silent myocardial ischemia [11, 14, 15]. The Action to Control Cardiovascular Risk in Diabetes (ACCORD) trial, which included more than 8,000 participants with type 2 DM (T2DM), revealed that the prevalence of CADN strongly predicts all-cause mortality and mortality from cardiovascular disease [16, 17]. CADN could increase mortality risk by promoting life-threatening events caused by drug side effects, hypoglycemia, hypokalemia, hypotension, and ischemia [18–20].

The earliest sign of CADN is reported to be a reduction in heart rate variability (HRV), which is detectable at the subclinical stage by deep respiration [21, 22]. The parasympathetic nerves are damaged, resulting in resting tachycardia because of the dominancy of sympathetic effects [23]. Additional symptoms, such as exercise intolerance, orthostatic hypotension, and an increasing limitation in HRV, are the manifestations of progressive damage to the autonomic balance [24]. Cardiac pain perception is often impaired by additional damage to sensory nerve fibers [10], leads to the delay of adequate treatment of myocardial ischemia [24].

4 Nerve Growth Factor and Cardiac Sympathetic Nerves

Nerve growth factor (NGF) is a prototypic member of the neurotrophin family, the members of which are crucial for the differentiation, survival, and synaptic activity of the peripheral sympathetic and sensory nervous systems [25–27]. NGF expression within innervated tissue corresponds to innervation density [28]. NGF expression is altered in diseased hearts, for example, following myocardial infarction (MI) and heart failure [29, 30]. NGF is upregulated following MI, resulting in the regeneration of cardiac sympathetic nerves and heterogeneous innervation [31]. In one study, NGF infusion after MI enhanced myocardial nerve sprouting, resulting in a marked increase in sudden cardiac death and a high incidence of ventricular tachycardia [32]. In another study, we demonstrated that NGF is upregulated in cardiac hypertrophy, leading to sympathetic hyperinnervation and rejuvenation [33]. Together, these results demonstrate that NGF has various crucial roles in the diseased heart.

5 NGF Downregulation Is Critical for Diabetic Neuropathy and Silent Ischemia

In contrast with sympathetic innervation, little is known about sensory innervation and its alteration in the diseased heart. Immunohistochemical studies using an antibody against calcitonin gene-related peptide (CGRP), a sensory nerve marker, demonstrated rich sensory innervation at epicardial sites and in the ventricular myocardium [34]. In a screen of several neurotrophic factors, we found that the development of cardiac sensory nerves parallels the production of NGF in the heart (Fig. 1) [10, 35]. Cardiac nociceptive sensory nerves that are immunopositive for CGRP, including those in the DRG and DH, are markedly retarded in NGF-deficient mice and rescued in mice overexpressing NGF specifically in the heart (Fig. 2). Thus, NGF synthesis in the heart is crucial for the development of the cardiac sensory nervous system. The cardiac sensory nervous system is responsible for pain perception and the initiation of the protective cardiovascular response during myocardial ischemia [36, 37]. Cardiac sensory nerve impairment causes silent myocardial ischemia, a major cause of sudden death in diabetic patients [38]. To investigate whether NGF is involved in diabetic neuropathy, diabetes was induced by streptozotocin (STZ) in wild-type (WT) and transgenic mice overexpressing NGF in the heart. Downregulation of NGF, CGRP-immunopositive cardiac sensory denervation, and atrophic changes in DRG were observed in STZ-diabetic WT mice, whereas these deteriorations were rescued in STZ-diabetic NGF-transgenic mice (Fig. 3). Cardiac sensory function, measured by myocardial ischemia-induced c-Fos expression in the DRG, was also downregulated in STZ-diabetic WT mice, but not in STZ-diabetic NGF-transgenic mice. In another study, direct gene transfer of NGF into diabetic rat hearts improved cardiac sensory innervation and function, as determined by the electrophysiological activity of cardiac afferent nerves during myocardial ischemia. Together, these findings indicate that diabetes-induced downregulation of NGF may lead to cardiac sensory neuropathy.

Vinik et al. reported a consistent association between CADN and the presence of silent ischemia, measured by exercise stress testing [39]. THe Asymptomatic Diabetics (DIAD) study, which included 1,123 patients with T2DM, reported that CADN was a strong predictor of silent ischemia and subsequent cardiovascular events [40]. Slow heart rate (HR) recovery after exercise was reported to reflect CADN and to be associated with silent myocardial ischemia [41]. The association between CADN and silent ischemia has important implications, because reduced appreciation of ischemic pain impairs timely recognition of myocardial ischemia or MI, thereby delaying appropriate treatment [42].

6 Sema3a and Cardiac Sympathetic Nerves

Sema3a is a class 3 secreted semaphorine that has been cloned and identified as a potent neural chemorepellent and a directional guidance molecule for nerve fibers [43, 44]. We have demonstrated that Sema3a produced by cardiomyocytes is crucial for cardiac sympathetic nerve development [45]. Sympathetic nerve density is reduced in the subepicardium and enhanced in the subendocardium of *Sema3a-deficient* (*Sema3a*$^{-/-}$) mice, resulting in disruption of the innervation gradient in ventricles from these mice. As a result, *Sema3a*$^{-/-}$ mice develop sinus bradycardia and abrupt sinus arrest consequent to sympathetic neural dysfunction (Fig. 4a). In contrast, overexpression of *Sema3a* specifically in the heart (*SemaTG* mice) resulted in reduced sympathetic innervation and attenuation of the epicardial to

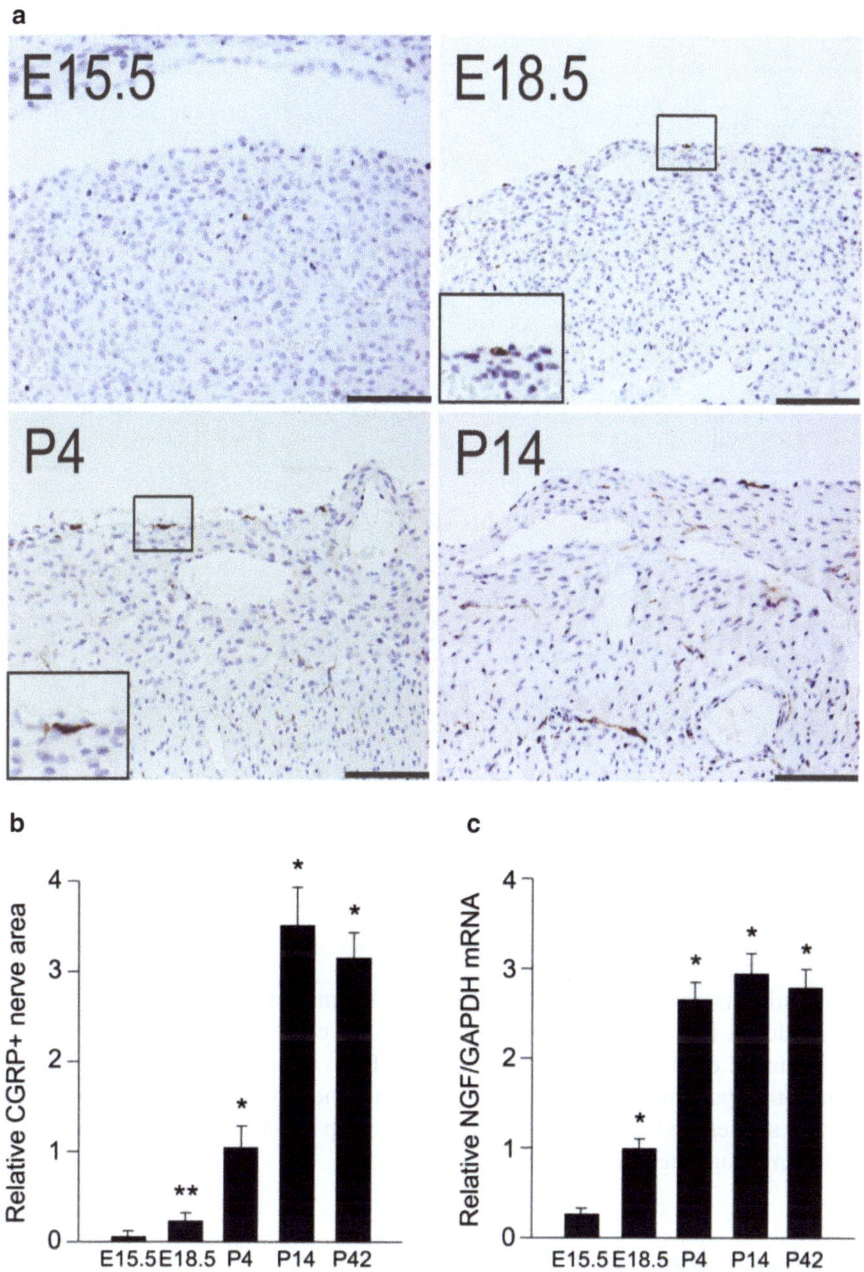

Fig. 1 Cardiac sensory innervation is increased with development coincident with nerve growth factor (NGF) expression in the heart. (**a**) Representative immunostaining for calcitonin gene-related peptide (CGRP) in murine hearts at *E15.5, 18.5, P4*, and *P14*. Sensory innervation was increased with development. (**b**) Quantitative analysis of the immunopositive nerve areas for CGRP. (**c**) Time-course of cardiac NGF expression was determined by quantitative RT-PCR

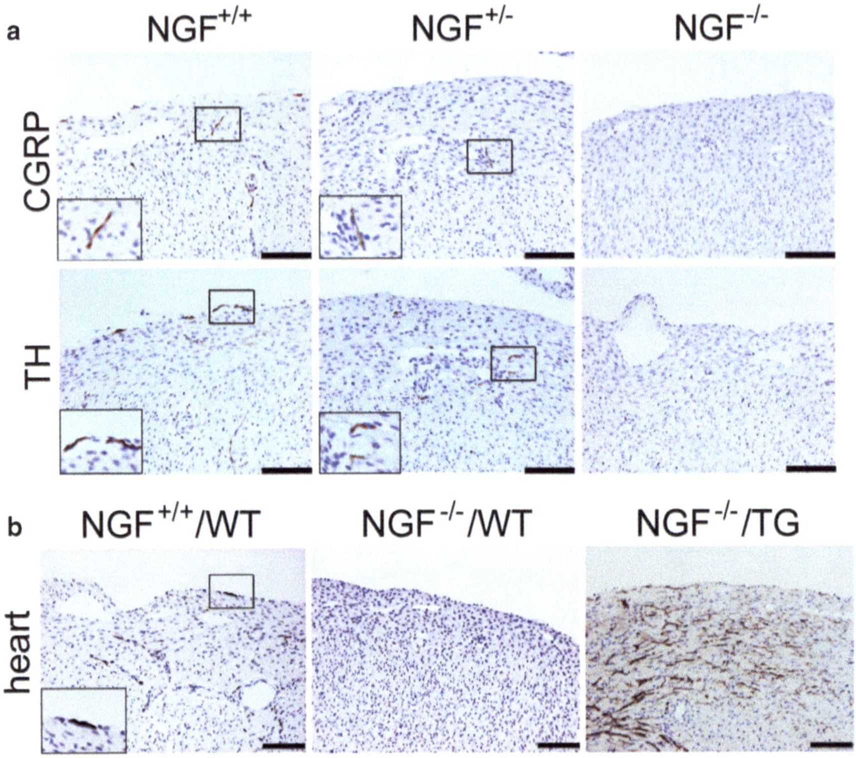

Fig. 2 The cardiac sensory nervous system is restarted in *NGF*-deficient mice. (**a**) Immunostaining for *CGRP* and throsine hydroxylase in $NGF^{-/-}$, $NGF^{+/-}$, and $NGF^{-/-}$ hearts at P4. Note that *CGRP*+ nerve endings were reduced in an *NGF* gene dosage-dependent manner. (**b**) Cardiac-specific overexpression of *NGF* rescues the defects of the cardiac sensory nervous system in *NGF*-deficient mice. Immunostaining for *CGRP* in the hearts of $NGF^{+/+}/WT$, $NGF^{-/-}/WT$, and $NGF^{-/-}/TG$ mice

endocardial innervation gradient. Consequently, sustained ventricular tachyarrhythmia was induced in *SemaTG* but not WT mice after epinephrine administration and programmed electrical stimulation (Fig. 4b). Thus, Sema3a-mediated sympathetic innervation patterning is crucial for the maintenance of an arrhythmia-free heart. Further studies are needed to clarify the relationship between DM and cardiac lethal arrhythmia mediated by Sema3a.

7 Conclusions

Cardiac autonomic nerve innervation is strictly controlled by the balance between NGF and Sema3a. CADN is a serious complication of DM and is an independent predictor of cardiovascular mortality, the result of a high risk of cardiac arrhythmia,

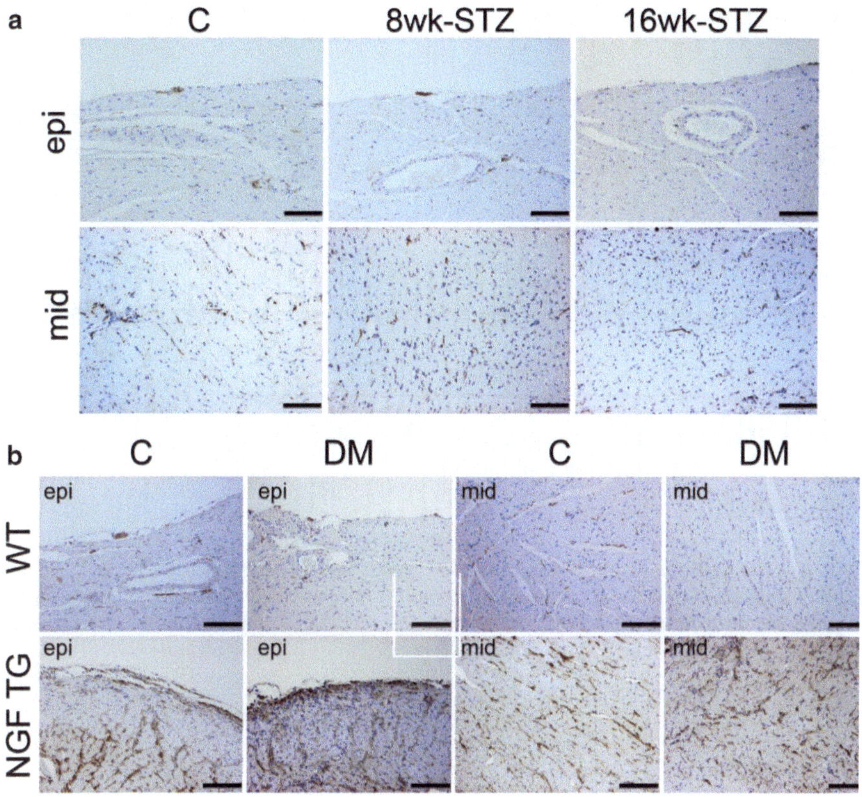

Fig. 3 Cardiac sensory innervation and NGF production are reduced in streptozotocin (STZ)-induced diabetic mice. (**a**) Immunostaining for CGRP in the hearts of control, 8-week-STZ, and 16-week-STZ mice. *Upper panels* show epicardial (*epi*) sites; *lower panels* show midventricular (*mid*) sites. Sensory innervation was significantly reduced in 16-week-STZ hearts. (**b**) Cardiac-specific overexpression of NGF rescues sensory denervation in the diabetic heart. Immunostaining for CGRP in the hearts of control and diabetic *WT* and *NGFTG* mice

silent myocardial ischemia, and sudden cardiac death. Abnormal sympathetic innervation may trigger cardiac lethal arrhythmia. Silent myocardial ischemia, characterized by loss of pain perception during ischemia, is a major complication of DM that can lead to sudden cardiac death. Downregulation of NGF decreases the density of cardiac sensory nerves, resulting in silent ischemia, a primary cause of sudden cardiac death in DM. Furthermore, Sema3a-mediated sympathetic innervation patterning is crucial for the maintenance of an arrhythmia-free heart. These findings can expand our understanding of the cardiac nervous system in patients with DM and may lead to further investigations that yield novel therapeutic targets to improve the prognosis of DM.

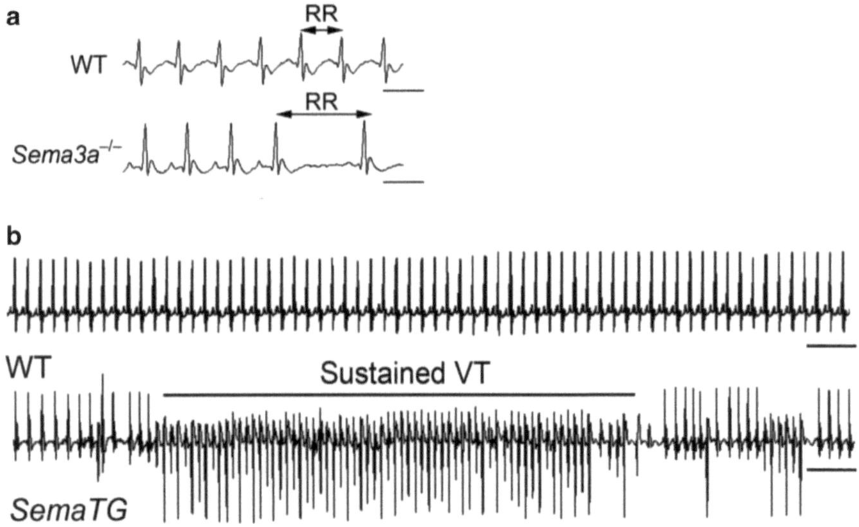

Fig. 4 ECG recordings from *WT* and *Sema3a*$^{-/-}$ mice. (**a**) The lengthened *RR* interval indicates abrupt sinus slowing in *Sema3a*$^{-/-}$ mice. (**b**) *SemaTG* mice are highly susceptible to induction of ventricular arrhythmia. Epinephrine administration revealed sustained ventricular tachycardia (*VT*) only in *SemaTG* mice

References

1. Ito M, Zipes DP (1994) Efferent sympathetic and vagal innervation of the canine right ventricle. Circulation 90:1459–1468
2. Crick SJ, Sheppard MN, Ho SY et al (1999) Localisation and quantitation of autonomic innervation in the porcine heart I: conduction system. J Anat 195:341–357
3. Chow LT, Chow SS, Anderson RH et al (1993) Innervation of the human cardiac conduction system at birth. Br Heart J 69:430–435
4. Hansson M, Kjorell U, Forsgren S (1998) Increased immunoexpression of atrial natriuretic peptide in the heart conduction system of the rat after cardiac sympathectomy. J Mol Cell Cardiol 30:2047–2057
5. Randall WC, Szentivanyi M, Pace JB et al (1968) Patterns of sympathetic nerve projections onto the canine heart. Circ Res 22:315–323
6. Crick SJ, Wharton J, Sheppard MN et al (1994) Innervation of the human cardiac conduction system. A quantitative immunohistochemical and histochemical study. Circulation 89: 1697–1708
7. Kanazawa H, Ieda M, Kimura K et al (2010) Heart failure causes cholinergic transdifferentiation of cardiac sympathetic nerves via gp130-signaling cytokines in rodents. J Clin Invest 120:408–421
8. Kimura K, Ieda M, Fukuda K (2012) Development, maturation, and transdifferentiation of cardiac sympathetic nerves. Circ Res 110:325–336
9. Ulphani JS, Cain JH, Inderyas F et al (2010) Quantitative analysis of parasympathetic innervation of the porcine heart. Heart Rhythm 7:1113–1119
10. Ieda M, Kanazawa H, Ieda Y et al (2006) Nerve growth factor is critical for cardiac sensory innervation and rescues neuropathy in diabetic hearts. Circulation 114:2351–2363

11. Ieda M, Fukuda K (2009) Cardiac innervation and sudden cardiac death. Curr Cardiol Rev 5:289–295
12. Hua F, Harrison T, Qin C et al (2004) c-Fos expression in rat brain stem and spinal cord in response to activation of cardiac ischemia-sensitive afferent neurons and electrostimulatory modulation. Am J Physiol Heart Circ Physiol 287:H2728–H2738
13. Schultz HD, Ustinova EE (1998) Capsaicin receptors mediate free radical-induced activation of cardiac afferent endings. Cardiovasc Res 38:348–355
14. Maser RE, Mitchell BD, Vinik AI et al (2003) The association between cardiovascular autonomic neuropathy and mortality in individuals with diabetes: a meta-analysis. Diabetes Care 26:1895–1901
15. Ieda M, Fukuda K (2009) New aspects for the treatment of cardiac diseases based on the diversity of functional controls on cardiac muscles: the regulatory mechanisms of cardiac innervation and their critical roles in cardiac performance. J Pharmacol Sci 109:348–353
16. Pop-Busui R, Evans GW, Gerstein HC et al (2010) Effects of cardiac autonomic dysfunction on mortality risk in the Action to Control Cardiovascular Risk in Diabetes (ACCORD) trial. Diabetes Care 33:1578–1584
17. Pop-Busui R (2012) What do we know and we do not know about cardiovascular autonomic neuropathy in diabetes. J Cardiovasc Transl Res 5:463–578
18. Landstedt-Hallin L, Englund A, Adamson U et al (1999) Increased QT dispersion during hypoglycaemia in patients with type 2 diabetes mellitus. J Intern Med 24:299–307
19. Robinson RT, Harris ND, Ireland RH et al (2004) Changes in cardiac repolarization during clinical episodes of nocturnal hypoglycaemia in adults with Type 1 diabetes. Diabetologia 47:312–315
20. Nordin C (2010) The case for hypoglycaemia as a proarrhythmic event: basic and clinical evidence. Diabetologia 53:1552–15561
21. Watkins PJ, Mackay JD (1980) Cardiac denervation in diabetic neuropathy. Ann Intern Med 92:304–307
22. Ewing DJ, Campbell IW, Clarke BF (1980) The natural history of diabetic autonomic neuropathy. Q J Med 49:95–108
23. Ewing DJ, Campbell IW, Clarke BF (1981) Heart rate changes in diabetes mellitus. Lancet 317:183–186
24. Schonauer M, Thomas A, Morbach S et al (2008) Cardiac autonomic diabetic neuropathy. Diab Vasc Dis Res 5:336–344
25. Brennan C, Rivas-Plata K, Landis SC (1999) The p75 neurotrophin receptor influences NT-3 responsiveness of sympathetic neurons in vivo. Nat Neurosci 2:699–705
26. Snider WD (1994) Functions of the neurotrophins during nervous system development: what the knockouts are teaching us. Cell 77:627–638
27. Lockhart ST, Turrigiano GG, Birren SJ (1997) Nerve growth factor modulates synaptic transmission between sympathetic neurons and cardiac myocytes. J Neurosci 17:9573–9582
28. Heumann R, Korsching S, Scott J et al (1984) Relationship between levels of nerve growth factor (NGF) and its messenger RNA in sympathetic ganglia and peripheral target tissues. EMBO J 3:3183–3189
29. Kanki H, Fukuda K, Okushi K et al (1999) Comparison of nerve growth factor mRNA expression in cardiac and skeletal muscle in streptozotocin-induced diabetic mice. Life Sci 65:2305–2313
30. Kimura K, Kanazawa H, Ieda M et al (2010) Norepinephrine-induced nerve growth factor depletion causes cardiac sympathetic denervation in severe heart failure. Auton Neurosci 156:27–35
31. Zhou S, Chen LS, Miyauchi Y et al (2004) Mechanisms of cardiac nerve sprouting after myocardial infarction in dogs. Circ Res 95:76–83
32. Cao JM, Chen LS, KenKnight BH et al (2000) Nerve sprouting and sudden cardiac death. Circ Res 86:816–821
33. Kimura K, Ieda M, Kanazawa H et al (2007) Cardiac sympathetic rejuvenation: a link between nerve function and cardiac hypertrophy. Circ Res 100:1755–1764

34. Zahner MR, Li DP, Chen SR et al (2003) Cardiac vanilloid receptor 1-expressing afferent nerves and their role in the cardiogenic sympathetic reflex in rats. J Physiol 551:515–523
35. Trupp M, Ryden M, Jornvall H et al (1995) Peripheral expression and biological activities of GDNF, a new neurotrophic factor for avian and mammalian peripheral neurons. J Cell Biol 130:137–148
36. Pan HL, Chen SR (2004) Sensing tissue ischemia: another new function for capsaicin receptors? Circulation 110:1826–1831
37. Wang L, Wang DH (2005) TRPV1 gene knockout impairs postischemic recovery in isolated perfused heart in mice. Circulation 112:3617–3623
38. Faerman I, Faccio E, Milei J et al (1977) Autonomic neuropathy and painless myocardial infarction in diabetic patients. Histologic evidence of their relationship. Diabetes 26:1147–1158
39. Vinik AI, Maser RE, Mitchell BD et al (2003) Diabetic autonomic neuropathy. Diabetes Care 26:1553–1579
40. Young LH, Wackers FJ, Chyun DA et al (2009) Cardiac outcomes after screening for asymptomatic coronary artery disease in patients with type 2 diabetes: the DIAD study: a randomized controlled trial. JAMA 301:1547–1555
41. Hage FG, Iskandrian AE (2011) Cardiovascular imaging in diabetes mellitus. J Nucl Cardiol 18:959–965
42. Vinik AI, Ziegler D (2007) Diabetic cardiovascular autonomic neuropathy. Circulation 115:387–397
43. Puschel AW, Adams RH, Betz H (1995) Murine semaphorin D/collapsin is a member of a diverse gene family and creates domains inhibitory for axonal extension. Neuron 14:941–948
44. Tanelian DL, Barry MA, Johnston SA et al (1997) Semaphorin III can repulse and inhibit adult sensory afferents in vivo. Nat Med 3:1398–1401
45. Ieda M, Kanazawa H, Kimura K et al (2007) Sema3a maintains normal heart rhythm through sympathetic innervation patterning. Nat Med 13:604–612

Liver and Fat in Type 2 Diabetes: New Insights and Clinical Relevance

Mukesh Nandave, Anup Ramdhave, and Ramesh K. Goyal

Abstract Among various organs such as the pancreas, kidney, liver, skeletal muscle, and adipose tissue responsible for control of blood glucose levels, the liver is emerging as one of the principal organs involved in insulin resistance associated with type 2 diabetes mellitus. The liver is involved both short- as well as long-term maintenance of glucose concentrations in the blood. In type 2 diabetes, impaired insulin-mediated suppression of glucose production and diminished glucose uptake ultimately causes an increase in postabsorptive glucose production. Type 2 diabetes associated with liver dysfunction and the vicious circle between liver, adipose tissue, and pancreas leads to various other diseases including nonalcoholic fatty liver disease, cardiovascular complications, and cancer. Despite current advances in pharmacotherapy for diabetes, attaining optimal glycemic control and preventing micro- and macrovascular diabetic complications has remained intangible and daunting. Novel therapeutic targets and their modulators, which include protein tyrosine phosphatase 1B inhibitors, glycogen phosphorylase inhibitors, glucokinase activators, diacylglycerol acyltransferase inhibitors, acetyl-CoA carboxylases inhibitors, and sirtuin activators, show promising results in preclinical and clinical studies. Adding new options with new mechanisms of action to the treatment armamentarium may eventually help to improve outcomes and reduce the burden of type 2 diabetes, which is only possible if we explore and understand the involvement of liver and fats in the pathogenesis of type 2 diabetes.

M. Nandave • A. Ramdhave
SPP School of Pharmacy and Technology Management, SVKM's NMIMS, V.L. Mehta Road, Vile Parle (West), Mumbai 400 056, India

R.K. Goyal (✉)
Institute of Life Science, Ahmedabad University, Opposite Gujarat University, Navrangpura, Ahmedabad 380009, India
e-mail: ramesh.goyal@ahduni.edu.in; goyalrk@rediffmail.com

B. Turan and N.S. Dhalla (eds.), *Diabetic Cardiomyopathy: Biochemical and Molecular Mechanisms*, Advances in Biochemistry in Health and Disease 9, DOI 10.1007/978-1-4614-9317-4_15,

Keywords Liver • Fats • Adipose tissue • Glucose metabolism • Insulin resistance • Diabetes mellitus

1 Introduction

Diabetes mellitus is an undoubtedly one of the most challenging health problems of the twenty-first century. The current estimates of the International Diabetes Federation (IDF) indicate that 366 million people had diabetes in 2011 and that by 2030 this will rise to 552 million globally. It is characterized by hyperglycemia or without insulin resistance. Insulin resistance, which is mainly associated with failure of various sensitive cells to respond to the normal actions of insulin (type 2 diabetes mellitus), accounts for more than 90 % of the diabetic population. In the early phase of type 2 diabetes, hyperglycemia is handled by the increased production of insulin by beta cells of the pancreatic islets of Langerhans. However, it may also cause hyperinsulinemia with decrease in beta-cell function, causing frank diabetes. The principal organs involved in insulin sensitivity are liver, muscle, and adipose tissue, which are mainly responsible for uptake of blood glucose (Fig. 1). Insulin resistance in these tissues is considered the most important cause of type 2 diabetes [2–5]. Recent studies have explored the significant role and involvement of

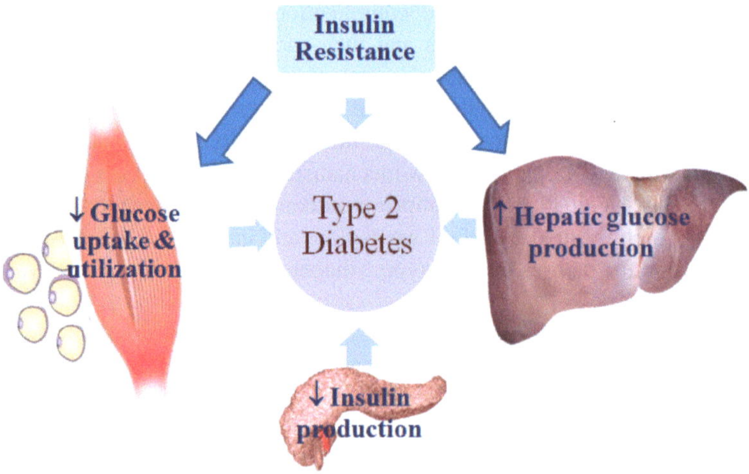

Fig. 1 Overview of the pathogenesis of type 2 diabetes contributed by insulin resistance in muscle, adipose tissue, and liver, as well as impaired insulin secretion by β-cells (Adapted from Baudry et al. [1])

liver in pathogenesis of type diabetes. The first study, in mice, demonstrated that inactivation of the insulin receptor gene in the liver leads to diabetes-like symptoms, indicating that insulin plays a critical role in liver metabolism regulation [3]. Another study in two different animal models reported that development of insulin resistance is associated with selective inactivation of the capacity of insulin to inhibit hepatic glucose production, retaining the ability of the hormone to stimulate fatty acid synthesis [3, 4]. In this chapter, we review the importance of glycemic control by insulin through the liver, exploring the liver as a target for insulin action and the possibility of discovering new drugs.

2 Physiology of Glucose Homeostasis

The maintenance of glucose homeostasis involves the integration of several major organs and is governed by insulin. In the postabsorptive state (12- to 16-h overnight fast), normal glucose level in the blood is maintained in the nondiabetic patient by the equilibrium between endogenous glucose production and glucose utilization. Approximately 2 mg/kg body weight of glucose is produced and used by the body every minute. The brain mainly contributes to glucose consumption, taking up ~1.1 mg/kg/min glucose (~60 % of total consumption). The splanchnic bed, erythrocytes, and other parenchymal organs contribute ~20–25 % to total glucose disposal, and the remainder is utilized primarily by skeletal muscle and, to a small extent, by adipose tissue. It is imperative to note that glucose utilization of brain, splanchnic bed, erythrocytes, and other parenchymal organs is insulin independent while only skeletal muscle and adipose tissue are insulin dependent for utilization of glucose [6]. Food ingestion, nutrient absorption, or increase in blood glucose causes stimulation of insulin release, which is a major cellular mechanism for disposal of glucose load. The signaling mechanism is a complex process involving the following stages: arrival of nutrient chyme to the intestine stimulates release of insulinotropic peptides, that is, glucose-dependent insulinotropic peptide, also called gastric inhibitory peptide (GIP) and glucagon-like peptide 1 (GLP-1), from specialized endocrine cells of the intestinal mucosa. GIP and GLP-1, also called incretins, are secreted in proportion to the nutrient load, which causes release of insulin from the β-cells. It is important to know that insulin secretion is much higher when the same amount of glucose is administered orally as compared to glucose administered intravenously, which can be attributed to the "incretin effect" [7].

The secreted insulin binds to the insulin receptor, a type of tyrosine kinase receptor, a dimer of α- and β-subunits. Binding of insulin to α-subunit activates its intrinsic tyrosine kinase activity, causing phosphorylation of β-subunit, which in turn activates various intracellular and intramembrane substrates. Intracellular substrates

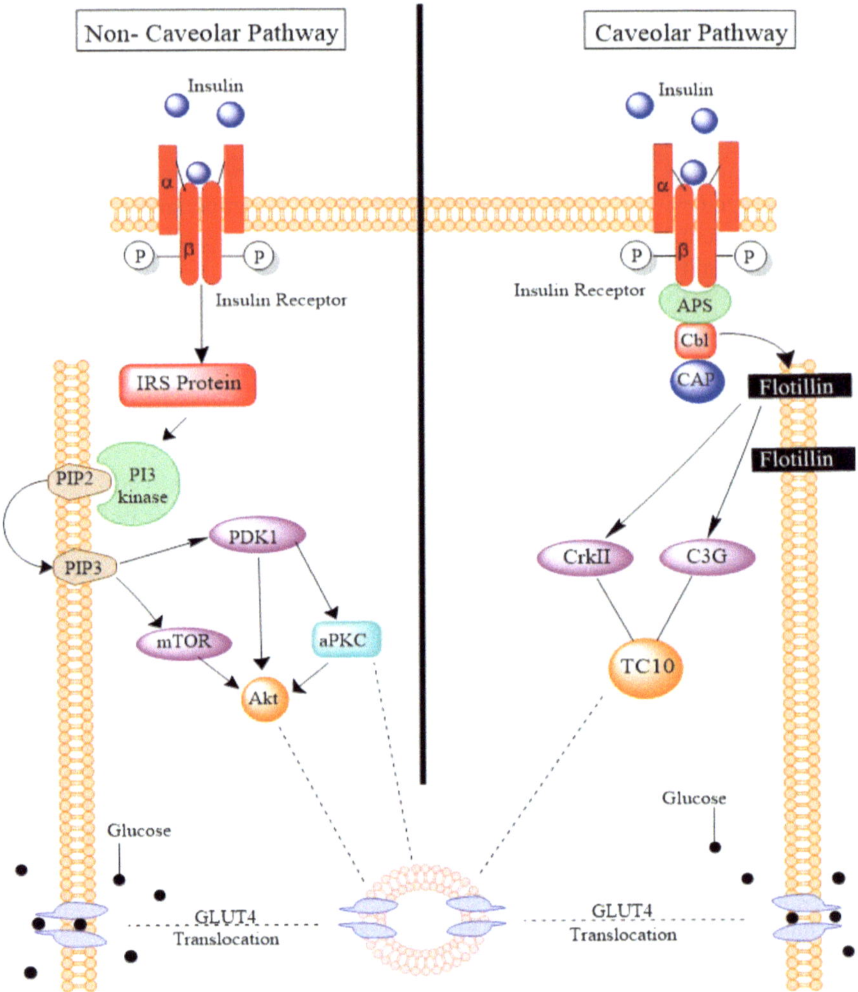

Fig. 2 Pathways of insulin signaling

such as insulin receptor substrate (IRS) protein, Gab-1, and Src-homology-2-containing protein (SHC), and intramembrane substrates such as Caveolin (CaV), adaptor protein (APS), and proto-oncogene Cbl, are activated. The key event is translocation of GLUT4 glucose transporter from intracellular vesicles to membrane, which is stimulated by both the caveolar and non-caveolar pathway [8]. There are 13 sugar transporter proteins (GLUT1–GLUT12, and HMIT) that catalyze hexose transport across cell membranes through an ATP-independent, facilitative diffusion mechanism. Of these, GLUT4, a 12-transmembrane domain-containing protein, is a key determinant of glucose homeostasis. A central role for GLUT4 in metabolism is strongly supported by a variety of genetically engineered mouse models [8].

In the noncaveolar pathway (Fig. 2), binding of insulin causes activation of IRS proteinn which further causes activation of PI3-kinase enzyme and leads to PIP3 formation in the membrane. PIP3 causes activation of PDK1 and mammalian target of rapamycin (mTOR), which in turn activates PKB, also known as Akt; the Akt cascade is also activated by a PKC. Activated Akt further induces glycogen synthesis through inhibition of glycogen synthase kinase 3 (GSK-3), protein synthesis via mTOR and downstream elements, and cell survival through inhibition of several pro-apoptotic agents (Bad, Forkhead family transcription factors, GSK-3). In the caveolar pathway (Fig. 2), insulin binding causes phosphorylation of β-subunit, which further causes phosphorylation of CAP and formation of the CAP-Cbl-APS complex. Next, the CAP-Cbl-APS complex moves to caveolin- and flotillin-enriched compartments, also termed lipid rafts, of the plasma membrane where flotillin forms a ternary complex with CAP, Cbl, and APS; this causes activation of CrkII with the guanine nucleotide exchange factor C3G, specifically activating TC10 and a small GTP-binding protein of the Rho family. Activation of these substrates of the noncaveolar and caveolar pathways causes translocation of GLUT4 glucose transporter from intracellular vesicle to membrane [7].

This translocation of GLUT4 glucose transporter causes intake of extracellular glucose. On entry of glucose into a cell, quick phosphorylation is caused by glucokinase, which converts it into glucose-6-phosphate, which further enters the glycolysis pathway. Insulin signaling also has growth and mitogenic effects mediated by the Akt cascade as well as by activation of the Ras/mitogen-activated protein kinase (MAPK) pathway. Insulin signaling also promotes fatty acid synthesis through activation of sterol regulatory element-binding proteins (SREBP-1C), upstream transcription factor 1 (USF1), and liver X receptor (LXR).

3 Role of Liver in Pathophysiology of Type 2 Diabetes Mellitus

Although the liver has a variety of functions, it plays a unique role in controlling carbohydrate metabolism by maintaining glucose concentrations in a normal range. The net glucose release is the result of two simultaneous ongoing pathways that include glycogenolysis and gluconeogenesis. Effects of insulin on hepatic glucose uptake occur rapidly, within the first 20 min after ingestion of meals, whereas stimulation of peripheral glucose uptake may require up to an hour to reach significant rates [7]. This delay may be caused quick access of insulin to hepatocytes via the hepatic portal circulation and liver sinusoids and the slower passage of insulin to the receptors on muscle and adipose tissues (Fig. 3).

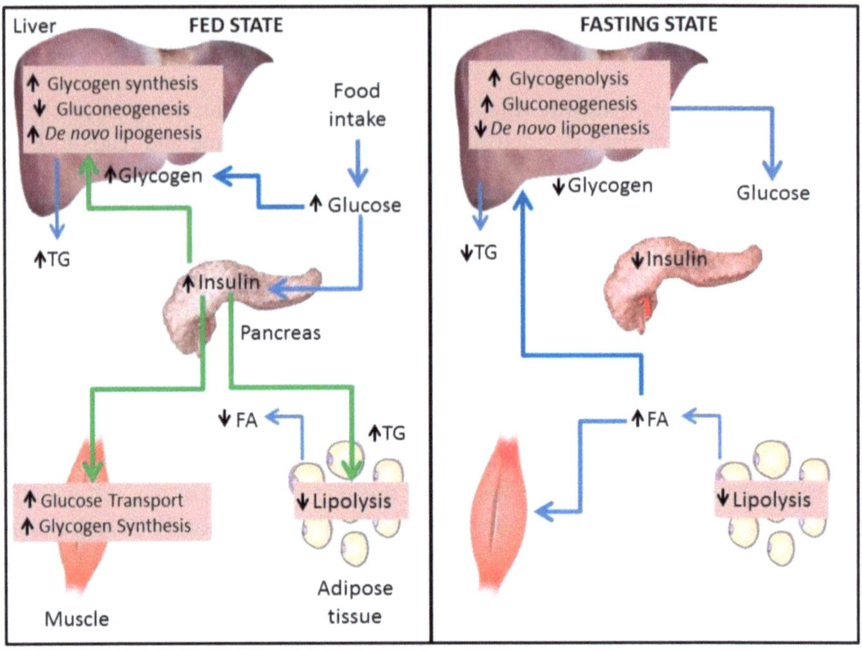

Fig. 3 Interplay between liver, pancreas, skeletal muscle, and white adipose tissue in nondiabetic person in fed and fasting states

4 Regulation of Hepatic Glucose Production

The rate of gluconeogenesis in liver is predominantly controlled by phosphoenol pyruvate-carboxykinase (PEPCK), fructose-1,6-bisphosphatase (FP_2ase), and glucose-6-phosphatase (G6Pase). PEPCK is known to catalyze the conversion of oxaloacetate to phosphoenolpyruvate (PEP), and G6Pase catalyzes the production of free glucose from glucose-6-phosphate (G6P). At the genetic level, insulin acting through its receptor inhibits gluconeogenesis by suppressing the expression of PEPCK and G6Pase whereas glucagon and glucocorticoids stimulate hepatic glucose production by inducing these genes [9].

In type 2 diabetes, there is insufficient insulin action to maintain a normal plasma level of glucose. Insulin action refers to the composite effect of plasma insulin concentration and insulin sensitivity of key target tissues. As discussed earlier, the major insulin-sensitive organs are skeletal muscles, adipose tissues, and liver. The failure of normal amounts of insulin to elicit the expected response is referred to as insulin resistance. In type 2 diabetes, hepatic glucose output is excessive in the fasting state and unacceptably suppressed after meals. Elevated levels of glucagon along with insulin resistance and insufficient insulin secretion lead to excessive hepatic gluconeogenesis and glycogenolysis, resulting in abnormally high glucose concentrations (Fig. 4). It is important to know that despite the ineffective insulin effects on insulin hepatic

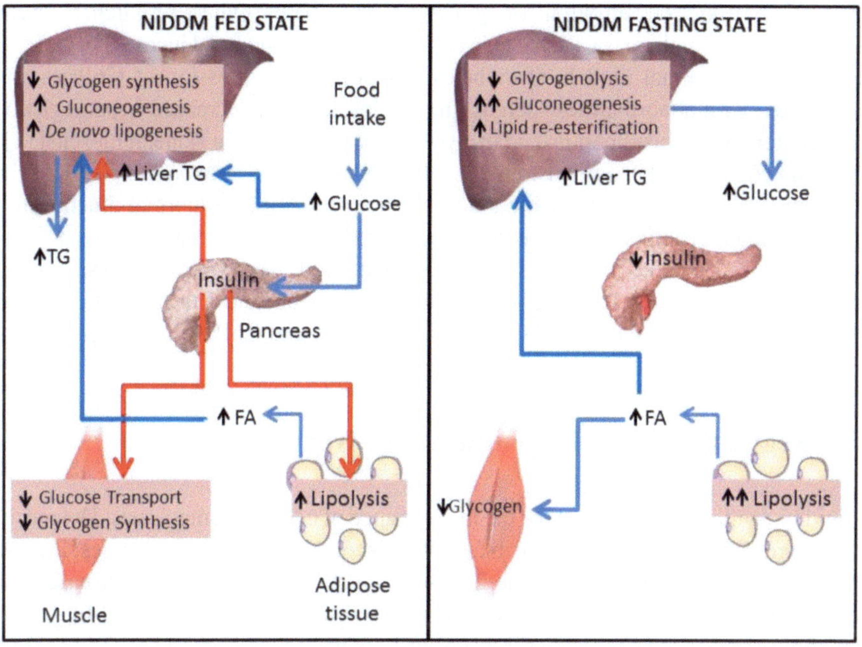

Fig. 4 Interplay between liver, pancreas, skeletal muscle, and white adipose tissue in type 2 diabetic person in fed and fasting states

glucose metabolism, the lipogenic effect of insulin in the liver is maintained by hyperinsulinemia, which contributes to hepatic steatosis that may further worsen to hepatocellular carcinoma [7].

The role of liver in type 2 diabetes was demonstrated by Michael et al. [3], in which they demonstrated that the loss of liver insulin receptors specifically causes hyperglycemia. Michael and colleagues used a targeted approach to knock out the gene encoding that insulin receptor specifically in the liver, resulting in liver-specific insulin receptor gene knockout (LIRKO) and severe insulin resistance in young (2-month-old) animals. It was further observed that elevated liver G6Pase and PEPCK expression were caused by the disruption of insulin action in the liver of LIRKO mice. This elevated expression leads to resistance to the blood glucose-lowering effect of insulin, severe glucose intolerance, and uncontrolled hepatic glucose production. The concentration of insulin was also found to escalate in LIRKO mice, mainly because of the failure of insulin receptor-mediated clearance of insulin by the liver and further, a sixfold increase in pancreatic β-cell mass. The resulting appearance of overt glucose intolerance was in contrast with the lack of effect of inactivating the insulin receptor gene in muscle (MIRKO). This finding clearly demonstrated that normal insulin signaling in the liver is important for handling glucose load.

At the cellular level, overexpression of G6Pase and PEPCK in liver leads to glucose intolerance, attributed to the deregulation of hepatic gluconeogenic genes [10–12]. In hepatocytes, equilibrium between G6Pase and glucokinase (GK) determines the concentration of glucose-6-phosphate. Other researchers have demonstrated that there is a decrease in hepatic GK expression and increase in G6Pase expression in animal models of diabetes [13–16]. Overexpression of G6Pase was associated with abnormalities such as glucose intolerance, decreased hepatic glycogen content, hyperinsulinemia, and triglyceride accumulation in skeletal muscle [11]. Sun and coauthors [10] demonstrated that transgenic mice overexpressing PEPCK in liver show increased basal hepatic glucose production but have normal glucose disposal during a hyperinsulinemic euglycemic clamp study compared to wild-type controls. At the genetic level, it was found that peroxisome proliferator-activated receptor-γ co-activator-1α (PGC-1α) and the hepatocyte nuclear factor-4α (HNF-4α) are crucial in the transcriptional regulation of PEPCK and G6Pase genes [17].

5 Regulation of Lipid Synthesis in Liver

Lipid biosynthesis is essential for the maintenance of cellular homeostasis. Defects in lipid synthesis or processing contribute to the development of many diseases, including obesity, insulin resistance, type 2 diabetes, nonalcoholic fatty liver disease, and cancer. Two synergistic pathways of acetyl-CoA carboxylase (ACC) and fatty acid synthase (FAS) are important in regulating lipogenic gene expression. The role of insulin is to induce GK gene expression, which is then essential for subsequent glucose phosphorylation [9]. Factors involved in the transcriptional effects of insulin are still largely unknown. The mechanisms that regulate lipid synthesis have come into focus recently with the discovery of SREBPs, which are membrane-bound transcription factors that control the rates of lipid synthesis in animal cells [18, 19]. Studies on transgenic and knockout mice suggest that SREBP-1 is involved in metabolism of glucose and fatty acid, whereas SREBP-2 is specific to cholesterol synthesis.

In type 2 diabetes and obesity, hepatic fatty acid synthesis is elevated when plasma insulin rises. The fatty acids are exported from the liver in lipoproteins, and they reach the extrahepatic organs where they increase insulin resistance and worsen the diabetic state [20, 21]. Shimomura and colleagues [21] demonstrated that transgenic mice expressing SREBP-1c in adipose tissue exhibit abnormal adipose differentiation, marked insulin resistance, and diabetes mellitus. In the streptozotocin-induced diabetes rat model, insulin stimulated lipid synthesis by selectively increasing hepatic SREBP-1c mRNA levels [22].

6 Role of the Adipocyte in the Pathogenesis of Type 2 Diabetes Mellitus

Adipocytes (lipocytes or fat cells) mainly contain adipose tissue of two types, white adipose tissue (WAT) and brown adipose tissue (BAT). The most important function of adipocytes is to store free fatty acids (FFAs) after food intake and release them during the fasting state to ensure sufficient energy status is maintained. In short, adipocytes store energy in the form of fat. Adipocytes also insulate and provide cushioning to the body. During the post-meal phase, triglycerides (TG) are hydrolyzed by lipoprotein lipase (LPL) and the resultant FFAs in blood are taken up in adipose tissue. This reserve of FFAs is mobilized by hormone-sensitive lipase (HSL)-mediated TG hydrolysis in adipocytes. Being a potent inhibitor of HSL as well as an activator of LPL, insulin primarily regulates the adipocyte fat content by enhancing FFA uptake and TG synthesis in adipocytes [23].

7 Dysfunctional Adipocytes and Obesity: A Bidirectional Relationship

7.1 Role of Adipocyte-Produced Adipocytokines in Obesity

Adipocytes and adipose tissue play multifaceted roles through production of wide array of hormones and cytokines involved in glucose metabolism (e.g., adiponectin, resistin), lipid metabolism (e.g., cholesteryl ester transfer protein, CETP), inflammation [e.g., tumor necrosis factor (TNF)-α, interleukin (IL)-6], coagulation (PAI-1), blood pressure (e.g., angiotensinogen, angiotensin II), and feeding behavior (leptin), thus affecting the metabolism and function of many organs and tissues including muscle, liver, vasculature, and brain. Thus, dysfunctional adipocytes are important in producing chronic low-grade pro-inflammation in obesity [24–27]. Increase in adipose tissue and adipocyte volume or size causes plasma adipocytokine levels to increase, except adiponectin, which decreases in obesity [28, 29]. Thus, increase in adipocyte size leads to increased secretion of adipocytokines that include IL-6, IL-8, monocyte chemo-attractant protein 1, and granulocyte colony-stimulating factor. Adipocyte hypertrophy causes an imbalance between pro- and antiinflammatory adipokines. In addition to adipose tissue, infiltration of pro-inflammatory macrophages is an important event and inflammatory source in obesity and type 2 diabetes [30]. For example, most of the tissue TNF-α and IL-6 production in adipose tissue is attributed to infiltrating macrophages. Cytokines produced by adipocytes or macrophages in adipose tissue that may lead to insulin resistance include TNF-α, IL-6, retinol binding protein 4 (RBP4), and lipocalin-2 (LCN2), leptin, and resistin [31].

7.2 Role of Adipocyte-Produced Hormones in Lipid Metabolism

Leptin, acylation-stimulating protein (ASP), and adiponectin are adipocyte hormones that are important in regulation of energy intake and expenditure and in lipid and carbohydrate metabolism. Leptin plays an important role in providing adaptation to reduced energy. Its deficiency or leptin gene mutation results in decreased level of circulating leptin, thereby leading to increased adiposity in the body. Its production is mainly regulated by insulin-mediated changes in adipocyte metabolism. Dietary fat and fructose do not cause increase in insulin release; however, they cause decreased leptin production, which explains the mechanism of a high-fat/high-sugar diet-related increase in energy intake and body weight [32–34].

ASP is a unique hormone and plays an important role in lipid metabolism regulation. In adipose tissue, ASP promotes glucose uptake and activity of diacyl glycerol acyl transferase (DGAT) while inhibiting the activity of hormone-sensitive lipase. Thus, in adipocytes, ASP increases the efficiency of triglyceride synthesis and storage, thereby leading to postprandial lipid clearance. One study demonstrated that exogenous ASP administration to mice increases the clearance of FFAs and TGs from the circulation after oral administration of fats [35, 36]. To support a role for ASP in lipid metabolism regulation in humans, one genetic study demonstrated that plasma ASP levels are directly related to genes controlling total cholesterol, LDL, and TG levels [37].

Adiponectin, a 30-kDa protein produced exclusively by mature adipocytes, is an antiinflammatory adipocytokine whose level decreases in obesity, insulin resistance, glucose intolerance, dyslipidemia, and atherosclerosis. Adiponectin mediates fatty acid metabolism and modulates energy homeostasis through the adiponectin receptors (AdipoR)-1 and -2 in peripheral tissues and brain. Adiponectin gene expression and production are reduced by TNF-α and IL-6 whereas insulin sensitizers and PPAR-γ agonists are known to increase adiponectin levels [38].

Leptin, ASP, and adiponectin production is affected by nutritional status. Therapeutic strategies that can modulate these hormones, the pathways that control their production, and receptors on which these hormones act are considered to be helpful in managing metabolic syndrome, hyperlipidemia, obesity, and insulin resistance.

8 Novel Approaches for the Treatment of Type 2 Diabetes and Obesity

Currently there are many novel therapeutic targets. This chapter discusses newer targets (Table 1) that have shown promising results for the treatment of obesity and type 2 diabetes.

8.1 HMG CoA Reductase Inhibitors

As discussed earlier, obesity has an important impact on predisposing risk factors for coronary heart disease, including dislipidemia, glucose intolerance, insulin resistance, and hypertension. HMG CoA reductase inhibitors are potent inhibitors

Table 1 Novel therapeutic targets in type 2 diabetes (T2DM) and obesity

Target	Mechanism	Drug (company)	Phase of development	Possible use
HMG CoA reductase	Competitive inhibition of HMG-CoA reductase	Atorvastatin (Pfizer)	USFDA approved	T2DM, other risk factors associated with cardio-vascular disease (CVD)
Lipase	Selective inhibition of gastric and pancreatic lipases	Orlistat (Roche)	USFDA approved	Obese patients with risks of hypertension, diabetes, dyslipidemia
		Cetilistat (Takeda)	NDA filed in Japan	Obese patients with T2DM and dyslipidemia
Protein tyrosine phosphatase 1B (PTP1B)	Inhibition of PTP1B	ISIS113715 (Isis Pharma)	Phase I	Obesity, T2DM
Glycogen phosphorylase	Inhibition of glycogen phosphorylase	CP-91149 (NA)	Preclinical	Obesity, T2DM
		PSN357 (OSI Pharma)	Preclinical	Obesity, T2DM
		GSK1362885 (GlaxoSmith-Kline)	Phase 1	Obesity, T2DM
Glucokinase	Activation and elevation of glucokinase (GK)	LY2599506 (Eli Lilly)	Phase 2	Obesity, T2DM
		LY2608204 (Eli Lilly)	Phase 1	Obesity, T2DM
		ZYGK1 (Cadila)	Phase 1	Obesity, T2DM
		ARRY-403 (Array Bio-Pharma)	Phase 1	Obesity, T2DM
Glucagon receptor	Antagonism of glucagon receptor	LY2409021 (Eli Lilly)	Phase 2	Obesity, T2DM
		MK-0893 (Merck)	Phase 2	Obesity, T2DM
Diacyl glycerol acyl transferase (DGAT-1)	Inhibition of DGAT-1 enzyme and blockade of triacylglycerol biosynthetic pathway	LCQ-908 (Novartis)	Phase 2	Obesity, T2DM
		PF-04620110 (Pfizer)	Phase 1	Obesity, T2DM
		AZD7687 (AstraZeneca)	Phase 1	Obesity, T2DM
Acetyl-CoA carboxylases (ACC)	Inhibition of ACC	CP-640186 (Pfizer)	Preclinical	Obesity, T2DM
Sirtuins	*Multifaceted activation of various* signaling pathways involved in regulation of energy utilization	Resveratrol (University of California)	Phase 2	Obesity, T2DM
		SRT2104 (SirtrisPharma)	Phase 2	Obesity, T2DM
		SRT2379 (SirtrisPharma)	Phase 1	Obesity, T2DM
		SRT3025 (SirtrisPharma)	Phase 1	Obesity, T2DM

of cholesterol biosynthesis that are used extensively to treat hypercholesterolemia. Statins exert an important beneficial effect of lowering lipid, which certainly attenuates the micro- and macrovascular complications associated with diabetes, but on the other hand statins also exhibit a pleiotropic effect, that is, a cholesterol-independent effect of statins. The pleiotropic effects of these drugs include direct beneficial effects on endothelial function, stabilizing atherosclerosis plaques, and improvement in insulin sensitivity [39].

The international guidelines of the American Diabetes Association (ADA) states a target LDL cholesterol of less than 100 mg/dl in individuals with diabetes and other cardiovascular risk factors. ADA provides guidelines for the use of statins in such diabetic individuals. The guidelines are based on numerous trials showing a benefit for statin therapy as both primary and secondary prevention of cardiovascular disease and mortality. Trials such as the Collaborative Atorvastatin Diabetes Study (CARDS) have proved beyond doubt that patients with type 2 diabetes and other risk factors for CVD should be treated with a statin [40]. Further, in trials such as the Stop Atherosclerosis in Native Diabetics Study (SANDS), which enrolled adults with type 2 diabetes ($n=499$), patients were treated to reach aggressive targets (LDL cholesterol of 70 mg/dl and systolic blood pressure of 115 mmHg) versus standard targets (LDL cholesterol of 100 mg/dl and systolic blood pressure of 130 mmHg). It was found that intimal medial thickness regressed in the aggressive group and progressed in the standard group [41].

8.2 Lipase Inhibitors

Orlistat is a gastrointestinal lipase inhibitor for obesity management that acts by inhibiting the absorption of dietary fats. Orlistat is indicated for obesity management in conjunction with a reduced-calorie diet. It is also indicated for obese patients with risk factors such as hypertension, diabetes, and dyslipidemia. Use of orlistat for diabetes associated with obesity was demonstrated by Hauptman [42]: in a series of 1- and 2-year randomized, placebo-controlled trials of obese subjects, treatment with orlistat in combination with a mildly calorie-restricted diet consistently produced significantly greater mean weight loss than diet alone. In addition, orlistat therapy resulted in significant improvements in several cardiovascular risk factors including serum total and low density lipoprotein cholesterol, serum insulin levels, systolic and diastolic blood pressure, and waist circumference. Furthermore, obese subjects with type 2 diabetes achieved a significantly greater decrease in body weight with orlistat compared with placebo, as well as significant improvements in HbA1c and fasting glucose levels. Among subjects with impaired glucose tolerance, orlistat compared with placebo reduced the proportion that developed type 2 diabetes. Conversely, orlistat increased the proportion of subjects who achieved a normalization of glucose tolerance. The only other lipase inhibitor currently in clinical development for the treatment of obesity is ATL-962 (Cetilistat). Although clinical trials for Cetilistat are still ongoing, the efficacy study demonstrates that Cetilistat 120 mg three times daily was superior to placebo in the primary endpoints, reduction of average body weight and also reduced HbA1c and LDL cholesterol levels.

8.3 Protein Tyrosine Phosphatase 1B Inhibitors

Protein tyrosine phosphatase 1B (PTP1B) is a prototypical member of the PTP family of enzymes. The involvement of PTPs in insulin action has been suggested ever since the discovery that vanadate, a nonselective inhibitor of PTPs, could mimic insulin activity [43]. PTP1B is a ubiquitously expressed enzyme known to dephosphorylate and control a multitude of signaling events during cell growth, differentiation, apoptosis, and cell movement. Insulin resistance is manifested in adipocytes, skeletal muscle, and hepatocytes by defects of insulin receptor and postreceptor signaling associated with the increase in activity or expression of protein tyrosine phosphatase (PTP)1B. This enzyme acts as a negative regulator of insulin signaling by causing termination of the RTK cascade and phosphorylated IRS 1, initiated when insulin binds to the IR. PTP1B is also a negative regulator of the leptin signaling pathway, contributing to leptin resistance [44].

The reported inhibitors can be divided into four general classes: difluoro methylene phosphonates, 2-carbomethoxybenzoic acids, 2-oxalylaminobenzoic acids, and lipophilic compounds. A series of these compounds can be found in the review by Johnson et al. [45]. A study with PTP1B-knockout mice demonstrated resistance to obesity and increased insulin sensitivity [46]. In a recent study in monkeys, inhibition of PTP1B with antisense oligonucleotides led to improved insulin sensitivity [47]. Along with peripheral tissues, neuronal PTP1B has also been implicated in controlling adiposity and leptin sensitivity [48]. Recently ISIS113715 has entered the clinical trial phase of drug development. Thus, PTP1B inhibition as a point of intervention has many attributes and could provide a new therapeutic option to patients with at-risk obesity or type 2 diabetes in the near future.

8.4 Glycogen Phosphorylase Inhibitors

The inappropriate overproduction of glucose by the liver as a result of glycogen breakdown is a contributor to hyperglycemia in type 2 diabetes. As the rate of glycogenolysis is regulated by glycogen phosphorylase, inhibition of this key enzyme may constitute a therapeutic option for the treatment of type 2 diabetes. Glycogen phosphorylase catalyzes the phosphorolytic cleavage of glycogen to produce glucose-1-phosphate, which is then isomerized to glucose-6-phosphate and enters the glycolytic pathway to produce glucose. Creating an inhibitor that specifically targets glycogen phosphorylase would then essentially decrease the amount of glucose produced by the liver [49].

Findings from a study by Martin and colleagues [50] show that CP-91149, a glycogen phosphorylase inhibitor, was able to reduce plasma glucose levels in mice. They also found that oral CP-91149 indirectly inhibits gluconeogenesis via the disruption of glucose/glycogen cycling, thereby improving glucose levels. In addition, CP-91149, which has been characterized in vitro and in vivo, suppressed glycogenolysis in both rat and human liver cells. While studying obese mice, the investigators found that a single 50-mg/kg oral dose of CP-91149 reduced plasma glucose concentrations to near-normal levels 3 h after administration.

PSN357 is another compound reported to inhibit glycogen phosphorylase and reduce blood glucose levels in animal models with diabetes. An increase in liver glycogen was also seen in the study; however, heart and skeletal muscle glycogen showed no changes from controls. In addition, a 9-day study in mice showed once-daily oral administration of PSN357 maintained antihyperglycemic efficacy throughout this period. Other endpoints observed in this study were a 57 % increase in liver glycogen and no changes in other glycogen deposits or in levels of plasma insulin or alanine aminotransferase. Thus, the development of glycogen phosphorylase inhibitors is promising in the near future.

8.5 Glucokinase Activators

Glucokinase is an enzyme that facilitates phosphorylation of glucose to glucose-6-phosphate. It determines the rate of glucose metabolism by regulating the amount of insulin produced from pancreatic beta cells. The rate of insulin release from the pancreas is directly proportional to the level of glucokinase, which ultimately depends on the level of blood glucose. In addition, glucokinase influences hepatic lipid metabolism and gluconeogenesis in the liver. Glucokinase has been found to function to a lesser degree in patients with type 2 diabetes [51].

A recent study reported that piragliatin (RO4389620) lowers plasma glucose both in the postabsorptive state and after a glucose challenge in patients with type 2 diabetes mellitus. In 15 human volunteers, piragliatin caused a significant dose-dependent reduction of glucose levels in both fasting and fed states. Thus, the glucokinase activator piragliatin had an acute glucose-lowering action in patients with mild type 2 diabetes, mainly mediated through a generalized enhancement of β-cell function and through fasting-restricted changes in glucose turnover [52]. LY2599506 and LY2608204 (developed by Eli Lilly), ZYGK1 (developed by Cadila), PSN105 and PSN010 (developed by OSI Pharmaceuticals), and ARRY-588 and ARRY-403 (developed by Array Biopharma) are some of the other glucokinase activators in various stages of clinical trials.

8.6 Glucagon-Receptor Antagonists

As discussed earlier, glucagon is secreted in response to falling glucose levels during the fasting period and is primarily a counter-regulatory hormone to insulin. It raises blood glucose by enhancing glycogenolysis and gluconeogenesis through the activation of the cAMP-dependent protein kinase cascade in the liver.

LY2409021 is a potent, selective antagonist of the glucagon receptor. It was tested in 23 healthy subjects and 9 patients with type 2 diabetes treated with diet and exercise. The study design was a randomized, double-blind, placebo (PBO)-controlled, crossover study to examine the safety, tolerability, pharmacokinetics, and pharmacodynamics of single escalating doses of LY2409021. A single dose of LY2409021 was found to reduce blood glucose in healthy subjects and patients with type 2 diabetes mellitus [53].

Bay 27-9955, a non-peptide compound, competitively blocks the interaction of glucagon with the human glucagon receptor. A clinical study on eight normal volunteers in a double-blind, placebo-controlled, crossover study revealed that a single dose of 200 mg was able to block the effect of exogenous glucagon [54].

In a phase II study of MK-0893, 342 type 2 diabetes patients were randomized to once-daily MK-0893 in four different dosages, metformin 1,000 mg twice daily, or placebo. At 12 weeks, treatment with MK-0893 resulted in significant dose-dependent reductions in fasting plasma glucose, ranging from 32 mg/dl with a 20-mg dose to 63 mg/dl with 80 mg, from a baseline of 180–193 mg/dl. Metformin reduced FPG by 37 mg/dl and placebo by just 2 mg/dl from baseline. For HbA_{1c}, reductions at 12 weeks ranged from 0.6 to 1.5 percentage points, compared with 0.8 percentage points with metformin and 0.5 with placebo (MK0893 Link).

8.7 Diacylglycerol Acyltransferase (DGAT1) Inhibitors

The enzyme catalyzing the final and committed step in the triacylglycerol biosynthetic pathway is acyl-CoA:diacylglycerolacyltransferase (DGAT). Transgenic overexpression of DGAT1 in adipose tissues resulted in whole-body insulin resistance and other metabolic disturbances, such as liver steatosis [55]. Thus, potent and selective DGAT1 inhibitors are now an area of interest for many industries.

Clinically, the most advanced DGAT-1 inhibitor is LCQ-908 (Novartis). In preclinical studies, Novartis and others have demonstrated that dosing of DGAT-1 inhibitors for 14–28 days caused weight loss, improved insulin sensitivity, improved lipid disposal following a triglyceride challenge, and reduced hepatic steatosis. The product is currently in phase II for the assessment of LCQ-908 when added to metformin in patients with type 2 diabetes mellitus and in phase III to determine long-term safety and tolerability and continued efficacy in lowering triglycerides of LCQ908 in subjects with familial chylomicronemia syndrome. A phase I study of AZD7687, a selective DGAT1 inhibitor developed by

AstraZeneca, revealed an attenuating effect of AZD7687 on postprandial TAG. However, dose- and diet-related gastrointestinal side effects may impact further development of DGAT1 inhibitors [56].

8.8 Acetyl-CoA Carboxylases (ACC) Inhibitors

Acetyl-CoA carboxylases 1 and 2 (ACC1 and ACC2) catalyze the synthesis of malonyl-CoA, the substrate for fatty acid synthesis and the regulator of fatty acid oxidation. These enzymes are highly regulated and are important in the energy metabolism of fatty acids [57]. CP-640186, a mixed inhibitor of ACC-1 and ACC-2, has been used to evaluate the effects of ACC inhibition on glycemic endpoints in diabetic ob/ob mice. Treatment of ob/ob mice with CP-640186 increased glucose and triglyceride levels and reduced insulin levels [58].

8.9 AMP-Activated Protein Kinase (AMPK) Activators

The growing realization that AMPK regulates the coordination of synthesis and storage of glucose and fatty acids and oxidation of glucose and fatty acids metabolic processes represents an attractive therapeutic target for intervention in many conditions of disordered energy balance. The net effect of AMPK activation is stimulation of hepatic fatty acid oxidation and ketogenesis, inhibition of cholesterol synthesis, lipogenesis, and triglyceride synthesis, inhibition of adipocyte lipolysis and lipogenesis, stimulation of skeletal muscle fatty acid oxidation and muscle glucose uptake, and modulation of insulin secretion by pancreatic β-cells. In skeletal muscle, AMPK is activated by contraction [59]. AMPK activators are being explored in areas such as type 2 diabetes, obesity, cancer, and Alzheimer disease [60].

8.10 Sirtuin Activators

The SIRT1 enzyme was the first mammalian sirtuin discovered and is the most studied of the seven human sirtuin family members. It has emerged as a major therapeutic target for the treatment of metabolic, inflammatory, and neurodegenerative diseases. SIRT1 activation has been shown to stimulate several key cellular signalling pathways involved in regulating energy utilization, including the synthesis of new mitochondria. Mitochondrial activity in metabolically active tissues, such as muscle, will increase metabolic rate, drive glucose metabolism and fatty acid oxidation, and thereby improve insulin sensitivity and enhance energy expenditure in multiple tissues. SIRT1 activation also plays a significant role on other key cellular regulators involved in metabolic and oxidative stress.

SIRT1-2104, SRT2379, and SRT3025 have been developed as the compounds targeting SIRT1. Development of SRT-501 was halted by GSK in late 2010 because of adverse effects in study participants. In an animal model of metabolic dysfunction including insulin resistance [a diet-induced obesity (DIO) mouse model], these new chemical entities (NCEs) discovered at Sirtris significantly improved insulin sensitivity and glucose levels in a similar manner observed in DIO mice genetically altered to have increased SIRT1 levels. DIO mice treated with these same NCE SIRT1 activators also had reduced weight gain, increased energy expenditure, improved insulin sensitivity, reversal of hepatic steatosis, and profiles of metabolites (metabolomic analysis) in tissues and plasma consistent with a reversal of the metabolic dysfunction induced by the high-fat diet.

9 Conclusions

It appears that although fat is important in the pathogenesis of type 2 diabetes, the role of the liver cannot be overlooked. The vicious circle of liver, adipose tissue, and pancreas leads to various other diseases including nonalcoholic fatty liver disease, cardiovascular complications, and cancer. Despite current advances in pharmacological therapies for diabetes, attaining and maintaining optimal glycemic control has remained daunting. Novel therapies such as protein tyrosine phosphatase 1B inhibitors, glycogen phosphorylase inhibitors, glucokinase activators, glucagon-receptor antagonists, diacylglycerol acyltransferase inhibitors, acetyl-CoA carboxylase inhibitors, and sirtuin activators are in various stages of development, showing promising results in clinical trials. Adding new options with new mechanisms of action to the treatment armamentarium may eventually help to improve outcomes and reduce the burden of type 2 diabetes: this is only possible if we explore and understand the involvement of liver and fats in pathogenesis of type 2 diabetes. However, it is prudent to remain optimistic as research in this field continues to evolve.

References

1. Baudry A, Leroux L, Jackerott M et al (2002) Genetic manipulation of insulin signaling, action and secretion in mice. Insights into glucose homeostasis and pathogenesis of type 2 diabetes. EMBO Rep 3:323–328
2. Korenblat KM, Fabbrini E, Mohammed BS et al (2008) Liver, muscle, and adipose tissue insulin action is directly related to intrahepatic triglyceride content in obese subjects. Gastroenterology 134:1369–1375
3. Michael MD, Kulkarni RN, Postic C et al (2000) Loss of insulin signalling in hepatocyes leads to severe insulin resistance and progressive hepatic dysfunction. Mol Cell 6:87–97
4. Shimomura I, Matsuda M, Hammer RE et al (2000) Decreased IRS-2 and increased SREBP-1c lead to mixed insulin resistance and sensitivity in livers of lipodystrophic and ob/ob mice. Mol Cell 6:77–86

5. Caro JF, Dohm LG, Pories WJ et al (1989) Cellular alterations in liver, skeletal muscle, and adipose tissue responsible for insulin resistance in obesity and type II diabetes. Diabetes Metab Rev 5:665–689
6. Consoli A (1992) Role of liver in pathophysiology of NIDDM. Diabetes Care 15:430–441
7. Powers AC, D'Alessio D (2011) Endocrine pancreas and pharmacotherapy of diabetes mellitus and hypoglycaemia. In: Brunton LL (ed) Goodman & Gilman's the pharmacological basis of therapeutics, 12th edn. McGraw-Hill, New York, pp 1258–1271
8. Huang S, Czech MP (2007) The GLUT4 glucose transporter. Cell Metab 5:237–252
9. Postic C, Dentin R, Girard J (2004) Role of the liver in the control of carbohydrate and lipid homeostasis. Diabetes Metab 30:398–408
10. Sun Y, Liu S, Ferguson S et al (2002) Phosphoenolpyruvate carboxykinase overexpression selectively attenuates insulin signalling and hepatic insulin sensitivity in transgenic mice. J Biol Chem 277:23301–23307
11. Trinh KY, O'Doherty RM, Anderson P et al (1998) Perturbation of fuel homeostasis caused by overexpression of the glucose-6-phosphatase catalytic subunit in liver of normal rats. J Biol Chem 273:31615–31620
12. Valera A, Pujol A, Pelegrin M et al (1994) Transgenic mice overexpressing phosphoenolpyruvate carboxykinase develop non-insulin-dependent diabetes mellitus. Proc Natl Acad Sci U S A 91:9151–9154
13. Massillon D, Barzilai N, Chen W et al (1996) Glucose regulates in vivo glucose-6-phosphatase gene expression in the liver of diabetic rats. J Biol Chem 271:9871–9874
14. Mithieux G, Vidal H, Zitoun C et al (1996) Glucose-6-phosphatase mRNA and activity are increased to the same extent in kidney and liver of diabetic rats. Diabetes 45:891–896
15. Barzilai N, Rossetti L (1993) Role of glucokinase and glucose-6-phosphatase in the acute and chronic regulation of hepatic glucose fluxes by insulin. J Biol Chem 268:25019–25025
16. Burchell A, Cain DI (1985) Rat hepatic microsomal glucose-6-phosphatase protein levels are increased in streptozotocin-induced diabetes. Diabetologia 28:852–856
17. Hanson RW, Reshef L (1997) Regulation of phosphoenolpyruvate carboxykinase (GTP) gene expression. Annu Rev Biochem 66:581–611
18. Wang X, Sato R, Brown MS et al (1994) SREBP-1, a membrane-bound transcription factor released by sterol-regulated proteolysis. Cell 77:53–62
19. Kim JB, Spiegelman BM (1996) ADD1/SREBP1 promotes adipocyte differentiation and gene expression linked to fatty acid metabolism. Genes Dev 10:1096–1107
20. McGarry JD (1998) Glucose-fatty acid interactions in health and disease. Am J Clin Nutr 67:500S–504S
21. Shimomura I, Hammer RE, Richardson JA et al (1998) Insulin resistance and diabetes mellitus in transgenic mice expressing nuclear SREBP-1c in adipose tissue: model for congenital generalized lipodystrophy. Genes Dev 12:3182–3194
22. Shimomura I, Bashmakov Y, Ikemoto S et al (1999) Insulin selectively increases SREBP-1c mRNA in the livers of rats with streptozotocin-induced diabetes. Proc Natl Acad Sci U S A 96:13656–13661
23. Hajer GR, Van Haeften TW, Visseren FL (2008) Adipose tissue dysfunction in obesity, diabetes, and vascular diseases. Eur Heart J 29:2959–2971
24. Ran J, Hirano T, Fukui T et al (2006) Angiotensin II infusion decreases plasma adiponectin level via its type 1 receptor in rats: an implication for hypertension-related insulin resistance. Metabolism 55:478–488
25. Engeli S, Bohnke J, Feldpausch M et al (2005) Activation of the peripheral endocannabinoid system in human obesity. Diabetes 54:2838–2843
26. Yamauchi T, Kamon J, Waki H et al (2001) The fat-derived hormone adiponectin reverses insulin resistance associated with both lipoatrophy and obesity. Nat Med 7:941–946
27. Chu NF, Spiegelman D, Hotamisligil GS et al (2001) Plasma insulin, leptin, and soluble TNF receptors levels in relation to obesity-related atherogenic and thrombogenic cardiovascular disease risk factors among men. Atherosclerosis 157:495–503

28. Wannamethee SG, Lowe GD, Rumley A et al (2007) Adipokines and risk of type 2 diabetes in older men. Diabetes Care 30:1200–1205
29. Skurk T, Alberti-Huber C, Herder C et al (2007) Relationship between adipocyte size and adipokine expression and secretion. J Clin Endocrinol Metab 92:1023–1033
30. Fernandez-Real JM, Pickup JC (2008) Innate immunity, insulin resistance and type 2 diabetes. Trends Endocrinol Metab 19:10–16
31. Esteve E, Ricart W, Fernández-Real JM (2009) Adipocytokines and insulin resistance: the possible role of lipocalin-2, retinol binding protein-4, and adiponectin. Diabetes Care 32:S362–S367
32. Havel PJ (2001) Peripheral signals conveying metabolic information to the brain: short-term and long- term regulation of food intake and energy homeostasis. Exp Biol Med 226:963–977
33. Havel PJ (2002) Control of energy homeostasis and insulin action by adipocyte hormones: leptin, acylation stimulating protein, and adiponectin. Curr Opin Lipidol 13:51–59
34. Havel PJ (2004) Update on adipocyte hormones: regulation of energy balance and carbohydrate/lipid metabolism. Diabetes 53:S143–S151
35. Cianflone K, Maslowska M, Sniderman AD (1999) Acylation stimulating protein (ASP), an adipocyte autocrine: new directions. Semin Cell Dev Biol 10:31–41
36. Cianflone K, Xia Z, Chen LY (2003) Critical review of acylation-stimulating protein physiology in humans and rodents. Biochim Biophys Acta 1609:127–143
37. Comuzzie AG, Cianflone K, Martin LJ et al (2001) Serum levels of acylation stimulating protein (ASP) show evidence of a pleiotropic relationship with total cholesterol, LDL, and triglycerides and preliminary evidence of linkage on chromosomes 5 and 17 in Mexican Americans (abstract). Obes Res 9:103S
38. Kadowaki T, Yamauchi T (2005) Adiponectin and adiponectin receptors. Endocr Rev 26:439–451
39. Danesh FR, Kanwar YS (2004) Modulatory effects of HMG-CoA reductase inhibitors in diabetic microangiopathy. FASEB J 18:805–815
40. Owen OG (2005) The collaborative atorvastatin diabetes study: preliminary results. Int J Clin Pract 59:121–123
41. Russell M, Fleg JL, Galloway WJ et al (2006) Examination of lower targets for low-density lipoprotein cholesterol and blood pressure in diabetes: the Stop Atherosclerosis in Native Diabetics Study (SANDS). Am Heart J 152:867–875
42. Hauptman J (2000) Orlistat: selective inhibition of caloric absorption can affect long-term body weight. Endocrine 13:201–206
43. Sekar N, Li J, Shechter Y (1996) Vanadium salts as insulin substitutes: mechanism of action, a scientific and therapeutic tool in diabetes mellitus research. Crit Rev Biochem Mol Biol 31:339–359
44. Morris DL, Rui L (2009) Recent advances in understanding leptin signaling and leptin resistance. Am J Physiol Endocrinol Metab 297:E1247–E1259
45. Johnson TO, Ermolieff J, Jirousek MR (2002) Protein tyrosine phosphatase 1B inhibitors for diabetes. Nat Rev Drug Discov 1:696–709
46. Swarbrick MM, Havel PJ, Levin AA et al (2009) Inhibition of protein tyrosine phosphatase-1B with antisense oligonucleotides improves insulin sensitivity and increases adiponectin concentrations in monkeys. Endocrinology 150:1670–1679
47. Martin TL, Alquier T, Asakura K et al (2006) Diet-induced obesity alters AMP kinase activity in hypothalamus and skeletal muscle. J Biol Chem 281:18933–18941
48. Ottanà R, Maccari R, Ciurleo R et al (2009) 5-Arylidene-2-phenylimino-4-thiazolidinediones as PTP1B and LMW-PTP inhibitors. Bioorg Med Chem Lett 17:1928–1937
49. Henke BR, Sparks SM (2006) Glycogen phosphorylase inhibitors. Mini Rev Med Chem 6:845–857
50. Martin WH, Hoover DJ, Armento SJ et al (1998) Discovery of a human liver glycogen phosphorylase inhibitor that lowers blood glucose in vivo. Proc Natl Acad Sci U S A 95: 1776–1781

51. Matschinsky FM, Porte D (2010) Glucokinase activators (GKAs) promise a new pharmacotherapy for diabetics. F1000 Med Rep 2:43
52. Bonadonna RC, Heise T, Arbet-Engels C et al (2010) Piragliatin (RO4389620), a novel glucokinase activator, lowers plasma glucose both in the postabsorptive state and after a glucose challenge in patients with type 2 diabetes mellitus: a mechanistic study. J Clin Endocrinol Metab 95:5028–5036
53. Kelly RP, Abu-Raddad EJ, Tham LS (2011) Single doses of the glucagon receptor antagonist LY2409021 reduce blood glucose in healthy subjects and patients with Type 2 Diabetes Mellitus (T2DM). American Diabetes Association abstract no. 1004-P
54. Petersen KF, Sullivan JT (2001) Effects of a novel glucagon receptor antagonist (Bay 27-9955) on glucagon-stimulated glucose production in humans. Diabetologia 44:2018–2024
55. Chen HC, Smith SJ, Ladha Z et al (2002) Increased insulin and leptin sensitivity in mice lacking acyl CoA:diacylglycerolacyltransferase 1. J Clin Invest 109:1049–1055
56. Denison H, Nilsson C, Kujacic M et al (2013) Proof of mechanism for the DGAT1 inhibitor AZD7687: results from a first-time-in-human single-dose study. Diabetes Obes Metab 15: 136–143
57. Wakil SJ, Abu-Elheiga LA (2009) Fatty acid metabolism: target for metabolic syndrome. J Lipid Res 50:S138–S143
58. Treadway JL, McPherson RK, Petras SF et al (2004) Effect of the acetyl-CoA carboxylase inhibitor CP-640186 on glycemic control in diabetic ob/ob mice. Pfizer Poster Presentation
59. Winder WW, Hardie DG (1999) AMP-activated protein kinase, a metabolic master switch: possible roles in type 2 diabetes. Am J Physiol Endocrinol Metab 277:E1–E10
60. Vingtdeux V, Chandakkar P, Zhao HC et al (2011) Novel synthetic small-molecule activators of AMPK as enhancers of autophagy and amyloid-β peptide degradation. FASEB J 25: 219–231
61. Lian Z, Li Y, Gao J et al (2011) A novel AMPK activator, WS070117, improves lipid metabolism discords in hamsters and HepG2 cells. Lipids Health Dis 10:67

Roles of PKC Isoforms in Development of Diabetes-Induced Cardiovascular Complications

Isil Ozakca and A. Tanju Ozcelikay

Abstract Diabetes-induced cardiovascular abnormalities are the major causes of mortality and morbidity in diabetic populations. Vascular complications of diabetes can be evaluated as microvascular anomalies leading to retinopathy, nephropathy, and neuropathy, and macrovascular anomalies causing atherosclerosis, coronary artery disease, and peripheral vascular disease. Independent of coronary artery disease and hypertension, cardiomyopathy is also an important abnormality that can occur in the diabetic heart. Hyperglycemia, hyperinsulinemia related to insulin resistance, and increased levels of free fatty acids and lipids seem to have prominent roles in the development of microvascular and macrovascular complications and diabetic cardiomyopathy. Several mechanisms can be implicated in these complications, including increased polyol pathway flux, enhanced nonenzymatic glycation, and intracellular formation of advanced glycation end products (AGEs), activation of protein kinase C (PKC) isoforms, and increased hexosamine pathway activity. The focus of this chapter is recent concepts regarding PKC isoform-specific activation mechanisms and actions that have implications for the development of PKC-targeted therapeutics in diabetic complications. The PKC family of serine/threonine kinases have been associated with a diverse array of biological responses in health and disease. In diabetes, activation of different isoforms of PKC is associated with many pathologies seen in the retina, kidneys, vasculature, and heart. Therefore, inhibition of PKC isoforms can be evaluated as a therapeutic target for preventing

I. Ozakca • A.T. Ozcelikay (✉)
Department of Pharmacology, Ankara University, Ankara 06100, Turkey
e-mail: ozcelikay@ankara.edu.tr

B. Turan and N.S. Dhalla (eds.), *Diabetic Cardiomyopathy: Biochemical and Molecular Mechanisms*, Advances in Biochemistry in Health and Disease 9,
DOI 10.1007/978-1-4614-9317-4_16, © Springer Science+Business Media New York 2014

of diabetic complications. In this regard, clinical trials using ruboxistaurin, a PKC-β isoform inhibitor, have promising results for treatment of diabetic retinopathy, nephropathy, and endothelial dysfunction.

Keywords PKC • Diabetes • Insulin resistance • Diabetic complications • PKC inhibitors

1 Introduction

The prevalence of diabetes mellitus is growing rapidly. According to International Diabetes Federation, the number of people with diabetes (20–79 years of age), which is 366 million in 2011, is estimated to reach 552 million worldwide by 2030.

The discovery of insulin by Banting and Best for the treatment of diabetes has prevented mortality caused by diabetic coma and ketoacidosis, thereby leading to an increase in life expectancy of diabetic patients. However, diabetes-induced chronic complications, especially related to cardiovascular system are currently the main reason for mortality and morbidity in diabetic patients. A common feature of these complications is abnormalities in the vasculature of the target organs include the eyes, kidneys, heart, lower limbs, and the nervous system. Cardiovascular complications caused by long-term diabetes can be evaluated as both microvascular anomalies, which are manifested as retinopathy, nephropathy, and neuropathy; and macrovascular anomalies that cause atherosclerosis, coronary artery disease, and peripheral vascular disease. In addition, diabetic cardiomyopathy, which occurs even in the absence of coronary artery disease and hypertension, is also an important diabetic complication leading to heart failure characterized as diastolic and systolic dysfunctions [1].

It is widely accepted that hyperglycemia is the most important metabolic abnormality, leading to development of diabetic vascular complications. Large prospective clinical studies demonstrated that better glucose control in patients with type 1 or type 2 diabetes can prevent microvascular complications [2, 3]. In addition, other metabolic factors such as insulin resistance associated with hyperinsulinemia and increased levels of free fatty acids and lipids seem to have important roles in the development of macrovascular complications and diabetic cardiomyopathy [1].

A great deal of evidence obtained from experimental and clinical studies has shown that insulin resistance and hyperglycemia can cause cardiovascular complications by several mechanisms, including increased polyol pathway flux [4], enhanced nonenzymatic glycation and intracellular formation of advanced glycation end products (AGEs) [5], activation of protein kinase C (PKC) isoforms [6], and increased hexosamine pathway activity [7]. It is suggested that all these mechanisms are activated by mitochondrial overproduction of reactive oxygen species (ROS) as a result of hyperglycemia and increased oxidation of fatty acids [8].

In this chapter, we specifically outline the association between cardiovascular complications and the activation of PKC in diabetes.

2 PKC Classification and Regulation

PKC was initially identified by Nishizuka and coworkers as a single molecular entity that is a cyclic nucleotide-independent, Ca^{2+}-dependent serine kinase in the mid- and late 1970s [9]. Today, it has been accepted that PKC is "a group of isoforms" that belongs to a family of serine/threonine kinases. By molecular cloning, nine different isoforms of PKC have been cloned and characterized [10]. PKC isoforms are subclassified on the basis of the membrane-targeting module that is located on the regulatory domain of the protein. The function of the membrane-targeting module is to control the subcellular localization of the enzyme [11]. The conventional PKC isoforms (cPKCs; α, the alternative spliced β1 and β2, and γ) are activated by diacylglycerol (DAG)/phorbol ester and Ca^{2+} as a result of C1 and C2 binding domains [12]. Novel PKCs (nPKCs; δ, ε, θ, and η) also contain C1 and C2 domains. However, compared to cPKCs, the change in the linear sequence of the C1 and C2 domains interfere with the binding of Ca^{2+}, so nPKCs are activated by DAG/phorbol ester, but not by Ca^{2+}. Atypical PKCs (aPKCs; ξ and ι/λ) contain only C1 domain; neither DAG/phorbol ester nor calcium is the activator of this isoform (Fig. 1). The activation of aPKC is maintained by protein–protein interactions and activation loop phosphorylation by phosphoinositide-dependent kinase 1 (PDK-1) [11].

The different types of PKC play an important role in a wide variety of cellular responses such as proliferation, hypertrophy, angiogenesis, regulation of endothelial permeability and blood flow, extracellular matrix (ECM) synthesis/turnover, cytokine activation, and leukocyte adhesion [13]. Diabetes or hyperglycemia can alter most of these cellular functions regulated by PKC.

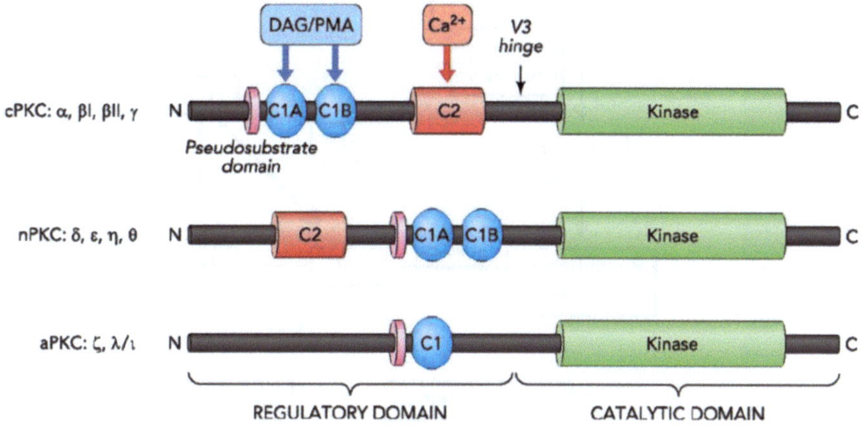

Fig. 1 Domain structure of protein kinase C (PKC) family enzymes showing the conserved pseudosubstrate motif NH_2-terminal to the *C1* domain, the *C2* domain, and the kinase domain and the more variable regions (Reproduced with permission from Physiology (Bethesda) 27:130–139, 2012)

3 Activation of DAG-PKC Pathways in Diabetes

Activation and translocation of PKC from the cytosol to the plasma membrane occur in response to a transient increase in DAG/phorbol esters, Ca^{2+}, or phosphatidylserine. Posttranslational phosphorylation of PKC also appears to constitute an important mechanism for regulating PKC translocation and isoform activity [14, 15]. Intracellular production of DAG is the main step leading to activation and translocation of PKC. DAG can be generated from phospholipase C-mediated hydrolysis of phosphoinositides, from the metabolism of phosphatidylcholine by phospholipase D or phospholipase C, through release of nonesterified fatty acids (NEFAs) from precursor lipids by the action of phospholipase A_2 and de novo synthesis of DAG from phosphatidic acid [13] (Fig. 2).

It has been demonstrated that diabetes and insulin resistance cause an increase in activation of DAG-PKC pathways in vascular (retina, aorta, heart, renal glomeruli) and nonvascular (liver, skeletal muscle) tissues, but not central nervous system or

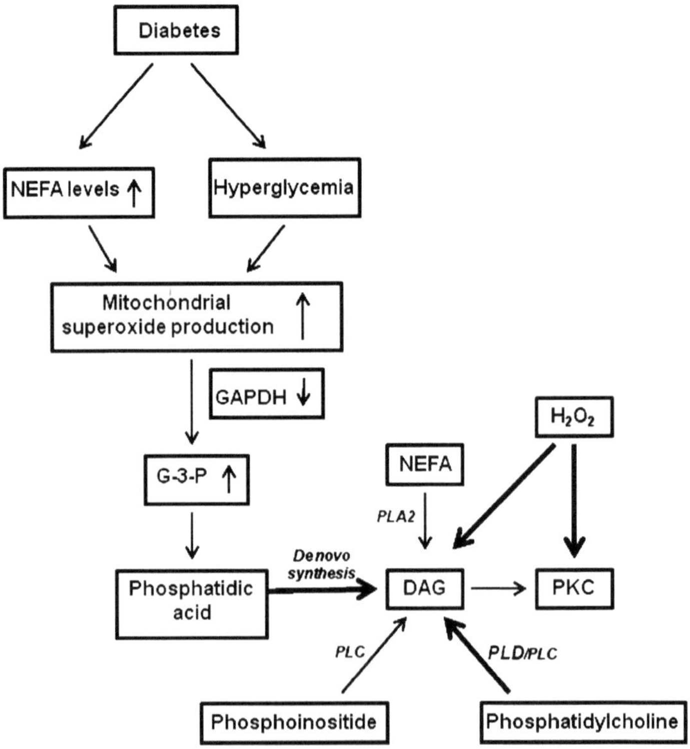

Fig. 2 Mechanisms of diacylglycerol (DAG)-PKC activation: the pathways activated in the diabetic state are shown as *bold arrows*. *G-3-P* glyceraldehyde-3-phosphate, *PLD* phospholipase D, *PLC* phospholipase C, *PLA2* phospholipase A2

peripheral nerves from diabetic rats, and in cultured vascular cells exposed to a high glucose level (22 mM) (see details in Geraldes and King [1]). The most prominent pathway responsible for enhanced DAG levels in cardiovascular tissues in diabetes is de novo synthesis from glyceraldehyde-3-phosphate as the result of inhibition of glyceraldehyde-3-phosphate dehydrogenase (GAPDH) activity caused by hyperglycemia and fatty acid oxidation-induced mitochondrial superoxide production. The inhibition of GAPDH activity, which is the main upstream event, is also responsible for activation of the other mechanisms including the polyol pathway, hexosamine pathway, and intracellular AGE formation [8]. Another path leading to hyperglycemia-induced DAG formation is hydrolysis of phosphatidylcholine by phospholipase D. PKCs can also be activated by oxidants such as H_2O_2, which can activate PKC in a manner related or unrelated to DAG.

The effects of PKC activation in diabetes can be classified as three mechanisms. (i) The increased permeability of albumin and other macromolecules through endothelial cells [16, 17] and neovascularization and endothelial permeability by vascular endothelial growth factor (VEGF)/vascular permeability factor [18]. These mechanisms can contribute to the effects of hyperglycemia on development of atherosclerosis and retinopathy [1, 19]. (ii) Vascular dysfunction by disruption of the balance between endogen vasoconstrictors and vasodilators [decreased bioavailability of nitric oxide (NO), decreased production of prostacyclin, increased production of endothelin 1 (ET-1) and thromboxan] [20–22] and by activating ROS production mediated by vascular NADPH oxidase [23]. The development of insulin resistance, retinopathy, and atherosclerosis can be a result of this mechanism [1]. (iii) The change and expansion of extracellular matrix and basement membrane thickening [13]. The accumulation of collagen, fibronectin, and laminin by increased transforming growth factor-beta 1 (TGF-β1) and connective tissue growth factor (CTGF) expression is suggested as a main signaling pathway of extracellular matrix change and membrane thickening [1, 13]. This mechanism can be contributed to the development of retinopathy, nephropathy, and cardiomyopathy in diabetes [1] (Fig. 3).

PKC isoforms can be detected in different tissues and cultured cells under normal conditions. Although the predominant isoforms are PKC-α and PKC-β2 in rat aorta, heart, glomeruli, and retina under physiological conditions, the major activated isoform(s) following exposure to elevated glucose or the diabetic state is PKC-β2 in aorta and heart [13], PKC-α, PKC-β, and PKC-ε in the retina [24, 25], and PKC-α, PKC-β, PKC-δ, PKC-ε, and PKC-ξ in the glomeruli [26, 27].

In addition, PKC-θ and PKC-ε lead to inhibition of insulin action by elevated DAG levels in target organs, which results in insulin resistance [28, 29].

On the other hand, the aPKC isoforms PKCξ and PKCλ contribute to insulin-stimulated glucose uptake and GLUT4 translocation in adipocytes and muscle in physiological conditions [30]. In contrast to other PKC isoforms that are increased in diabetes, decreased activation of aPKCs has been reported in skeletal muscle of type 2 diabetic humans and rodents [31].

In the following sections, specific PKC isoforms related to diabetes-induced cardiovascular complications are discussed in detail.

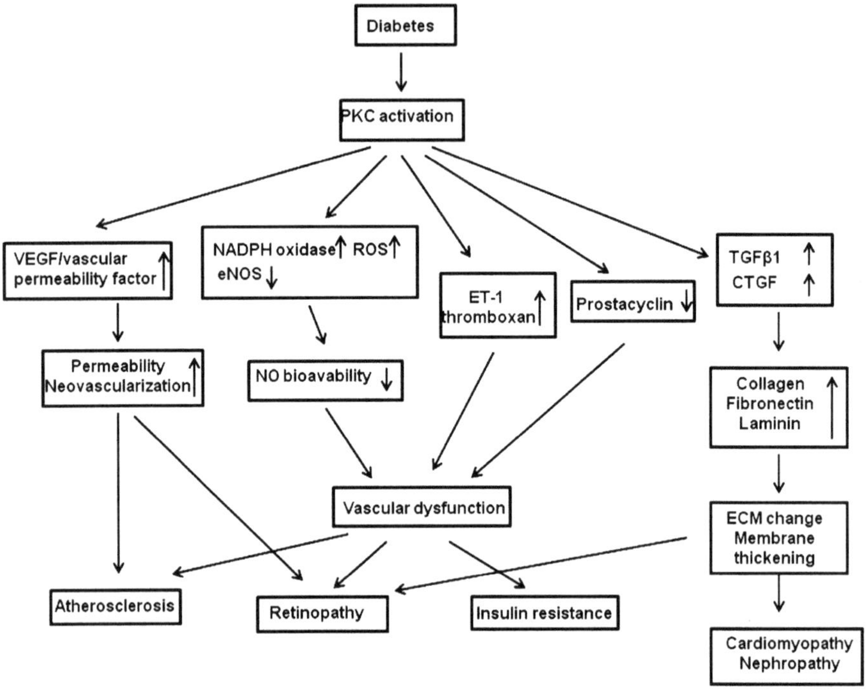

Fig. 3 Schematic representation of diabetic complications induced by PKC activation

4 PKC-β

There is evidence that chronic hyperglycemia-induced PKC activation can contribute to cardiovascular mortality in the presence of diabetic cardiomyopathy [32]. The changes in activated PKC isoforms in the diabetic state with a tissue-dependent manner indicate the possible role of PKC-β2 compared to other isoforms in diabetic cardiomyopathy leading to heart failure. Similarly, PKC-β expression and activity are also shown to be increased in animal models of heart failure [33, 34] and human heart failure [35]. However, studies in transgenic animal models of PKC-β emphasized that PKC-β activity depends on the expression ratio, specificity of expression, and stage of development. Although in PKC-β overexpressed neonatal cardiomyocytes, PKC-β activation caused to sudden death associated with increased L-type calcium channel activity, and calcium cycling abnormalities [33, 36], progressive ventricular hypertrophy and diastolic dysfunction were induced without any histological impairment in adult cardiomyocytes that have conditional PKC-β overexpression [33]. In addition, contrary to increased contractility in cardiomyocytes isolated from conditionally expressing PKC-β hearts [37], in transgenic mice with cardiac-specific overexpression of PKC-β, cellular and functional changes leading to cardiomyopathy were detected [38]. In a recent study, it has been shown that modulation of contractile function by PKC-β2 upregulation is associated with activation of phosphatase(s) and signaling pathways, which modulate the phosphorylation of a number of Ca^{2+} handling and myofilament targets [39].

The observation that in the myocardium of transgenic PKC-β2 mice increased expression of CTGF, which may contribute to the cardiac fibrosis, suggested the role of CTGF in the development of cardiomyopathy [40]. In addition to CTGF, TNF-β1 may play a role in extracellular matrix accumulation in diabetes [41]. It is well known that PKC activation can induce the synthesis of type IV collagen and fibronectin [42, 43] and transcription factors [42, 44–46]. In PKC-β-knockout mice, increased diabetes-induced CTGF, TGF-β1, and extracellular protein expression were significantly reduced [47]. In early stages of diabetes, enhanced myocardial NADPH oxidase activation is found to be parallel with increased myocardial PKC-β2 activation and myocardial dysfunction [48]. In addition, treatment with ruboxistaurin (RBX) improved the cardiac performance compared to N-acetylcysteine, which is an antioxidant, indicating that the action mechanism of RBX is independent of antioxidant activity. Furthermore, RBX treatment inhibited PKC-β2 activity by reducing translocation of the enzyme to membrane and phosphorylation at Ser660 [48]. In diabetic mice that are complicated with a model of myocardial infarction, pharmacological inhibition of PKC-β impaired the angiogenic response induced by hyperglycemia [49].

Previous studies associate high-glucose concentrations or diabetes with increased activity of PKC-β [1] and show its link to increased intracellular DAG concentrations in several tissues and cell culture systems [50]. In cardiomyocytes, free fatty acid-mediated inhibition of basal and insulin-stimulated glucose oxidation and the detrimental effects of free fatty acids were prevented with RBX treatment [1]. Furthermore, RBX and captopril, an ACE inhibitor, improved diastolic function and increased glucose utilization [51] and attenuated myocyte hypertrophy and collagen deposition [52] in diabetic rat hearts. Together, these results suggest that both ACE and PKC-β inhibitors regulate metabolic gene expression directly in the myocardium and consequently improve cardiac function and metabolism in diabetes. In addition, there is evidence that chronic hyperglycemia-induced PKC activation can contribute to cardiovascular mortality in the presence of diabetic cardiomyopathy [32].

In diabetic patients, besides cardiomyopathy, diabetic macrovasculopathy can contribute to an increased risk of cardiac mortality. In membranous fraction of aortic smooth muscle cells, hyperglycemia activated the PKC-β2 isoform but not PKC-α [53]. Compared with other isoforms, PKC-β1/2 exhibited significant increases in the membrane fraction in all vascular tissues [26].

Although PKC isoforms mediate certain insulin responses, chronic PKC activation in diabetes leads to a loss of insulin responsiveness caused by inhibitory Ser/Thr phosphorylations on the insulin receptor itself or its downstream signaling partners [54, 55]. The loss of insulin action on vascular endothelium can lead to endothelial dysfunction, which is an indicator of atherosclerosis [56]. In different models of hyperglycemic state and diabetes, endothelial insulin resistance impairs endothelial nitric oxide synthase (eNOS) activation, resulting in reduction of endothelium-dependent NO-mediated vasodilation, and promotes atherogenesis [57–59]. Insulin-induced eNOS stimulation was disturbed by the activation of PKC-β [60], and inhibition of PKC-β activation by RBX restored the vascular function in diabetic rats [61]. Additionally, PKC-β inhibition with RBX treatment significantly inhibited the development of endothelial dysfunction induced by acute hyperglycemia and improved flow-mediated dilation in diabetic patients [62, 63]. Recently, consistent with these studies, higher PKC-β levels, which were observed in endothelial cells from diabetic patients, were associated with reduced endothelial function and inhibition of PKC-β activity ameliorated the insulin response in endothelial cells and reduced basal eNOS phosphorylation in diabetes [64].

Different from type 1 diabetes, in type 2 diabetes, independent of hyperglycemia, abnormalities including increased cytokine expression and circulating levels of free fatty acids can cause vascular insulin resistance. Dysfunctional endothelium displays activation of vascular NADPH oxidase, uncoupling of eNOS, increased expression of ET-1, and induction of adhesion molecules [47]. By using transgenic ($PKC\beta^{-/-}/ApoE^{-/-}$ double-knockout mice) and pharmacological (treatment with RBX) approaches, the link between PKC-β and atherosclerosis was recently demonstrated [65] as PKC-β depletion or inhibition leads to decreased atherosclerosis.

Diabetes-mediated increases in renal hypertrophy, glomerular hyperfiltration, extracellular matrix production, and ROS were prevented by the deletion of the PKC-β gene or treatment with RBX [28, 47, 61, 66].

Treatment with RBX (0.1–10 mg/kg) improved retinal hemodynamic abnormalities and retinal blood flow [52] and ET-1 mRNA expression [67] in streptozotocin-diabetic rats. The importance of the PKC-β isoform in the development of diabetic retinopathy has been shown by several studies that are discussed in the last part of this chapter.

5 PKC-α

PKC-α activation occurs through ligand activation of G-protein-coupled receptors, including α-adrenoceptors [68], angiotensin II receptors [69], and ET-1 receptors [70]. This activation results in hydrolysis of a membrane phospholipid, phosphatidylinositol-4,5-bisphosphate (PIP2), by phospholipase C. The PIP2-derived second messengers are DAG and inositol-1,4,5-triphosphate (IP3). IP3 contributes to PKC-α activation through opening of IP3-gated Ca^{2+} transport channels in the endoplasmic reticulum membrane, thereby increasing $[Ca^{2+}]i$.

Increased expression and/or activity of PKC-α has been shown in cardiac hypertrophy and heart failure models [34, 71–73]. In addition, the effects of PKC-α activation on cardiomyocytes are revealed with the genetic animal models of PKC-α. In PKC-α knockout mice, contractile performance was found to be increased, and in transgenic mice overexpressing PKC-α, contractility was found to be reduced [74]. The mechanism associated with PKC-α-dependent changes on contractility has been attributed to reduced sarcoplasmic reticulum Ca^{2+} loading and

Ca^{2+} transients as a result of enhancement of protein phosphatase I activity [74]. Additional mechanisms including different sarcomeric proteins [75, 76] and phosphorylation of G protein-coupled receptor kinase 2 [77] and L-type Ca^{2+} channel subunits [78] may contribute to contractility changes in response to PKC-α activation. However, as already mentioned, in vasculature and heart, one of the predominant PKC isoform is PKC-α, the state of activated PKC isoform change into PKC-β in diabetes. Thus, the detrimental effect of PKC-α is limited only to glomeruli in which the PKC-α isoform is activated in hyperglycemic conditions.

Pathophysiologically, PKC activation can play a pivotal role in the development of hyperglycemia-induced inflammation and endothelial dysfunction. Activation with DAG/phorbol ester and Ca^{2+} in a human hepatic cell line increased low density lipoprotein (LDL) mRNA expression [79], LDL binding, and PKC-α translocation [80]. Activation of PKC-α seems to be required for LDL receptor upregulation and the resulting decrease in serum LDL [81]. It has been also shown that uncoupling of eNOS (increased O_2^- production by eNOS) observed in oxLDL-treated human endothelial cells can be associated with inactivation of PKC-α as a result of dephosphorylation of Thr495 of the enzyme [82]. However, PKC-α activation is observed in monocyte-mediated LDL oxidation [83, 84] and increased monocyte adherence and vascular endothelial permeability [18]. Accumulation of oxLDL results in the formation of foam cells, generation of atherosclerotic lesions, and ultimately dysfunction of endothelium [81]. In diabetic nephropathy, inhibition of NADPH oxidase prevents AGEs-mediated damage in the PKC-α-dependent pathway [85]. Additionally, PKC-α-deficient diabetic mice were found to be resistant to the development of albuminuria and glomerular hyperfiltration stimulated by VEGF [86].

6 PKC-δ

PKC-δ, a member of the novel PKCs, has also been involved in vascular inflammation [87]. By using three different in vivo models (liver-specific overexpression, whole-body knockout, and liver-specific reduction in PKC-δ), PKC-δ has been explored as a potential key genetic modifier of hepatic insulin sensitivity and hepatic lipid accumulation [88]. Also, the correlation between increased levels of PKC-δ expression in livers of obese and type 2 diabetic patients and hyperglycemia and hypertriglyceridemia indicates the role of this enzyme in the development of metabolic syndrome in humans [88]. Also, PKC-δ inhibition induced CTGF expression and potentially cardiac fibrosis in the presence of angiotensin II [89]. Interestingly, in the myocardial ischemia–reperfusion model it has been shown that PKC-δ activation coordinates proapoptotic pathways [90].

It has been recently shown that in retinal lysates of diabetic mice there is a correlation between PKC-δ levels and pericyte apoptosis in vitro and formation of acellular capillaries [91]. Furthermore, retinal PKC-δ activation of diabetic mice resulted in platelet-derived growth factor (PDGF) resistance. Pharmacological inhibition of PKC-δ in pericytes and genetic deletion of *Prkcd* inhibit nuclear factor kappa B (NF-κB) activation and improve PDGF signaling. Based on these findings, increased reactive oxygen species (ROS) production and NF-κB activity and a decrease in the survival signaling pathway of PDGF are mechanisms that can be suggested for the hyperglycemia-mediated effects of PKCδ [91].

7 DAG-PKC Pathway as Therapeutic Target and PKC Inhibitors

There is clear evidence to support the hypothesis that PKC activation plays a critical role in development of diabetic cardiovascular complications. Several members of PKC isoforms, especially PKC-β, seem to be involved in this process. Therefore, PKC isoforms can considered an excellent therapeutic target. On this pathway, two mechanisms including preventing the rise of enzyme substrate (inhibition of DAG accumulation) and blocking the enzyme activity (inhibition of PKC activity) can be suggested as possible mechanisms for a therapeutic effect. The only candidate for the inhibition of DAG accumulation is α-tocopherol (vitamin E), which acts by stimulation of DAG kinase [92]. Additionally, it has been shown that α-tocopherol can modulate PKC-α activity by preventing the transient activation induced by oxidation [93, 94]. It has been reported that α-tocopherol pretreatment of endothelial cells inhibits vascular cell adhesion molecule 1 (VCAM-1)-induced oxidative activation of PKCα [95]. However, during VCAM-1 activation of PKCα, in addition to oxidative activation of PKCα, there is also generation of calcium [96] and consumption of the PKCα cofactor DAG [94], suggesting a contribution of both oxidative activation of PKC-α and cofactor-dependent activation of PKC-α during VCAM-1 signaling in endothelial cells. Therefore, α-tocopherol may inhibit VCAM-1 signaling by functioning both as an antioxidant and as an antagonist of PKC-α [97].

The rationale of treatment with PKC inhibitors is to provide more effective inhibition on the DAG-PKC pathway than treatment with α-tocopherol. However, the disadvantage of PKC inhibitors is lack of selectivity because of common pathways of signaling enzymes as action mechanism. As can be seen in general PKC isoform inhibitors such as PKC-412, in vivo toxic and severe side effects of nonselective PKC inhibitors can be a result of interacting with other ATP-binding kinases [1, 13]. As mentioned in the section "PKC Classification and Regulation," the enzyme has a regulatory domain that is a binding site for phospholipids and phorbol ester and a catalytic domain that is a binding site for substrate or ATP. Thus, suitable specific PKC isoform inhibitors should target one of these domains [1]. The specific PKC-β inhibitor, RBX, which belongs to the bisindolylmaleimide class, is designed as a target for the catalytic domain of the enzyme [61, 66]. The IC_{50} of RBX to inhibit PKC-β1 and PKC-β2 are 4.7 and 5.9 nM, respectively, and for other PKC isoforms the inhibitory constants are higher than 300 nM [98]; this compound is the most studied PKC inhibitor in cellular, animal, and especially human studies. The effectiveness of RBX has been evaluated in diabetic patients who have retinopathy, and in a lesser extent renal and endothelium dysfunction. Phase I studies were performed to determine the effective dose of RBX for preventing the decrease in retinal blood flow that is an early indicator of diabetic retinopathy, and results of these studies showed that the effective RBX dose for oral use is 32 mg/day [99]. Phase II and phase III studies of RBX were performed in late stages of diabetic retinopathy. Although the results of the PKC-Diabetic Retinopathy Study (PKC-DRS) and PKC-Diabetic Macular Edema Study (PKC-DMES) were not as satisfying as expected [100, 101], it has been shown that the 32 mg/day RBX treatment for 1 month provides an improvement in the strength of vision that was destroyed because of diabetic retinopathy in the PKC-Diabetic Retinopathy Study 2 (PKC-DRS2) [102, 103]. The clinical outcomes of phase II study in type II diabetic patients with high albuminuria treated with ACE inhibitors or angiotensin receptor blockers (ARBs) in addition to 32 mg/day RBX were a decrease in albumin-to-creatinine ratio, which is a marker for albuminuria and maintenance of glomerular filtration rate [104]. The positive effects of RBX (32 mg/day) on blood flow have been determined in hyperglycemia-induced endothelium dysfunction [62] and in flow-mediated dilation in type II diabetic patients [63]. The effect of RBX treatment (32 and 64 mg) for 1 year was evaluated in patients with diabetic peripheral neuropathy [105]. In this study, the results had been scored by the neuropathy total symptoms score-6 (NTSS-6) and vibration detection threshold [105]. The improvement has only been observed in patients treated with 64 mg/day RBX with an NTSS-6 score above 6 [105]. It is obvious that further clinical studies are required to determine whether RBX can effectively improve retinal, renal, and endothelial dysfunction in type 1 and 2 diabetic patients.

8 Conclusions

PKC activation is one of the mechanisms leading to diabetic complications that affect the retina, kidneys, vasculature, and heart. Some mechanisms can be responsible for DAG-PKC activation in diabetes (shown in Fig. 2). In light of accumulated evidence, the common activated PKC isoforms in hyperglycemia and diabetes are PKC-β, PKC-α, and PKC-δ in heart, aorta, glomeruli, and retina. Compared to the nonselective PKC inhibitors that have in vivo toxic and severe side effects, the selective PKC-β inhibitor RBX can be a promising drug for the treatment of diabetic complications.

References

1. Geraldes P, King GL (2010) Activation of protein kinase C isoforms and its impact on diabetic complications. Circ Res 106:1319–1331
2. The Diabetes Control and Complications Trial Research Group (1993) The effect of intensive treatment of diabetes on the development and progression of long-term complications in insulin-dependent diabetes mellitus. N Engl J Med 329:977–986

3. UK Prospective Diabetes Study (UKPDS) Group (1998) Intensive blood-glucose control with sulphonylureas or insulin compared with conventional treatment and risk of complications in patients with type 2 diabetes (UKPDS 33). Lancet 352:837–853
4. Obrosova IG, Minchenko AG, Vasupuram R et al (2003) Aldose reductase inhibitor fidarestat prevents retinal oxidative stress and vascular endothelial growth factor overexpression in streptozotocin-diabetic rats. Diabetes 52:864–871
5. Wendt T, Harja E, Bucciarelli L et al (2006) Rage modulates vascular inflammation and atherosclerosis in a murine model of type 2 diabetes. Atherosclerosis 185:70–77
6. Koya D, King GL (1998) Protein kinase C activation and the development of diabetic complications. Diabetes 47:859–866
7. Brownlee M (2001) Biochemistry and molecular cell biology of diabetic complications. Nature 414:813–820
8. Giacco F, Brownlee M (2010) Oxidative stress and diabetic complications. Circ Res 107:1058–1070
9. Inoue M, Kishimoto A, Takai Y et al (1977) Studies on a cyclic nucleotide-independent protein kinase and its proenzyme in mammalian tissues. II. Proenzyme and its activation by calcium-dependent protease from rat brain. J Biol Chem 252:7610–7616
10. Freeley M, Kelleher D, Long A (2011) Regulation of protein kinase C function by phosphorylation on conserved and non-conserved sites. Cell Signal 23:753–762
11. Steinberg SF (2012) Cardiac actions of protein kinase C isoforms. Physiology (Bethesda) 27:130–139
12. Steinberg SF (2008) Structural basis of protein kinase C isoform function. Physiol Rev 88:1341–1378
13. Das Evcimen N, King GL (2007) The role of protein kinase C activation and the vascular complications of diabetes. Pharmacol Res 55:498–510
14. Newton AC (1995) Protein kinase C: structure, function and regulation. Minireview. J Biol Chem 270:28495–28498
15. Pears C, Stabel S, Cazaubon S et al (1992) Studies on the phosphorylation of protein kinase C-α. Biochem J 283:515–518
16. Lynch JJ, Ferro TJ, Blumenstock FA et al (1990) Increased endothelial albumin permeability mediated by protein kinase C activation. J Clin Invest 85:1991–1998
17. Wolf BA, Williamson JR, Easom RA et al (1991) Diacylglycerol accumulation and microvascular abnormalities induced by elevated glucose levels. J Clin Invest 87:31–38
18. Xia P, Aiello LP, Ishii H et al (1996) Characterization of vascular endothelial growth factor's effect on the activation of protein kinase C, its isoforms, and endothelial cell growth. J Clin Invest 98:2018–2026
19. Harhaj NS, Felinski EA, Wolpert EB et al (2006) VEGF activation of protein kinase C stimulates occludin phosphorylation and contributes to endothelial permeability. Invest Ophthalmol Vis Sci 47:5106–5115
20. Hink U, Li H, Mollnau H et al (2001) Mechanisms underlying endothelial dysfunction in diabetes mellitus. Circ Res 88:e14–e22
21. Cardillo C, Campia U, Bryant MB et al (2002) Increased activity of endogenous endothelin in patients with type 2 diabetes mellitus. Circulation 106:1783–1787
22. Cosentino F, Eto M, De Paolis P et al (2003) High glucose causes upregulation of cyclooxygenase-2 and alters prostanoid profile in human endothelial cells: role of protein kinase C and reactive oxygen species. Circulation 107:1017–1023
23. Inoguchi T, Li P, Umeda F et al (2000) High glucose level and free fatty acid stimulate reactive oxygen species production through protein kinase C-dependent activation of NAD(P)H oxidase in cultured vascular cells. Diabetes 49:1939–1945
24. Inoguchi T, Battan R, Handler E et al (1992) Preferential elevation of protein kinase C isoform beta II and diacylglycerol levels in the aorta and heart of diabetic rats: differential reversibility to glycemic control by islet cell transplantation. Proc Natl Acad Sci USA 89:11059–11063

25. Shiba T, Inoguchi T, Sportsman JR et al (1993) Correlation of diacylglycerol level and protein kinase C activity in rat retina to retinal circulation. Am J Physiol 265:E783–E793
26. Babazono T, Kapor-Drezgic J, Dlugosz JA et al (1998) Altered expression and subcellular localization of diacylglycerol-sensitive protein kinase C isoforms in diabetic rat glomerular cells. Diabetes 47:668–676
27. Koya D, Jirousek MR, Lin YW et al (1997) Characterization of protein kinase C beta isoform activation on the gene expression of transforming growth factor-beta, extracellular matrix components, and prostanoids in the glomeruli of diabetic rats. J Clin Invest 100:115–126
28. Samuel VT, Petersen KF, Shulman GI (2010) Lipid-induced insulin resistance: unravelling the mechanism. Lancet 375:2267–2277
29. Idris I, Gray S, Donnelly R (2001) Protein kinase C activation: isozyme-specific effects on metabolism and cardiovascular complications in diabetes. Diabetologia 44:659–673
30. Farese RV (2002) Function and dysfunction of aPKC isoforms for glucose transport in insulin-sensitive and insulin resistant states. Am J Physiol Endocrinol Metab 283:E1–E11
31. Farese RV, Sajan MP, Standaert ML (2005) Atypical protein kinase C in insulin action and insulin resistance. Biochem Soc Trans 33:350–353
32. Davidoff AJ, Davidson MB, Carmody MW et al (2004) Diabetic cardiomyocyte dysfunction and myocyte insulin resistance: role of glucose-induced PKC activity. Mol Cell Biochem 262:155–163
33. Bowman JC, Steinberg SF, Jiang TR et al (1997) Expression of protein kinase Cβ in the heart causes hypertrophy in adult mice and sudden death in neonates. J Clin Invest 100:2189–2195
34. Gu X, Bishop SP (1994) Increased protein-kinase-C and isozyme redistribution in pressure-overload cardiac-hypertrophy in the rat. Circ Res 75:926–931
35. Bowling N, Walsh RA, Song GJ et al (1999) Increased protein kinase C activity and expression of Ca^{2+}-sensitive isoforms in the failing human heart. Circulation 99:384–391
36. Alden KJ, Goldspink PH, Ruch SW et al (2002) Enhancement of L-type Ca^{2+} current from neonatal mouse ventricular myocytes by constitutively active PKC-β II. Am J Physiol Cell Physiol 282:C768–C774
37. Huang L, Wolska BM, Montgomery DE et al (2001) Increased contractility and altered Ca^{2+} transients of mouse heart myocytes conditionally expressing PKCβ. Am J Physiol Cell Physiol 280:C1114–C1120
38. Wakasaki H, Koya D, Schoen FJ et al (1997) Targeted overexpression of protein kinase CβII isoform in myocardium causes cardiomyopathy. Proc Natl Acad Sci USA 94:9320–9325
39. Hwang H, Robinson DA, Stevenson TK et al (2012) PKCβII modulation of myocyte contractile performance. J Mol Cell Cardiol 53:176–186
40. Way KJ, Isshiki K, Suzuma K et al (2002) Expression of connective tissue growth factor is increased in injured myocardium associated with protein kinase C beta2 activation and diabetes. Diabetes 51:2709–2718
41. Sharma K, Ziyadeh FN (1995) Hyperglycemia and diabetic kidney disease. The case for transforming growth factor-beta as a key modulator. Diabetes 44:1139–1146
42. Chiu R, Boyle WJ, Meek J et al (1988) The c-Fos protein interacts with c-Jun/AP-1 to stimulate transcription of AP-1 responsive genes. Cell 54:541–552
43. Edwards AS, Faux MC, Scott JD et al (1999) Carboxyl-terminal phosphorylation regulates the function and subcellular localization of protein kinase C betaII. J Biol Chem 274:6461–6468
44. Messina JL, Standaert ML, Ishizuka T et al (1992) Role of protein kinase C in insulin's regulation of c-fos transcription. J Biol Chem 267:9223–9228
45. Franza BR Jr, Rauscher FJ 3rd, Josephs SF et al (1988) The Fos complex and Fos-related antigens recognize sequence elements that contain AP-1 binding sites. Science 239:1150–1153
46. Zhang J, Wang L, Petrin J et al (1993) Characterization of site-specific mutants altered at protein kinase C beta 1 isosyme autophosphorylation sites. Proc Natl Acad Sci USA 90:6130–6134

47. Ohshiro Y, Ma RC, Yasuda Y et al (2006) Reduction of diabetes-induced oxidative stress, fibrotic cytokine expression, and renal dysfunction in protein kinase Cbeta-null mice. Diabetes 55:3112–3120
48. Liu Y, Lei S, Gao X et al (2012) PKCβ inhibition with ruboxistaurin reduces oxidative stress and attenuates left ventricular hypertrophy and dysfunction in rats with streptozotocin-induced diabetes. Clin Sci (Lond) 122:161–173
49. Ikeda A, Matsushita S, Sakakibara Y (2012) Inhibition of protein kinase C β ameliorates impaired angiogenesis in type I diabetic mice complicating myocardial infarction. Circ J 76:943–949
50. Rask-Madsen C, King GL (2005) Proatherosclerotic mechanisms involving protein kinase C in diabetes and insulin resistance. Arterioscler Thromb Vasc Biol 25:487–496
51. Arikawa E, Ma RC, Isshiki K et al (2007) Effects of insulin replacements, inhibitors of angiotensin, and PKCbeta's actions to normalize cardiac gene expression and fuel metabolism in diabetic rats. Diabetes 56:1410–1420
52. Connelly KA, Kelly DJ, Zhang Y et al (2009) Inhibition of protein kinase C in diabetic cardiomyopathy. Circ Heart Fail 2:129–137
53. Inoguchi T, Xia P, Kunisaki M et al (1994) Insulin's effect on protein kinase C and diacylglycerol induced by diabetes and glucose in vascular tissues. Am J Physiol 267:E369–E379
54. Mima A, Ohshiro Y, Kitada M et al (2011) Glomerular-specific protein kinase C-beta-induced insulin receptor substrate-1 dysfunction and insulin resistance in rat models of diabetes and obesity. Kidney Int 79:883–896
55. Taniguchi CM, Emanuelli B, Kahn CR (2006) Critical nodes in signalling pathways: insights into insulin action. Nat Rev Mol Cell Biol 7:85–96
56. Rask-Madsen C, King GL (2007) Mechanisms of disease. Endothelial dysfunction in insulin resistance and diabetes. Nat Clin Pract Endocrinol Metab 3:46–56
57. Ding Y, Vaziri ND, Coulson R et al (2000) Effects of simulated hyperglycemia, insulin, and glucagon on endothelial nitric oxide synthase expression. Am J Physiol Endocrinol Metab 279:E11–E17
58. Kearney MT, Duncan ER, Kahn M et al (2008) Insulin resistance and endothelial cell dysfunction: studies in mammalian models. Exp Physiol 93:158–163
59. Rask-Madsen C, Li Q, Freund B et al (2010) Loss of insulin signaling in vascular endothelial cells accelerates atherosclerosis in apolipoprotein E null mice. Cell Metab 11:379–389
60. Naruse K, Rask-Madsen C, Takahara N et al (2006) Activation of vascular protein kinase C-beta inhibits Akt-dependent endothelial nitric oxide synthase function in obesity-associated insulin resistance. Diabetes 55:691–698
61. Ishii H, Jirousek MR, Koya D et al (1996) Amelioration of vascular dysfunctions in diabetic rats by an oral PKC beta inhibitor. Science 272:728–731
62. Beckman JA, Goldfine AB, Gordon MB et al (2002) Inhibition of protein kinase Cbeta prevents impaired endothelium-dependent vasodilation caused by hyperglycemia in humans. Circ Res 90:107–111
63. Mehta NN, Sheetz M, Price K et al (2009) Selective PKC beta inhibition with ruboxistaurin and endothelial function in type-2 diabetes mellitus. Cardiovasc Drugs Ther 23:17–24
64. Tabit CE, Shenouda SM, Holbrook M et al (2013) Protein kinase C-β contributes to impaired endothelial insulin signaling in humans with diabetes mellitus. Circulation 127:86–95
65. Harja E, Chang JS, Lu Y et al (2009) Mice deficient in PKCβ and apolipoprotein E display decreased atherosclerosis. FASEB J 23:1081–1091
66. Koya D, Haneda M, Nakagawa H et al (2000) Amelioration of accelerated diabetic mesangial expansion by treatment with a PKC beta inhibitor in diabetic db/db mice, a rodent model for type 2 diabetes. FASEB J 14:439–447
67. Yokota T, Ma RC, Park JY et al (2003) Role of protein kinase C on the expression of platelet-derived growth factor and endothelin-1 in the retina of diabetic rats and cultured retinal capillary pericytes. Diabetes 52:838–845
68. Rohde S, Sabri A, Kamasamudran R et al (2000) The alpha(1)-adrenoceptor subtype- and protein kinase C isoform-dependence of Norepinephrine's actions in cardiomyocytes. J Mol Cell Cardiol 32:1193–1209

69. Dixon BS, Sharma RV, Dickerson T et al (1994) Bradykinin and angiotensin II: activation of protein kinase C in arterial smooth muscle. Am J Physiol 266:C1406–C1420
70. Sugden PH (2003) An overview of endothelin signaling in the cardiac myocyte. J Mol Cell Cardiol 35:871–886
71. Jalili T, Takeishi Y, Song G et al (1999) PKC translocation without changes in Galphaq and PLC-beta protein abundance in cardiac hypertrophy and failure. Am J Physiol Heart Circ Physiol 277:H2298–H2304
72. Bayer AL, Heidkamp MC, Patel N et al (2003) Alterations in protein kinase C isoenzyme expression and autophosphorylation during progression of pressure overload-induced left ventricular hypertrophy. Mol Cell Biochem 242:145–152
73. Simonis G, Briem SK, Schoen SP et al (2007) Protein kinase C in the human heart: differential regulation of the isoforms in aortic stenosis or dilated cardiomyopathy. Mol Cell Biochem 305:103–111
74. Braz JC, Gregory K, Pathak A et al (2004) PKC-α regulates cardiac contractility and propensity toward heart failure. Nat Med 10:248–254
75. Kooij V, Boontje N, Zaremba R et al (2010) Protein kinase Cα and ξ phosphorylation of troponin and myosin binding protein C reduce Ca^{2+} sensitivity in human myocardium. Basic Res Cardiol 105:289–300
76. Sumandea MP, Pyle WG, Kobayashi T et al (2003) Identification of a functionally critical protein kinase C phosphorylation residue of cardiac troponin T. J Biol Chem 278:35135–35144
77. Malhotra R, D'Souza KM, Staron ML et al (2010) G q-mediated activation of GRK2 by mechanical stretch in cardiac myocytes: the role of protein kinase C. J Biol Chem 285:13748–13760
78. Yang L, Liu G, Zakharov SI et al (2005) S1928 is a common site for Cav1.2 phosphorylation by protein kinase C isoforms. J Biol Chem 280:207–214
79. Auwerx JH, Chait A, Deeb SS (1989) Regulation of the low density lipoprotein receptor and hydroxymethylglutaryl coenzyme A reductase genes by protein kinase C and a putative negative regulatory protein. Proc Natl Acad Sci USA 86:1133–1137
80. Kumar A, Chambers TC, Cloud-Heflin BA et al (1997) Phorbol ester-induced low density lipoprotein receptor gene expression in HepG2 cells involves protein kinase C-mediated p42/44 MAP kinase activation. J Lipid Res 38:2240–2248
81. Churchill E, Budas G, Vallentin A et al (2008) PKC isozymes in chronic cardiac disease: possible therapeutic targets? Annu Rev Pharmacol Toxicol 48:569–599
82. Fleming I, Mohamed A, Galle J et al (2005) Oxidized low-density lipoprotein increases superoxide production by endothelial nitric oxide synthase by inhibiting PKCα. Cardiovasc Res 65:897–906
83. Cathcart MK, McNally AK, Morel DW et al (1989) Superoxide anion participation in human monocyte-mediated oxidation of low-density lipoprotein and conversion of low-density lipoprotein to a cytotoxin. J Immunol 142:1963–1969
84. Cathcart MK, Morel DW, Chisolm GM 3rd (1985) Monocytes and neutrophils oxidize low density lipoprotein making it cytotoxic. J Leukoc Biol 38:341–350
85. Thallas-Bonke V, Thorpe SR, Coughlan MT et al (2008) Inhibition of NADPH oxidase prevents advanced glycation end product-mediated damage in diabetic nephropathy through a protein kinase C-alpha dependent pathway. Diabetes 57:460–469
86. Menne J, Park JK, Boehne M et al (2004) Diminished loss of proteoglycans and lack of albuminuria in protein kinase C-alpha-deficient diabetic mice. Diabetes 53:2101–2109
87. Wong SL, Lau CW, Wong WT et al (2011) Pivotal role of protein kinase Cdelta in angiotensin II-induced endothelial cyclooxygenase-2 expression: a link to vascular inflammation. Arterioscler Thromb Vasc Biol 31:1169–1176
88. Bezy O, Tran TT, Pihlajamaki J et al (2011) PKCdelta regulates hepatic insulin sensitivity and hepatosteatosis in mice and humans. J Clin Invest 121:2504–2517
89. He Z, Way KJ, Arikawa E et al (2005) Differential regulation of angiotensin II-induced expression of connective tissue growth factor by protein kinase C isoforms in the myocardium. J Biol Chem 280:15719–15726

90. Churchill EN, Mochly-Rosen D (2007) The roles of PKCdelta and epsilon isoenzymes in the regulation of myocardial ischaemia/reperfusion injury. Biochem Soc Trans 35:1040–1042
91. Geraldes P, Hiraoka-Yamamoto J, Matsumoto M et al (2009) Activation of PKC-delta and SHP-1 by hyperglycemia causes vascular cell apoptosis and diabetic retinopathy. Nat Med 15:1298–1306
92. Kunisaki M, Bursell SE, Umeda F et al (1994) Normalization of diacylglycerol-protein kinase C activation by vitamin E in aorta of diabetic rats and cultured rat smooth muscle cells exposed to elevated glucose levels. Diabetes 43:1372–1377
93. Venugopal SK, Devaraj S, Yang T et al (2002) α-Tocopherol decreases superoxide anion release in human monocytes under hyperglycemic conditions via inhibition of protein kinase C-α. Diabetes 51:3049–3054
94. Abdala-Valencia H, Cook-Mills JM (2006) VCAM-1 signals activate endothelial cell protein kinase Cα via oxidation. J Immunol 177:6379–6387
95. Berdnikovs S, Abdala-Valencia H, McCary C et al (2009) Isoforms of vitamin E have opposing immunoregulatory funcitons during inflammation by regulating leukocyte recruitment. J Immunol 182:4395–4405
96. Cook-Mills JM, Johnson JD, Deem TL et al (2004) Calcium mobilization and Rac1 activation are required for VCAM-1 (vascular cell adhesion molecule-1) stimulation of NADPH oxidase activity. Biochem J 378:539–547
97. McCary CA, Yoon Y, Panagabko C et al (2012) Vitamin E isoforms directly bind PKCα and differentially regulate activation of PKCα. Biochem J 441:189–198
98. Tran K, Proulx PR, Chan AC (1994) Vitamin E suppresses DAG level in thrombin-stimulated endothelial cells through an increase of DAG kinase activity. Biochem Biophys Acta 1212:193–202
99. Aiello LP, Clermont A, Arora V et al (2006) Inhibition of PKC beta by oral administration of ruboxistaurin is well tolerated and ameliorates diabetes-induced retinal hemodynamic abnormalities in patients. Invest Ophthalmol Visl Sci 47:86–92
100. PKC-DRS Study Group (2005) The effect of ruboxistaurin on visual loss in patients with moderately severe to very severe nonproliferative diabetic retinopathy: initial results of the Protein Kinase C Beta Inhibitor Diabetic Retinopathy Study (PKC-DRS) multicenter randomized clinical trial. Diabetes 54:2188–2197
101. PKC-DMES Study Group (2007) Effect of ruboxistaurin in patients with diabetic macular edema: thirty-month results of the randomized PKC-DMES clinical trial. Arch Ophthalmol 125:318–324
102. Aiello LP, Davis MD, Girach A et al (2006) Effect of ruboxistaurin on visual loss in patients with diabetic retinopathy. Ophthalmology 113:2221–2230
103. Aiello LP, Vignati L, Sheetz MJ et al (2011) Oral protein kinase C β inhibition using ruboxistaurin: efficacy, safety, and causes of vision loss among 813 patients (1,392 eyes) with diabetic retinopathy in the Protein Kinase C β Inhibitor-Diabetic Retinopathy Study and the Protein Kinase C β Inhibitor-Diabetic Retinopathy Study 2. Retina 31:2084–2094
104. Tuttle KR, Bakris GL, Toto RD et al (2005) The effect of ruboxistaurin on nephropathy in type 2 diabetes. Diabetes Care 28:2686–2690
105. Vinik AI, Bril V, Kempler P, Litchy WJ, Tesfaye S, Price KL, Bastyr EJ III (2005) Treatment of symptomatic diabetic peripheral neuropathy with the protein kinase C beta-inhibitor ruboxistaurin mesylate during a 1-year, randomized, placebo-controlled, double-blind clinical trial. Clin Ther 27:1164–1180

Calcium-Handling Proteins in Diabetic Cardiomyopathy

Vijayan Elimban, Adriana Frota P. Pinto, and Naranjan S. Dhalla

Abstract Diabetes mellitus increases the risk of cardiomyopathy independently of underlying comorbidities, and heart failure is a major cause of death in diabetic patients. The development of this distinct cardiomyopathy in both type 1 and type 2 diabetes is associated with complex and multifactorial cellular and molecular perturbations. It is widely recognized that cardiac dysfunction in chronic diabetes involves hormonal imbalance, oxidative stress, proteases activation, defects in Ca^{2+} cycling, and varying degrees of subcellular remodeling of organelles.

Ca^{2+} -handling abnormalities in diabetic cardiomyocytes have primarily been attributed to changes in the sarcolemmal Na^+–Ca^{2+} exchanger, L-type Ca^{2+} channel, Na^+–K^+ ATPase, and Na^+–H^+ exchanger proteins as well as Ca^{2+}-release channels and Ca^{2+}-pump proteins embedded in the sarcoplasmic reticulum. Intracellular Ca^{2+} overload has been implicated in the impairment of excitation–contraction coupling as a result of alterations in Ca^{2+}-entry, Ca^{2+}-removal, Ca^{2+}-uptake, and Ca^{2+}-release processes in the diabetic heart. These observations are consistent with the view that defects in Ca^{2+}-handling proteins play a critical role in the pathogenesis of cardiac dysfunction during the development of diabetic cardiomyopathy.

V. Elimban (✉)
Institute of Cardiovascular Sciences and Department of Physiology, University of Manitoba, St. Boniface Hospital Research, Winnipeg, MB R2H 2A6, Canada

Institute of Cardiovascular Sciences, St. Boniface Hospital Research, 351 Tache Avenue, Winnipeg, MB R2H 2A6, Canada
e-mail: velimban@sbrc.ca

A.F.P. Pinto • N.S. Dhalla
Institute of Cardiovascular Sciences and Department of Physiology, University of Manitoba, St. Boniface Hospital Research, Winnipeg, MB R2H 2A6, Canada

B. Turan and N.S. Dhalla (eds.), *Diabetic Cardiomyopathy: Biochemical and Molecular Mechanisms*, Advances in Biochemistry in Health and Disease 9,
DOI 10.1007/978-1-4614-9317-4_17,

Keywords Diabetic heart • Diabetic cardiomyopathy • Calcium cycling proteins • Sarcolemma remodeling • Sarcoplasmic reticulum remodeling • Cardiac dysfunction • Na^+–Ca^{2+} exchange • Na^+–K^+ ATPase • Ca^{2+}-pump ATPase • Ca^{2+}-release channels

1 Introduction

Cardiovascular disease is the leading cause of death in the diabetic population. Although diabetic cardiomyopathy is associated with several comorbidities including atherosclerosis, hypertension, coronary artery disease, and valvular malfunction, it has been demonstrated that chronic diabetes impairs ventricular function independently of other risk factors [1, 2]. This distinct diabetic cardiomyopathy is characterized by reduced diastolic compliance and rate of myocardial relaxation as well as a decrease in absolute force development [3, 4]. The exact underlying pathological mechanisms are not clear; however, several studies have suggested that cardiac dysfunction in chronic diabetes is intimately associated with varying degrees of defects in subcellular organelles such as sarcolemma (SL), sarcoplasmic reticulum (SR), mitochondria (MT), myofibrils (MF), and extracellular matrix (ECM) [3, 5, 6]. Remodeling of these components in the diabetic heart primarily occurs in response to hormonal imbalance, oxidative stress, activation of different proteases, changes in gene expression, and metabolic shift caused by increased levels of cholesterol and fatty acids. It is worthwhile to note that remodeling of SL and SR along with altered calcium metabolism has been shown to be an early sign in the process for the development of diabetic cardiomyopathy [7–9].

It is well known that intracellular Ca^{2+} is a major regulator of excitation–contraction coupling, and multiple aspects of calcium handling are considered to underlie the subcellular mechanisms responsible for the impaired cardiac contraction and relaxation in diabetic cardiomyopathy [6]. Indeed, several studies have reported the occurrence of intracellular Ca^{2+} overload in diabetic cardiomyocytes [3, 7, 10]. This alteration have been mostly attributed to the SL and SR remodeling, leading to depressed SL Na^+–Ca^{2+} exchanger activity, decreased SR Ca^{2+}-pump ATPase (SERCA2a) activity, reduced SR Ca^{2+} load, and Ca^{2+}-release channel (ryanodine receptor) dysfunction [11–13]. It is pointed out that the inward Ca^{2+} current is the critical initiator of the contractile and relaxation cycle in the heart. Cardiac depolarization opens L-type Ca^{2+} channels in the SL membrane and allows the entry of Ca^{2+} into cardiomyocytes. This transient increase in cytoplasmic Ca^{2+} concentration triggers Ca^{2+} release from SR, mainly through the Ca^{2+}-release channel or ryanodine receptor2 (RyR2) and by inositol triphosphate receptors (InsP3R) to a lesser extent. This event, described as calcium-induced calcium release (CICR), is crucial for excitation–contraction coupling in cardiac muscle [14, 15]. Following the opening of a RyR2 cluster on the SR, Ca^{2+} sparks are generated; this local, rapid, and brief elevation in $[Ca^{2+}]_i$ elevates cytosolic-free Ca^{2+} by tenfold or more and initiates contraction. The relaxation of cardiac muscle occurs upon lowering the concentration of free Ca^{2+} by intracellular SR uptake via SERCA2a as well as SL efflux via the Na^+–Ca^{2+} exchanger and the SL Ca^{2+} pump in the SL membrane [6, 14]. Although the MT and nucleus are also known to accumulate a significant amount of Ca^{2+} in cardiomyocytes, their role in the regulation of cytoplasmic concentration of free Ca^{2+} during the contraction and relaxation processes is not well established [6, 14]. This chapter is therefore focused on discussion regarding the status Ca^{2+}-handling proteins in SL and SR during the development of diabetic cardiomyopathy.

2 SL Defects in Diabetic Heart

Alterations in SL L-type Ca^{2+} channels, Na^+–Ca^{2+} exchanger, Na^+–K^+ ATPase, and Na^+–H^+ exchanger proteins, which are involved in Ca^{2+} handling directly or indirectly, have been shown to occur in diabetic cardiomyopathy [2, 6]. L-type Ca^{2+} channels are voltage-gated channels mostly located in the transverse tubules in proximity with RyR in SR, thereby suggesting the existence of a physical coupling between both Ca^{2+}-entry and Ca^{2+}-release channels [16]. SL Ca^{2+} channels in cardiomyocytes are modulated by several pathways including calmodulin (CaM), β-adrenergic receptors, phosphatidylinositol-3-kinase (PI3K), protein kinase A (PKA), and protein kinase C (PKC) [17, 18]. Although most of the calcium for cardiac contraction is provided by the SR, the activity of L-type Ca^{2+} channels is of critical importance for heart function. For instance, genetic mutation of SL Ca^{2+} channels leading to their impaired function has been linked with short QT syndrome, arrhythmia, and sudden death [19]. One of the early alterations detected in diabetic hearts was the prolongation of the ventricular action potential, which was attributed mainly to depressed transient outward K^+ current and to L-type Ca^{2+} current [3, 20]. Experimental investigations in diabetic animals have revealed an unaltered [21, 22] or decreased [23–26] SL Ca^{2+}-channel density. These disparities in results seems to reflect differences in models used, especially regarding the progression of disease, because alteration of the Ca^{2+} current has been shown to occur only in later stages of diabetes. The reduced Ca^{2+}-channel density has been attributed to decreased levels of protein content [23, 25], depressed cell-surface expression [24–26], and changes in the phosphorylation status [23]. Lu et al. [24], after a series of investigations using type 1 (*Ins2*^{Akita} rats) and type 2 (*db/db* rats) [23] diabetes models, have reported decreased Ca^{2+}-current density in both groups of diabetic animals, although the reduction was more intense in *db/db* than in *Ins2*^{Akita} myocytes as compared to nondiabetic cells. Because reduced phosphorylation status of the L-type Ca^{2+} channel was observed, it was hypothesized that Ca^{2+}-current alteration could be related to a lack of insulin in type 1 diabetes and downregulation of the Akt pathway.

Following intracellular infusion of phosphatidylinositol-3,4,5-trisphosphate (PIP3), a second messenger produced by PI3K, and consequently because of stimulation of the Akt pathway, depression in Ca^{2+}-current density was fully restored in *Ins2*^{Akita} myocytes in contrast with the partial restoration seen in *db/db* myocytes. The reduced levels of SL Ca^{2+}-channel protein in the *db/db* cardiomyocytes were not seen in *Ins2*^{Akita} cardiomyocytes, thereby leading to the hypothesis that hyperglycemia in combination with obesity and insulin resistance in type 2 diabetes could cause more damage to SL Ca^{2+}-channel function than hyperglycemia and lack of insulin in type 1 diabetes [23, 24]. It is important to highlight that another study has revealed that the activation of the PI3K-dependent Akt signaling pathway by insulin-like growth factor 1 restored L-type Ca^{2+} channels function in type 1 diabetic animals [27]. Taken together, these data suggest that insulin may have a positive inotropic effect and could explain how insulin resistance can affect heart function in several pathological states [3]. Despite the fact that most of the investigations support the idea that L-type Ca^{2+}-channel activity is not impaired in cardiac hypertrophy, Ca^{2+} transients trigged by Ca^{2+}-channel current have shown to be desynchronized, presenting a decreased amplitude and slow kinetics. These findings support the view that the intermolecular failure state would also apply to SL Ca^{2+} channels, the SR Ca^{2+}-release channels, considering that the Ca^{2+} current becomes less effective in triggering SR Ca^{2+} release in the diabetic heart [16, 28–30].

It has become clear that Na^+–Ca^{2+} exchanger 1 (NCX1) is the major SL protein for extruding Ca^{2+} that enters the cardiac cell via SL Ca^{2+} channels [31, 32]. This exchanger promotes the influx of 3 Na^+ for the extrusion of each Ca^{2+}, and its activity is controlled by both internal and external Na^+ and Ca^{2+} levels as well as by the membrane potential. Under certain pathological conditions, NCX1 also works in the reverse mode, contributing to the development of intracellular Ca^{2+} overload in cardiomyocytes. It is noteworthy that the direction and amplitude of NCX1 current relies on the activity of SL Na^+–K^+-ATPase, which is responsible for maintaining the intracellular Na^+ concentration at a low level [31, 32]. In type 1 diabetes, both depressed NCX activity [10, 33] and expression [10, 34, 35] were observed in the heart. It has been suggested that NCX1 dysfunction is related to alterations in the phospholipid composition of SL and reduced stimulation of the transporter by protein kinase C [36]. Furthermore, marked depression in SL Na^+–K^+ ATPase activity in insulin-dependent diabetes animals is considered to stimulate the NCX activity in a reverse mode to normalize the cytosolic Na^+ concentration [37–39]. Depressed SL activity of Ca^{2+}-pump ATPase was also reported in the diabetic heart [40, 41]. Consequently, a net gain of Ca^{2+} would occur as a result of the impaired efflux and increased Ca^{2+} entry, leading to intracellular Ca^{2+} overload as well as mechanical and electrical dysfunction in diabetic cardiomyocytes. On the other hand, in some studies involving type 2 diabetes, the NCX1 activity was either increased [25] or unchanged [42, 43], and no difference in mRNA level or protein content was detected [36, 42, 43]. Thus, the role of NCX in the etiology of cardiomyocyte dysfunction is complex, and changes in its expression or activity are viewed as compensatory or causal, depending upon the stage and severity of diabetes.

SL Na^+–K^+ ATPase plays a key role in maintenance of the resting membrane potential in cardiac cells by removing intracellular Na^+ in exchange for extracellular K^+. It has been demonstrated that Na^+–K^+ ATPase dysfunction in diabetic cardiomyopathy is related to downregulation of its subunit expression as well as alteration in the enzyme kinetics [37, 39]. The activity of this enzyme may also be influenced by alterations in composition of SL membrane observed in diabetes [44]. The abnormality in Na^+–K^+

ATPase activity in the diabetic heart results in cytosolic Ca^{2+} overload involving the NCX exchanger. It is important to emphasize that treatment of diabetic animals with insulin upregulates the expression of Na^+–K^+ ATPase and improves cardiac function [45]. Moreover, antioxidant agents, including vitamin E [46] and fish oil containing n-3 fatty acids [47], were able to attenuate and even prevent the diabetic-induced changes in SL Na^+–K^+ ATPase and cardiac dysfunction. These observations suggest the role of the observed depression in Na^+–K^+ in Ca^{2+}-handling abnormalities in cardiomyocytes during the development of diabetic cardiomyopathy.

Another integral SL protein, Na^+–H^+ exchanger (NHE), is involved in intracellular Ca^{2+} modulation. NHE-1, which isoform is mostly expressed in cardiac cells, regulates intracellular pH by exchanging one intracellular H^+ ion for an extracellular Na^+ ion. In addition, NHE-1 participates in the regulation of Na^+ fluxes and cell volume. Although emerging evidence supports NHE-1 involvement in diabetic cardiomyopathy, the results are controversial, and its potential role has not been established [48]. The NHE-1 activity has been shown to be decreased in isolated cardiomyocytes as well as the SL membranes of the diabetic heart [49, 50]. In another study, the reduced activity of NHE-1 in diabetes has been considered responsible for resistance of diabetic hearts to ischemia–reperfusion injury [51]. An increase in NHE-1 activity in cardiomyocytes of the Goto-Kakizaki rat model of type 2 diabetes has also been detected [52]. It has been suggested that intracellular acidification in cardiac cells stimulates the Akt signaling pathway, which could represent a likely mechanism that mediates the myocardial hypertrophy observed in the diabetic animals. In addition, chronic treatment with cariporide, a NHE-1-selective inhibitor, has been shown to prevent the phenotype of hypertrophy [52]. It is worth noting that some studies have also indicated that chronic administration of NHE-1-selective inhibitors may prevent vascular hypertrophy in diabetic rats [53] and also attenuate or even reverse the development of cardiac hypertrophy and its progression to heart failure in different animal models [54–56]. Thus, the observed alterations in SL Na^+–H^+ exchanger in diabetes can be seen to indirectly affect the Ca^{2+} handling by cardiomyocytes and participate in the development of diabetic cardiomyopathy.

3 SR Changes in Diabetic Heart

Several studies have revealed that different Ca^{2+}-handling proteins embedded in the SR membrane become abnormal during the development of diabetic cardiomyopathy [2, 6, 46]. SR channel or RyR is a key component in Ca^{2+} handling and excitation–contraction coupling in the heart. Cardiac cells express mostly the RyR2 isoform, which is regulated by proteins such as calmodulin (CaM), Ca^{2+}-CaM-dependent kinase (CaMKII), and PKA [57]. Following the opening of a RyR2 cluster on the SR, Ca^{2+} sparks are generated and result in local, rapid, and brief elevation in cytosolic-free Ca^{2+} by tenfold or more and trigger cardiac contraction. It has been demonstrated that RyR2 function in diabetic cardiomyocytes is compromised, becoming leaky to Ca^{2+} during diastole and accounting for a reduced SR Ca^{2+} load.

In addition, a leaky RyR would promote Ca^{2+} accumulation in the cytosol, resulting in increased SL NCX activity to remove the intracellular excess Ca^{2+} in exchange for Na^+. Consequently, the increased Na^+ influx would induce cell membrane depolarization, thereby leading to extrasystolic depolarizations and development of premature beats [58–60]. It has been suggested that these abnormalities may be linked to reduced levels of FKBP12.6 and increased activity of PKA [25, 61]. It should be mentioned that FKBP 12.6 is an accessory protein that plays a role in coordinating the opening and closing of individual RyRs in an array. The hyperphosphorylation of RyR2 by PKA leads to the dissociation of FKBP 12.6 and increasing the open probability of the RyR2 receptor [43]. This increased phosphorylation at Ser2809 and Ser2814 of RyR2 is also observed in stress/exercise-induced cardiac arrhythmias, sudden death, and catecholaminergic ventricular tachycardia [62, 63].

In a model of diabetic cardiomyopathy, Bidasee et al. [58] have reported that RyR2 proteins of 6-week streptozotocin (STZ)-induced diabetes rats bound less [^{3}H]ryanodine in comparison to control, although the affinity of this specific ligand and protein expression of the receptor remained unchanged in comparison to control. In a later study using 6- and 8-week STZ-induced diabetes rats [64], they also observed impaired binding ability of RyR2 to [^{3}H]ryanodine, which was even more pronounced in 8-week STZ-induced diabetes cardiomyocytes. In addition, 8-week STZ-induced diabetes rats showed a decrease in RyR2 expression (mRNA and protein). In both studies [58, 64], 2 weeks of insulin treatment initiated after 4 and 6 weeks of untreated diabetes was able to minimize the loss in function and expression of RyR2. Taken together, the findings indicate that the loss of functional integrity of the receptor precedes reduction in its expression and that the severity depends on the duration of untreated disease. The underlying mechanisms for RyR2 dysfunction remain unclear, but it has been shown that it could be caused by oxidative stress, nonenzymatic glycation reactions, and increased formation of disulfide bonds between adjacent sulfhydryl groups of the receptor [65–67].

The InsP3R plays a minor role in excitation–contraction coupling compared to the RyR in ventricular cardiomyocytes, but in atrial myocytes InsP3Rs are much more numerous and coexist with RyR on the SR, suggesting a prominent role in atrial contraction [68]. Several studies have shown that the InsP3R pathway is involved in progression of heart failure and delayed after depolarizations arrhythmias [69, 70]. In an experiment involving animals with obesity and type 2 diabetes, InsP3R expression was unaltered in ventricles from *ob/ob* mice [71], but in other diabetes studies it was shown to be decreased in diabetic rats [72] and in the atrium from diabetic patients [73]. The existing data indicate that altered InsP3R signaling may account for impaired Ca^{2+} handling and arrhythmogenesis in diabetic cardiomyopathy. However, the precise role of InsP3R in such pathological conditions requires further study.

Most of the intracellular Ca^{2+} is stored in SR via SERCA, which transfers two Ca^{2+} ions from the cytosol to the lumen at the expense of the hydrolysis of one ATP molecule. SERCA2a, the isoform predominately expressed by cardiomyocytes, is regulated by phosphorylation of a SR protein, phospholamban (PLB) [32, 74]. In its dephosphorylated form, PLB interacts with the pump, reducing its affinity for Ca^{2+}. However, when phosphorylated by PKC or CAMK, PLB is not able to inhibit

SERCA2a activity [75, 76]. SR function in diabetic cardiomyocytes has been shown to be compromised, presenting a reduced Ca^{2+} uptake that could explain the prolonged cardiac relaxation observed. As a consequence, SR calcium storage declines, resulting in reduced systolic calcium release and therefore a weaker cardiac contraction [74]. In this regard, some investigations with STZ-induced type 1 diabetes rats have reported decreased protein content and SERCA2a pumping dysfunction, which might be partly associated with an upregulation of activity and inhibitory PLB expression [35, 77]. In addition, it was proposed that products from advanced glycosylation reactions would form irreversible crosslinks within many proteins, leading to impairment of SERCA2a activity in diabetes [78, 79]. Thus far, it has been difficult to establish a general conclusion regarding myocardial SERCA2a and PLB changes in type 2 diabetes, most likely because of data limitation and ambiguity, especially taking in consideration the differences in animals models used.

Several studies using different type 2 diabetes animal models observed compromised SERCA2a function [25, 42, 43, 80]. SERCA2a expression was shown to be downregulated in Otsuka Long-Evans Tokushima fatty rats [80] and *db/db* mice [43] but unaltered in sucrose (SU)-fed rats [42]. Increased protein level of inhibitory PLB was only detected by one study [43]. Furthermore, Fredersdorf et al. [81] evaluated cardiac function and protein expression of Zucker diabetic fatty (ZDF) rats in the early stages of type 2 diabetes. They were able to demonstrate that animals in transition from insulin resistance to type 2 diabetes developed significant myocardial hypertrophy initially characterized by an increased systolic function and an intense SR Ca^{2+} uptake. In addition, myocardial expression of SERCA2a was markedly elevated and PLB expression was depressed. These changes were attributed to Akt signaling pathway activation induced by high levels of insulin, thereby supporting the view that upregulation of myocardial SERCA2a expression may be seen as a feedback mechanism in handling volume overload in the early phase of diabetes type 2. Taken together, the conflicting results regarding gene and protein expressions for SERCA2a and PLB can be explained by differences in the duration and severity of diabetes in various studies. Nonetheless, these observations are consistent with the view that alterations in SR function and SR remodeling occur in the diabetic heart [74]. Moreover, the critical role of SERCA2a in excitation–relaxation coupling is reinforced with the evidence that upregulation of its expression is able to reverse contractile dysfunction and abnormal calcium flux in established diabetic cardiomyopathy [82–84].

4 Mechanisms of SL and SR Alterations in the Diabetic Heart

It has been suggested that hyperglycemia along with metabolic shift, as a result of the hormonal imbalance caused by elevated plasma levels of catecholamines and angiotensin II, leads to oxidative stress and contributes to diabetic injury to multiple organs, especially the cardiac muscle [2, 4, 46]. The shift in myocardial metabolism, marked by decreased use of glucose and excessive utilization of long-chain fatty acids as an energy substrate, intensify the production of reactive oxygen species (ROS) that damage the respiratory and oxidative phosphorylation activities of mitochondria, contributing to decreased myocardial efficiency [14, 85]. In addition, there is experimental evidence to suggest that mitochondria under several pathological conditions can act as a Ca^{2+} sink [86, 87]. Although this mechanism initially seems to play an important compensatory role in Ca^{2+} regulation by preventing or delaying intracellular Ca^{2+} overload in cardiomyocytes, it also accounts for the development of oxidative stress at late stages of diabetes. The generation of ROS can lead to leakage of toxic proteins through opening of mitochondrial pores and further damage of cardiomyocytes [65, 88]. Another mechanism of oxidative stress is mediated by advanced glycation end products (AGE), which are able to activate signaling pathways that induce ROS production, and its accumulation is related to structural and functional alterations of proteins in chronic diabetic tissues. It is worthwhile to note that hyperglycemia can impair and decrease the antioxidant system capacity in the heart and other organs in diabetes [85, 89, 90]. Thus, both the intense generation of ROS and reduced antioxidant capacity contribute significantly to oxidative stress and therefore myocardial damage in chronic diabetes.

It is now well established that genomic alterations lead to myocardial dysfunction in diabetic cardiomyopathy. Numerous studies have also been relating the diabetic state with activation of proteases and changes in signal transduction pathways, including PKC, PKA, CaM kinase, and mitogen-activated protein kinase, contributing to subcellular remodeling [91]. With respect to Ca^{2+} cycling, downregulation of SERCA2a expression, as well as its promoter activity were reported. Some investigations also detected reduced protein levels of SL Ca^{2+} channels, NCX1, Na^+–K^+ ATPase, and RyR2 [74, 92]. These alterations have been attributed to an increased nuclear O-GlcN acylation, as a result of oxidative stress induced by hyperglycemia and enhanced activity of the PKC signaling pathway [93]. Moreover, genomic alterations also seem to underlie myosin dysfunction [94, 95]. In models of diabetic cardiomyopathy, abnormal myosin isozyme distribution, shift in myosin content from V1 to V3, and increased troponin I phosphorylation via the PKC pathway have been detected. Taken together, this could contribute to the decrease in Ca^{2+} sensitivity of myofilaments [96–101].

5 Conclusions

From the foregoing discussion it can be appreciated that diabetes is a complex pathology and that a wide variety of mechanisms contributes to cardiac dysfunction. The hormonal imbalance along with metabolic shift enhances oxidative stress, which leads to several abnormalities including activation of proteases, increased intracellular concentration of free Ca^{2+}, and alterations in cardiac gene expression (Fig. 1). Intracellular Ca^{2+} overload has been implicated not only in the process of excitation–contraction impairment but also in subcellular remodeling of organelles in cardiac cells. This event has been attributed to decreased SR Ca^{2+} load, depressed SERCA2a activity, and RyR2 dysfunction as well as changes in L-type Ca^{2+} channels. Abnormalities of SL proteins such as SL NCX, Na^+–K^+ ATPase, NHE-1, and

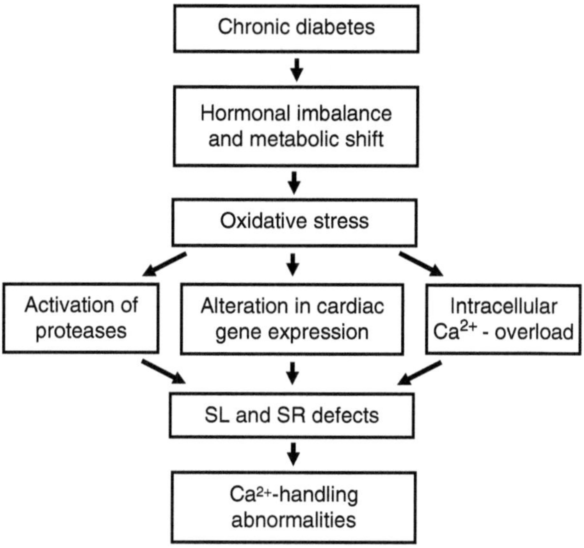

Fig. 1 Role of hormonal imbalance in sarcolemma (*SL*) and sarcoplasmic reticulum (*SR*) defects and subsequent Ca^{2+}-handling abnormalities in chronic diabetes

Ca^{2+}-pump ATPase have also been shown to be involved in diabetic cardiomyopathy. Molecular targeting approaches to revert or even attenuate alterations in proteins associated with Ca^{2+} handling hold promise as a new therapeutic modality. In addition, recent data have suggested that the insulin signaling pathway and Ca^{2+} regulatory processes are clearly interrelated, although many of these relationships are yet to be defined. Thus, further in-depth studies regarding the interactions between these pathways should lay the foundations for the design of new therapeutic approaches for diabetic heart disease.

Acknowledgments This work was supported by the St. Boniface Hospital Research Foundation, and A.F.P. Pinto was supported by the National Council for Scientific and Technological Development (CNPq), Brazil.

References

1. Go AS, Mozaffarian D, Roger VL et al (2013) Heart disease and stroke statistics—2013 update: a report from the American Heart Association. Circulation 127:e6–e245
2. Schaffer SW (1991) Cardiomyopathy associated with noninsulin-dependent diabetes. Mol Cell Biochem 107:1–20
3. Lebeche D, Davidoff AJ, Hajjar RJ (2008) Interplay between impaired calcium regulation and insulin signaling abnormalities in diabetic cardiomyopathy. Nat Clin Pract Cardiovasc Med 5:715–724
4. Hayat SA, Patel B, Khattar RS, Malik RA (2004) Diabetic cardiomyopathy: mechanisms, diagnosis and treatment. Clin Sci (Lond) 107:539–557
5. Boudina S, Abel ED (2007) Diabetic cardiomyopathy revisited. Circulation 115:3213–3223
6. Dhalla NS, Takeda N, Rodriguez-Leyva D, Elimban V (2013) Mechanisms of subcellular remodeling in heart failure due to diabetes. Heart Fail Rev. doi:10.1007/s10741-013-9385-8

7. Ligeti L, Szenczi O, Prestia CM et al (2006) Altered calcium handling is an early sign of streptozotocin-induced diabetic cardiomyopathy. Int J Mol Med 17:1035–1043
8. Fein FS, Sonnenblick EH (1994) Diabetic cardiomyopathy. Cardiovasc Drugs Ther 8:65–73
9. Op den Buijs J, Miklós Z, Van Riel NAW et al (2005) β-Adrenergic activation reveals impaired cardiac calcium handling at early stage of diabetes. Life Sci 76:1083–1098
10. Hattori Y, Matsuda N, Kimura J et al (2000) Diminished function and expression of the cardiac Na^+- Ca^{2+} exchanger in diabetic rats: implication in Ca^{2+} overload. J Physiol 527(Pt 1):85–94
11. Ren J, Davidoff AJ (1997) Diabetes rapidly induces contractile dysfunctions in isolated ventricular myocytes. Am J Physiol 272:H148–H158
12. Bai S, Sun J, Wu H et al (2012) Decrease in calcium-sensing receptor in the progress of diabetic cardiomyopathy. Diabetes Res Clin Pract 95:378–385
13. Kralik P, Ye G, Metreveli N et al (2005) Cardiomyocyte dysfunction in models of type 1 and type 2 diabetes. Cardiovasc Toxicol 5:285–292
14. Dhalla NS, Saini HK, Tappia PS et al (2007) Potential role and mechanisms of subcellular remodeling in cardiac dysfunction due to ischemic heart disease. J Cardiovasc Med (Hagerstown) 8:238–250
15. Halling DB, Aracena-Parks P, Hamilton SL (2005) Regulation of voltage-gated Ca^{2+} channels by calmodulin. Sci STKE 2005:re15
16. Xu M, Zhou P, Xu S-M et al (2007) Intermolecular failure of L-type Ca^{2+} channel and ryanodine receptor signaling in hypertrophy. PLoS Biol 5:e21
17. Petrovic MM, Vales K, Putnikovic B et al (2008) Ryanodine receptors, voltage-gated calcium channels and their relationship with protein kinase A in the myocardium. Physiol Res 57:141–149
18. Shaw RM, Colecraft HM (2013) L-type calcium channel targeting and local signalling in cardiac myocytes. Cardiovasc Res 98:177–186
19. Splawski I, Timothy KW, Sharpe LM et al (2004) Ca(V)1.2 calcium channel dysfunction causes a multisystem disorder including arrhythmia and autism. Cell 119:19–31
20. Shimoni Y, Firek L, Severson D, Giles W (1994) Short-term diabetes alters K^+ currents in rat ventricular myocytes. Circ Res 74:620–628
21. Lacombe VA, Viatchenko-Karpinski S, Terentyev D et al (2007) Mechanisms of impaired calcium handling underlying subclinical diastolic dysfunction in diabetes. Am J Physiol 293:R1787–R1797
22. Lengyel C, Virág L, Bíró T et al (2007) Diabetes mellitus attenuates the repolarization reserve in mammalian heart. Cardiovasc Res 73:512–520
23. Lu Z, Ballou LM, Jiang Y-P et al (2011) Restoration of defective L-type Ca^{2+} current in cardiac myocytes of type 2 diabetic db/db mice by Akt and PKC-ι. J Cardiovasc Pharmacol 58:439–445
24. Lu Z, Jiang Y-P, Xu X-H et al (2007) Decreased L-type Ca^{2+} current in cardiac myocytes of type 1 diabetic akita mice due to reduced phosphatidylinositol 3-kinase signaling. Diabetes 56:2780–2789
25. Pereira L, Matthes J, Schuster I et al (2006) Mechanisms of $[Ca^{2+}]i$ transient decrease in cardiomyopathy of db/db type 2 diabetic mice. Diabetes 55:608–615
26. Malhotra A, Sanghi V (1997) Regulation of contractile proteins in diabetic heart. Cardiovasc Res 34:34–40
27. Sun H, Kerfant B-G, Zhao D et al (2006) Insulin-like growth factor-1 and PTEN deletion enhance cardiac L-type Ca^{2+} currents via increased PI3Kα/PKB signaling. Circ Res 98:1390–1397
28. Gómez AM, Valdivia HH, Cheng H et al (1997) Defective excitation-contraction coupling in experimental cardiac hypertrophy and heart failure. Science 276:800–806
29. Bénitah J-P, Kerfant BG, Vassort G et al (2002) Altered communication between L-type calcium channels and ryanodine receptors in heart failure. Front Biosci 7:e263–e275
30. Shao C-H, Rozanski GJ, Patel KP, Bidasee KR (2007) Dyssynchronous (non-uniform) Ca^{2+} release in myocytes from streptozotocin-induced diabetic rats. J Mol Cell Cardiol 42:234–246

31. Eisner DA, Choi HS, Diaz ME et al (2000) Integrative analysis of calcium cycling in cardiac muscle. Circ Res 87:1087–1094
32. Bers DM (2008) Calcium cycling and signaling in cardiac myocytes. Annu Rev Physiol 70:23–49
33. Chattou S, Diacono J, Feuvray D (1999) Decrease in sodium-calcium exchange and calcium currents in diabetic rat ventricular myocytes. Acta Physiol Scand 166:137–144
34. Wold LE, Ceylan-Isik AF, Fang CX et al (2006) Metallothionein alleviates cardiac dysfunction in streptozotocin-induced diabetes: role of Ca^{2+} cycling proteins, NADPH oxidase, poly(ADP-Ribose) polymerase and myosin heavy chain isozyme. Free Radic Biol Med 40:1419–1429
35. Choi KM, Zhong Y, Hoit BD et al (2002) Defective intracellular Ca^{2+} signaling contributes to cardiomyopathy in type 1 diabetic rats. Am J Physiol Heart Circ Physiol 283:H1398–H1408
36. Schaffer SW, Ballard-Croft C, Boerth S, Allo SN (1997) Mechanisms underlying depressed Na^+-Ca^{2+} exchanger activity in the diabetic heart. Cardiovasc Res 34:129–136
37. Pierce GN, Dhalla NS (1983) Sarcolemmal Na^+–K^+-ATPase activity in diabetic rat heart. Am J Physiol 245:C241–C247
38. Kjeldsen K, Braendgaard H, Sidenius P et al (1987) Diabetes decreases Na^+–K^+ pump concentration in skeletal muscles, heart ventricular muscle, and peripheral nerves of rat. Diabetes 36:842–848
39. Golfman L, Dixon IM, Takeda N et al (1998) Cardiac sarcolemmal Na^+–Ca^{2+} exchange and Na^+–K^+ATPase activities and gene expression in alloxan-induced diabetes in rats. Mol Cell Biochem 188:91–101
40. Schaffer SW, Allo S, Punna S, White T (1991) Defective response to cAMP-dependent protein kinase in non-insulin-dependent diabetic heart. Am J Physiol 261:E369–E376
41. Allo SN, Lincoln TM, Wilson GL et al (1991) Non-insulin-dependent diabetes-induced defects in cardiac cellular calcium regulation. Am J Physiol 260:C1165–C1171
42. Wold LE, Dutta K, Mason MM et al (2005) Impaired SERCA function contributes to cardiomyocyte dysfunction in insulin resistant rats. J Mol Cell Cardiol 39:297–307
43. Belke DD, Swanson EA, Dillmann WH (2004) Decreased sarcoplasmic reticulum activity and contractility in diabetic db/db mouse heart. Diabetes 53:3201–3208
44. Stubbs CD, Smith AD (1984) The modification of mammalian membrane polyunsaturated fatty acid composition in relation to membrane fluidity and function. Biochim Biophys Acta 779:89–137
45. Vér Á, Szántó I, Bányász T et al (1997) Changes in the expression of Na^+–K^+-ATPase isoenzymes in the left ventricle of diabetic rat hearts: effect of insulin treatment. Diabetologia 40:1255–1262
46. Dhalla NS, Liu X, Panagia V, Takeda N (1998) Subcellular remodeling and heart dysfunction in chronic diabetes. Cardiovasc Res 40:239–247
47. Gerbi A, Barbey O, Raccah D et al (1997) Alteration of Na, K-ATPase isoenzymes in diabetic cardiomyopathy: effect of dietary supplementation with fish oil (n-3 fatty acids) in rats. Diabetologia 40:496–505
48. Chen S, Khan ZA, Karmazyn M, Chakrabarti S (2007) Role of endothelin-1, sodium hydrogen exchanger-1 and mitogen activated protein kinase (MAPK) activation in glucose-induced cardiomyocyte hypertrophy. Diabetes Metab Res Rev 23:356–367
49. Pierce GN, Ramjiawan B, Dhalla NS, Ferrari R (1990) Na^+/H^+exchange in cardiac sarcolemmal vesicles isolated from diabetic rats. Am J Physiol 258:H255–H261
50. Le Prigent K, Lagadic-Gossmann D, Feuvray D (1997) Modulation by pH^0 and intracellular Ca^{2+} of Na^+–H^+ exchange in diabetic rat isolated ventricular myocytes. Circ Res 80:253–260
51. Khandoudi N, Bernard M, Cozzone P, Feuvray D (1990) Intracellular pH and role of Na^+–H^+ exchange during ischaemia and reperfusion of normal and diabetic rat hearts. Cardiovasc Res 24:873–878
52. Darmellah A, Baetz D, Prunier F et al (2007) Enhanced activity of the myocardial Na^+–H^+ exchanger contributes to left ventricular hypertrophy in the Goto–Kakizaki rat model of type 2 diabetes: critical role of Akt. Diabetologia 50:1335–1344

53. Jandeleit-Dahm K, Hannan KM, Farrelly CA et al (2000) Diabetes-induced vascular hypertrophy is accompanied by activation of Na^+–H^+ exchange and prevented by Na^+–H^+ exchange inhibition. Circ Res 87:1133–1140
54. Kusumoto K, Haist JV, Karmazyn M (2001) Na^+/H^+ exchange inhibition reduces hypertrophy and heart failure after myocardial infarction in rats. Am J Physiol Heart Circ Physiol 280:H738–H745
55. Baartscheer A, Hardziyenka M, Schumacher CA et al (2008) Chronic inhibition of the Na^+–H^+-exchanger causes regression of hypertrophy, heart failure, and ionic and electrophysiological remodelling. Br J Pharmacol 154:1266–1275
56. Vial G, Dubouchaud H, Couturier K et al (2008) Na^+–H^+ exchange inhibition with cariporide prevents alterations of coronary endothelial function in streptozotocin-induced diabetes. Mol Cell Biochem 310:93–102
57. Meissner G (2004) Molecular regulation of cardiac ryanodine receptor ion channel. Cell Calcium 35:621–628
58. Bidasee KR, Dinçer ÜD, Besch HR (2001) Ryanodine receptor dysfunction in hearts of streptozotocin-induced diabetic rats. Mol Pharmacol 60:1356–1364
59. Lanner JT (2012) Ryanodine receptor physiology and its role in disease. Adv Exp Med Biol 740:217–234
60. Dincer UD, Araiza A, Knudson JD et al (2006) Dysfunction of cardiac ryanodine receptors in the metabolic syndrome. J Mol Cell Cardiol 41:108–114
61. Netticadan T, Temsah RM, Kent A et al (2001) Depressed levels of Ca^{2+}-cycling proteins may underlie sarcoplasmic reticulum dysfunction in the diabetic heart. Diabetes 50:2133–2138
62. Lehnart SE, Mongillo M, Bellinger A et al (2008) Leaky Ca^{2+} release channel/ryanodine receptor 2 causes seizures and sudden cardiac death in mice. J Clin Invest 118:2230–2245
63. Turan B, Vassort G (2011) Ryanodine receptor: a new therapeutic target to control diabetic cardiomyopathy. Antioxid Redox Signal 15:1847–1861
64. Bidasee K, Nallani K, Henry B et al (2003) Chronic diabetes alters function and expression of ryanodine receptor calcium-release channels in rat hearts. Mol Cell Biochem 249:113–123
65. Dhalla NS, Temsah RM, Netticadan T (2000) Role of oxidative stress in cardiovascular diseases. J Hypertens 18:655–673
66. Bidasee KR, Nallani K, Besch HR, Dincer UD (2003) Streptozotocin-induced diabetes increases disulfide bond formation on cardiac ryanodine receptor (RyR2). J Pharmacol Exp Ther 305:989–998
67. Wolff SP, Jiang ZY, Hunt JV (1991) Protein glycation and oxidative stress in diabetes mellitus and ageing. Free Radic Biol Med 10:339–352
68. Hund TJ, Ziman AP, Lederer WJ, Mohler PJ (2008) The cardiac IP3 receptor: uncovering the role of "the other" calcium-release channel. J Mol Cell Cardiol 45:159–161
69. Mackenzie L, Bootman MD, Laine M et al (2002) The role of inositol 1,4,5-trisphosphate receptors in Ca^{2+} signalling and the generation of arrhythmias in rat atrial myocytes. J Physiol 541:395–409
70. Zima AV, Blatter LA (2004) Inositol-1,4,5-trisphosphate-dependent Ca^{2+} signalling in cat atrial excitation-contraction coupling and arrhythmias. J Physiol 555:607–615
71. Fauconnier J, Lanner JT, Zhang S-J et al (2005) Insulin and inositol 1,4,5-trisphosphate trigger abnormal cytosolic Ca^{2+} transients and reveal mitochondrial Ca^{2+} handling defects in cardiomyocytes of ob/ob mice. Diabetes 54:2375–2381
72. Zhou B-Q, Hu S-J, Wang G-B (2006) The analysis of ultrastructure and gene expression of sarco/endoplasmic reticulum calcium handling proteins in alloxan-induced diabetic rat myocardium. Acta Cardiol 61:21–27
73. Guner S, Arioglu E, Tay A et al (2004) Diabetes decreases mRNA levels of calcium-release channels in human atrial appendage. Mol Cell Biochem 263:143–150
74. Dhalla NS, Rangi S, Zieroth S, Xu Y-J (2012) Alterations in sarcoplasmic reticulum and mitochondrial functions in diabetic cardiomyopathy. Exp Clin Cardiol 17:115–120

75. Dutta K, Carmody MW, Cala SE, Davidoff AJ (2002) Depressed PKA activity contributes to impaired SERCA function and is linked to the pathogenesis of glucose-induced cardiomyopathy. J Mol Cell Cardiol 34:985–996
76. Davidoff AJ, Davidson MB, Carmody MW et al (2004) Diabetic cardiomyocyte dysfunction and myocyte insulin resistance: role of glucose-induced PKC activity. Mol Cell Biochem 262:155–163
77. Zhong Y, Ahmed S, Grupp IL, Matlib MA (2001) Altered SR protein expression associated with contractile dysfunction in diabetic rat hearts. Am J Physiol Heart Circ Physiol 281:H1137–H1147
78. Bidasee KR, Nallani K, Yu Y et al (2003) Chronic diabetes increases advanced glycation end products on cardiac ryanodine receptors/calcium-release channels. Diabetes 52:1825–1836
79. Bidasee KR, Zhang Y, Shao CH et al (2004) Diabetes increases formation of advanced glycation end products on sarco(endo)plasmic reticulum Ca^{2+}-ATPase. Diabetes 53:463–473
80. Abe T, Ohga Y, Tabayashi N et al (2002) Left ventricular diastolic dysfunction in type 2 diabetes mellitus model rats. Am J Physiol Heart Circ Physiol 282:H138–H148
81. Fredersdorf S, Thumann C, Zimmermann WH et al (2012) Increased myocardial SERCA expression in early type 2 diabetes mellitus is insulin dependent: in vivo and in vitro data. Cardiovasc Diabetol 11:57
82. Trost SU, Belke DD, Bluhm WF et al (2002) Overexpression of the sarcoplasmic reticulum Ca^{2+}-ATPase improves myocardial contractility in diabetic cardiomyopathy. Diabetes 51:1166–1171
83. Vetter R, Rehfeld U, Reissfelder C et al (2002) Transgenic overexpression of the sarcoplasmic reticulum Ca^{2+} ATPase improves reticular Ca^{2+} handling in normal and diabetic rat hearts. FASEB J 16:1657–1659
84. Golfman L, Dixon IC, Takeda N et al (1999) Differential changes in cardiac myofibrillar and sarcoplasmic reticular gene expression in alloxan-induced diabetes. Mol Cell Biochem 200:15–25
85. Cai L, Kang YJ (2001) Oxidative stress and diabetic cardiomyopathy: a brief review. Cardiovasc Toxicol 1:181–193
86. Boudina S, Abel ED (2006) Mitochondrial uncoupling: a key contributor to reduced cardiac efficiency in diabetes. Physiology (Bethesda) 21:250–258
87. An D, Rodrigues B (2006) Role of changes in cardiac metabolism in development of diabetic cardiomyopathy. Am J Physiol Heart Circ Physiol 291:H1489–H1506
88. Turrens JF (2003) Mitochondrial formation of reactive oxygen species. J Physiol 552: 335–344
89. Chen Y, Saari JT, Kang YJ (1994) Weak antioxidant defenses make the heart a target for damage in copper-deficient rats. Free Radic Biol Med 17:529–536
90. Yan SD, Schmidt AM, Anderson GM et al (1994) Enhanced cellular oxidant stress by the interaction of advanced glycation end products with their receptors/binding proteins. J Biol Chem 269:9889–9897
91. Müller AL, Dhalla NS (2012) Role of various proteases in cardiac remodeling and progression of heart failure. Heart Fail Rev 17:395–409
92. Dhalla NS, Rangi S, Babick AP et al (2012) Cardiac remodeling and subcellular defects in heart failure due to myocardial infarction and aging. Heart Fail Rev 17:671–681
93. Clark RJ, McDonough PM, Swanson E et al (2003) Diabetes and the accompanying hyperglycemia impairs cardiomyocyte calcium cycling through increased nuclear O-GlcNAcylation. J Biol Chem 278:44230–44237
94. Belin RJ, Sumandea MP, Allen EJ et al (2007) Augmented protein kinase C-induced myofilament protein phosphorylation contributes to myofilament dysfunction in experimental congestive heart failure. Circ Res 101:195–204
95. Shimoni Y, Liu X-F (2003) Role of PKC in autocrine regulation of rat ventricular K^+ currents by angiotensin and endothelin. Am J Physiol Heart Circ Physiol 284:H1168–H1181
96. Rupp H, Elimban V, Dhalla NS (1989) Diabetes-like action of intermittent fasting on sarcoplasmic reticulum Ca^{2+} pump ATPase and myosin isoenzymes can be prevented by sucrose. Biochem Biophys Res Commun 164:319–325

97. Dillmann WH (1982) Influence of thyroid hormone administration on myosin ATPase activity and myosin isoenzyme distribution in the heart of diabetic rats. Metabolism 31:199–204
98. Afzal N, Pierce GN, Elimban V et al (1989) Influence of verapamil on some subcellular defects in diabetic cardiomyopathy. Am J Physiol 256:E453–E458
99. Dillmann WH (1980) Diabetes mellitus induces changes in cardiac myosin of the rat. Diabetes 29:579–582
100. Yu JZ, Rodrigues B, McNeill JH (1997) Intracellular calcium levels are unchanged in the diabetic heart. Cardiovasc Res 34:91–98
101. Pierce GN, Russell JC (1997) Regulation of intracellular Ca^{2+} in the heart during diabetes. Cardiovasc Res 34:41–47

Abnormalities in ATP Production and Utilization in Diabetic Cardiomyopathy

Naranjan S. Dhalla, Arthur H. Cunha-Volpato, and Yan-Jun Xu

Abstract For the heart to produce mechanical force, two cellular components are essential: myofibrils, which are responsible for generating contractile activity, and mitochondria, which provide most of the required energy. Diabetic cardiomyopathy is associated with defects in both mitochondria and myofibrils, indicating that changes in energy production and energy utilization are of foremost relevance in the etiology of cardiac dysfunction in chronic diabetes. Several elements including hyperglycemia, hyperlipidemia, and changes in the level of different hormones contribute directly or indirectly to contractile impairment in this multifactorial diabetic disorder. Metabolic imbalance, characterized by excessive fatty acid oxidation, Ca^{2+} overload, and oxidative stress are considered to reduce mitochondrial phosphorylation activity and impair the mitochondrial electron-transport chain in the diabetic heart. These subcellular alterations result in reduced level of adenosine triphosphate (ATP) in the diabetic heart, limiting cardiomyocyte contractile ability. Altered gene expression and excessive proteolytic activity caused by intracellular Ca^{2+} overload and oxidative stress in chronic diabetes promote changes in both composition and structure of myofibrils; this myofibril remodeling, characterized by diminished energy consumption and insensitivity to Ca^{2+}, further impairs heart function in diabetic cardiomyopathy.

Keywords Diabetic cardiomyopathy • Mitochondria • Myofibrils • ATP • Metabolic abnormalities • Ca^{2+} overload • Oxidative stress • Cardiac dysfunction

---

N.S. Dhalla (✉)
Institute of Cardiovascular Sciences and Departments of Physiology, Faculty of Medicine, University of Manitoba, St. Boniface Hospital Research, Winnipeg, MB, Canada

Institute of Cardiovascular Sciences, St. Boniface Hospital, 351 Tache Avenue, Winnipeg, MB 2RH 2A6, Canada
e-mail: nsdhalla@sbrc.ca

A.H. Cunha-Volpato • Y.-J. Xu
Institute of Cardiovascular Sciences and Departments of Physiology, Faculty of Medicine, University of Manitoba, St. Boniface Hospital Research, Winnipeg, MB, Canada

B. Turan and N.S. Dhalla (eds.), *Diabetic Cardiomyopathy: Biochemical and Molecular Mechanisms*, Advances in Biochemistry in Health and Disease 9,
DOI 10.1007/978-1-4614-9317-4_18, © Springer Science+Business Media New York 2014

1 Introduction

Diabetes mellitus is becoming a more relevant disease, as the global prevalence of diabetes was estimated to be 2.8 % in 2000 and is expected to increase more than 50 % until reaching 4.4 % in 2030 [1]. Diabetes is associated to a wide range of clinical manifestations related to either insulin deficiency in type 1 diabetes or insulin insensitivity in type 2 diabetes [2, 3]. This abnormal metabolic state is mainly characterized by, but not limited to, hyperglycemia (elevated glucose levels). Other hormonal and metabolic abnormalities, especially hyperlipidemia, are known to contribute to diabetes-related cardiovascular complications [4–9]. Although the name diabetes mellitus, in opposition to insipidus, is a remnant of a time when tasting urine was an acceptable medical procedure, the idea of a defective cardiac phenotype as a result of diabetes is a relatively new concept. Although diabetes mortality is primarily attributed to cardiovascular complications [10], the concept of diabetic cardiomyopathy was neglected for a long time, mainly because of the confounding effect of chronic diabetes on heart function by atherosclerosis and hypertension [9, 11]. It was in 1972, with the description of four diabetic patients showing congestive heart failure in the absence of coronary artery atherosclerosis by Rubler et al. [12], that the term diabetic cardiomyopathy was first used. Diabetic cardiomyopathy was defined as a cardiac dysfunction that occurs because of chronic diabetes, independently of coronary artery disease [13]. Patients with diabetes, in the absence of atherosclerosis, were found to suffer from ventricular dysfunction including shortened left ventricular ejection time, longer pre-ejection period, and also elevated end-diastolic pressure [14]. Some epidemiological studies also revealed that diabetes shows increased risk of heart failure even when atherosclerosis and hypertension risks are taken into account [15]. Animal models for diabetes also attest to the harmful effects of diabetes on heart function. Streptozotocin-induced diabetes in rats is associated with reduced heart rate, lower peak ventricular pressure- and also impaired left ventricular contractions and relaxations [16–21]. Further animal studies have also shown that diabetes, in conjunction with hypertension, leads to congestive heart failure [22, 23].

Although the molecular and cellular mechanisms of the cardiac dysfunction in diabetic cardiomyopathy are not completely understood [24], it is clear that an imbalance between energy production, in the form of adenosine triphosphate (ATP), and energy consumption is a key factor that contributes to the development of this pathological disorder [13, 24, 25]. The major players of the high-energy phosphate production and utilization cycle, in the cardiomyocytes, are the mitochondria (MT) and myofibrils (MF), the subcellular components responsible for the phosphorylation of adenosine diphosphate (ADP) into ATP and the hydrolysis of ATP by ATPase activity, respectively. Herein, this review focuses on the mechanisms by which chronic diabetes affects MT and MF functions, and also the consequences of the resulting energetic imbalance on cardiomyocytes and heart function.

2 Defects in Energy Production

In cardiac tissue, as in most tissues, MT are accountable for most of the ATP production, and thus any deterioration of these organelles may lead to a state of energy restriction and consequently of cellular deficiency. In fact, MT are responsible, under normal conditions, for more than 95 % of all myocardial ATP synthesis [26]; cardiomyocytes have a rather limited ATP pool that would be consumed in approximately 10 s without continuous mitochondrial activity [26]. To maintain a stable ATP content, and proper myocardial function, energy consumption and production have to be tightly coupled in the cardiac muscle. Proper MT function is, therefore, essential for cardiac function as MT phosphorylation is commonly compromised in several different types of cardiac disorders [26–28], including diabetic cardiomyopathy [29–31]. MT dysfunction is mainly credited to metabolic alterations in diabetes, resulting in increased free fatty acid (FFA) utilization. Oxidative stress and Ca^{2+} overload are also relevant to the process that leads to MT damage and energy deficiency in the diabetic heart. These events are depicted in Fig. 1.

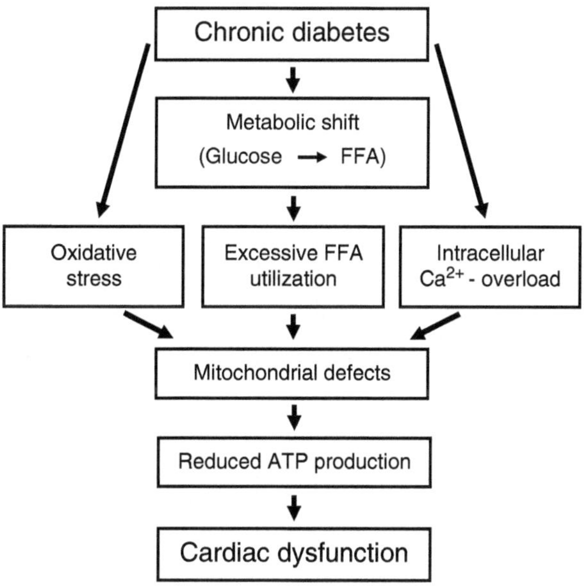

Fig. 1 Mechanisms by which chronic diabetes leads to mitochondrial defect, reduced ATP reserves, and cardiac dysfunction

3 Defects in Energy Utilization

Cardiomyocytes require energy, in the form of ATP, to produce MF contractions and the mechanical force that ultimately allows the heart to pump blood. Because cardiomyocyte contractions are controlled, in both intensity and rhythm, by changes in intracellular Ca^{2+} concentration, a precise MF response to Ca^{2+} is crucial to overall heart performance. MF is the functional unit of the cardiac muscle, being composed mainly of actin and myosin; myosin is the protein that is actively responsible for muscular contractions, as it possesses ATPase activity. MF contractions are regulated by the troponin–tropomyosin complex (TnTm): Ca^{2+} binds to the TnTm complex, which exposes the myosin ATP-binding site, and this allows actin–myosin interactions to occur, resulting in the shortening of the muscle fiber. This contractile process requires energy in the form of ATP and represents about 60–70 % of all ATP consumption of the myocardium under normal conditions [26]. As a matter of fact, diabetic cardiomyopathy is associated with MF abnormalities that result in defects in the energy utilization process [13, 32–34]. Two main factors can be seen promoting cardiac contractile dysfunction: defects in energy utilization and abnormal MF regulation. Several cellular abnormalities involved in defects in energy utilization and regulation of MF in diabetic heart are depicted in Fig. 2.

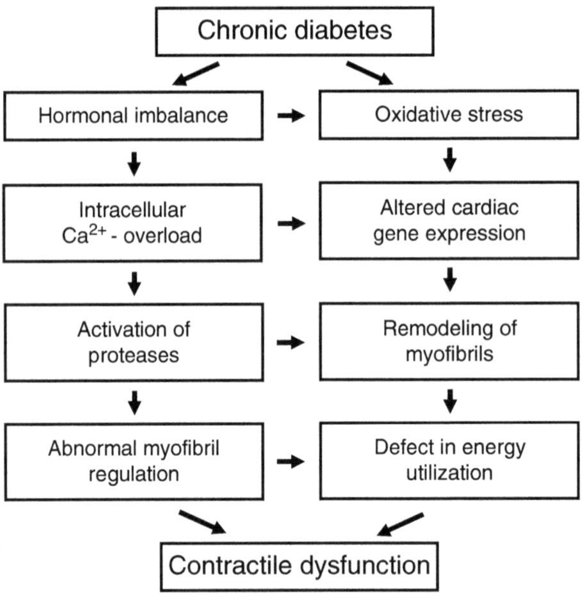

Fig. 2 Role of hormonal imbalance in diabetic cardiomyopathy in promoting myofibrillar dysfunction, as a consequence of oxidative stress, Ca^{2+} overload, and the activation of proteolytic enzymes

Abnormal energy utilization is associated with MF remodeling, a process that is related to changes in gene expression, oxidative stress, intracellular Ca^{2+}, and activation of proteolytic enzymes [13, 21, 32, 34, 35]. Several animal studies have reported decreased MF Ca^{2+}-dependent ATPase activity in chronic diabetes [17, 19, 21, 36, 37], and MF insensibility to Ca^{2+} is most likely involved in diabetic cardiomyopathy [13]. Several factors interfere with myofibril function, involving both functional and regulatory enzymes [36–39]. Chronic diabetes, in animals, is associated with the prevalence of myosin heavy-chain (MHC) isoenzyme β over isoenzyme α [34, 40–48]. This shift in expression of MHC isoforms could explain the depressed ATPase activity in other animals; however, the relevance of this mechanism is uncertain in human hearts, in which the β-isoform is normally predominant over the α isoform [49, 50]. On the other hand, the phosphorylation of the myosin light chain (MLC) by the myosin light-chain kinase (MLCK) is a factor that could explain cardiac dysfunction in human diabetic cardiomyopathy, because MLC phosphorylation is related to increased MF sensitivity to Ca^{2+} [51]. However, MLC, MLCK, and MLC phosphorylation were shown to be significantly reduced in diabetic rat heart homogenate, and these changes were partially reversed by insulin treatment [38]. The activation of proteolytic enzymes, seen in cardiac dysfunction [35, 52–60], can also participate in the development of MF dysfunction. Intracellular Ca^{2+} overload and oxidative stress are related to the activation of proteases [61–63] that would lead to degradation of cardiomyocyte MF proteins in diabetic cardiomyopathy.

Chronic diabetes is also known to affect cardiac function through impaired MF regulation mainly caused by TnTm abnormalities [13]. The TnTm complex is formed by tropomyosin and three subunits of troponin, TnC, TnI, and TnT, that are responsible for Ca^{2+} binding, ATPase inhibition, and myosin binding, respectively. There is evidence that the phosphorylation of TnI and TnT by protein kinase A (PKA) and C (PKC) contribute to myofibrillar insensitivity to Ca^{2+}, leading, in the given order, to reduced ATPase response to Ca^{2+} and reduced myosin–actin interactions [13]. These findings are supported by the fact that impaired actomyosin ATPase activity of diabetic animals could be partially normalized in the presence of TnTm extracted from healthy animals [64].

4 Mechanisms of Alterations in Mitochondria and Myofibrils

The primary energy source of cardiomyocytes are fatty acids, accounting for approximately 60–70 % of the myocardial substrate [65]. The participation of FFA in cardiac muscle energy metabolism is known to be even more expressive in a chronic diabetic state [66, 67]. Although lipids are essential for heart function, from both an energetic and structural point of view, excessive FFA uptake by myocytes is known to have deleterious effects on cardiac function, supporting the concept of 'lipid paradox' [65]. Studies conducted with both animal and human tissues have concluded that sarcolemmal (SL) glucose transporter (GLUT) 1, GLUT4, and sodium–glucose-linked transporter (SLGT) 1 expression is reduced in diabetic hearts [68–76]. In fact, glucose uptake is impaired in diabetic hearts by insulin deficiency or insensitivity [77–79]. The opposite is true for FFA uptake: high plasma levels of FFA not only increase cardiomyocyte FFA uptake but also depress glucose utilization by the diabetic heart [26, 28, 80]. Increased MT fatty acid oxidation, in chronic diabetes, is credited to the upregulation of the perioxisome proliferator-activated receptor α [71, 81–84]. Such an elevation in the rate of fatty acid oxidation, if maintained for a long period of time, is believed to impair MT oxidative phosphorylation, causing damage to the electron-transport chain and depressing MT Mg^{2+} ATPase [29–31]. In addition to the reduction of ATP reserves and depressed cardiac function [30, 31], MT dysfunction is associated with the formation of MT pores, leaking of MT proteins, and cellular dysfunction [9, 13, 24, 30, 85, 86]. Triglyceride synthesis is also remarkably upregulated in diabetic cardiomyopathy, leading to a phenomenon called lipotoxicity that is intimately related to cellular damage [9, 87]. Enhanced FFA uptake is associated with the accumulation of lipid droplets in the myocardium during the development of diabetic cardiomyopathy [9, 86].

Diabetes is intimately associated with defects in several metabolic pathways that are responsible for a marked increase in intracellular Ca^{2+} concentration [13, 18, 24, 73, 86, 88–95]. Ca^{2+}-handling defects in chronic diabetes are known to be related to SL and sarcoplasmic reticulum (SR) alterations that favor the occurrence of intracellular Ca^{2+} overload [96–99]. In this regard, MT are known to function as Ca^{2+} sinks, in the event of Ca^{2+} overload, in an attempt to maintain the normal cytoplasmic level of free Ca^{2+} [24, 73, 100, 101]. Although this mechanism is intended to be protective in nature, excessive MT Ca^{2+} uptake depresses MT phosphorylation activity [13]. Different drugs that are capable of attenuating Ca^{2+} overload, such as Ca^{2+} channel blockers and angiotensin-receptor antagonists, have been shown to ameliorate cardiac dysfunction in diabetic cardiomyopathy [40, 95, 102, 103].

It is becoming clear that oxidative stress contributes to the development of several diabetic complications [104–108], including diabetic cardiomyopathy [13, 85, 109–113]. Hormonal imbalance in diabetes, marked by elevated levels of angiotensin II, catecholamines, and endothelins, plays a significant role in promoting oxidative stress and cardiac dysfunction [13, 114]. Damaged MT [13, 115, 116] and advanced protein glycation, caused by hyperglycemia, are also known to result in the development of oxidative stress in chronic diabetes [117, 118]. Initially, cardiomyocytes cope with increased oxidative stress by boosting their natural antioxidant defenses. Elevated activities of superoxide dismutase, glutathione peroxidase, and other antioxidant enzymes have been reported in the diabetic rat heart [119–121]. Most likely, these initial compensatory mechanisms are eventually exhausted, resulting in oxidative stress and leading to subcellular remodeling and myocardial cell damage [13]. Some reports have indicated that antioxidants are effective in preventing diabetic cardiomyopathy and cardiac dysfunction [13, 122, 123] and thus provide evidence regarding the relevance of oxidative stress in the pathophysiology of diabetic cardiomyopathy (Fig. 2).

5 Conclusions

From the foregoing discussion, it is apparent that a defect occurs in the process of energy production that leads to depression of ATP stores in the diabetic heart, primarily because of a shift in the balance between the utilization of glucose and the utilization of FFA as substrates by the myocardium. Excessive utilization of FFA for a prolonged period is considered to impair MT function with respect to oxidative phosphorylation and the electron-transport system. MT defects also includes opening of MT pores for leakage of cytoplasmic proteins that lead to the development of myocardial cell damage in the form of diabetic cardiomyopathy. Abnormalities in myocardial metabolism also promotes the occurrence of oxidative stress and intracellular Ca^{2+} overload, which results in the activation of different proteases and defects in gene expression in the diabetic myocardium. During this process of subcellular remodeling caused by diabetes, MF become defective in respect to their ability to utilize ATP as well as sensitivity to Ca^{2+} for the generation of contractile force. Thus, not only is cardiac dysfunction the result of defects in energy production by MT, but also defects in energy utilization by MF play a critical role during the development of diabetic cardiomyopathy.

Acknowledgments The work referred in this article was supported by the infrastructure provided by the St. Boniface Hospital Research Foundation. A.H. Cunha-Volpato was supported by the National Council for Scientific and Technological Development (CNPq) of Brazil.

References

1. Wild S, Roglic G, Green A et al (2004) Global prevalence of diabetes: estimates for the year 2000 and projections for 2030. Diabetes Care 27:1047–1053
2. Regan TJ, Ahmed S, Haider B et al (1994) Diabetic cardiomyopathy: experimental and clinical observations. N Engl J Med 91:776–778
3. Notkins AL (1979) The causes of diabetes. Sci Am 241(5):62–73
4. Christlieb AR (1973) Diabetes and hypertensive vascular disease. Am J Cardiol 32:592–606
5. Regan TJ (1983) Congestive heart failure in the diabetic. Annu Rev Med 34:161–168
6. Kannel WB, McGee DL (1979) Diabetes and cardiovascular disease. JAMA 241:2035–2038
7. Shah S (1980) Cardiomyopathy in diabetes mellitus. Angiology 31:502–504
8. Factor SM, Okun EM, Minase T (1980) Capillary microaneurysms in the human diabetic heart. N Engl J Med 302:384–388
9. Dhalla NS, Pierce GN, Innes IR, Beamish RE (1985) Pathogenesis of cardiac dysfunction in diabetes mellitus. Can J Cardiol 1:263–281
10. Garcia MJ, McNamara PM, Gordon T, Kannel WB (1974) Morbidity and mortality in diabetics in the Framingham population. Sixteen year follow-up study. Diabetes 23:105–111
11. Fein FS, Sonnenblick EH (1985) Diabetic cardiomyopathy. Prog Cardiovasc Dis 25:255–270
12. Rubler S, Dlugash J, Yuceoglu YZ et al (1972) New type of cardiomyopathy associated with diabetic glomerulosclerosis. Am J Cardiol 30:595–602

13. Dhalla NS, Liu X, Panagia V et al (1998) Subcellular remodeling and heart dysfunction in chronic diabetes. Cardiovasc Res 40:239–247
14. Ahmed SS, Jaferi G, Narang RM, Regan TJ (1975) Preclinical abnormality of left ventricular function in diabetes mellitus. Am Heart J 89:153–158
15. Boudina S, Albel ED (2007) Diabetic cardiomyopathy revisited. Circulation 115:3213–3223
16. Fein FS, Malhotra A, Miller GB et al (1984) Diabetic cardiomyopathy in rats: mechanical and biochemical response to different insulin doses. Am J Physiol 247:H817–H823
17. Fein FS, Strobeck JE, Malhotra A, Scheuer J (1981) Reversibility of diabetic cardiomyopathy with insulin in rats. Circ Res 49:1251–1261
18. Ganguly PK, Pierce GN, Dhalla KS, Dhalla NS (1983) Defective sarcoplasmic reticular calcium transport in diabetic cardiomyopathy. Am J Physiol 244:E528–E535
19. Pierce GN, Dhalla NS (1981) Cardiac myofibrillar ATPase activity in diabetic rats. J Mol Cell Cardiol 13:1063–1069
20. Vadlamudi RVSV, Rodgers RL, McNeill JH (1982) The effect of chronic alloxan- and streptozotocin-induced diabetes on isolated rat heart performance. Can J Physiol Pharmacol 60:902–911
21. Takeda N, Dixon IMC, Hata T et al (1996) Sequence of alterations in subcellular organelles during the development of heart dysfunction in diabetes. Diabetes Res Clin Pract 30(Suppl):S113–S122
22. Factor SM, Minase T, Sonnenblick EH (1980) Clinical and morphological features of human hypertensive-diabetic cardiomyopathy. Am Heart J 99:446–458
23. Factor SM, Minase T, Bhan R et al (1983) Hypertensive diabetic cardiomyopathy in the rat: ultrastructural features. Virchows Arch 398:305–317
24. Dhalla NS, Rangi S, Zieroth S, Xu YJ (2012) Alterations in sarcoplasmic reticulum and mitochondrial functions in diabetic cardiomyopathy. Exp Clin Cardiol 17:115–120
25. Dhalla NS, Takeda N, Rodriguez-Leyva D, Elimban V (2013) Mechanisms of subcellular remodeling in heart failure due to diabetes. Heart Fail Rev. doi:10.1007/s10741-013-9385-8
26. Stanley WC, Recchia FA, Lopaschuk GD (2005) Myocardial substrate metabolism in the normal and failing heart. Phys Rev 85:1093–1129
27. Dhalla NS, Das PK, Sharma GP (1978) Subcellular basis of cardiac contractile failure. J Mol Cell Cardiol 10:363–385
28. Rodrigues B, Cam MC, McNeill JH (1995) Myocardial substrate metabolism: implications in diabetic cardiomyopathy. J Mol Cell Cardiol 27:169–179
29. Kuo TH, Moore KH, Giacomelli F et al (1983) Defective oxidative metabolism of heart mitochondria from genetically diabetic mice. Diabetes 32:781–787
30. Pierce GN, Dhalla NS (1985) Heart mitochondrial function in chronic experimental diabetes in rats. Can J Cardiol 1:48–54
31. Flarsheim CE, Grupp IL, Matlib MA (1996) Mitochondrial dysfunction accompanies diastolic dysfunction in diabetic rat heart. Am J Physiol Heart Circ Physiol 271:H192–H202
32. Dillman WH (1989) Diabetes and thyroid-hormone-induced changes in cardiac function and their molecular basis. Annu Rev Med 40:373–394
33. Malhotra A, Sanghi V (1997) Regulation of contractile proteins in diabetic heart. Cardiovasc Res 34:34–40
34. Machackova J, Barta J, Dhalla NS (2005) Molecular defects in cardiac myofibrillar proteins due to thyroid hormone imbalance and diabetes. Can J Physiol Pharmacol 83:1071–1091
35. Müller AL, Dhalla NS (2012) Role of various proteases in cardiac remodeling and progression of heart failure. Heart Fail Rev 17:395–409
36. Dillmann WH (1980) Diabetes mellitus induces changes in cardiac myosin of the rat. Diabetes 29:579–582
37. Pierce GN, Dhalla NS (1985) Mechanisms of the defect in cardiac myofibrillar function during diabetes. Am J Physiol 248:E170–E175
38. Liu X, Takeda N, Dhalla NS (1997) Myosin light-chain phosphorylation in diabetic cardiomyopathy in rats. Metabolism 46:71–75

39. Liu X, Takeda N, Dhalla NS (1996) Troponin I phosphorylation in heart homogenate from diabetic rat. Biochim Biophys Acta 1316:78–84
40. Afzal N, Pierce GN, Elimban V et al (1989) Influence of verapamil on some subcellular defects in diabetic cardiomyopathy. Am J Physiol 256:E453–E458
41. Rupp H, Elimban V, Dhalla NS (1989) Diabetes-like action of intermittent fasting on sarcoplasmic reticulum Ca^{2+}-pump ATPase and myosin isoenzymes can be prevented by sucrose. Biochem Biophys Res Commun 164:319–325
42. Golfman L, Dixon IMC, Takeda N, Chapman D, Dhalla NS (1999) Differential changes in cardiac myofibrillar and sarcoplasmic reticular gene expression in alloxan-induced diabetes. Mol Cell Biochem 200:15–25
43. Depre C, Young ME, Ying J, Ahuja HS, Han Q, Garza N et al (2000) Streptozotocin-induced changes in cardiac gene expression in the absence of severe contractile dysfunction. J Mol Cell Cardiol 32:985–996
44. Dillmann WH (1982) Influence of thyroid hormone administration on myosin ATPase activity and myosin isoenzyme distribution in the heart of diabetic rats. Metabolism 31:199–204
45. Malhotra A, Penpargkul S, Fein FS et al (1981) The effect of streptozotocin-induced diabetes in rats on cardiac contractile proteins. Circ Res 49:1243–1250
46. Malhotra A, Mordes JP, McDermott L, Schaible TF (1985) Abnormal cardiac biochemistry in spontaneously diabetic Bio-Breeding/Worcester rat. Am J Physiol 249:H1051–H1055
47. Schaffer SW, Mozaffari MS, Artman M, Wilson GL (1989) Basis for myocardial mechanical defects associated with noninsulin-dependent diabetes. Am J Physiol 256:E25–E30
48. Takeda N, Nakamura I, Hatanaka T, Ohkubo T, Nagano M (1988) Myocardial mechanical and myosin isoenzyme alterations in streptozotocin-diabetic rats. Jpn Heart J 29:455–463
49. Mercadier JJ, Bouveret P, Gorza L et al (1983) Myosin isoenzymes in normal and hypertrophied human ventricular myocardium. Circ Res 53:52–62
50. Caforio ALP, Grazzini M, Mann JM et al (1992) Identification of α- and β-cardiac myosin heavy chain isoforms as major autoantigens in dilated cardiomyopathy. Circulation 85: 1734–1742
51. Sweeney HL, Stull JT (1986) Phosphorylation of myosin in permeabilized mammalian cardiac and skeletal muscle cells. Am J Physiol 250:C657–C660
52. Ali MAM, Schulz R (2009) Activation of MMP-2 as a key event in oxidative stress injury to the heart. Front Biosci 14:699–716
53. Huang Y, Wang KK (2001) The calpain family and human disease. Trends Mol Med 7:355–362
54. Goll DE, Thompson VF, Li H et al (2003) The calpain system. Physiol Rev 83:731–801
55. Spinale FG (2007) Myocardial matrix remodeling and the matrix metalloproteinases: influence on cardiac form and function. Physiol Rev 87:1285–1342
56. Schulz R (2007) Intracellular targets of matrix metalloproteinase-2 in cardiac disease: rationale and therapeutic approaches. Annu Rev Pharmacol Toxicol 47:211–242
57. Müller AL, Hryshko LV, Dhalla NS (2013) Extracellular and intracellular proteases in cardiac dysfunction due to ischemia–reperfusion injury. Int J Cardiol 164:39–47
58. Müller AL, Freed D, Hryshko L, Dhalla NS (2012) Implications of protease activation in cardiac dysfunction and development of genetic cardiomyopathy in hamsters. Can J Physiol Pharmacol 90:995–1004
59. Kandasamy AD, Chow AK, Ali MAM, Schulz R (2010) Matrix metalloproteinase-2 and myocardial oxidative stress injury: beyond the matrix. Cardiovasc Res 85:413–423
60. León H, Baczkó I, Sawicki G et al (2008) Inhibition of matrix metalloproteinases prevents peroxynitrite-induced contractile dysfunction in the isolated cardiac myocyte. Br J Pharmacol 153:676–683
61. Singh RB, Dhalla NS (2010) Ischemia-reperfusion-induced changes in sarcolemmal Na^+/K^+-ATPase are due to the activation of calpain in the heart. Can J Physiol Pharmacol 88:388–397
62. Singh RB, Hryshko L, Freed D, Dhalla NS (2012) Activation of proteolytic enzymes and depression of the sarcolemmal Na^+/K^+-ATPase in ischemia-reperfused heart may be mediated through oxidative stress. Can J Physiol Pharmacol 90:249–260

63. Müller AL, Freed D, Dhalla NS (2013) Activation of proteases and changes in Na^{+}-K^{+}-ATPase subunits in hearts subjected to ischemia-reperfusion. J Appl Physiol 114:351–360
64. Malhotra A, Lopez MC, Nakouzi A (1995) Troponin subunits contribute to altered myosin ATPase activity in diabetic cardiomyopathy. Mol Cell Biochem 151:165–172
65. Dhalla NS, Elimban V, Rupp H (1992) Paradoxical role of lipid metabolism in heart function and dysfunction. Mol Cell Biochem 116(1–2):3–9
66. Mazumder PK, O'Neill BT, Roberts MW et al (2004) Impaired cardiac efficiency and increased fatty acid oxidation in insulin-resistant *ob/ob* mouse hearts. Diabetes 53:2366–2374
67. Herrero P, Peterson LR, McGill JB et al (2006) Increased myocardial fatty acid metabolism in patients with type 1 diabetes mellitus. J Am Coll Cardiol 47:598–604
68. Berger J, Biswa C, Vicario PP et al (1989) Decreased expression of the insulin-responsive glucose transporter in diabetes and fasting. Nature 340:70–72
69. Camps M, Castello A, Munoz P et al (1992) Effect of diabetes and fasting on GLUT-4 (muscle/fat) glucose-transporter expression in insulin-sensitive tissues. Heterogeneous response in heart, red and white muscle. Biochem J 282:765–772
70. Dahl J, Herman WH, Hicks RJ et al (1993) Myocardial glucose uptake in patients with insulin-dependent diabetes mellitus assessed quantitatively by dynamic positron emission tomography. Circulation 88:395–404
71. Sivitz WI, DeSautel SL, Kayano T et al (1989) Regulation of glucose transporter messenger RNA in insulin-deficient states. Nature 340:72–74
72. Stanley WC, Lopaschuk GD, McCormack JG (1997) Regulation of energy substrate metabolism in the diabetic heart. Cardiovasc Res 34:25–33
73. Dhalla NS, Elimban V (2013) Mechanisms of sarcolemmal defects in cardiac dysfunction due to chronic diabetes. In: Kimchi A (ed) Proceedings of the 17th world congress on heart disease, Medimond, Bologna (in press)
74. Cook SA, Varela-Carver A, Mongillo M, Kleinert C, Khan MT, Leccisotti L et al (2010) Abnormal myocardial insulin signalling in type 2 diabetes and left-ventricular dysfunction. Eur Heart J 31:100–111
75. Osborn BA, Daar JT, Laddaga RA, Romano FD, Paulson DJ (1997) Exercise training increases sarcolemmal GLUT-4 protein and mRNA content in diabetic heart. J Appl Physiol 82:828–834
76. Banerjee SK, McGaffin KR, Pastor-Soler NM, Ahmad F (2009) SGLT1 is a novel cardiac glucose transporter that is perturbed in disease states. Cardiovasc Res 84:111–118
77. Kaestner KH, Flores-Riveros JR, McLenithan JC et al (1991) Transcriptional repression of the mouse insulin-responsive glucose transporter (GLUT4) gene by cAMP. Proc Natl Acad Sci U S A 88:1933–1937
78. Kainulainen H, Breiner M, Schurmann A et al (1994) In vivo glucose uptake and glucose transporter proteins GLUT1 and GLUT4 in heart and various types of skeletal muscle from streptozotocin-diabetic rats. Biochim Biophys Acta 1225:275–282
79. Stroedter D, Schmidt T, Bretzel RG, Federlin K (1995) Glucose metabolism and left ventricular dysfunction are normalized by insulin and islet transplantation in mild diabetes in the rat. Acta Diabetol 32:235–243
80. Randle PJ, Garland PB, Hales CN et al (1963) The glucose fatty-acid cycle. Its role in insulin sensitivity and the metabolic disturbances of diabetes mellitus. Lancet 1:785–789
81. Finck BN, Kelly DP (2002) Peroxisome proliferator-activated receptor alpha (PPARalpha) signalling in the gene regulatory control of energy metabolism in the normal and diseased heart. J Mol Cell Cardiol 34:1249–1257
82. Banke NH, Wende AR, Leone TC et al (2010) Preferential oxidation of triacylglyceride-derived fatty acids in heart is augmented by the nuclear receptor PPAR alpha. Circ Res 107:233–241
83. Duncan JG, Fong JL, Medeiros DM et al (2007) Insulin-resistant heart exhibits a mitochondrial biogenic response driven by the peroxisome proliferator-activated receptor-alpha/PGC-1alpha gene regulatory pathway. Circulation 115:909–917

84. Finck BN, Lehman JJ, Leone TC, Welch MJ, Bennett MJ, Kovacs A et al (2002) The cardiac phenotype induced by PPAR-alpha overexpression mimics that caused by diabetes mellitus. J Clin Invest 109:121–130
85. Dhalla NS, Temsah RM, Netticadan T (2000) Role of oxidative stress in cardiovascular diseases. J Hypertens 18:655–673
86. Schaffer SW (1991) Cardiomyopathy associated with non-insulin dependent diabetes. Mol Cell Biochem 107:1–20
87. Yagyu H, Chen G, Yokoyama M et al (2003) Lipoprotein lipase (LpL) on the surface of cardiomyocytes increases lipid uptake and produces a cardiomyopathy. J Clin Invest 111: 419–426
88. Dhalla NS, Golfman LS, Elimban V, Takeda N (1996) Remodelling of subcellular organelles during the development of diabetic cardiomyopathy. In: Chatham JC, Fodder JR, McNeill JH (eds) The heart in diabetes. Kluwer, Boston, pp 100–142
89. Penpargkul S, Fein F, Sonnenblick EH, Scheuer J (1981) Depressed cardiac sarcoplasmic reticular function from diabetic rats. J Mol Cell Cardiol 13:303–309
90. Lopaschuk GD, Tahiliani AG, Vadlamudi RV, Katz S, McNeill JH (1984) Cardiac sarcoplasmic reticulum function in insulin- or carnitine-treated diabetic rats. Am J Physiol 245: H969–H976
91. Lopaschuk GD, Katz S, McNeill JH (1984) The effect of alloxan- and streptozotocin-induced diabetes on calcium transport in rat cardiac sarcoplasmic reticulum. The possible involvement of long chain acylcarnitines. Can J Physiol Pharmacol 61:439–448
92. Lopaschuk GD, Eibschutz B, Katz S, McNeill JH (1984) Depression of calcium transport in sarcoplasmic reticulum from diabetic rats: lack of involvement by specific regulatory mediators. Gen Pharmacol 15:1–5
93. Yu Z, Tibbits GF, McNeill JH (1994) Cellular functions of diabetic cardiomyocytes: contractility, rapid-cooling contracture, and ryanodine binding. Am J Physiol 266:H2082–H2089
94. Choi KM, Zhong Y, Hoit BD, Grupp IL, Hahn H, Dilly KW et al (2002) Defective intracellular Ca^{2+} signaling contributes to cardiomyopathy in type 1 diabetic rats. Am J Physiol Heart Circ Physiol 283:H1398–H1408
95. Yaras N, Bilginoglu A, Vassort G, Turan B (2007) Restoration of diabetes-induced abnormal local Ca^{2+} release in cardiomyocytes by angiotensin II receptor blockade. Am J Physiol Heart Circ Physiol 292:H912–H920
96. Lagadic-Gossmann D, Buckler KJ, Le Prigent K, Feuvray D (1996) Altered Ca^{2+} handling in ventricular myocytes isolated from diabetic rats. Am J Physiol 270:H1529–H1537
97. Ishikawa T, Kajiwara H, Kurihara S (1999) Alterations in contractile properties and Ca^{2+} handling in streptozotocin-induced diabetic rat myocardium. Am J Physiol 277:H2185–H2194
98. Allo SN, Lincoln TM, Wilson GL, Green FJ, Watanabe AM, Schaffer SW (1991) Non-insulin-dependent diabetes-induced defects in cardiac cellular calcium regulation. Am J Physiol 260:C1165–C1171
99. Ligeti L, Szenczi O, Prestia CM, Szabo C, Horvath K, Marcsek ZL et al (2006) Altered calcium handling is an early sign of streptozotocin-induced diabetic cardiomyopathy. Int J Mol Med 17:1035–1043
100. Dhalla NS, Pierce GN, Panagia V et al (1982) Calcium movements in relation to heart function. Basic Res Cardiol 77:117–139
101. Dhalla NS, Wang X, Beamish RE (1996) Intracellular calcium handling in normal and failing hearts. Exp Clin Cardiol 1:7–20
102. Afzal N, Ganguly PK, Dhalla KS et al (1988) Beneficial effects of verapamil in diabetic cardiomyopathy. Diabetes 37:936–942
103. Liu X, Suzuki H, Sethi R et al (2006) Blockade of the renin-angiotensin system attenuates sarcolemma and sarcoplasmic reticulum remodeling in chronic diabetes. Ann N Y Acad Sci 1084:1141–1154
104. Frank RN (1991) On the pathogenesis of diabetic retinopathy. A 1990 update. Ophthalmology 98:586–593

105. Wohaieb SA, Godin DV (1987) Alterations in free radical tissue-defense mechanisms in streptozotocin-induced diabetes in rat. Effects of insulin treatment. Diabetes 36:1014–1018
106. Thompon KH, Godin DV, Lee M (1992) Tissue antioxidant status in streptozotocin-induced diabetes in rats. Effects of dietary manganese deficiency. Biol Trace Elem Res 35:213–224
107. Giugliano D, Ceriello A, Paolisso G (1995) Diabetes mellitus, hypertension and cardiovascular disease: which role for oxidative stress? Metabolism 44:363–368
108. Fonseca VA, Stone A, Munshi M et al (1997) Oxidative stress in diabetic macrovascular disease: does homocysteine play a role? South Med J 90:903–906
109. Baynes JW, Thorpe SR (1999) Role of oxidative stress in diabetic complications: a new perspective on an old paradigm. Diabetes 48:1–9
110. Giardino I, Fard AK, Hatchell DL et al (1998) Aminoguanidine inhibits reactive oxygen species formation, lipid peroxidation, and oxidant-induced apoptosis. Diabetes 47:1114–1120
111. Ohuwa T, Sato Y, Naoi M (1995) Hydroxyl radical formation in diabetic rats induced by streptozotocin. Life Sci 56:1789–1798
112. Boudina S, Abel ED (2010) Diabetic cardiomyopathy, causes and effects. Rev Endocr Metab Disord 11:31–39
113. Kowluru RA, Engerman RL, Kern TS (2000) Diabetes-induced metabolic abnormalities in myocardium: effect of antioxidant therapy. Free Radic Res 32:67–74
114. Wolff SP, Jiang ZY, Hunt JV (1991) Protein glycation and oxidative stress in diabetes mellitus and ageing. Free Radic Biol Med 10:339–359
115. Taniguchi N, Kaneto H, Asahi M et al (1996) Involvement of glycation and oxidative stress in diabetic macroangiopathy. Diabetes 45:S81–S83
116. Turrens JF (2003) Mitochondrial formation of reactive oxygen species. J Physiol 552: 335–344
117. Brownlee M (1995) Advanced protein glycosylation in diabetes and aging. Annu Rev Med 46:223–234
118. Peiro C, Lafuente N, Matesanz N et al (2001) High glucose induces cell death of cultured human aortic smooth muscle cells through the formation of hydrogen peroxide. Br J Pharmacol 133:967–974
119. Kakkar R, Kalra J, Mantha V, Prasad K (1995) Lipid peroxidation and activity of antioxidant enzymes in diabetic rats. Mol Cell Biochem 151:113–119
120. Kakkar R, Mantha SV, Kalra J, Prasad K (1996) Time course study of oxidative stress in aorta and heart of diabetic rat. Clin Sci 91:441–448
121. Mak DHF, Ip SP, Li PC, Poon MKT, Ko KM (1996) Alterations in tissue glutathione antioxidant system in streptozotocin-induced diabetic rats. Mol Cell Biochem 162:153–158
122. Ye G, Metreveli NS, Donthi RV et al (2004) Catalase protects cardiomyocyte function in models of type 1 and type 2 diabetes. Diabetes 53:1336–1343
123. Cai L, Wang Y, Zhou G et al (2006) Attenuation by metallothionein of early cardiac cell death via suppression of mitochondrial oxidative stress results in a prevention of diabetic cardiomyopathy. J Am Coll Cardiol 48:1688–1697

Part III
Therapeutic Aspects of Diabetic Cardiomyopathy

The Next Generation of Diagnostic Biomarkers for Type 2 Diabetes

Samarjit Das and Tengku Ain Kamalden

Abstract Diabetes is a major cause of morbidity and mortality in both the United States and Asia. The greatest public health impact of diabetes is its vascular complications, which include both microvascular (retinopathy, nephropathy, and neuropathy) and macrovascular (ischemic heart disease, cerebrovascular disease, peripheral vascular disease) processes. Given that the diabetes epidemic continues to grow worldwide, there is a clear need for improvements in the management of the disease and its complications. The identification of biomarkers and pathogenic determinants of progression not only could provide tools physicians could use to monitor disease progression but could also give important insights into the mechanisms behind diabetic vascular complications, thus potentially opening new avenues for treatment.

The roles played by circulating miRNAs in diabetes mellitus (DM) and its vascular complications, both as biomarkers of disease and as critical contributors to disease progression, are not very clear. Recently, researchers have identified powerful roles for circulating miRNAs and their potential therapeutic implications, partly by advances in various technologies with which researchers can identify circulating miRNAs (next-generation sequencing, RNA Seq, digital qPCR, etc.) and then successfully isolate and study them. Circulating miRNAs play a role in the metabolic, inflammatory, and antiangiogenic pathways in type 2 diabetes (T2D).

S. Das, Ph.D. (✉)
Department of Cardiac-Pathology, Johns Hopkins University, Baltimore, MD, USA

Department of Pathology, Cardiac Division, Johns Hopkins University, Baltimore, MD 21205, USA
e-mail: sdas11@jhmi.edu

T.A. Kamalden, D. Phil
University of Malaya Eye Research Centre Faculty of Medicine, University of Malaya, Kuala Lumpur, Malaysia

B. Turan and N.S. Dhalla (eds.), *Diabetic Cardiomyopathy: Biochemical and Molecular Mechanisms*, Advances in Biochemistry in Health and Disease 9,
DOI 10.1007/978-1-4614-9317-4_19, © Springer Science+Business Media New York 2014

Keywords Diabetes mellitus • Type 2 diabetes • MicroRNA (miRNA) • Circulating miRNA • Biomarkers

1 Introduction

Diabetes mellitus (DM) is a metabolic disorder caused by defects in insulin secretion, insulin action, or both [1–3]. The disease involves disturbances in carbohydrate, protein, and fat metabolism and chronic hyperglycemia from insulin deficiency [4–7]. This disease is the most common endocrine disorder and remains a global health problem. More than 150 million people are diagnosed with diabetes globally, and it is estimated that by the year 2025 this number will be doubled [8–10]. Long-term complications such as retinopathy, neuropathy, nephropathy, and cardiovascular disease arise when the disease is uncontrolled or has progressed [11–14]. There are two major classifications of DM. In type 1, patients have little or no endogenous insulin secretary capacity [15]. The two main forms of clinical type 1 diabetes are type 1a, which is caused by immunological destruction of pancreatic cells, resulting in insulin deficiency, and type 1b (idiopathic), in which there is no evidence of autoimmunity [16]. Type 1b is more prevalent among individuals of African and Asian origin [17]. Type 2 is the most common form of diabetes and is characterized by disorders of insulin secretion and insulin resistance [18]. In Western countries, the disease affects up to 7 % of the population [19, 20], and globally, it affects 5–7 % of the world's population [19, 20]. This prevalence is underestimated because many cases, perhaps as much as 50 % in some populations, remain undiagnosed, especially in developing countries [15].

Type 2 diabetes (T2D) is the more common form of DM across the globe: almost 5–7 % of the world's population is now afflicted [8, 9, 20]. The prevalence of T2D varies considerably, ranging from less than 1 % to more than 50 %, depending on the country [15]. The incidence of T2D is higher in urban centers compared to rural areas [9, 21, 22]. Interestingly, T2D is associated with a more "Westernized" lifestyle, marked by decreased physical activity and obesity [23–25]. Recent data highlight an alarming trend with an increased incidence of T2D [25, 26]. In some countries, the numbers of children diagnosed with T2D is beginning to outnumber those diagnosed with type 1 [27].

2 Importance of Early Detection of Diabetes Mellitus

DM is a significant cause of morbidity worldwide as a result of its myriad macro- and microvascular complications. Diabetic complications usually arise from long-standing disease and are aggravated by poor glycemic control, hypertension, and hyperlipidemia. Of the 366 million people affected by this disease worldwide [1], about a third have either vision-threatening retinopathy [2] or nephropathy [3] or both.

Nearly half the patients suffering from T2D have at least one macrovascular complication such as cardiovascular, cerebrovascular, or peripheral artery disease; the majority have polyvascular disease [28]. The Framingham study showed that there was a two- to threefold increased incidence of cardiovascular disease among men and women with diabetes [29]. In contrast to type 1, where the diagnosis is often made at the time of the disease onset, T2D has a long asymptomatic phase, which may continue for as long as 12 years before the appearance of the earliest symptoms [30]. During this undiagnosed and untreated 'silent' preclinical phase, early changes to the vascular system, especially endothelial dysfunction, gradually occur. Retinopathy is a common microvascular diabetic complication and is one of the earliest clinical signs. As many as 29 % of newly diagnosed T2D patients have retinopathy at the time of diagnosis [31]. The presence of diabetic retinopathy increases the risk for developing coronary heart disease [13] and ischemic stroke [14], partially explainable by the effects of diabetic microangiopathy on larger vessels, where it has been shown that there is an increased angiogenesis of vaso vasorum on the walls of larger arteries [32]. The development of diabetic complications may not be completely prevented but can be delayed with early and appropriate treatment with good glycemic control. Thus, early detection and treatment of DM significantly delays the onset of micro- and macrovascular complications and reduces morbidity and mortality. Nowadays more studies are focused on identifying specific biomarkers to predict the risk of developing DM [33, 34] as well as development of early complications such as retinopathy.

3 Biomarkers

Biomarkers, are detectable and quantifiable biological differences or parameters in the body that indicate normal physiological processes and pathological processes or changes related to pharmacological therapeutics. Biomarkers may precede the onset of any early signs of clinical disease and hence are important as a diagnostic tool. Biomarkers are usually detected in the blood, urine, or tissues and are used as a measure of physiological status of the body's responses to pharmacological treatment. As shown in Table 1, in recent years many groups have worked to identify early-stage biomarkers for diabetes.

3.1 Circulating miRNAs

MicroRNAs (miRNAs) are small (~19–22 nt), stable noncoding RNAs that critically modulate posttranscriptional gene regulation by binding at the 3′-untranslated region (or UTR) of corresponding mRNAs [43]. Because of this binding at the 3′-UTR of mRNAs, miRNA either block translation or cause message degradation through RNA-induced silencing complex (RISC)-mediated events [43]. The biogenesis of a

Table 1 Biomarkers for type 1 and and type 2 diabetes mellitus (DM)

Type of DM	Complications	Source	Biomarkers	References
1	Retinopathy	Blood	E-selectin, TNF-α, sICAM-1	[35]
2	Nephropathy	Peritoneal dialysate	APOA4, AZGP1, EIF4GI, HLA-A, ALB, AMBP, APOA1, IGHG1, RBP4, HP	[36]
2	Unspecified	Urine	HINT1, CLU, EPRS	[37]
2	Unspecified	Plasma	Sfrp5	[38]
2	Unspecified	Blood	AGEs, AOPPs, oxLDL, NOX	[39]
2	Unspecified	Serum	B2M	[40]
2	Coronary artery disease	Serum	Fetuin-A	[41]
1	Retinopathy	Blood	hsCRP, ICAM-1	[42]

miRNA is to be transcribed and processed by RNase II (Drosha) in the nucleus from primary miRNA (pri-miRNA) transcript, then transported to the cytoplasm by exportin 5, to be spliced by Dicer before formation of the RISC, by association with aragonate 2 (Ago 2) and target mRNA [43]. miRNAs can be found in various biofluids, including blood and urine, and have been reported as specific biomarkers in a variety of nonneoplastic diseases. Several studies have shown that the stability of circulating miRNAs come from the formation of complexes between miRNA and some proteins, including Ago 2 [44–46]. Circulating exosomes or other microvesicles have also been shown to play a protective role for miRNAs, shielding them from the high level of nucleases in the blood (Fig. 1) [47].

More recently miRNAs are emerging as key players in the development of various diabetic complications, such as nephropathy [48–50], retinopathy [51–53], and cardiomyopathy [54–56]. miRNAs can provide opportunities to explore possible new treatment modalities for DM.

4 Circulating miRNAs and Diabetes

Mitchell et al. [57] showed for the first time that miRNAs can be found in the blood and can act as a potential biomarker for cancer detection. This finding led to the further realization that miRNAs could be secreted actively by different cell types or in response to tissue injury, and these miRNAs could subsequently enter freely into the bloodstream. These blood-based miRNAs acquire stability in the blood in many ways [44–47]. This interesting phenomenon has caught researcher attention. Multiple groups have started to look at the use of plasma or serum miRNAs as biomarkers for disease, both neoplastic and nonneoplastic. A number of well-characterized studies on circulating miRNA and cardiovascular complications have been published [58–61]. Although this burgeoning field has produced studies for various disease conditions, only a handful of publications on DM [54, 62–65] have

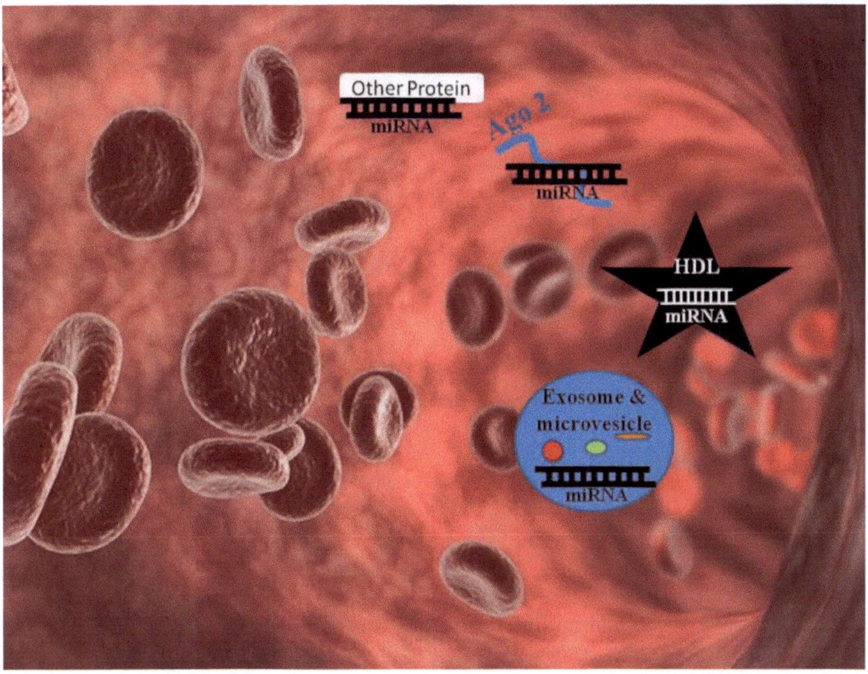

Fig. 1 *MicroRNAs (miRNAs) in circulating system.* Circulating miRNAs acquire protection from nucleases present in the blood by multiple mechanisms, including the formation of complexes with proteins (such as Ago 2) or by residing in a vesicle such as an exosome or HDL

surfaced in the past few years. We have summarized here each of the target miRNAs from these five different studies, and Fig. 2 shows all the candidate miRNAs that may have an important role in DM. Although most of these studies are in animal models, Zampetaki et al. [63] presented a very compelling case for the use of miRNA diagnostic biomarkers for T2D in their study for which blood from T2D patients was collected and analyzed.

5 Conclusions

It is not clear how miRNAs are being released into the circulation or how they are taken up into organs from the circulatory system. However, it is well documented that changes in miRNAs in the circulation can have profound effects on organ function. This field is at its very early stages and is evolving constantly. More than 5,000 hits were found on a "Goggle Patent" search engine for keywords such as "biomarker" and "miRNAs," suggesting that researchers are beginning to embrace circulating miRNAs as a potential biomarker for DM.

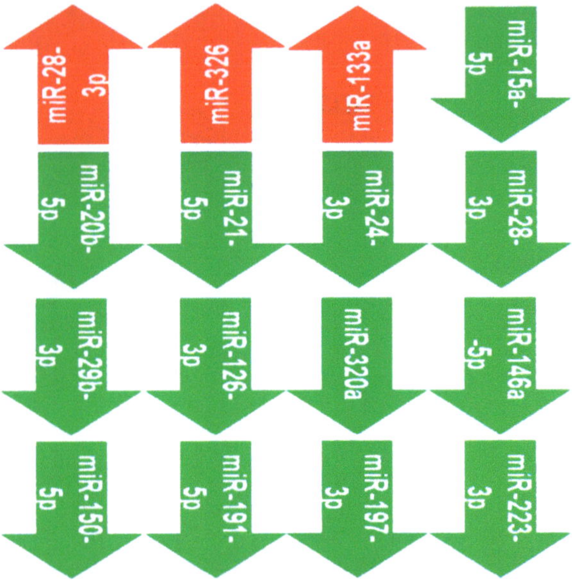

Fig. 2 Published evidence of miRNA targets between nondiabetic and type 2 diabetes (T2D) groups. Through a literature search, we have identified five published manuscripts to date [54, 62–65] that compared diabetic samples with nondiabetic samples. Among these 5, only 1 used human T2D samples and compared with nondiabetic patient samples [63]. However, in this study, miRNAs were not isolated from fresh patient sera and the patient population was not well characterized. Growing evidence highlights the importance of utilizing fresh serum when performing a miRNA isolation protocol

References

1. Whiting DR, Guariguata L, Weil C, Shaw J (2011) IDF diabetes atlas: global estimates of the prevalence of diabetes for 2011 and 2030. Diabetes Res Clin Pract 94:311–321
2. Saaddine JB, Honeycutt AA, Narayan KM et al (2008) Projection of diabetic retinopathy and other major eye diseases among people with diabetes mellitus: United States, 2005–2050. Arch Ophthalmol 126:1740–1747
3. Schernthaner G, Schernthaner GH (2013) Diabetic nephropathy: new approaches for improving glycemic control and reducing risk. J Nephrol. doi:10.5301/jn.5000281. [Epub ahead of print]
4. Kumar PJ, Clark M (2002) Textbook of clinical medicine. Saunders, London
5. Mellitus ECotDaCoD (1997) Report of the expert committee on the diagnosis and classification of diabetes mellitus. Committee Reports & Consensus Statements. Diabetes Care 20:1183–1197
6. Beverley B, Eschwège E (2003) The diagnosis and classification of diabetes and impaired glucose tolerance. In: Textbook of Diabetes 1 Ed: John C Pickup and Gareth Williams Third edition; Chapter 2, pp 2.1-2.11.
7. Johnson JA, et al (2002) Decreased mortality associated with the use of metformin compared with sulfonylurea monotherapy in type 2 diabetes. Diabetes Care 25:2244–2248

8. Amos AF, McCarty DJ, Zimmet P (1997) The rising global burden of diabetes and its complications: estimates and projections to the year 2010. Diabet Med 14(Suppl 5):S1–S85
9. King H, Aubert RE, Herman WH (1998) Global burden of diabetes, 1995–2025: prevalence, numerical estimates, and projections. Diabetes Care 21:1414–1431
10. Zimmet P (2000) Globalization, coca-colonization and the chronic disease epidemic: can the Doomsday scenario be averted? J Intern Med 247:301–310
11. Foong AW, Saw SM, Loo JL et al (2007) Rationale and methodology for a population-based study of eye diseases in Malay people: the Singapore Malay Eye Study (SiMES). Ophthalmic Epidemiol 14:25–35
12. Cheung N, Mitchell P, Wong TY (2010) Diabetic retinopathy. Lancet 376:124–136
13. Cheung N, Wang JJ, Klein R et al (2007) Diabetic retinopathy and the risk of coronary heart disease: the Atherosclerosis Risk in Communities Study. Diabetes Care 30:1742–1746
14. Cheung N, Rogers S, Couper DJ et al (2007) Is diabetic retinopathy an independent risk factor for ischemic stroke? Stroke 38:398–401
15. Bastaki S (2005) Diabetes mellitus and its treatment. Int J Diabetes Metab 13:111–134
16. McLarty DG, Athaide I, Bottazzo GF et al (1990) Islet cell antibodies are not specifically associated with insulin-dependent diabetes in Tanzanian Africans. Diabetes Res Clin Pract 9:219–224
17. Ahren B, Corrigan CB (1985) Intermittent need for insulin in a subgroup of diabetic patients in Tanzania. Diabet Med 2:262–264
18. DeFronzo RA (1997) Pathogenesis of NIDDM. Wiley, Chichester
19. Group RoWHOS Report of a WHO Study Group (1994) Prevention of diabetes mellitus. In: World Health Organ Tech Rep Ser. 844:1–100
20. Harris MI, Flegal KM, Cowie CC et al (1998) Prevalence of diabetes, impaired fasting glucose, and impaired glucose tolerance in U.S. adults. The Third National Health and Nutrition Examination Survey, 1988–1994. Diabetes Care 21:518–524
21. Zimmet P (1982) Type 2 (non-insulin-dependent) diabetes—an epidemiological overview. Diabetologia 22:399–411
22. Christacopoulos P (1975) The prevalence of diabetes mellitus and non-diabetic glycosuria in a rural population in Greece. Diabetologia 11:A-335
23. Bloomgarden ZT (1996) American Diabetes Association annual meeting 1996: the etiology of type II diabetes, obesity, and the treatment of type II diabetes. Diabetes Care 19:1311–1315
24. Eriksson J, Saloranta C, Widen E et al (1991) Non-esterified fatty acids do not contribute to insulin resistance in persons at increased risk of developing type 2 (non-insulin-dependent) diabetes mellitus. Diabetologia 34:192–197
25. Glaser NS (1997) Non-insulin-dependent diabetes mellitus in childhood and adolescence. Pediatr Clin N Am 44:307–337
26. Scott CR, Smith JM, Cradock MM, Pihoker C (1997) Characteristics of youth-onset noninsulin-dependent diabetes mellitus and insulin-dependent diabetes mellitus at diagnosis. Pediatrics 100:84–91
27. Cockram CS (2000) Diabetes mellitus: perspective from the Asia-Pacific region. Diabetes Res Clin Pract 50(Suppl 2):S3–S7
28. Papa G, Degano C, Iurato MP et al (2013) Macrovascular complication phenotypes in type 2 diabetic patients. Cardiovasc Diabetol 12:20
29. Kannel WB, McGee DL (1979) Diabetes and cardiovascular disease. The Framingham Study. JAMA 241:2035–2038
30. Harris MI, Klein R, Welborn TA, Knuiman MW (1992) Onset of NIDDM occurs at least 4–7 yr before clinical diagnosis. Diabetes Care 15:815–819
31. Dolben J, Owens D, Young S, Vora J, Atiea J, Dean J, Luzio S (1988) Retinopathy at presentation in type 2 (non-insulin dependent) diabetic patients. Diabet Med 5(Suppl 2):20
32. Arcidiacono MV, Traveset A, Rubinat E et al (2013) Microangiopathy of large artery wall: a neglected complication of diabetes mellitus. Atherosclerosis 228:142–147
33. Meigs JB, Hu FB, Rifai N, Manson JE (2004) Biomarkers of endothelial dysfunction and risk of type 2 diabetes mellitus. JAMA 291:1978–1986

34. Meigs JB (2009) Multiple biomarker prediction of type 2 diabetes. Diabetes Care 32:1346–1348
35. Roy MS, Janal MN, Crosby J, Donnelly R (2013) Inflammatory biomarkers and progression of diabetic retinopathy in African Americans with type 1 diabetes. Invest Ophthalmol Vis Sci 54:5471–5480
36. Yang MH, Wang HY, Lu CY et al (2013) Proteomic profiling for peritoneal dialysate: differential protein expression in diabetes mellitus. BioMed Res Int 2013:642964
37. Chu L, Fu G, Meng Q et al (2013) Identification of urinary biomarkers for type 2 diabetes using bead-based proteomic approach. Diabetes Res Clin Pract 101:187–193
38. Lu YC, Wang CP, Hsu CC et al (2013) Circulating secreted frizzled-related protein 5 and wingless-type MMTV integration site family member 5a levels in patients with type 2 diabetes mellitus. Diabetes Metab Res Rev 29:551–556
39. Gradinaru D, Borsa C, Ionescu C, Margina D (2013) Advanced oxidative and glycoxidative protein damage markers in the elderly with type 2 diabetes. J Proteomics 92:313–322
40. Cheung CL, Lam KS, Cheung BM (2013) Serum beta-2 microglobulin predicts mortality in people with diabetes. Eur J Endocrinol 169:1–7
41. Zhao ZW, Lin CG, Wu LZ et al (2013) Serum fetuin-A levels are associated with the presence and severity of coronary artery disease in patients with type 2 diabetes. Biomarkers 18:160–164
42. Muni RH, Kohly RP, Lee EQ et al (2013) Prospective study of inflammatory biomarkers and risk of diabetic retinopathy in the diabetes control and complications trial. JAMA Ophthalmol 131:514–521
43. Das S, Ferlito M, Kent OA et al (2012) Nuclear miRNA regulates the mitochondrial genome in the heart. Circ Res 110:1596–1603
44. Arroyo JD, Chevillet JR, Kroh EM et al (2011) Argonaute2 complexes carry a population of circulating microRNAs independent of vesicles in human plasma. Proc Natl Acad Sci U S A 108:5003–5008
45. Vickers KC, Palmisano BT, Shoucri BM et al (2011) MicroRNAs are transported in plasma and delivered to recipient cells by high-density lipoproteins. Nat Cell Biol 13:423–433
46. Wang K, Zhang S, Weber J et al (2010) Export of microRNAs and microRNA-protective protein by mammalian cells. Nucleic Acids Res 38:7248–7259
47. Hunter MP, Ismail N, Zhang X et al (2008) Detection of microRNA expression in human peripheral blood microvesicles. PLoS One 3:e3694
48. Kato M, Zhang J, Wang M et al (2007) MicroRNA-192 in diabetic kidney glomeruli and its function in TGF-beta-induced collagen expression via inhibition of E-box repressors. Proc Natl Acad Sci U S A 104:3432–3437
49. Kato M, Wang L, Putta S et al (2010) Post-transcriptional up-regulation of Tsc-22 by Ybx1, a target of miR-216a, mediates TGF-beta-induced collagen expression in kidney cells. J Biol Chem 285:34004–34015
50. Deshpande SD, Putta S, Wang M et al (2013) Transforming growth factor-beta induced cross talk between p53 and a microRNA in the pathogenesis of diabetic nephropathy. Diabetes 62:3151–3162
51. Kovacs B, Lumayag S, Cowan C, Xu S (2011) MicroRNAs in early diabetic retinopathy in streptozotocin-induced diabetic rats. Invest Ophthalmol Vis Sci 52:4402–4409
52. Feng B, Chen S, McArthur K et al (2011) miR-146a-mediated extracellular matrix protein production in chronic diabetes complications. Diabetes 60:2975–2984
53. McArthur K, Feng B, Wu Y et al (2011) MicroRNA-200b regulates vascular endothelial growth factor-mediated alterations in diabetic retinopathy. Diabetes 60:1314–1323
54. Feng B, Chen S, George B et al (2010) miR133a regulates cardiomyocyte hypertrophy in diabetes. Diabetes Metab Res Rev 26:40–49
55. Shen E, Diao X, Wang X et al (2011) MicroRNAs involved in the mitogen-activated protein kinase cascades pathway during glucose-induced cardiomyocyte hypertrophy. Am J Pathol 179:639–650
56. Greco S, Fasanaro P, Castelvecchio S et al (2012) MicroRNA dysregulation in diabetic ischemic heart failure patients. Diabetes 61:1633–1641

57. Mitchell PS, Parkin RK, Kroh EM et al (2008) Circulating microRNAs as stable blood-based markers for cancer detection. Proc Natl Acad Sci U S A 105:10513–10518
58. Wang GK, Zhu JQ, Zhang JT, Li Q, Li Y, He J, Qin YW, Jing Q (2010) Circulating microRNA: a novel potential biomarker for early diagnosis of acute myocardial infarction in humans. Eur Heart J 31:659–666
59. Ji X, Takahashi R, Hiura Y, Hirokawa G, Fukushima Y, Iwai N (2009) Plasma miR-208 as a biomarker of myocardial injury. Clin Chem 55:1944–1949
60. Fichtlscherer S, De Rosa S, Fox H, Schwietz T, Fischer A, Liebetrau C, Weber M, Hamm CW, Röxe T, Müller-Ardogan M, Bonauer A, Zeiher AM, Dimmeler S (2010) Circulating microRNAs in patients with coronary artery disease. Circ Res 107:677–684
61. Tijsen AJ, Creemers EE, Moerland PD, de Windt LJ, van der Wal AC, Kok WE, Pinto YM (2010) MiR423-5p as a circulating biomarker for heart failure. Circ Res 106:1035–1039
62. Sebastiani G, Grieco FA, Spagnuolo I et al (2011) Increased expression of microRNA miR-326 in type 1 diabetic patients with ongoing islet autoimmunity. Diabetes Metab Res Rev 27: 862–866
63. Zampetaki A, Kiechl S, Drozdov I et al (2010) Plasma microRNA profiling reveals loss of endothelial miR-126 and other microRNAs in type 2 diabetes. Circ Res 107:810–817
64. Balasubramanyam M, Aravind S, Gokulakrishnan K et al (2011) Impaired miR-146a expression links subclinical inflammation and insulin resistance in Type 2 diabetes. Mol Cell Biochem 351:197–205
65. Chavali V, Tyagi SC, Mishra PK (2012) MicroRNA-133a regulates DNA methylation in diabetic cardiomyocytes. Biochem Biophys Res Commun 425:668–672

PDE-5 Inhibitors in Protection of Diabetic Heart

Saisudha Koka and Rakesh C. Kukreja

Abstract Cardiovascular diseases are the leading cause of death and disability among diabetic patients. Diabetes-induced cardiovascular complications are primarily caused by impaired NO bioavailability, eventually leading to endothelial dysfunction and a sequela of cardiovascular disorders. Phosphodiesterase-5 (PDE-5) inhibitors are class erectile dysfunction (ED) drugs that are shown to induce powerful cardiovascular benefits against ischemia–reperfusion injury, myocardial infarction, heart transplantation, cardiac hypertrophy, heart failure, and doxorubicin-induced cardiotoxicity. The use of PDE-5 inhibitors, including sildenafil (Viagra), vardenafil (Levitra), and tadalafil (Cialis), represent a potential therapeutic strategy to reduce the incidence of heart diseases in diabetic patients because these compounds prevent damage to the vascular endothelium by upregulating eNOS, iNOS, and increased NO production. This review provides new insights into the potential benefits of PDE-5 inhibitors for diabetic patients and discusses the multiple molecular mechanisms by which these drugs protect the diabetic hearts.

Keywords PDE-5 inhibitors • Diabetes • NO signaling • ROS • Oxidative stress • Cardioprotection

S. Koka
Department of Internal Medicine, Division of Cardiology, VCU Pauley Heart Center, Virginia Commonwealth University, Richmond, VA 23298-0204, USA

R.C. Kukreja, Ph.D. (✉)
Division of Cardiology, Pauley Heart Center, Virginia Commonwealth University, Box 980204, 1101 East Marshall Street, Room 7-020D, Richmond, VA 23298, USA
e-mail: rakesh@vcu.edu

B. Turan and N.S. Dhalla (eds.), *Diabetic Cardiomyopathy: Biochemical and Molecular Mechanisms*, Advances in Biochemistry in Health and Disease 9,
DOI 10.1007/978-1-4614-9317-4_20, © Springer Science+Business Media New York 2014

1 Introduction

Diabetes is a major public health concern affecting nearly 170 million people worldwide [1]. Its incidence is dramatically on the rise, with a global estimate of more than 300 million diabetic patients by 2025 [1, 2]. Diabetes is associated with increased risk for a wide variety of long-term health problems including macrovascular/microvascular complications that induce damage to major organs such as brain, heart, kidney, eye, peripheral vascular system, and nervous system. Among the various diabetes-related complications, heart diseases and associated cardiovascular complications continue to be the leading cause of morbidity and mortality, accounting for nearly 80 % of the deaths in the diabetic population [3–6]. Both type 1 and type 2 diabetes have been closely linked to cardiovascular disease (CVD). Diabetic patients exhibit increased risk for multiple cardiovascular complications such as endothelial dysfunction, coronary atherosclerosis, microangiopathy, and hypertension [7, 8]. They also exhibit enhanced risk for cardiac dysfunction that occurs independently of other risk factors such as hypertension and coronary artery disease. Several factors exaggerate the risk of heart failure and stroke in diabetic patients, such as hypertension, insulin resistance, hyperinsulinemia, hyperamilinemia, dyslipidemia, and coagulation system disorder and hyperhomosisteinemia. It is estimated that two of three diabetes patients develop heart failure and eventually die of myocardial infarction or stroke [9, 10]. In addition, diabetics are known to experience worse outcomes following acute myocardial infarction, coronary angioplasty, and cardiac surgery. The CVD treatment accounts for a large part of the healthcare costs attributable to management of diabetic complications, averaging nearly 50 % of the total healthcare expenditure of diabetic patients [11]. These concerns all denote the overwhelming significance of CVD in diabetic patients [12].

2 Pathophysiology of Cardiovascular Disease in Diabetic Patients

Cardiovascular risk actually begins with insulin resistance, a condition that occurs well before diabetes mellitus. Reduced nitric oxide (NO) bioavailability and enhanced reactive oxygen species (ROS) formation within the vascular wall results in an imbalance between NO and ROS, contributing to impaired insulin utilization and thus leading to insulin resistance [13]. Insulin resistance causes a number of micro- and macrovascular insults leading to retinopathy, nephropathy, and painful neuropathy, and eventually to more adverse complications such as atherosclerosis, coronary artery disease, and cerebrovascular disease. The increased incidence of these complications has been attributed to higher levels of inflammatory cytokines, chronic hyperglycemia leading to formation of advanced glycation end products (AGEs), and elevated levels of oxidative stress that lead to endothelial dysfunction. Vascular NO is critical for normal vasodilatation and endothelial function, and impairment of NO bioavailability is known to cause endothelial dysfunction [14]. In a large meta-analysis, it was reported that endothelial dysfunction is a significant independent risk factor for cardiac death, myocardial infarction, stroke, and the need for coronary revascularization [15]. A number of clinical studies have shown that hyperglycemia and increased AGEs are key factors in potentiating vascular inflammation and increasing levels of ROS and oxidative stress [16, 17]. This vascular milieu of elevated inflammation, impaired NO bioavailability, and oxidative stress plays an integral role in the progression of atherosclerosis and subsequently acute coronary syndromes, culminating in significant morbidity and mortality of the diabetic patient [18].

In recent years, it is known that diabetes may influence heart muscles independently in addition to early atherosclerosis of the coronary artery, which causes ischemic heart disease [19]. This condition is referred as diabetic cardiomyopathy, a disease process that affects the myocardium in diabetic patients, causing a wide range of structural abnormalities that eventually lead to left ventricular hypertrophy and diastolic and systolic dysfunction [20]. The cardiomyopathy associated with diabetes is a unique myopathic state that appears to be independent of macrovascular/microvascular disease and contributes significantly to CVD morbidity and mortality in diabetic patients, especially those with coexistent hypertension or coronary artery disease with resulting synergistic adverse effects. At the onset of diabetes, the heart undergoes short-term physiological adaptation to metabolic alterations, but prolonged hyperglycemic conditions induce degenerative changes that eventually culminate in irreversible pathological remodeling. Morphological changes include thickening of capillary basement membrane, proliferation of small arterioles, myocyte atrophy, accumulation of ground matrix, and cardiac fibrosis [21, 22] in the diabetic myocardium. Clinically, diabetic patients manifest abnormalities in diastolic left ventricular function, which is caused by changes such as interstitial fibrosis process, collagen formation, reduced ventricular elasticity, and hypertrophy of heart muscle cells. At the cellular level, disruption of calcium release from cytoplasm, changes of troponin T structure, and increased activity of pyruvate kinase appear [19]. These changes cause distraction of heart muscle contraction and relaxation, as well as elevation of end-diastolic pressure, that cause diabetic cardiomyopathy [20, 23].

The pathogenesis of CVD in diabetes is multifactorial, including increased oxidative stress, disturbances in glucose and fatty acid metabolism, mitochondrial dysfunction, alterations of vasoactive factors such as endothelin-1 and NO, abnormalities in intracellular calcium homeostasis, altered transcription of genes encoding for contractile and structural proteins, autonomic dysfunction, and abnormal expression of growth factors and their receptors [24, 25]. Moreover, there is increase in inflammatory cytokines, apoptosis of cardiac muscles, dysregulation of the renin-angiotensin system, hyperglycemia-induced activation of protein kinase C isoforms, and alterations in gene expression induced by miRNAs [26–28]. Additionally, diabetic conditions are known to impede key cardioprotective signaling pathways and blunt a number of cardioprotective modalities [29–31].

3 Role of NO-cGMP Signaling in Diabetes-Induced Cardiovascular Damage

High levels of glucose cause biochemical changes in the endothelial cells that are reminiscent of early molecular alterations in the target organs of diabetes [32, 33]. In vitro studies revealed that acute exposure of endothelial cells to high glucose significantly lowers NO production [34]. In addition to its reduced level under hyperglycemic conditions, NO may also be sequestered by glucose-induced oxidative stress, which generates an imbalance in the counter-activity of NO and endothelins, which are major vasoactive factors produced by endothelial cells [32, 33, 35]. Endothelium-derived NO activates soluble guanylyl cyclase in vascular smooth muscle cells, resulting in enhanced cyclic guanosine monophosphate (cGMP) concentrations and vasorelaxation. Downregulation of the NO-cGMP pathway has been implicated in the pathogenesis of diabetes-induced cardiovascular complications [36, 37]. Consistent with these observations, type 2 diabetic patients have impaired NO synthesis and decreased expression of eNOS and iNOS in skeletal muscle [38]. Furthermore, nitric oxide synthase (NOS) enzymes have been suggested to play important roles in acetylcholine-induced paradoxical vasoconstriction in atherosclerotic coronary arteries [39]. Recent animal studies with NOS inhibitors and eNOS gene deficiency suggest that the NO signaling pathway may regulate and promote glucose uptake in myocytes and enhance muscle glucose utilization as well [40–43]. Moreover, insulin-stimulated muscle glucose uptake and endothelial NO production was blocked using the eNOS inhibitor, L-NMMA (*NG*-monomethyl-L-arginine) [44]. eNOS knockout mice had decreased oxygen consumption, increased weight gain, and were resistant to insulin [45]. Furthermore, genetic variations of eNOS gene influenced energy expenditure, severity of glucose intolerance, and risk of developing type 2 diabetes [46]. A recent epidemiological study provided evidence of a strong correlation between the risk factors associated with metabolic syndrome (i.e., obesity, elevated fasting glucose levels, dyslipidemia, hypertension) and urinary cGMP excretion, suggesting that a reduction of NO bioactivity concurs with these cardiovascular risk factors [47].

4 PDE-5 Inhibitors in the Treatment of Diabetes-Associated Complications

Phosphodiesterase 5 (PDE-5) inhibitors including sildenafil (Viagra), vardenafil (Levitra), and tadalafil (Cialis) are erectile dysfunction (ED) drugs that reduce damage to the penile vascular endothelium by upregulating eNOS, iNOS, and increased NO production [48–50]. PDE-5 inhibitors are commonly used among the diabetic patient population because ED is a major and prevalent vascular complication in diabetes [51]. ED is present in 32 % of insulin-dependent diabetics and 46 % of non-insulin-dependent diabetics. Chronic administration of PDE-5 inhibitors has been associated with increased persistent vascular and endothelial function by increasing the level of endothelial cGMP generated by activation of eNOS [52]. Sildenafil dilates epicardial coronary arteries, improves endothelial dysfunction, and inhibits platelet activation in patients with coronary artery disease [53] and acutely enhances flow-mediated vasodilation in patients with heart failure [54]. Therefore, inhibiting cGMP degradation by sildenafil might be a rational approach to treat patients with diabetes, coronary artery disease, or heart failure [55]. In support of this notion, previous reports demonstrated that sildenafil and vardenafil improved vasorelaxation through enhanced endogenous NO signaling in streptozotocin-induced diabetic rats [56, 57]. In streptozotocin-induced type 1 diabetic rats, 14 days of treatment with sildenafil improved vasorelaxation, and long-term administration of the PDE-5 inhibitor DA-8159 prevented ED and preserved endothelial function through enhanced endogenous NO signaling [56, 58]. The PDE-5 inhibitor vardenafil improved cardiovascular dysfunction in experimental diabetes mellitus [57]. Diabetic rats treated with vardenafil showed a tendency toward higher dP/dt_{max} and dP/dt_{min}, without reaching statistical significance. The load-independent, PV-loop-derived contractility indices (E_{max}, PRSW, and dP/dt_{max}-EDV) were significantly improved in the vardenafil treatment group. In a clinical study, chronic (alternate-day) administration of tadalafil in men with ED of any etiology had improved endothelial function as indicated by marked changes in serum markers of endothelial function [59]. Furthermore, both acute and chronic administration of sildenafil improved endothelial function in patients with type 2 diabetes [60, 61].

Pioneering studies from our laboratory have shown that PDE-5 inhibitors restore NO signaling and protect against myocardial ischemia–reperfusion (I/R) injury in normoglycemic mice [52, 62–64]. Several other investigators have also demonstrated the cardioprotective effects of PDE-5 inhibitors in different models of ischemic injury [65–67]. Moreover, PDE-5 inhibitors attenuate cardiac dysfunction following myocardial infarction and doxorubicin-induced cardiomyopathy [68–71]. Using proteomic analysis, our laboratory has recently demonstrated that chronic tadalafil treatment (28-day treatment) modulates cardiac proteins, specifically those associated with cytoskeletal rearrangement such as myosin light chain-2, myosin light chain-4, myosin heavy chain-α, and myosin binding protein-C, which contribute to contractile dysfunction [72]. These data suggest that tadalafil therapy may downregulate cytoskeletal contractile proteins associated with cardiac remodeling and heart failure. A recent study from our laboratory has shown that tadalafil, similar to other PDE-5 inhibitors in nondiabetic models, significantly reduces infarct size following I/R in the diabetic heart and attenuates necrosis and apoptosis following simulated ischemia and reoxygenation in isolated ventricular cardiomyocytes [73]. Moreover, tadalafil therapy in type 2 diabetic mice ameliorated circulating inflammatory cytokines and chemokines while improving fasting glucose levels and reducing infarct size following I/R injury in the diabetic heart [73]. Tadalafil treatment also improved PKG activity in the cardiomyocytes of diabetic mice compared to vehicle-treated controls [73]. These studies provided evidence that PDE-5 inhibitors restore NO signaling and induce cardioprotective effects through several cellular and molecular mechanisms.

5 PDE-5 Inhibitors in Attenuation of Oxidative Stress in the Diabetic Myocardium

In diabetes, multiple hyperglycemia-induced pathways causing oxidative stress including oxidized low lipoprotein, AGEs/RAGEs, and heme oxygenase pathways [74, 75] have been reported. In addition, increased mitochondrial superoxide and ROS production lead to apoptosis of cardiomyocytes. Depletion of proteins involved in electron-chain transport and increased expression of proteins involved in β-oxidation causing increased superoxide production, mitochondrial uncoupling, and reduced ATP generation in diabetic myocardium were reported earlier [76, 77]. Enhanced oxidative stress can modify cellular components (including proteins) to elicit damage and alter gene transcription of specific vasoactive and cardioprotective factors, leading to structural and functional defects in the diabetic myocardium [35, 78]. Oxidative stress is a major cause of reduced endothelial NO bioavailability in diabetes and is involved in the pathogenesis and progression of diabetic tissue damage. Increased ROS generation and impaired antioxidant defenses could both contribute to oxidative stress. Many studies have shown that ROS generation increases in both type 1 and type 2 diabetes [79–82]. PDE-5 inhibitors may suppress oxidative stress. Importantly, we previously demonstrated that PDE-5 inhibitors sildenafil and vardenafil reduced myocardial infarct size when administered at reperfusion following ischemia [83], a well-established model in which ROS have been widely implicated in causing reperfusion injury. Recent studies have also shown that sildenafil inhibits superoxide formation in cultured corpus cavernosal smooth muscle cells derived from rabbit penis [84, 85]. Chronic treatment with tadalafil in diabetic mice was shown to improve redox signaling by enhancing the antioxidant enzyme glutathione-*S*-transferase kappa-1 (GSKT-1) and downregulating redox regulatory chaperones, heat shock protein 8, and 75-kDa glucose regulatory protein [72]. Moreover, tadalafil-treated diabetic mice had significantly lowered plasma levels of GSSG/GSH, suggesting reduction of oxidative stress [72]. Using a combined physiological and biochemical approach we recently demonstrated that chronic treatment with tadalafil attenuated oxidative stress induced in type 2 diabetic hearts [86]. Tadalafil treatment protected the diabetic mice hearts from I/R injury, consistent with our previous studies that demonstrated PDE-5 inhibitors induce powerful cardioprotective effects against in vivo myocardial I/R injury in normoglycemic mice [64, 71]. Tadalafil also exhibited beneficial effects on the systemic abnormalities induced by diabetes. Tadalafil treatment improved the metabolic status of mice as evidenced by slight decrease in body weight and blood glucose coupled with significant reductions in hyperinsulinemia and hypertriglyceridemia. Furthermore, tadalafil treatment decreased ROS production in isolated ventricular myocytes following simulated ischemia and reoxygenation (Fig. 1).

Multiple sources of endogenously generated ROS have been implicated in oxidative damage of the diabetic vasculature and heart [87–90]. Importantly, increased expression of NAD(P)H oxidase proteins has been observed in the vasculature from animal models of diabetes or from diabetic patients [87, 91, 92]. In this recent study,

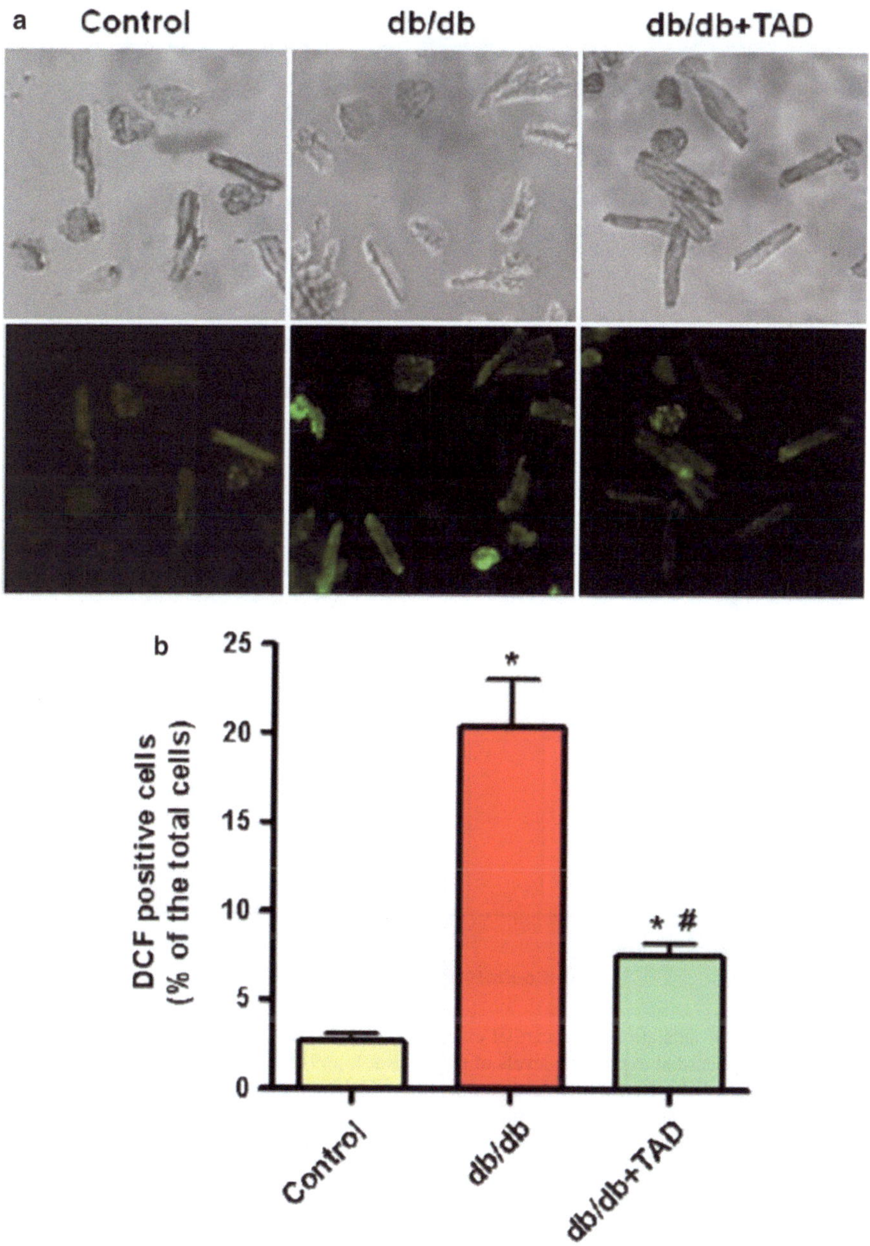

Fig. 1 Effect of tadalafil on reactive oxygen species (ROS) generation in adult cardiomyocytes following simulated ischemia and reoxygenation: representative images with chlorofluorescein (DCF) staining in isolated cardiomyocytes. (**a**) Bright-field images (*top*) and green fluorescent images (*bottom*). ×200. (**b**) ROS production in cardiomyocytes quantified and expressed as percent (%) for DCF-positive cells among total cells. Data are mean ± SEM ($n = 4$/group). *$P < 0.05$ versus db/db mouse cardiomyocytes

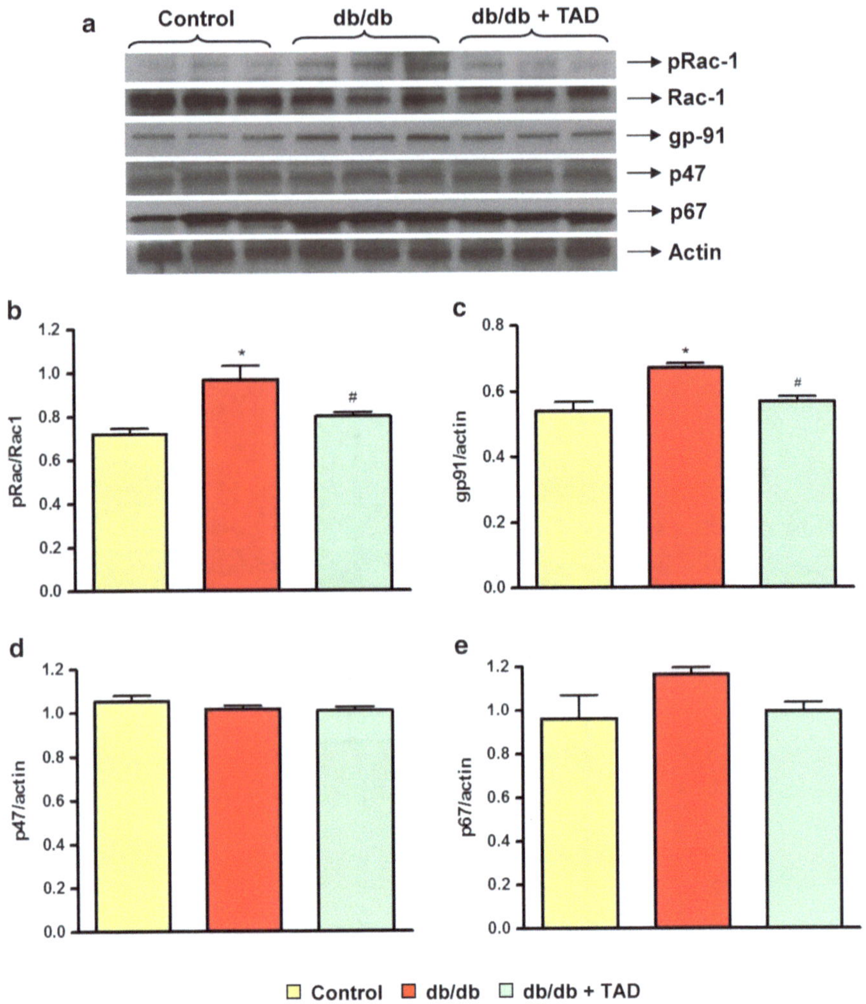

Fig. 2 Effect of tadalafil on myocardial expression of NAD(P)H oxidase subunits. (**a**) Western blots show myocardial expression levels of pRac1, Rac1, gp91^{phox} p47^{phox} p67^{phox}, and representative actin bands. (**b**) Densitometric quantification of protein expression of pRac1 normalized against Rac1. (**c**) Densitometric quantification of protein expression of gp91^{phox} normalized against actin. (**d**) Densitometric quantification of protein expression of p47^{phox} normalized against actin. (**e**) Densitometric quantification of protein expression of p67^{phox} normalized against actin. Data are mean ± SEM (n = 6/group). *P < 0.05 versus control; $^{\#}P$ < 0.05 versus db/db

we also tested the effect of tadalafil treatment on NAD(P)H oxidase, a major enzyme involved in oxidative stress [86]. The NAD(P)H oxidase enzyme complex consists of the membrane subunit cytochrome *b* 558 (p22 phox and gp91 phox), multiple cytoplasmic subunits (p67 phox and p47phox), and the small G protein Rac-1. We showed that tadalafil treatment attenuated expression of NAD(P)H oxidase subunits pRac-1 and gp91phox in type 2 diabetic hearts (Fig. 2) and NAD(P)H oxidase activity (Fig. 3a).

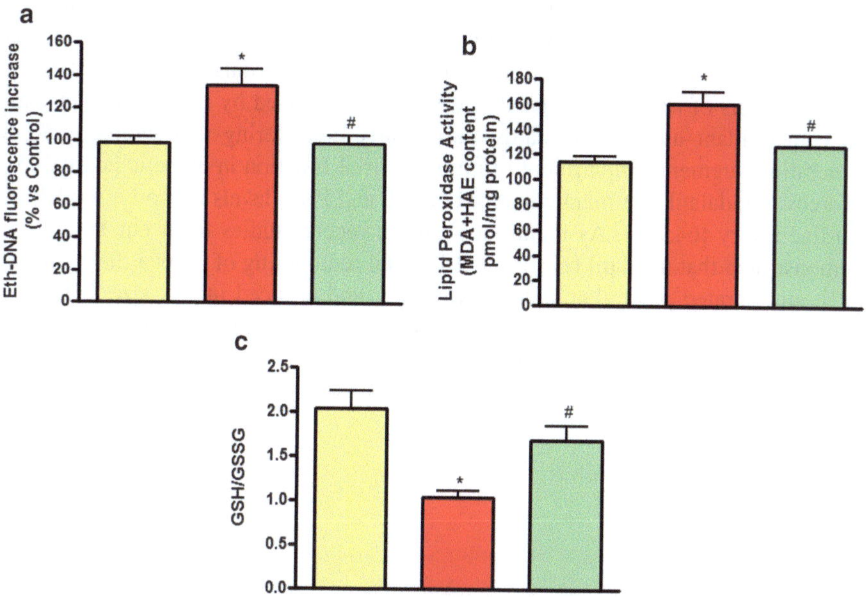

Fig. 3 Effect of tadalafil on NAD(P)H oxidase activity, lipid peroxidation and glutathione levels in diabetic hearts. NAD(P)H-dependent activity represented as percent increases in ethidium fluorescence compared to control (**a**), myocardial lipid peroxidation (**b**), and GSH/GSSG ratio (**c**) in control, db/db, and tadalafil (TAD)-treated db/db groups. Data are mean ± SEM ($n = 6$/group).*$P < 0.05$ versus control; $^{\#}P < 0.05$ versus db/db

Enhanced lipid peroxidation in diabetes is an autocatalytic mechanism leading to oxidative destruction of cellular membranes in the heart. Cardiac lipid peroxidation activity in the db/db mice was increased by 41.2 % as compared to the control group ($n = 6$/group, $P < 0.05$, Fig. 3b). Tadalafil treatment significantly attenuated the enhanced lipid peroxidation in the diabetic mice. Moreover, tadalafil treatment enhanced the GSH/GSSG ratio in the myocardium of db/db mice (Fig. 3c). Reduced glutathione is a major intracellular redox buffer and functions as a direct free radical scavenger.

In type 2 diabetes, ROS are involved in insulin resistance, via its regulatory effects on mitochondrial function [93]. Maintenance of mitochondrial membrane potential ($\Delta\Psi_m$) is necessary for production of energy (ATP) and preservation of cellular homeostasis. It has been demonstrated that maintenance of $\Delta\Psi_m$ is a critical primary determinant of myocyte survival [94]. We measured dissipation of $\Delta\psi_m$ of ventricular cardiomyocytes following simulated ischemia and reoxygenation injury by JC-1 staining. Cardiomyocytes from untreated diabetic mice exhibited loss of $\Delta\psi_m$ whereas control and tadalafil-treated diabetic mice demonstrated preserved $\Delta\psi_m$ and intact mitochondrial membrane. Similar preservation of $\Delta\psi_m$ was observed following simulated ischemia/reoxygenation in nondiabetic cardiomyocytes treated with sildenafil [49]. The mechanism by which tadalafil preserves $\Delta\psi_m$ in diabetic cardiomyocytes is not clear, although it may be mediated in part through opening of mitochondrial ATP-sensitive potassium channels, which appear to be the effectors of cardioprotection with PDE-5 inhibitors as reported previously [63, 83]. The protective effects of tadalafil in diabetic heart may be caused by maintenance of balance in oxidant–antioxidant status, particularly considering that increased ROS generation augments impairment of mitochondrial function in diabetic hearts [95]. Moreover, tadalafil treatment enhanced cGMP and PKG levels in mouse models of cardiac injury [64, 69]. As mentioned earlier, recent studies from our laboratory demonstrated that tadalafil reversed detrimental remodeling of myocardial proteins [72], ameliorated pro-inflammatory cytokines, and reduced infarct size following I/R injury while upregulating PKG activity in isolated cardiomyocytes of diabetic mice [73]. In this respect, tadalafil is attractive because this drug can modulate multiple molecular targets of cardioprotection as compared to antioxidants, which only reduce oxidative stress.

6 Conclusions

PDE-5 inhibitors have been successfully used by millions of people worldwide for treatment of male erectile dysfunction. Several basic science studies now demonstrate that sildenafil and other PDE-5 inhibitors have a protective effect in clinical scenarios including ischemia/reperfusion injury, myocardial infarction, heart transplantation, cardiac hypertrophy, heart failure, doxorubicin-induced cardiotoxicity, Duchenne muscular dystrophy, and stem-cell preconditioning [96]. Today, nearly 100 clinical trials with PDE-5 inhibitors have been completed or are ongoing that focus on the potential cardiovascular benefits [97]. Our studies provide new insights into the potential role of PDE-5 inhibitors in type 2 diabetic heart (Fig. 4). We believe these drugs may protect the diabetic heart through multiple redundant mechanisms. First, PDE-5 inhibitors could attenuate ROS generation and decrease the accumulation of oxidized glutathione (GSSG) in the diabetic through inhibition of NADPH oxidase [86]. Second, initiation of cGMP-dependent PKG signaling by PDE-5 inhibitors may increase PI3/Akt phosphorylation, which could increase eNOS phosphorylation leading to enhanced NO bioavailability and activation of PKG through soluble guanylate cyclase (sGC)-dependent formation of cGMP. The ROS-dependent inhibition of AMPK and PI3/Akt phosphorylation may reduce NO formation through disruption of eNOS phosphorylation. The increased NO bioavailability and formation of cGMP/PKG may reverse or attenuate mitochondrial ROS by increased synthesis of the putative mitochondrial antioxidant enzyme GSTK1. Finally, PDE-5 inhibition could increase PKG-dependent phosphorylation of glycogen synthase 3β (GSK-3β), which has a role in inhibition of MPTP [96] and therefore apoptosis in the diabetic heart following I/R. Our results have also shown that PDE-5 inhibition attenuated detrimental alterations in proteins involved in the cytoskeletal structure, cardiac contractility, and redox regulatory mechanisms of the diabetic myocardium. We propose that tadalafil treatment could be pursued as a

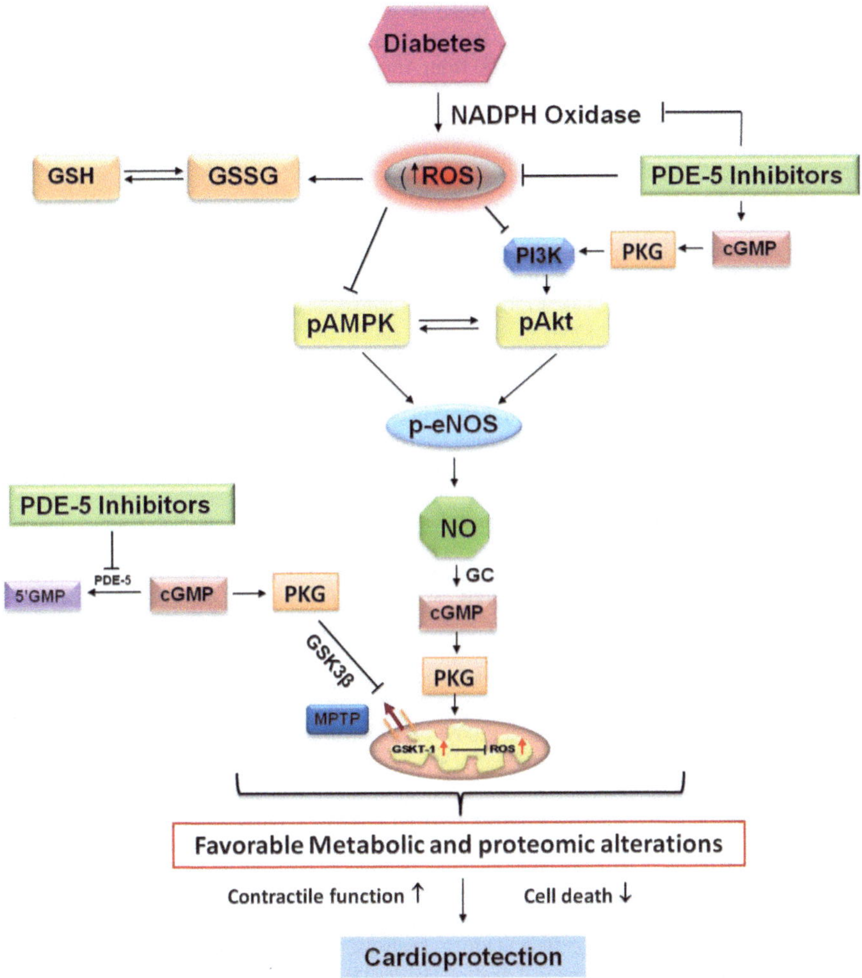

Fig. 4 Proposed cardioprotective pathways by phosphodiesterase 5 (PDE-5) inhibition in diabetes

novel therapeutic approach in protection against diabetes-induced cardiomyopathy. In fact, a recent clinical study suggested that chronic treatment with sildenafil caused an anti-remodeling effect in patients with early features of diabetic cardiomyopathy, such as left ventricular concentric hypertrophy associated with altered myocardial contraction dynamics [98]. Moreover, it is tempting to speculate the use of PDE-5 inhibitors may be therapeutically beneficial in protection against other oxidative stress-induced organ damage during diabetes, including nephropathy and retinopathy. These findings have potential clinical significance in the current scenario as PDE-5 inhibitors are now serving as first-line therapeutics in the treatments of ED in diabetic patients.

Acknowledgments This work was supported by grants from the National Institutes of Health (HL51045, HL79424, HL93685) to Dr. Rakesh C Kukreja. Dr. Saisudha Koka is supported by postdoctoral fellowship (11POST7400028) from American Heart Association.

References

1. King H, Aubert RE, Herman WH (1998) Global burden of diabetes, 1995–2025: prevalence, numerical estimates, and projections. Diabetes Care 21:1414–1431
2. Zimmet P, Alberti KG, Shaw J (2001) Global and societal implications of the diabetes epidemic. Nature 414:782–787
3. Garcia MJ, McNamara PM, Gordon T et al (1974) Morbidity and mortality in diabetics in the Framingham population. Sixteen year follow-up study. Diabetes 23:105–111
4. Tschoepe D, Roesen P (1998) Heart disease in diabetes mellitus: a challenge for early diagnosis and intervention. Exp Clin Endocrinol Diabetes 106:16–24
5. Kannel WB, Hjortland M, Castelli WP (1974) Role of diabetes in congestive heart failure: the Framingham study. Am J Cardiol 34:29–34
6. Malmberg K, Ryden L (1988) Myocardial infarction in patients with diabetes mellitus. Eur Heart J 9:259–264
7. Fein FS (1990) Diabetic cardiomyopathy. Diabetes Care 13:1169–1179
8. Iltis I, Kober F, Dalmasso C et al (2005) Noninvasive characterization of myocardial blood flow in diabetic, hypertensive, and diabetic-hypertensive rats using spin-labeling MRI. Microcirculation 12:607–614
9. Lopaschuk GD, Rebeyka IM, Allard MF (2002) Metabolic modulation: a means to mend a broken heart. Circulation 105:140–142
10. Maier C, Clodi M, Neuhold S et al (2009) Endothelial markers may link kidney function to cardiovascular events in type 2 diabetes. Diabetes Care 32:1890–1895
11. Caro JJ, Ward AJ, O'Brien JA (2002) Lifetime costs of complications resulting from type 2 diabetes in the U.S. Diabetes Care 25:476–481
12. Grossman E, Messerli FH (1996) Diabetic and hypertensive heart disease. Ann Intern Med 125:304–310
13. Ayala JE, Bracy DP, Julien BM et al (2007) Chronic treatment with sildenafil improves energy balance and insulin action in high-fat-fed conscious mice. Diabetes 57:1025–1033
14. Deyoung L, Chung E, Kovac JR et al (2012) Daily use of sildenafil improves endothelial function in men with type 2 diabetes. J Androl 33:176–180
15. Lerman A, Zeiher AM (2005) Endothelial function: cardiac events. Circulation 111:363–368
16. Wen Y, Skidmore JC, Porter-Turner MM et al (2002) Relationship of glycation, antioxidant status and oxidative stress to vascular damage in diabetes. Diabetes Obes Metab 4:305–308
17. Soloman H, Man JW, Jackson G (2003) Erectile dysfunction and the cardiovascular patient: endothelial dysfunction is the common denominator. Heart 89:251–254
18. Versari A (2009) Endothelial dysfunction as a target for prevention of cardiovascular disease. Diabetes Care 32:S314–S321
19. Shahab A (2006) Why does diabetes mellitus increase the risk of cardiovascular disease? Acta Med Indones 38:33–41
20. Hayat SA, Patel B, Khattar RS et al (2004) Diabetic cardiomyopathy: mechanisms, diagnosis and treatment. Clin Sci (Lond) 107:539–557
21. Bella JN, Devereux RB, Roman MJ et al (2001) Separate and joint effects of systemic hypertension and diabetes mellitus on left ventricular structure and function in American Indians (the Strong Heart Study). Am J Cardiol 87:1260–1265
22. Fein FS, Sonnenblick EH (1985) Diabetic cardiomyopathy. Prog Cardiovasc Dis 27:255–270
23. Bauters C, Lamblin N, Mc Fadden EP et al (2003) Influence of diabetes mellitus on heart failure risk and outcome. Cardiovasc Diabetol 2:1

24. Aronow WS, Ahn C (1999) Incidence of heart failure in 2,737 older persons with and without diabetes mellitus. Chest 115:867–868
25. Liu JE, Palmieri V, Roman MJ et al (2001) The impact of diabetes on left ventricular filling pattern in normotensive and hypertensive adults: the Strong Heart Study. J Am Coll Cardiol 37:1943–1949
26. Spector KS (1998) Diabetic cardiomyopathy. Clin Cardiol 21:885–887
27. Tziakas DN, Chalikias GK, Kaski JC (2005) Epidemiology of the diabetic heart. Coron Artery Dis 16:S3–S10
28. Feng B, Chen S, George B et al (2010) miR133a regulates cardiomyocyte hypertrophy in diabetes. Diabetes Metab Res Rev 26:40–49
29. Zhu SG, Xi L, Kukreja RC (2011) Type 2 diabetic obese db/db mice are refractory to myocardial ischemic postconditioning in vivo: potential role for Hsp20, F(1)-ATPase δ, and Echs1. J Cell Mol Med 15:2512–2524
30. Ye Y, Perez-Polo JR, Aguilar D et al (2011) The potential effects of anti-diabetic medications on myocardial ischemia-reperfusion injury. Basic Res Cardiol 106:925–952
31. Ghaboura N, Tamareille S, Ducluzeau PH et al (2011) Diabetes mellitus abrogates erythropoietin-induced cardioprotection against ischemic-reperfusion injury by alteration of the RISK/GSK-3 beta signaling. Basic Res Cardiol 106:147–162
32. Khan ZA, Chakrabarti S (2003) Endothelins in chronic diabetic complications. Can J Physiol Pharmacol 81:622–634
33. Khan ZA, Chakrabarti S (2006) Therapeutic targeting of endothelial dysfunction in chronic diabetic complications. Recent Pat Cardiovasc Drug Discov 1:167–175
34. Giugliano D, Marfella R, Coppola L et al (1997) Vascular effects of acute hyperglycemia in humans are reversed by L-arginine. Evidence for reduced availability of nitric oxide during hyperglycemia. Circulation 95:1783–1790
35. Khan ZA, Farhangkhoee H, Chakrabarti S (2006) Towards newer molecular targets for chronic diabetic complications. Curr Vasc Pharmacol 4:45–57
36. Lin KY, Ito A, Asagami T et al (2002) Impaired nitric oxide synthase pathway in diabetes mellitus: role of asymmetric dimethylarginine and dimethylarginine dimethylaminohydrolase. Circulation 106:987–992
37. Munzel T, Daiber A, Ullrich V (2005) Vascular consequences of endothelial nitric oxide synthase uncoupling for the activity and expression of the soluble guanylyl cyclase and the cGMP-dependent protein kinase. Arterioscler Thromb Vasc Biol 25:1551–1557
38. Torres SH, De Sanctis JB, de L Briceño M (2004) Inflammation and nitric oxide production in skeletal muscle of type 2 diabetic patients. J Endocrinol 181:419–427
39. Ludmer PL, Selwyn AP, Shook TL et al (1986) Paradoxical vasoconstriction induced by acetylcholine in atherosclerotic coronary arteries. N Engl J Med 315:1046–1051
40. Duplain H, Burcelin R, Sartori C et al (2001) Insulin resistance, hyperlipidemia, and hypertension in mice lacking endothelial nitric oxide synthase. Circulation 104:342–345
41. Shankar RR, Wu Y, Shen HQ et al (2000) Mice with gene disruption of both endothelial and neuronal nitric oxide synthase exhibit insulin resistance. Diabetes 49:684–687
42. Roy D, Perreault M, Marette A (1998) Insulin stimulation of glucose uptake in skeletal muscles and adipose tissue in vivo is NO dependent. Am J Physiol 274:E692–E699
43. Baron AD, Zhu JS, Marshall S et al (1995) Insulin resistance after hypertension induced by the nitric oxide synthase inhibitor L-NMMA in rats. Am J Physiol 269:E709–E715
44. Higaki Y, Hirshman MF, Fujii N et al (2001) Nitric oxide increases glucose uptake through a mechanism that is distinct from the insulin and contraction pathways in rat skeletal muscle. Diabetes 50:241–247
45. Cook S, Hugli O, Egli M et al (2003) Clustering of cardiovascular risk factors mimicking the human metabolic syndrome X in eNOS null mice. Swiss Med Wkly 133:360–363
46. Franks PW, Luan J, Barroso I et al (2005) Variation in the eNOS gene modifies the association between total energy expenditure and glucose intolerance. Diabetes 54:2795–2801
47. Cui R, Iso H, Pi J et al (2007) Metabolic syndrome and urinary cGMP excretion in the general population. Atherosclerosis 190:423–428

48. Behr-Roussel D, Gorny D, Mevel K et al (2005) Chronic sildenafil improves erectile function and endothelium-dependent cavernosal relaxations in rats: lack of tachyphylaxis. Eur Urol 47:87–91
49. Das A, Xi L, Kukreja RC (2005) Phosphodiesterase-5 inhibitor sildenafil preconditions adult cardiac myocytes against necrosis and apoptosis. Essential role of nitric oxide signaling. J Biol Chem 280:12944–12955
50. Wallis RM (1999) The pharmacology of sildenafil, a novel and selective inhibitor of phosphodiesterase (PDE) type 5. Nippon Yakurigaku Zasshi 114:22P–26P
51. Vickers MA, Satyanarayana R (2002) Phosphodiesterase type 5 inhibitors for the treatment of erectile dysfunction in patients with diabetes mellitus. Int J Impot Res 14:466–471
52. Salloum F, Yin C, Xi L (2003) Sildenafil induces delayed preconditioning through inducible nitric oxide synthase-dependent pathway in mouse heart. Circ Res 92:595–597
53. Halcox JP, Nour KR, Zalos G et al (2002) The effect of sildenafil on human vascular function, platelet activation, and myocardial ischemia. J Am Coll Cardiol 40:1232–1240
54. Hryniewicz K, Dimayuga C, Hudaihed A et al (2005) Inhibition of angiotensin-converting enzyme and phosphodiesterase type 5 improves endothelial function in heart failure. Clin Sci (Lond) 108:331–338
55. Gross GJ (2005) Sildenafil and endothelial dysfunction in humans. Circulation 111:721–723
56. Schafer A, Fraccarollo D, Pfortsch S et al (2008) Improvement of vascular function by acute and chronic treatment with the PDE-5 inhibitor sildenafil in experimental diabetes mellitus. Br J Pharmacol 153:886–893
57. Radovits T, Bomicke T, Kokeny G et al (2009) The phosphodiesterase-5 inhibitor vardenafil improves cardiovascular dysfunction in experimental diabetes mellitus. Br J Pharmacol 156:909–919
58. Choi SM, Kim JE, Kang KK (2006) Chronic treatment of DA-8159, a new phosphodiesterase type V inhibitor, attenuates endothelial dysfunction in stroke-prone spontaneously hypertensive rat. Life Sci 78:1211–1216
59. Aversa A, Greco E, Bruzziches R et al (2007) Relationship between chronic tadalafil administration and improvement of endothelial function in men with erectile dysfunction: a pilot study. Int J Impot Res 19:200–207
60. DeSouza C, Parulkar A, Lumpkin D et al (2002) Acute and prolonged effects of sildenafil on brachial artery flow-mediated dilatation in type 2 diabetes. Diabetes Care 25:1336–1339
61. Aversa A, Bruzziches R, Vitale C et al (2007) Chronic sildenafil in men with diabetes and erectile dysfunction. Expert Opin Drug Metab Toxicol 3:451–464
62. Kukreja RC, Ockaili R, Salloum F et al (2003) Sildenafil-induced cardioprotection in rabbits. Cardiovasc Res 60:700–701
63. Ockaili R, Salloum F, Hawkins J et al (2002) Sildenafil (Viagra) induces powerful cardioprotective effect via opening of mitochondrial K(ATP) channels in rabbits. Am J Physiol Heart Circ Physiol 283:H1263–H1269
64. Salloum FN, Chau VQ, Hoke NN et al (2009) Phosphodiesterase-5 inhibitor, tadalafil, protects against myocardial ischemia/reperfusion through protein-kinase g-dependent generation of hydrogen sulfide. Circulation 120:S31–S36
65. Elrod JW, Greer JJ, Lefer DJ (2007) Sildenafil-mediated acute cardioprotection is independent of the NO/cGMP pathway. Am J Physiol Heart Circ Physiol 292:H342–H347
66. Nagy O, Hajnal A, Parratt JR et al (2004) Sildenafil (Viagra) reduces arrhythmia severity during ischaemia 24 h after oral administration in dogs. Br J Pharmacol 141:549–551
67. Sesti C, Florio V, Johnson EG et al (2007) The phosphodiesterase-5 inhibitor tadalafil reduces myocardial infarct size. Int J Impot Res 19:55–61
68. Fisher PW, Salloum F, Das A et al (2005) Phosphodiesterase-5 inhibition with sildenafil attenuates cardiomyocyte apoptosis and left ventricular dysfunction in a chronic model of doxorubicin cardiotoxicity. Circulation 111:1601–1610
69. Koka S, Das A, Zhu SG et al (2010) Long-acting phosphodiesterase-5 inhibitor tadalafil attenuates doxorubicin-induced cardiomyopathy without interfering with chemotherapeutic effect. J Pharmacol Exp Ther 334:1023–1030

70. Koka S, Kukreja RC (2010) Attenuation of doxorubicin-induced cardiotoxicity by Tadalafil: a long acting phosphodiesterase-5 inhibitor. Mol Cell Pharmacol 2:173–178
71. Salloum FN, Abbate A, Das A et al (2008) Sildenafil (Viagra) attenuates ischemic cardiomyopathy and improves left ventricular function in mice. Am J Physiol Heart Circ Physiol 294:H1398–H1406
72. Koka S, Xi L, Kukreja RC (2012) Chronic treatment with long acting phosphodiesterase-5 inhibitor tadalafil alters proteomic changes associated with cytoskeletal rearrangement and redox regulation in type II diabetic hearts. Basic Res Cardiol 107:249
73. Varma A, Das A, Hoke NN et al (2012) Anti-inflammatory and cardioprotective effects of tadalafil in diabetic mice. PLoS One 7:e45243
74. Bucala R, Tracey KJ, Cerami A (1991) Advanced glycosylation products quench nitric oxide and mediate defective endothelium-dependent vasodilatation in experimental diabetes. J Clin Invest 87:432–438
75. Farhangkhoee H, Khan ZA, Mukherjee S et al (2003) Heme oxygenase in diabetes-induced oxidative stress in the heart. J Mol Cell Cardiol 35:1439–1448
76. Boudina S, Abel ED (2007) Diabetic cardiomyopathy revisited. Circulation 115:3213–3223
77. Shen X, Zheng S, Metreveli NS et al (2006) Protection of cardiac mitochondria by overexpression of MnSOD reduces diabetic cardiomyopathy. Diabetes 55:798–805
78. Zanetti M, Sato J, Katusic ZS et al (2000) Gene transfer of endothelial nitric oxide synthase alters endothelium-dependent relaxations in aortas from diabetic rabbits. Diabetologia 43:340–347
79. Barouch LA, Berkowitz DE, Harrison RW et al (2003) Disruption of leptin signaling contributes to cardiac hypertrophy independently of body weight in mice. Circulation 108:754–759
80. Cai L, Li W, Wang G et al (2002) Hyperglycemia-induced apoptosis in mouse myocardium: mitochondrial cytochrome C-mediated caspase-3 activation pathway. Diabetes 51:1938–1948
81. Wold LE, Ren J (2004) Streptozotocin directly impairs cardiac contractile function in isolated ventricular myocytes via a p38 MAP kinase-dependent oxidative stress mechanism. Biochem Biophys Res Commun 318:1066–1071
82. Zhou YT, Grayburn P, Karim A et al (2000) Lipotoxic heart disease in obese rats: implications for human obesity. Proc Natl Acad Sci U S A 97:1784–1789
83. Salloum FN, Takenoshita Y, Ockaili RA et al (2007) Sildenafil and vardenafil but not nitroglycerin limit myocardial infarction through opening of mitochondrial K(ATP) channels when administered at reperfusion following ischemia in rabbits. J Mol Cell Cardiol 42:453–458
84. Hotston MR, Jeremy JY, Bloor J et al (2007) Sildenafil inhibits the up-regulation of phosphodiesterase type 5 elicited with nicotine and tumour necrosis factor-alpha in cavernosal vascular smooth muscle cells: mediation by superoxide. BJU Int 99:612–618
85. Koupparis AJ, Jeremy JY, Muzaffar S et al (2005) Sildenafil inhibits the formation of superoxide and the expression of gp47 NAD[P]H oxidase induced by the thromboxane A2 mimetic, U46619, in corpus cavernosal smooth muscle cells. BJU Int 96:423–427
86. Koka S, Das A, Salloum FN et al (2013) Phosphodiesterase-5 inhibitor tadalafil attenuates oxidative stress and protects against myocardial ischemia/reperfusion injury in type 2 diabetic mice. Free Radic Biol Med 60:80–88
87. Guzik TJ, Mussa S, Gastaldi D et al (2002) Mechanisms of increased vascular superoxide production in human diabetes mellitus: role of NAD(P)H oxidase and endothelial nitric oxide synthase. Circulation 105:1656–1662
88. Kuroda J, Ago T, Matsushima S et al (2010) NADPH oxidase 4 (Nox4) is a major source of oxidative stress in the failing heart. Proc Natl Acad Sci U S A 107:15565–15570
89. Kuroda J, Sadoshima J (2010) NADPH oxidase and cardiac failure. J Cardiovasc Transl Res 3:314–320
90. Shao D, Oka S, Brady CD et al (2012) Redox modification of cell signaling in the cardiovascular system. J Mol Cell Cardiol 52:550–558
91. Hink U, Li H, Mollnau H et al (2001) Mechanisms underlying endothelial dysfunction in diabetes mellitus. Circ Res 88:E14–E22

92. Kim YK, Lee MS, Son SM et al (2002) Vascular NADH oxidase is involved in impaired endothelium-dependent vasodilation in OLETF rats, a model of type 2 diabetes. Diabetes 51:522–527
93. Newsholme P, Haber EP, Hirabara SM et al (2007) Diabetes associated cell stress and dysfunction: role of mitochondrial and non-mitochondrial ROS production and activity. J Physiol 583:9–24
94. Akao M, O'Rourke B, Teshima Y et al (2003) Mechanistically distinct steps in the mitochondrial death pathway triggered by oxidative stress in cardiac myocytes. Circ Res 92:186–194
95. Sack MN (2009) Type 2 diabetes, mitochondrial biology and the heart. J Mol Cell Cardiol 46:842–849
96. Kukreja RC, Salloum FN, Das A (2012) Cyclic guanosine monophosphate signaling and phosphodiesterase-5 inhibitors in cardioprotection. J Am Coll Cardiol 59:1921–1927
97. http://clinicaltrials.gov/
98. Giannetta E, Isidori AM, Galea N et al (2012) Chronic inhibition of cGMP phosphodiesterase 5A improves diabetic cardiomyopathy: a randomized, controlled clinical trial using magnetic resonance imaging with myocardial tagging. Circulation 125:2323–2333

Restoration of Angiogenesis: A Promising Therapeutic Strategy in Diabetic Cardiomyopathy

Veeranjaneyulu Addepalli and Dipti Gatne

Abstract Diabetic cardiomyopathy involves vascular endothelial cell dysfunction with structural abnormalities related to hyperglycemia and insulin resistance leading to cardiac complications. The cardiac disorder in diabetic condition is characterized by alteration in myocardial vascular endothelial growth factor (VEGF) and VEGF receptor expression levels, which are well-known regulators of angiogenesis. Hence, promoting angiogenesis in heart can be of therapeutic importance to increase blood flow and reduce vasoconstriction to prevent tissue ischemia and myocyte necrosis. Apart from VEGF, fibroblast growth factor, platelet-derived growth factor, hepatocyte growth factor, and placental growth factor may be involved in defective intracellular signaling in diabetes. Therefore, targeting disturbed intracellular signaling with various strategies to restore angiogenesis using small molecules may be a promising strategy in the management of diabetic cardiomyopathy.

Keywords Angiogenesis • Diabetic cardiomyopathy • VEGF • Gene therapy

1 Introduction

Diabetes is manifested by insulin deficiency or tissue resistance to insulin leading to hyperglycemia. As the incidence of diabetes is drastically increased worldwide, treating the disease and its associated complications with new and effective therapeutic avenues has gained major attention.

V. Addepalli (✉)
Department of Pharmacology, SPP School of Pharmacy and Technology Management, SVKM's NMIMS, V. L. Mehta Road, Mumbai, Maharashtra 400056, India
e-mail: veeranjaneyulu.addepalli@nmims.edu

D. Gatne
Departments of Pharmaceutical Analysis, SPP School of Pharmacy and Technology Management, SVKM's NMIMS, V. L. Mehta Road, Mumbai, Maharashtra 400056, India

B. Turan and N.S. Dhalla (eds.), *Diabetic Cardiomyopathy: Biochemical and Molecular Mechanisms*, Advances in Biochemistry in Health and Disease 9,
DOI 10.1007/978-1-4614-9317-4_21, © Springer Science+Business Media New York 2014

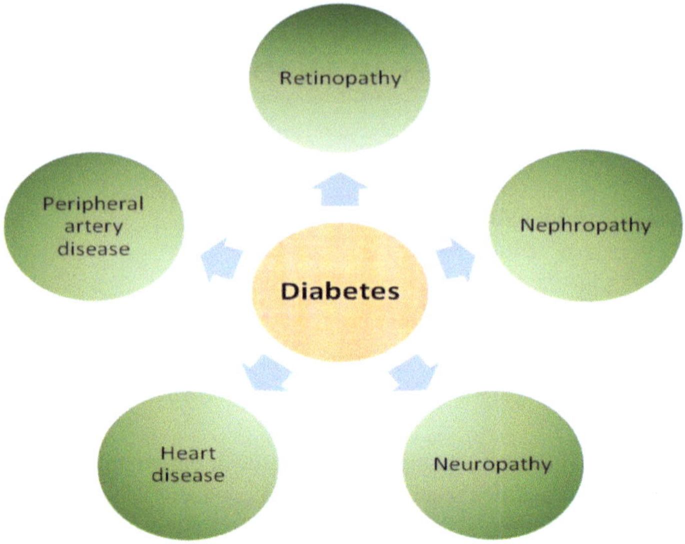

Fig. 1 Complications related to the diabetic condition

The syndrome is associated with diverse complications such as retinopathy, neuropathy, nephropathy, and atherosclerosis (Fig. 1). Hyperglycemia is believed to be the key determinant in these chronic diabetic complications [1, 2].

In heart muscles, lack of oxygen leads to ischemia and risk of such ischemic heart disease (IHD) as a result of inadequate blood supply increases with factors such as diabetes. The cardiovascular complications in diabetes have played a major role in mortality and morbidity. Some of these cardiovascular complications include accelerated atherosclerosis, disregulation of neovascularization, defective arteriogenesis, endothelial cell dysfunction, hypertension, cardiomyopathy, and congestive heart failure [3].

2 Diabetic Cardiomyopathy

Diabetic cardiomyopathy involves ventricular dysfunction in the absence of hypertension, coronary artery, and valvular heart disease, which increases the risk of heart failure. Hyperglycemia, insulin resistance, abnormal fatty acid metabolism, increased apoptosis, cardiac autonomic neuropathy, and local renin-angiotensin-aldosterone system (RAAS) overactivation are the important mechanisms of diabetic cardiomyopathy. It is characterized by microvascular pathology, which may lead to ischemia and contribute to further adverse events mediated by vascular endothelial growth factor (VEGF).

Metabolic abnormalities of diabetes produce large amounts of superoxide in the myocardium, leading to activation of a few major pathways, namely, protein kinase C activation, hexosamine pathway, advanced glycation end (AGE) products and expression of AGE receptors, and sarcoplasmic reticular Ca^{2+}-ATPase responsible

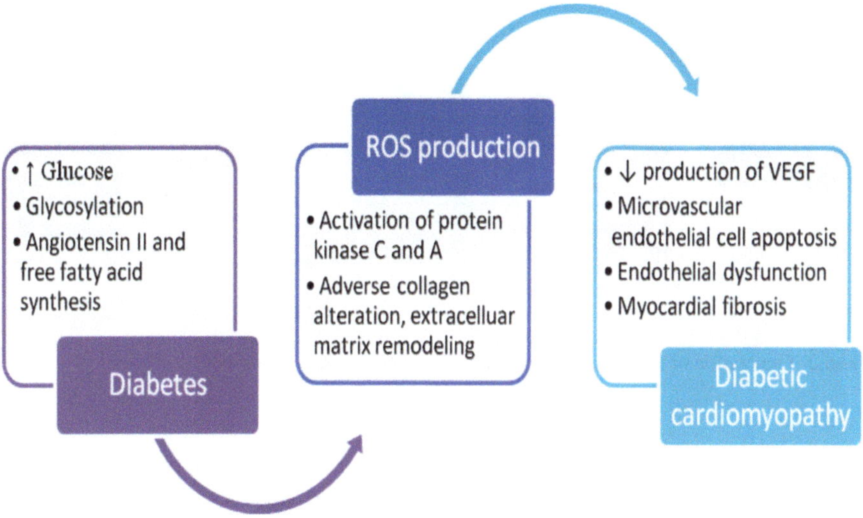

Fig. 2 Etiology of diabetic cardiomyopathy

for complications such as diabetic cardiomyopathy. A few microvascular abnormalities such as interstitial fibrosis that are influenced by VEGF exist without hypertension and coronary atherosclerosis in diabetic cardiomyopathy. The superoxide also inactivates crucial anti-atherosclerotic enzymes, endothelial nitric oxide synthase, and prostacyclin synthase, which lead to enhanced generation of reactive oxygen species (ROS). These species produce defective angiogenesis as a response to ischemia, activate pro-inflammatory pathways, and cause persistent expression of pro-inflammatory genes, causing long-lasting epigenetic change [4–15].

Diabetes involves elevated levels of ROS and decreased activity of the antioxidant defense system, which creates oxidative stress and related damage in the diabetic vasculature (Fig. 2). Hyperglycemia can induce production of ROS in mitochondria from various sources by exacerbating glucose oxidation and leads to increased oxidative stress. This oxidative stress contributes to the development of diabetic cardiomyopathy in the insulin-sensitive heart [4, 14].

Because ROS initiate several cell-signaling pathways, they contribute to diabetes-mediated endothelial damage and develop many complications such as DNA damage, accelerated apoptosis, and hyperglycemia-mediated cell injury [16–18].

3 Impaired Angiogenesis in Diabetic Cardiomyopathy

Angiogenesis or neovascularization involves endothelial cell migration and proliferation, breakdown of extracellular matrix, accumulation of pericytes and macrophages, smooth muscle cell proliferation and migration, and formation of new vasculature. Many growth factors such as VEGF and fibroblast growth factor (FGF) are involved in generating angiogenic response in ischemic conditions.

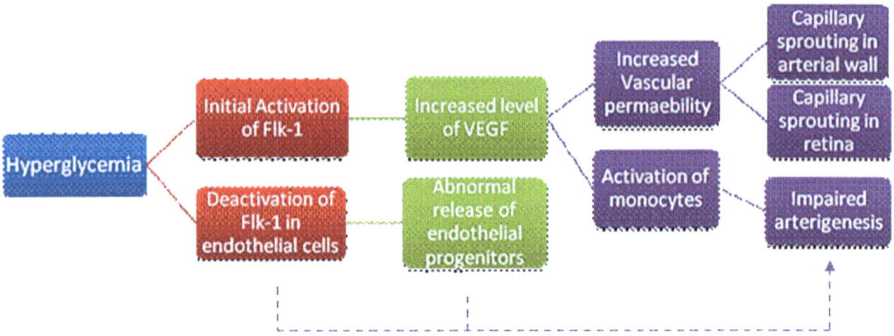

Fig. 3 Generation of impaired arteriogenesis in hyperglycemic condition

Diabetes is closely associated with impaired neovascularization related to vascular occlusion. The abnormalities associated with neovascularization in diabetic condition include enhanced angiogenesis in retina (retinopathy) and in vessel walls (unstable atherosclerosis plaque), impaired angiogenesis in wound healing (skin ulcer), impaired release of endothelial progenitor cells from bone marrow with defective functions, reduced VEGF and VEGF receptor expression in myocardium, and increased production of angiostatin, an angiogenesis inhibitor, which have significant roles in increased cardiovascular risks in diabetes [9, 19–22].

A few studies have shown reduced expression of VEGFR-1 (Flt-1) and VEGFR-2 (Flk-1), reduced Flk-1 phosphorylation, and activation of serine-threonine protein kinase Akt-1 and endothelial nitric oxide synthase (eNOS) in myocardium of diabetic patients [23, 24]. The expression of mRNA and VEGF, with its two receptors (VEGFR-1 and VEGFR-2), was found to be decreased by 40–70 % in myocardium of diabetic animals. The ventricles of diabetic patients also showed twofold reduction in VEGF and VEGFR-2 expression, both responsible for inadequate collateral formation in myocardium of diabetic and insulin-resistant conditions [23]. Akt-1 is a major regulator of cell functions and Flk-1 regulates cell proliferation. Akt-1 activates eNOS to stimulate nitric oxide production for cell proliferation and inhibition of apoptosis. Together, Akt-1 and VEGFR-2 act toward maintenance of the intact vasculature [17]. It is also important to note that in contrast to myocardial conditions, expression of VEGF and its receptors are reported to increase twofold in retina and glomeruli of diabetic animals, which explains the capillary leakage and enhanced neovascularization in the retina in the diabetic condition [23] (Fig. 3).

4 Restoration of Angiogenesis: A Therapeutic Approach

Treatment of coronary artery diseases involves medical therapy and revascularization procedures such as percutaneous transluminal angioplasty and coronary artery bypass surgery. An alternative therapy is required if the patient cannot be treated with these strategies. Restoring angiogenesis can be one of those therapies for development of collateral blood vessels [25].

Therapeutic angiogenesis with growth factor- or cell-based therapy is aimed at restoration of perfusion to chronically ischemic myocardium [26, 27].

5 Gene Therapy in Ischemic Conditions

Clinical trials have suggested gene therapy for treatment of ischemic conditions to induce angiogenesis. Administration of genes that encode angiogenic growth factors induces therapeutic angiogenesis and act as an alternative to traditional revascularization methods.

Two types of therapeutic genes have been used. Angiogenic growth factors such as VEGF, fibroblast growth factor (FGF), and hepatocyte growth factor (HGF), and antiapoptotic genes such as Bcl-2, heme oxygenase-1 (HO-1), and small interfering RNAs (siRNAs) have been used in gene therapy of ischemic heart disease (IHD). Growth factors have been found to increase blood capillary density in ischemic area and protect cells from death. Because expression levels of growth factors are low in ischemic tissues, the genes can increase the expression and thereby the effect of growth factors. Antiapoptotic genes protect ischemic myocardial cells from death [28, 29].

However, development of efficient and safe gene carriers and effective therapeutic genes for treatment remains a challenge. Viral and nonviral vectors have been used to deliver therapeutic genes. Although viral vectors efficiently transfer the genes, many safety issues such as toxicity, oncogenicity, and cellular immune response are of major concern. Nonviral carriers such as naked plasmid DNA and several biodegradable synthetic chemical vectors such as cationic polymers, cationic lipids, genetically modified stem cells, liposomes, inorganic nanoparticles, dendrimers, and nonviral RNA interference (RNAi) have been employed in the studies. One study reported replenishment of myocardial expression of VEGF using naked DNA gene therapy via direct intramyocardial injection of plasmid DNA encoding VEGF in a diabetic rat model with progressive cardiomyopathy [30].

There are some serious side effects related to gene therapy, especially nonspecific gene expression, leaking of genes into unwanted cells, delivery of genes to specific tissues, and overproliferation of endothelial cells in normal tissues.

6 Administration of Growth Factors

Replenishment of cardiac-specific growth factors in local areas is seen to restore microvascular homeostasis, angiogenesis, and cytoprotection to protect myocardial tissues from ischemic or metabolic injury [31].

A clinical phase I study evaluated efficacy, safety, and tolerability of intracoronary administration of FGF in severe ischemic heart disease. Patients with diabetes but with no retinopathy were involved in the study. The efficacy data showed improvement in quality of life, exercise tolerance, and myocardial perfusion in the ischemic area. A few side effects such as hypotension, proteinuria, and myocardial infarction were noted in some patients [25].

Treatment with granulocyte colony-stimulating factor (G-CSF) in nonischemic diabetic cardiomyopathy showed improvement in myocardial infarction in an animal model. Recovery of diastolic function and significant decrease in interstitial fibrosis were observed, with reduction in transforming growth factor (TGF)-β immunoreactivity, in G-CSF-treated animals. Microscope studies showed less disruption of myofilaments and decreased collagen deposition [32, 33].

The angiogenic cytokines may benefit the damaged heart through direct effects on the myocardium and indirectly by stimulating progenitor cells [34, 35].

7 Angiogenic Cell Precursors

Adult stem cells were used in a study as angiogenic cell precursors involving patients suffering from nonischemic dilated cardiomyopathy, some of whom also suffered from diabetes mellitus. The cell precursors were collected from the patients' own blood to avoid any immunological complications and intramyocardially injected into the left ventricle. Improvement in quality of life was observed in the health and physical condition of treated patients, showing a benefit of this therapy [36].

8 Small Molecule Angiogenic Activators

Statins have shown favorable effects such as improved endothelial function, coronary angiogenesis, and reduction of ventricular remodeling in patients with diabetes. In these patients, statin therapy showed a remarkably decreased risk of cardiovascular events [37]. Resveratrol, which is found abundantly in red wine, has upregulated pro-angiogenic VEGF expression, modified glucose metabolism, and decreased oxidative stress and thus gave a therapeutic option for favorable modification of cardiovascular risk factors associated with diabetes [38].

Activation of peroxisome proliferator-activated receptors (PPAR), a group of nuclear hormone receptors that regulate lipid and glucose metabolism, has shown stimulation of angiogenesis, which may suggest PPAR agonists as new molecules for therapeutic angiogenesis [39]. A study involving pioglitazone showed cardioprotective effect with improvement in endothelial dysfunction, increase in endothelial progenitor cells (EPC), enhancement of EPC differentiation, survival, and function, and enhancement of angiogenesis in diabetic patients [40].

Curcumin, a yellow pigment isolated from turmeric, is well known for its antioxidant, antiinflammatory, and antimicrobial activities. Curcumin showed beneficial effects in treatment of diabetic cardiomyopathy by modifying fibrosis, oxidative stress, inflammation, cell death, and Akt/GSK-3b signalling pathway [41].

9 Erythropoietin

Erythropoietin protects the tissues of the ischemic heart to prevent vascular and tissue damage. It also stimulates production of endothelial progenitor cells, VEGF expression, and neovascularization in the ischemic heart. Erythropoietin decreases myocardial interstitial fibrosis to improve cardiac function. A study with erythropoietin in diabetic rats showed improvement in cardiac contractility and reduction in diastolic dysfunction. There was significant increase in endothelial progenitor cells, capillary density, and VEGF expression in left ventricular myocardial tissues in diabetic rats [19].

10 Conclusion

Diabetic cardiomyopathy progresses to mortality and morbidity in patients with diabetes. Elucidation of the mechanisms responsible for the pathological changes can suggest ways to develop target-oriented therapy for diabetic cardiomyopathy.

Apart from well-known therapeutic strategies, there are new upcoming treatments for diabetes-related cardiac complications. Bone morphogenetic proteins (BMPs) and growth differentiation factors (GDFs) are reported to regulate tissue growth and differentiation [42]. These members of the transforming growth factor-β (TGF-β) superfamily may provide potential therapeutic development. Glycogen synthase kinase-3β protein regulates myocardial apoptosis, modifying Akt phosphorylation, which can also give a new treatment approach [43]. The endothelin system [44, 45] and hypoxia-inducible factor (HIF) signaling pathway [46] in cardiovascular systems may be therapeutic targets for cardioprotection. New animal models, clinical trial designs, and combination therapy to address organ-specific and glucose-mediated approaches are required for the future.

References

1. Kolluru G, Bir SC, Kevil CG (2012) Endothelial dysfunction and diabetes: effects on angiogenesis, vascular remodeling, and wound healing. Int J Vasc Med. doi:10.1155/2012/918267
2. Calcutt NA, Cooper ME, Kern TS et al (2009) Therapies for hyperglycaemia-induced diabetic complications: from animal models to clinical trials. Nat Rev Drug Discov 8:417–429

3. Baliga V, Sapsford R (2009) Diabetes mellitus and heart failure: an overview of epidemiology and management. Diab Vasc Dis Res 6:164–171
4. Han B, Baliga R, Huang H et al (2009) Decreased cardiac expression of vascular endothelial growth factor and redox imbalance in murine diabetic cardiomyopathy. Am J Physiol Heart Circ Physiol 297:H829–H835
5. Acar E, Ural D, Bildirici U et al (2011) Diabetic cardiomyopathy. Anadolu Kardiyol Derg 11:732–737
6. Hayat S, Patel B, Khattar R et al (2004) Diabetic cardiomyopathy: mechanisms, diagnosis and treatment. Clin Sci 107:539–557
7. Abel ED (2005) Myocardial insulin resistance and cardiac complications of diabetes. Curr Drug Targets Immune Endocr Metabol Disord 5:219–226
8. Pacher P, Szabo C (2006) Role of peroxynitrite in the pathogenesis of cardiovascular complications of diabetes. Curr Opin Pharmacol 6:136–141
9. He Z, Rask-Madsen C, King GL (2003) Managing heart disease: mechanisms of cardiovascular complications in diabetes and potential new pharmacological therapies. Eur Heart J 5:B51–B57
10. Miki T, Yuda S, Kouzu H et al (2012) Diabetic cardiomyopathy: pathophysiology and clinical features. Heart Fail Rev. doi:10.1007/s10741-012-9313-3
11. Liu H, Yu S, Zhang H et al (2012) Angiogenesis impairment in diabetes: role of methylglyoxal-induced receptor for advanced glycation end products, autophagy and vascular endothelial growth factor receptor 2. PLoS One. doi:10.1371/journal.pone.0046720
12. Farhangkhoee H, Khan Z, Kaur H et al (2006) Vascular endothelial dysfunction in diabetic cardiomyopathy: pathogenesis and potential treatment targets. Pharmacol Ther 111:384–399
13. Voulgari C, Papadogiannis D, Tentolouris N (2010) Diabetic cardiomyopathy: from the pathophysiology of the cardiac myocytes to current diagnosis and management strategies. Vasc Health Risk Manag 6:883–903
14. Schaffer S, Jong C, Mozaffari M (2012) Role of oxidative stress in diabetes-mediated vascular dysfunction: unifying hypothesis of diabetes revisited. Vascul Pharmacol 57:139–149
15. Giacco F, Brownlee M (2010) Oxidative stress and diabetic complications. Circ Res 107:1058–1070
16. Roe N, Ren J (2012) Nitric oxide synthase uncoupling: a therapeutic target in cardiovascular diseases. Vascul Pharmacol 57:168–172
17. Simons M (2005) Angiogenesis, arteriogenesis, and diabetes paradigm reassessed? J Am Coll Cardiol 46:835–837
18. Xu J, Zhou Q, Xu W et al (2012) Endoplasmic reticulum stress and diabetic cardiomyopathy. Exp Diabetes Res. doi:10.1155/2012/827971
19. Lu J, Yao Y, Dai Q et al (2012) Erythropoietin attenuates cardiac dysfunction by increasing myocardial angiogenesis and inhibiting interstitial fibrosis in diabetic rats. Cardiovasc Diabetol. doi:10.1186/1475-2840-11-105
20. Sasso F, Torella D, Carbonara O et al (2005) Increased vascular endothelial growth factor expression but impaired vascular endothelial growth factor receptor signaling in the myocardium of type 2 diabetic patients with chronic coronary heart disease. Diabetes paradigm reassessed? J Am Coll Cardiol 46:827–834
21. Xu L, Kanasaki K, Kitada M et al (2012) Diabetic angiopathy and angiogenic defects. Fibrogenesis Tissue Repair. doi:10.1186/1755-1536-5-13
22. Shiojima I, Walsh K (2006) Regulation of cardiac growth and coronary angiogenesis by the Akt/PKB signaling pathway. Genes Dev 20:3347–3365
23. Chou E, Suzuma I, Way K et al (2002) Decreased cardiac expression of vascular endothelial growth factor and its receptors in insulin-resistant and diabetic states: a possible explanation for impaired collateral formation in cardiac tissue. Circulation 105:373–379
24. Khazaei M, Fallahzadeh A, Sharifi M et al (2011) Effects of diabetes on myocardial capillary density and serum angiogenesis biomarkers in male rats. Clinics 66:1419–1424

25. Laham R, Chronos N, Pike M et al (2000) Intracoronary basic fibroblast growth factor (FGF-2) in patients with severe ischemic heart disease: results of a phase I open-label dose escalation study. J Am Coll Cardiol 36:2132–2139
26. Boodhwani M, Sellke F (2009) Therapeutic angiogenesis in diabetes and hypercholesterolemia: influence of oxidative stress. Antioxid Redox Signal 11:1945–1959
27. Chachques J, Duarte F, Herreros J et al (2003) Cellular myogenic and angiogenic therapy for patients with cardiac or limb ischemia. Basic Appl Myol 13:29–37
28. Park H, Yang F, Cho S (2012) Nonviral delivery of genetic medicine for therapeutic angiogenesis. Adv Drug Deliv Rev 64:40–52
29. Kim H, Rhim T, Lee M (2011) Regulatory systems for hypoxia-inducible gene expression in ischemic heart disease gene therapy. Adv Drug Deliv Rev 63:678–687
30. Yoon Y, Uchida S, Masuo O et al (2005) Progressive attenuation of myocardial vascular endothelial growth factor expression is a seminal event in diabetic cardiomyopathy restoration of microvascular homeostasis and recovery of cardiac function in diabetic cardiomyopathy after replenishment of local vascular endothelial growth factor. Circulation 111:2073–2085
31. Gonzalez A, Ravassa S, Beaumont J et al (2011) New targets to treat the structural remodeling of the myocardium. J Am Coll Cardiol 58:1833–1843
32. Lim Y, Joe J, Jang K et al (2011) Effects of granulocyte-colony stimulating factor (G-CSF) on diabetic cardiomyopathy in Otsuka Long-Evans Tokushima fatty rats. Cardiovasc Diabetol. doi:10.1186/1475-2840-10-92
33. Krzeminski T, Nozynski J, Grzyb J et al (2005) Angiogenesis and cardioprotection after TNFβ-inducer-Tolpa peat preparation treatment in rat hearts after experimental myocardial infarction in vivo. Vascul Pharmacol 43:164–170
34. Khurana R, Simons M, Martin J et al (2005) Role of angiogenesis in cardiovascular disease: a critical appraisal. Circulation 112:1813–1824
35. Beohar N, Rapp J, Pandya S et al (2010) Rebuilding the damaged heart—the potential of cytokines and growth factors in the treatment of ischemic heart disease. J Am Coll Cardiol 56:1287–1297
36. Arom K, Ruengsakulrach P, Belkin M et al (2009) Intramyocardial angiogenic cell precursors in nonischemic dilated cardiomyopathy. Asian Cardiovasc Thorac Ann 17:382–388
37. Riou B (2008) Diabetic cardiomyopathy and anesthesia. Anesthesiology 108:524–530
38. Robich M, Osipov R, Chu L et al (2011) Resveratrol modifies risk factors for coronary artery disease in swine with metabolic syndrome and myocardial ischemia. Eur J Pharmacol 664:45–53
39. Park C, Kim H, Lim J et al (2009) Vascular endothelial growth factor inhibition by dRK6 causes endothelial apoptosis, fibrosis, and inflammation in the heart via the Akt/eNOS axis in db/db mice. Diabetes 58:2666–2676
40. Ozawa T, Oda H, Oda M et al (2009) Improved cardiac function after sirolimus-eluting stent placement in diabetic patients by pioglitazone: combination therapy with statin. J Cardiol 53:402–409
41. Yu W, Wu J, Cai F et al (2012) Curcumin alleviates diabetic cardiomyopathy in experimental diabetic rats. PLoS One. doi:10.1371/journal.pone.0052013
42. Tobin J, Celeste A (2006) Bone morphogenetic proteins and growth differentiation factors as drug targets in cardiovascular and metabolic diseases. Drug Discov Today 11:405–411
43. Gurusamya N, Watanabea K, Maa M et al (2006) Glycogen synthase kinase 3b together with 14-3-3 protein regulates diabetic cardiomyopathy: effect of losartan and tempol. FEBS Lett 580:1932–1940
44. Rodriguez-Pascuala F, Busnadiego O, Lagares D et al (2011) Role of endothelin in the cardiovascular system. Pharmacol Res 63:463–472
45. Ergul A (2011) Endothelin-1 and diabetic complications: focus on the vasculature. Pharmacol Res 63:477–482
46. Ong S, Hausenloy D (2012) Hypoxia-inducible factor as a therapeutic target for cardioprotection. Pharmacol Ther 136:69–81

The Angiotensin-Converting Enzyme 2/Angiotensin-(1-7)/Mas Receptor Axis: A Potential Target for Treating Diabetic Cardiovascular Disease

Ibrahim F. Benter, Mariam H.M. Yousif, Jasbir S. Juggi, and Saghir Akhtar

Abstract Activation of the angiotensin-converting enzyme (ACE)/angiotensin II (Ang II)/angiotensin type-1 receptor (AT_1) axis in the heart and vasculature plays an important role in the development of diabetic complications. The angiotensin-converting enzyme 2/angiotensin-(1-7)/Mas receptor axis has been shown to oppose the actions of Ang II. Our results have shown that chronic treatment with Ang-(1-7) led to correction of the altered responses to norepinephrine or endothelin-1 and carbachol in the mesenteric bed, carotid, and renal arteries of diabetic rats and improved recovery of left ventricular function from 40 min of global ischemia. We showed that the beneficial effects of Ang-(1-7) on the diabetic vasculature involve inhibition of the detrimental EGFR/ERK1/2/p38MAP kinase pathway. Ang-(1-7) treatment inhibited cardiac NADPH oxidase (NOX) and cardiac nuclear factor-κB (NF-κB) activity and also suppressed the expression of several pro-inflammatory genes involved in the NF-κB signaling pathway, including complement component 3 (C3), interleukin 1-beta (Il-1β), interleukin 6 (Il-6), NACHT-containing protein (Nalp12), and caspase 1 (Casp1). We conclude that activation of the Ang-(1-7)-mediated signal transduction pathway appears to be an important therapeutic strategy to reduce cardiovascular events in diabetic patients.

Keywords Diabetes • Cardiomyopathy • Angiotensin II • EGFR • ACE2 • Mas • NOX • NF-κB • Heart • Ischemia

---

I.F. Benter (✉) • M.H.M. Yousif • S. Akhtar
Department of Pharmacology and Toxicology, Faculty of Medicine, Kuwait University, P.O. Box: 24923, Safat 13110, Kuwait
e-mail: ibenter@hsc.edu.kw; ifbenter@gmail.com

J.S. Juggi
Faculty of Allied Health, Kuwait University, Safat, Kuwait

B. Turan and N.S. Dhalla (eds.), *Diabetic Cardiomyopathy: Biochemical and Molecular Mechanisms*, Advances in Biochemistry in Health and Disease 9,
DOI 10.1007/978-1-4614-9317-4_22,

1 Introduction

Diabetes mellitus is a chronic debilitating disease characterized by high blood glucose levels whose prevalence is increasing rapidly to near-epidemic proportions [1, 2]. The latest global estimates suggesting that by the year 2030 there will be more than 550 million people with this debilitating disease [3, 4]. Diabetic patients tend to have a three- to eightfold higher risk of developing cardiovascular dysfunction, and complications of the heart and micro- and macrovasculature are major contributors to increased morbidity and mortality [5, 6]. The exact underlying mechanisms for the development of diabetes-induced cardiovascular complications are poorly understood and may involve multiple signaling pathways that are affected by hyperglycemia [7].

Although numerous factors contribute to the development of diabetes-induced cardiovascular disease, it is well established that activation of angiotensin-converting enzyme (ACE)/angiotensin II (Ang II)/angiotensin type-1 receptor (AT_1) axis in the heart and vasculature plays an important role in the development of diabetic complications [7, 8]. In the diabetic heart and vasculature, Ang II-mediated signaling is associated with oxidative damage, inflammation, fibrogenesis, myocyte apoptosis, and impaired myocardial contractility [7–9]. The importance of the ACE/Ang II/AT_1 axis in the development and progression of cardiovascular pathology associated with diabetes is supported by findings showing that inhibition of Ang II synthesis or activity can prevent or slow the progression of diabetes-induced cardiovascular complications. ACE inhibitors and AT_1 blockers have become an integral part of any therapeutic strategy to reduce cardiovascular events in patients with diabetes [8, 9].

Our studies showing that angiotensin-(1-7) [Ang-(1-7)] is a vasodilator peptide with antihypertensive effects have introduced the concept that the renin-angiotensin-aldosterone system (RAAS) also has a protective arm [10–13]. Ang-(1-7) is formed from Ang II by ACE-2, a homologue of ACE, and ACE2 is the primary enzyme responsible for the metabolism of Ang II in the murine cardiovascular system [14, 15]. Reduced metabolism of Ang II by ACE2 can result in elevated levels of Ang II, reduced levels of Ang-(1-7), and end-organ damage [16, 17]. Genetic deficiency of ACE2 is associated with cardiac dysfunction, increased tissue and circulating levels of Ang II, and decreased Ang-(1-7) levels [18]. Ang-(1-7) produces its vasodilatory, antifibrotic, antihypertrophic, and antiarrhythmic actions by activating the Mas receptor [19–22]. Ang-(1-7) activates the Mas receptor to induce release of vasodilatory prostaglandins and nitric oxide [21, 22]. Mas knockout mice show pronounced impairment of cardiac and renal function [23, 24].

2 Angiotensin-(1-7) Treatment Can Attenuate Diabetes-Induced Cardiovascular Dysfunction

Diabetes mellitus is associated with alterations in vascular structure and function that contribute to the development of cardiovascular complications. Endothelial dysfunction, enhanced vascular responsiveness to vasoconstrictors, and attenuated

Table 1 The effect of angiotension (Ang)-(1-7) and AVE0991 on postischemic recovery in global cardiac contractility and hemodynamics

	Left ventricular developed pressure (mmHg)			Coronary flow (ml min^{-1})		
Groups studied	Perfusion	Reperfusion	% R	Perfusion	Reperfusion	% R
Control	51±9	19±3	39±4	6.6±0.6	2.1±0.3	32±5
Diabetic (D)	86±14	24±9	28±4[a]	8.5±0.6	1.5±0.2	18±4[a]
D+Ang-(1-7)	115±12	44±3	42±7[b]	9.5±0.5	3.4±0.3	36±3[b]
D+AVE0991	94±10	48±8	51±8[b]	7.9±0.6	3.1±0.4	41±4[b]

Data were computed at 30 min reperfusion and expressed as mean±SEM (n=6). % R=% Recovery=(reperfusion/perfusion) ×100
[a]Value significantly different from control, $p<0.05$
[b]Value significantly different from diabetes, $p<0.05$

responses to vasodilators are common in diabetic patients and animals. We studied the effect of chronic treatment (4 weeks) with Ang-(1-7) (24 μg/kg/h i.p.) or Ang-(1-7) non-peptide analogue AVE0991 (24 μg/kg/h i.p.) on vascular reactivity, cardiac recovery from ischemia–reperfusion (I/R), and proteinuria in streptozotocin-treated rats (diabetes) [25]. Our results have shown that chronic treatment with Ang-(1-7) or AVE0991 led to correction of the altered responses to norepinephrine or endothelin-1 and carbachol in the mesenteric bed, carotid, and renal arteries of diabetic rats, indicating that activation of Ang-(1-7) receptors can produce protection against diabetes-induced vascular dysfunction [25]. In the diabetic animals, recovery of left ventricular function from 40 min of global ischemia was significantly impaired in isolated perfused hearts from diabetic animals compared to control animals (Table 1). Hearts from Ang-(1-7)- or AVE0991-treated diabetic rats recovered from ischemia with P_{max} and coronary flow values similar to those of control animals (Table 1). Induction of diabetes also resulted in significant increase in urine protein (231±2 mg/24 h) in diabetic animals compared to controls (88±6 mg/24 h). Treatment of diabetic animals with Ang-(1-7) or AVE0991 resulted in a significant reduction in urine protein compared to controls (183±16 and 149±15 mg/24 h, respectively) [25].

3 Beneficial Effects of Ang-(1-7) in Diabetic Vasculature Are Partly Mediated via Inhibition of EGFR Transactivation

The epidermal growth factor receptor (EGFR) tyrosine kinase is an important central transducer and signaling hub for cellular signaling pathways that regulate growth, differentiation, migration, survival, and apoptosis. Recent reports suggest that it is also an important mediator of cardiovascular dysfunction (for review, see Akhtar and Benter [26]). In the normal and pathological states, EGFR, a 175-kDa receptor tyrosine kinase, can be activated by several different epidermal growth factor (EGF)-like ligands including EGF and heparin-binding (HB)-EGF to induce receptor clustering and autophosphorylation. This step subsequently leads to activation of multiple downstream signaling pathways such as the mitogenic Ras/Raf/ extracellular signal-regulated kinase 1/2 (ERK1/2), the p38 mitogen-activated protein (MAP) kinase, or the PI3-kinase/Akt survival pathways that regulate cell growth, proliferation, and differentiation [27]. Alternatively, transactivation of EGFR can occur via G protein-coupled receptors (GPCRs), such as members of RAAS including Ang II, aldosterone, and endothelin [28, 29]. We have recently shown that hyperglycemia-induced transactivation of EGFR in vascular smooth muscle cells (VSMC) is mediated via the nonreceptor tyrosine kinase, src, through its phosphorylation at tyrosine 416 [30]. Also, enhanced EGFR phosphorylation and signaling via ERK1/2 and p38 MAP kinase pathways appear to be important mediators of vascular dysfunction associated with hyperglycemia or diabetes [30]. In experimental diabetes, chronic or acute treatment with AG1478, a selective inhibitor of EGFR tyrosine kinase, significantly corrected vascular dysfunction in the mesenteric bed, the renal vasculature, and the carotid artery [31–34]. Gene expression profiling of the mesenteric vasculature showed that the correction in vascular dysfunction achieved by AG1478 was attained by blocking the upregulation of the majority (~85 %) of the 1,100+ genes whose expression had been altered in the diabetic mesenteric bed vasculature [33]. Thus, EGFR was proposed to be an important early detrimental pathway mediating vascular dysfunction, both in an experimental model of type 1 diabetes [31, 32] and subsequently in a model of type 2 diabetes [35, 36].

Given the facts that (a) diabetes-induced vascular dysfunction was also prevented by chronic Ang-(1-7) treatment [25] in a manner analogous to that also reported by us for AG1478, a selective inhibitor of EGFR signaling [31, 32], and that (b) Ang II signaling, including its transactivation of EGFR, is enhanced in the diabetes state, we hypothesised that Ang-(1-7) may counter-regulate Ang II signaling via inhibition of EGFR transactivation in the diabetic vasculature. In a report by Akhtar et al. [30] in which this hypothesis was tested, we first showed that diabetes or high glucose can induce Ang II-mediated transactivation of EGFR via a src-dependent pathway that subsequently, through signaling cascades involving p38 MAP kinase and ERK1/2, leads to vascular complications [30]. Second, we showed that src-dependent EGFR transactivation can be inhibited by Ang-(1-7) acting through its Mas receptor, representing a novel mechanism of action for inhibiting EGFR and for counter-regulating the actions of Ang II. Thus, the beneficial effects of Ang-(1-7) on the diabetic vasculature involve inhibition, at least in part, of the detrimental EGFR/ERK1/2/p38MAP kinase pathway [30]. Our findings lead us to speculate that this could represent a general mechanism by which Ang-(1-7) exerts its beneficial effects in many conditions, especially where enhanced EGFR signaling has been implicated.

4 Angiotensin-(1-7) Inhibits Renal NF-κB and Inflammatory Response Genes

We examined the influence of Ang-(1-7), both endogenous and after chronic treatment with the peptide, on ischemia–reperfusion (I/R)-induced cardiac dysfunction and cardiac nuclear factor-κB (NF-κB) activity in streptozotocin-treated spontaneously hypertensive rats (diabetic SHR). In isolated perfused hearts, recovery of left ventricular function from 40 min of global ischemia was significantly improved in Ang-(1-7)- or captopril-treated diabetic SHR and worsened in animals treated with A779, a Mas receptor antagonist [37]. The beneficial effect of captopril on cardiac recovery was reduced when coadministered with A779. Cardiac NF-κB activity was higher in SHR as compared to WKY and also higher in diabetic SHR as compared to SHR (Table 2). Treatment with Ang-(1-7) decreased NF-κB activity in both SHR and diabetic SHR whereas treatment with A779 slightly increased NF-κB activity (Table 2). The captopril-induced decrease in NF-κB activity was partially prevented by coadministration of A779 (Table 2). Real-time PCR-based gene array analysis of cardiac tissue revealed that Ang-(1-7) or captopril treatment reduced expression of the Toll-like receptor 2 (Tlr2), interleukin-1 receptor kinase (Irak1), and inhibitor of NF-κB kinase-b subunit (Ikbkb) (Table 3). Ang-(1-7) or captopril treatment also suppressed the expression of several pro-inflammatory genes involved in the NF-κB signaling pathway, including complement component 3 (C3), interleukin 1-beta (Il-1β), interleukin 6 (Il-6), NACHT-containing protein (Nalp12), and caspase 1 (Casp1) (Table 3) [37].

Table 2 Effect of Ang-(1-7) on cardiac NF-κB DNA-binding activity

Groups studied	NF-κB activity
1. WKY	7.0 ± 0.5
2. SHR	25.0 ± 0.9^{a}
3. SHR-Ang-(1-7)	$3.6 \pm 0.3^{a,b}$
4. Diabetic SHR	$33.5 \pm 1.3^{a,b}$
5. Diabetic SHR-Ang-(1-7)	$3.5 \pm 0.4^{a,b,c}$
6. Diabetic SHR-A779	$38.7 \pm 2.3^{a,b,c}$
7. Diabetic SHR-captopril	$10.5 \pm 0.6^{a,b,c}$
8. Diabetic SHR-captopril-A779	$21.6 \pm 0.8^{a,b,c,d}$

Data are expressed as mean ± SEM ($n = 5$)
[a]Value significantly different from WKY, $p < 0.05$
[b]Value significantly different from SHR, $p < 0.05$
[c]Value significantly different from diabetic SHR, $p < 0.05$
[d]Value significantly different from diabetic SHR-captopril, $p < 0.05$

Table 3 The effect of Ang-(1-7) on cardiac gene expression in the NF-κB signaling pathway

	WKY	SHR	SHR-Ang-(1-7)	Diabetic SHR	Diabetic SHR+Ang-(1-7)	Diabetic SHR+captopril
Activation of the NF-κB pathway						
NM_198769 Tlr2	1	1.52	0.99	1.52	0.88	0.83
NM_053355 Ikbkb/AIM-1	1	1.58	0.87	1.92	0.66	0.64
XM_343844 Irak1	1	2.01	1.03	2.35	0.69	0.77
Inflammatory response genes						
NM_016994 C3	1	2.45	0.70	2.85	0.66	0.63
NM_031512 Il-1β	1	1.85	1.21	2.68	0.54	0.64
NM_012589 Il-6	1	1.30	0.92	2.58	0.52	0.61
XM_218181 Nalp12	1	2.60	1.40	2.09	0.26	0.33
NM_012762 Casp1	1	2.45	0.90	3.16	0.55	0.68

All the values indicate increase (above 1) or decrease (below 1) compared to WKY

5 Endogenous Angiotensin-(1-7) Inhibits NOX

We previously showed that Ang-(1-7) inhibits activation of NADPH oxidase (NOX) and prevents diabetes-induced attenuation in peroxisome proliferator-activated receptor (PPAR)-γ and catalase activities in diabetic hypertensive rats [38, 39]. In a subsequent study, we reported on the effects of chronically inhibiting endogenous Ang-(1-7) formation with DX600, a selective angiotensin-converting enzyme-2 (ACE2) inhibitor, on renal and cardiac NOX activity, vascular reactivity, and cardiac function in a model of type 1 diabetes [40]. The contribution of endogenous Ang-(1-7) to the protective effects of losartan and captopril and that of prostaglandins to the cardiovascular effects of exogenous Ang-(1-7) were also examined [40]. Chronic treatment with DX600 exacerbated diabetes-induced increase in cardiac and renal NOX activity. Diabetes-induced abnormal vascular reactivity to ET-1 and cardiac dysfunction were improved by treatment with Ang-(1-7) and worsened by treatment with DX600 or A779, a Mas receptor antagonist [40]. Ang-(1-7)-mediated improvement in cardiac recovery or vascular reactivity was attenuated by indomethacin. Captopril- and losartan-induced improvement in cardiovascular function was attenuated when these drugs were coadministered with A779. Ang-(1-7)-mediated decrease in renal NOX activity was prevented by indomethacin. Losartan also decreased renal NOX activity that could be attenuated with A779 cotreatment. From these studies, we concluded that endogenous Ang-(1-7) inhibits diabetes-induced cardiac/renal NOX activity and end-organ damage, and likely mediates the actions of captopril and losartan. Furthermore, prostaglandins were important intermediaries in the beneficial effects of Ang-(1-7) in diabetes. Combining either losartan or captopril with Ang-(1-7) had additional beneficial effects in preventing diabetes-induced cardiac dysfunction, and this may represent a novel therapeutic strategy [40].

6 Conclusion

In summary, we showed that Ang-(1-7) and its analogue AVE0991 can attenuate the development of diabetes-induced cardiovascular dysfunction without correcting hyperglycemia (Fig. 1). This finding implies that although certain signaling pathways may be triggered by hyperglycemia/diabetes, their subsequent blockade via other mechanisms can, at least partially, prevent end-organ dysfunction. We conclude that activation of the Ang-(1-7)-mediated signal transduction pathway appears to be an important therapeutic strategy to reduce cardiovascular events in diabetic patients and may underlie a portion of the beneficial effects of ACE inhibition or AT_1 blockade.

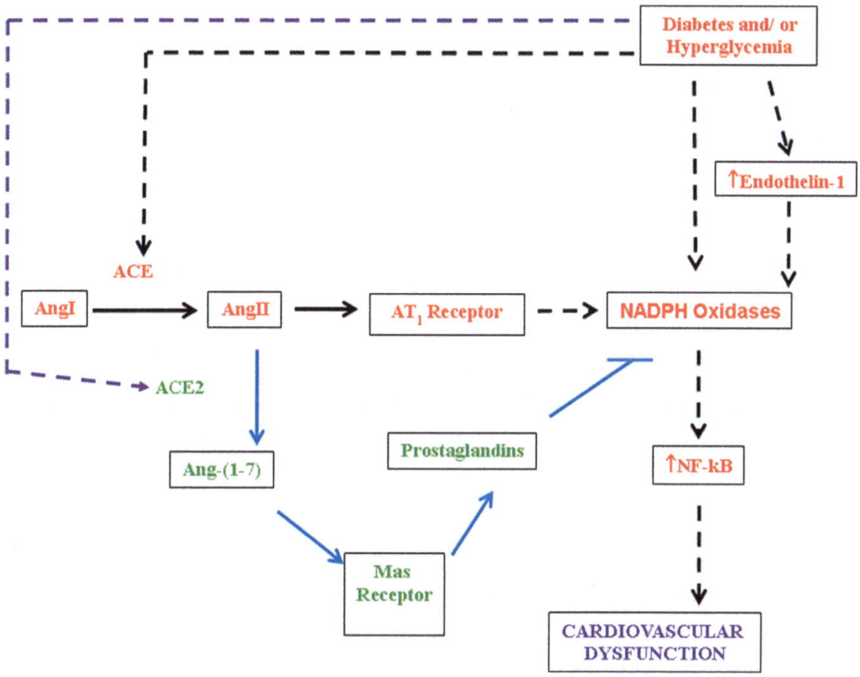

Fig. 1 Schematic model summarising our findings on the role of angiotensin-(1-7) signaling in diabetes-induced cardiovascular dysfunction

Acknowledgments This work was supported by a grant from Kuwait University Research Sector (MR05/09).

References

1. Yau JW, Rogers SL, Kawasaki R et al (2012) Global prevalence and major risk factors of diabetic retinopathy. Diabetes Care 35:556–564
2. Scully T (2012) Diabetes in numbers. Nature 485:S2–S3
3. Whiting DR, Guariguata L, Shaw J (2011) IDF diabetes atlas: global estimates of the prevalence of diabetes for 2011 and 2030. Diabetes Res Clin Pract 94:311–321
4. Von-Lueder TG, Krum H (2012) RAAS inhibitors and cardiovascular protection in large scale trials. Cardiovasc Drugs Ther 2:171–179
5. Forbes JM, Cooper ME (2013) Mechanisms of diabetic complications. Physiol Rev 93:137–188
6. Patel VB, Bodiga S, Basu R et al (2012) Loss of angiotensin-converting enzyme-2 exacerbates diabetic cardiovascular complications and leads to systolic and vascular dysfunction: a critical role of the angiotensin II/AT1 receptor axis. Circ Res 110:1322–1335
7. Aneja A, Tang WH, Bansilal S (2008) Diabetic cardiomyopathy: insights into pathogenesis, diagnostic challenges, and therapeutic options. Am J Med 121:748–757
8. Boudina S, Abel ED (2007) Diabetic cardiomyopathy revisited. Circulation 115:3213–3223
9. Fang ZY, Prins JB, Marwick TH (2004) Diabetic cardiomyopathy: evidence, mechanisms, and therapeutic implications. Endocr Rev 25:543–567
10. Benter IF, Diz DI, Ferrario CM (1993) Cardiovascular actions of angiotensin-(1-7). Peptides 14:679–684
11. Benter IF, Diz DI, Morris M, Ferrario CM (1995) Antihypertensive actions of angiotensin-(1-7) in spontaneously hypertensive rats. Am J Physiol Heart Circ Physiol 269:H313–H319
12. Benter IF, Diz DI, Ferrario CM (1995) Pressor and reflex sensitivity is altered in spontaneously hypertensive rats treated with angiotensin-(1-7). Hypertension 26:1138–1144
13. Benter IF, Yousif MHM, Anim JT et al (2006) Angiotensin-(1-7) prevents development of severe hypertension and end-organ damage in spontaneously hypertensive rats-treated with L-NAME. Am J Physiol Heart Circ Physiol 290:H684–H691
14. Crackower MA, Sarao R, Oudit GY et al (2002) Angiotensin converting enzyme 2 is an essential regulator of heart function. Nature 417:822–828
15. Tikellis C, Pickering R, Tsorotes D (2012) Interaction of diabetes and ACE2 in the pathogenesis of cardiovascular disease in experimental diabetes. Clin Sci (Lond) 123:519–529
16. Kassiri Z, Zhong J, Guo D et al (2009) Loss of angiotensin-converting enzyme 2 accelerates maladaptive left ventricular remodeling in response to myocardial infarction. Circ Heart Fail 2:446–455
17. Stewart JA, Lazartigues E, Lucchesi PA (2008) The angiotensin converting enzyme 2/Ang-(1-7) axis in the heart: a role for MAS communication? Circ Res 103:1197–1199
18. Nakamura K, Koibuchi N, Nishimatsu A et al (2008) Candesartan ameliorates cardiac dysfunction observed in angiotensin-converting enzyme 2-deficient mice. Hypertens Res 31: 1953–1961
19. Tallant EA, Clark MA (2003) Molecular mechanisms of inhibition of vascular growth by angiotensin-(1-7). Hypertension 42:574–579
20. Tallant EA, Ferrario CM, Gallagher PE (2005) Angiotensin-(1-7) inhibits growth of cardiac myocytes through activation of the Mas receptor. Am J Physiol Heart Circ Physiol 289:H1560–H1566
21. Santos RAS, Simões e Silva AC, Maric C et al (2003) Angiotensin-(1-7) is an endogenous ligand for the G protein-coupled receptor Mas. Proc Natl Acad Sci USA 100:8258–8263

22. Santos RAS, Ferreira AJ, Pinheiro SV et al (2005) Angiotensin-(1-7) and its receptor as a potential targets for new cardiovascular drugs. Expert Opin Invest Drugs 14:1019–1031
23. Santos RAS, Castro CH, Gava E et al (2006) Impairment of in vitro and in vivo heart function in angiotensin-(1-7) receptor MAS knockout mice. Hypertension 47:996–1002
24. Pinheiro SV, Simões e Silva AC, Sampaio WO et al (2004) Nonpeptide AVE 0991 is an angiotensin-(1-7) receptor Mas agonist in the mouse kidney. Hypertension 44:490–496
25. Benter IF, Yousif MH, Cojocel C et al (2007) Angiotensin-(1-7) prevents diabetes-induced cardiovascular dysfunction. Am J Physiol Heart Circ Physiol 292:H666–H672
26. Akhtar S, Benter IF (2013) The role of epidermal growth factor receptor in diabetes-induced cardiac dysfunction. Bioimpacts 3:5–9
27. Avraham R, Yarden Y (2011) Feedback regulation of EGFR signalling: decision making by early and delayed loops. Nat Rev Mol Cell Biol 12:104–117
28. Higuchi S, Ohtsu H, Suzuki H et al (2007) Angiotensin II signal transduction through the AT1 receptor: novel insights into mechanisms and pathophysiology. Clin Sci (Lond) 112:417–428
29. Smith NJ (2011) Determination of the exact molecular requirements for type 1 angiotensin receptor epidermal growth factor receptor transactivation and cardiomyocyte hypertrophy. Hypertension 57:973–980
30. Akhtar S, Yousif MH, Benter IF (2012) Angiotensin-(1-7) inhibits epidermal growth factor receptor transactivation via a Mas receptor-dependent pathway. Br J Pharmacol 165: 1390–1400
31. Benter IF, Yousif MH, Griffiths SM et al (2005) Epidermal growth factor receptor tyrosine kinase-mediated signalling contributes to diabetes-induced vascular dysfunction in the mesenteric bed. Br J Pharmacol 145:829–836
32. Benter IF, Yousif MH, Hollins AJ et al (2005) Diabetes-induced renal vascular dysfunction is normalized by inhibition of epidermal growth factor receptor tyrosine kinase. J Vasc Res 42:284–291
33. Benter IF, Benboubetra M, Hollins AJ et al (2009) Early inhibition of EGFR signaling prevents diabetes-induced up-regulation of multiple gene pathways in the mesenteric vasculature. Vasc Pharmacol 51:236–245
34. Yousif MH, Benter IF, Akhtar S (2005) The role of tyrosine kinase-mediated pathways in diabetes-induced alterations in responsiveness of rat carotid artery. Auton Autacoid Pharmacol 25:69–78
35. Belmadani S, Palen DI, Gonzalez-Villalobos RA et al (2008) Elevated epidermal growth factor receptor phosphorylation induces resistance artery dysfunction in diabetic db/db mice. Diabetes 57:1629–1637
36. Galán M, Kassan M, Choi SK et al (2012) A novel role for epidermal growth factor receptor tyrosine kinase and its downstream endoplasmic reticulum stress in cardiac damage and microvascular dysfunction in type 1 diabetes mellitus. Hypertension 60:71–80
37. Al-Maghrebi M, Benter IF, Diz DI (2009) Endogenous angiotensin-(1-7) reduces cardiac ischemia-induced dysfunction in diabetic hypertensive rats. Pharmacol Res 59:263–268
38. Benter IF, Yousif MH, Dhaunsi GS et al (2008) Angiotensin-(1-7) prevents activation of NADPH oxidase and renal vascular dysfunction in diabetic hypertensive rats. Am J Nephrol 28:25–33
39. Dhaunsi GS, Yousif MH, Akhtar S et al (2010) Angiotensin-(1-7) prevents diabetes-induced attenuation in PPAR-gamma and catalase activities. Eur J Pharmacol 638:108–114
40. Yousif MH, Dhaunsi GS, Benter IF (2012) Characterization of angiotensin-(1-7) effects on the cardiovascular system in an experimental model of type-1 diabetes. Pharmacol Res 66: 269–275

Targeting Matrix Metalloproteinase 2 and 9 for Treatment of Cardiovascular Dysfunction of Diabetes

Lokesh Kumar Bhatt and Veeranjaneyulu Addepalli

Abstract Cardiovascular complications in the diabetic population are the foremost cause of death and impose a huge economic burden. Despite recent advances in our understanding of diabetic complications, there is an unmet need for treatment. Matrix metalloproteinases (MMPs) are zinc-dependent enzymes that are involved in the remodeling of several components of the extracellular matrix. Upregulation of MMP-2 and MMP-9 in diabetes leads to disruption of extracellular matrix and thus causes complications of diabetes. Matrix metalloproteinase 2 and 9 can be potential targets to treat cardiovascular complications of diabetes.

Keywords Matrix metalloproteinase 2 • Matrix metalloproteinase 9 • Extracellular matrix • Diabetic cardiovascular complications

1 Introduction

Diabetes mellitus refers to a number of disorders that share the common feature of elevated blood glucose levels. The classification accepted by the World Health Organization (WHO) and the American Diabetes Association (ADA) [1] combines both clinical stages of hyperglycemia and the etiological types. Two main subtypes of diabetes are type 1, either autoimmune or idiopathic, and type 2, attributable to insulin resistance, insulin secretion defects, or both. Although diabetes has been known for centuries, the etiology and pathogenesis of this disease are still

L.K. Bhatt
Department of Pharmacology, Dr. BhanubenNanavati College of Pharmacy, Vile Parle (W), Mumbai, Maharashtra 400056, India

V. Addepalli (✉)
Department of Pharmacology, SPP School of Pharmacy and Technology Management, SVKM's NMIMS, V. L. Mehta Road, Mumbai, Maharashtra 400056, India
e-mail: veeranjaneyulu.addepalli@nmims.edu

B. Turan and N.S. Dhalla (eds.), *Diabetic Cardiomyopathy: Biochemical and Molecular Mechanisms*, Advances in Biochemistry in Health and Disease 9, DOI 10.1007/978-1-4614-9317-4_23,

incompletely defined. Hallmarks of diabetes are chronic hyperglycemia with disturbances of carbohydrate, fat, and protein metabolism that results from defects in insulin secretion, insulin action, or both [2]. It is characterized by hyperglycemia, particularly during fasting conditions. Several causes of diabetes mellitus have been identified; however, the etiology and pathogenesis are not clearly understood. A huge economic burden is imposed on healthcare systems by diabetes. Diabetes accounts for 11.6 % of the total healthcare expenditure in the world [3].

Diabetes is strongly accompanied by both microvascular and macrovascular complications that include ischemic heart disease, peripheral vascular disease, and cerebrovascular disease (macrovascular), and retinopathy, nephropathy, and neuropathy (microvascular). These complications results in organ and tissue damage in approximately one third to one half of people with diabetes. Expectancy of life for people with diabetes has been estimated to be as much as 10 years shorter than for people without diabetes [4]. Similarly, diabetes imposes substantial demands on healthcare systems because medical expenditures for people with diabetes are up to three times greater than for those without diabetes, largely because of macrovascular complications.

In the management of micro- and macrovascular complications, a substantial economic burden is created by diabetes mellitus [5]. The microvascular complications in diabetes include retinopathy, nephropathy, and neuropathy. The first signs of these complications in type 1 patients may develop even in adolescence, particularly if insulin treatment has been inadequate [6]. In type 2 patients, similar complications occur later in life and are frequently present at the time of diagnosis. The injurious effects of hyperglycemia are separated into macrovascular complications (coronary artery disease, peripheral arterial disease, and stroke) and microvascular complications (diabetic nephropathy, neuropathy, and retinopathy). These complications not only are inconvenient but also severely impact quality of life [6].

Two different clinical studies, the DCCT (Diabetes Control and Complications Trial) and the UKPDS (U.K. Prospective Diabetes Study), established that hyperglycemia is the initiating cause of the diabetic tissue damage that we see clinically [7, 8]. The process of damage in complications of diabetes is modified both by genetic determinants of individual susceptibility and by independent accelerating factors such as hypertension. On the basis of many laboratory findings as well as clinical trials, four main hypotheses have been suggested to explain how hyperglycemia causes diabetic complications: (1) increased polyol pathway flux; (2) increased advanced glycation end-product (AGE) formation; (3) activation of protein kinase C (PKC) isoforms; and (4) increased hexosamine pathway flux [9].

2 Matrix Metalloproteinases

Matrix metalloproteinases (MMPs) are a group of 20 zinc-dependent enzymes (Table 1) that are involved in the remodeling of several components of the extracellular matrix (ECM). MMPs play a role in several physiological processes, including bone remodeling and organogenesis, and also act in the reorganization of tissues

Table 1 Different types of matrix metalloproteinases and their substrates

MMP subtype (enzyme)	Substrate
Gelatinases	
MMP-2 (gelatinase A)	Gelatin, collagens (I, IV, V, VII, X, XI, XIV), elastin, fibronectin, laminin, MMP-1, MMP-9, MMP-13
MMP-9 (gelatinase B)	Gelatin, collagens (IV, V, VII, X, XIV), elastin, fibronectin, plasminogen
Collagenases	
MMP-1 (collagenase-1)	Collagens (I, II, III, VII, VIII, X), gelatin, proteoglycans, MMP-2, MMP-9
MMP-8 (collagenase-2)	Collagens (I, II, III, V, VII, VIII, X), gelatin, proteoglycans
MMP-13 (collagenase-3)	Collagens (I, II, III, IV, IX, X, XIV), gelatin, MMP-9
MMP-18 (collagenase-4)	Not known
Stromelysins	
MMP-3 (stromelysin-1)	Collagens (III, IV, V, IX, X) gelatin, fibronectin, laminin, MMP-1, MMP-7, MMP-8, MMP-9, MMP-13
MMP-10 (stromelysin-2)	Collagenes (III, IV, V), gelatin, casein, MMP-1, MMP-8
MMP-11 (stromelysins-3)	Gelatin, collagen IV, fibronectin, casein, proteoglycans
Matrilysins	
MMP-7 (matrilysin-1)	Collagens (IV, X), gelatin, fibronectin, laminin, MMP-1, MMP-2, MMP-9, MMP-9/TIMP-1 complex
MMP-26 (mattrilysin-2)	Collagen IV, fibronectin, gelatin, pro-MMP-9, fibrinogen
Membrane type	
MMP-14 (MT1-MMP)	Collagens (I, II, III), gelatin, casein, elastin, fibronectin, laminin, MMP-2, MMP-13
MMP-15 (MT2-MMP)	Gelatin, fibronectin, laminin, MMP-2
MMP-16 (MT3-MMP)	Collagen III, gelatin, casein, fibronectin, MMP-2
MMP-17 (MT4-MMP)	Gelatin, pro-MMP-2
MMP-24 (MT5-MMP)	Proteoglycans, pro-MMP-2, collagen 1, gelatin, fibronectin, laminin
MMP-25 (MT6-MMP)	Collagen IV, pro-MMP-8, pro-MMP-9
Other MMPs	
MMP-12 (macrophage metalloelastase)	Collagen IV, gelatin, elastin, fibronectin, casein, fibrinogen, plasminogen, MMP-2
MMP-19, MMP-20	Aggrecan, cartilage oligomeric matrix protein

MMP matrix metalloproteinases, *MT* membrane type

during pathological conditions such as inflammation, wound healing, and invasion by cancer cells [10, 11]. In normal cells, MMPs are responsible for remodeling of the ECM, including basement membranes. MMPs contain zinc ions that are coordinated by three histidine residues in their active site. Although all MMPs possess different primary structures, they are composed of shared modules known as protein domains. The catalytic domain of this enzyme possesses a zinc-binding consensus sequence, a characteristic sequence shared with other metalloproteinase families. Another characteristic feature of the MMP is activation by the cysteine switch [12, 13]. When cells produce this enzyme, most of it is secreted in a latent pro-form. In the latent form a cysteine sulfydryl group in the amino-terminal pro-domain interacts with the zinc ion and blocks the active site. Exclusion of the pro-peptide from the active site of the enzyme leads to its activation. The amount of MMP secretion depends on biological context; for example, some constitutive or homeostatic MMP genes hold simple promoter enhancer regions with *cis*-acting elements for basal transcription whereas others have complex promoter regions. Different agonists can regulate MMP expression. The levels of expression of these inducible or inflammatory MMPs are determined by the biological milieu. In triple-helical collagens, a specific scissile bond is cleaved by the interstitial collagenase (MMP-1), neutrophil collagenase (MMP-8), and collagenase 3 (MMP-13) [14].

2.1 Regulation of Matrix Metalloproteinases

Activities of MMPs are tightly controlled. The proteolytic activity of MMPs is regulated at three levels, as follows.

Gene Transcription

Transcriptional activation of MMPs is induced by inflammatory cytokines, growth factors, and tumor promoters such as interleukin 1b (IL-1b), IL-6, platelet-derived growth factor, epidermal growth factor, tumor necrosis factor-α, and macrophage colony-stimulating factor [15, 16]. Hyperglycemia, thrombin, reactive oxygen species (ROS), oxidized low density lipoprotein (ox-LDL), CD40 ligand, and hypoxia are other factors that control its expression [17, 18].

Activation

Activation of MMPs by protease is required because MMPs are secreted as latent enzymes (especially membrane-type MMPs) [17, 19]. Activation of MMPs takes place in two steps that lead to exposure of the catalytic site. In the first step, the cysteine–zinc interaction is disrupted by an activator, followed by the second step in which the pro-peptide domain of MMPs is removed. However, pro-MMP-2 is an exception of this two-step process [19].

Inhibition by Tissue Inhibitors of Matrix Metalloproteinases (TIMPs)

Active or latent forms of MMPs in a molecular 1:1 ratio are inhibited by inherent inhibitors of MMPs [20]. The net proteolytic activity is determined by the equilibrium between active MMPs and their inhibitors [21].

3 Hyperglycemia and Matrix Metalloproteinases

Various studies using cell lines have shown that production of MMPs, their expression, and their activity can be modulated by glucose concentration. Activity and expression of MMP-1, MMP-2, and macrophage-derived MMP-9 may be enhanced by hyperglycemic cultures of endothelial cells (ECs) [22]. On the other hand, MMP-3 expression and protein levels decrease with high glucose concentration while TIMP-1 concentrations remain unaffected. Thus, degradation of the ECM occurs because of irregular expression of MMPs and TIMPs. ECM disruption may enhance migration of monocytes and vascular smooth muscle cells (VSMC). This increase in migration may further exacerbate atherosclerosis in diabetics, as indicated by a previous study [23] showing that synthesis of MMP-2 and MMP-9 in ECs was enhanced by long-term incubation with high glucose concentrations. Genetic analysis of MMP/TIMP promoters demonstrated they host response elements that are capable of binding several transcription factors [24]. Results from earlier studies suggest that high glucose levels stimulate these factors [25, 26]. Thus, it is conceivable that hyperglycemia positively regulates these growth and transcription factors followed by enhanced MMP gene transcription. However, the effect of genetic variation is not clear. In one study, polymorphism of the AP-1 response element was found in the promoter region of the MMP-12 gene. This polymorphism links with higher MMP-12 activity in diabetic patients with coronary heart disease [27]. Results suggest that hyperglycemia, and oxidative stress in addition to genetic stimuli, regulate MMP gene transcription [22].

3.1 Glycation Products and MMPS

Evidence suggests that advanced glycation end products (AGEs) and their receptors (RAGEs) control several signaling pathways that are involved in vascular dysfunction [28]. Collagen deficiency, macrophage accumulation in the shoulder area of the plaque, and MMP colocalization with inflammatory cells are the main feature of a plaque [29]. Inflammatory cells are stimulated by susceptible atherosclerotic plaques and COX-2/PGES and release MMPs, which then leads to plaque vulnerability [30]. Within the diabetic atherosclerotic plaque, MMPs may cause increased frequency of vascular events. Diabetes likely shares common pathways of synthesis of MMPs and plaque deterioration, followed by atherosclerosis. Studies showed that COX-2 inhibitors decrease MMP expression [31, 32]. MMP activity may also be influenced by aspirin (a nonselective COX inhibitor) [31]. In diabetes, synthesis of prostacyclin (PGI2) may decrease [33]. Diabetes may also lead to several irregularities of other mediators of inflammation [33], which can influence the activity of MMPs [34, 35].

4 Matrix Metalloproteinases and Diabetic Vascular Complications

4.1 Diabetic Microvascular Complications and MMPs

Apart from the role of MMPs in diabetic macrovascular complications, these changes may play a role in DM-related microvascular complications. One study [36] provides a justification for the association between microangiopathic complications, such as retinopathy, nephropathy, and peripheral neuropathy, and MMP-2 in patients with DM1. Increased concentrations of activated MMP-2 and MMP-9 were seen in retinas from diabetic patients in another study [37], and in the basement membranes of new nonfunctional retinal capillaries MMP-2 and MMP-9 levels were also increased [37]. In proliferative diabetic retinopathy patients, a large proportion of specimens from epiretinal membranes were stained for MMP-1, -2, -3, and -9 as much as for TIMP-1, -2, and -3 [38]. These processes may lead to retinal neovascularization.

In diabetic nephropathy, thickening of mesangial expansion and glomerular basement membranes is the histopathological hallmark [39]. Because accretion of ECM within the glomerulus leads to renal dysfunction in diabetes, homeostasis of the ECM is important [40]. Disproportion between synthesis and degradation of the ECM is recognized by the amount and composition of mesangial matrix. Diabetes stimulates synthesis of mesangial matrix components such as collagen type IV, fibronectin, and laminin [41]. Further, hyperglycemia restrains the degradation process of ECM, which leads to accumulation of ECM in the glomerular mesangium [42]. In addition, AGE formation in glomeruli leaves the ECM less susceptible to degradation [43]. There is good support for the hypothesis that decrease in MMP activity of mesangial cells controls ECM degradation and may be involved in the pathogenesis of diabetic nephropathy [39, 44]. It is suggested that hyperglycemia regulates MMP gene expression through different mechanisms including protein kinase C (PKC) agonists, cytokines, transforming growth factor (TGF)-β, and nuclear factor-κB (NF-κB). The mechanisms by which hyperglycemia induces MMP expression are not clear. For example, in diabetics glycation of the ECM decreases MT1-MMP expression but increases MMP-2, TIMP-1, and TIMP-2 expression [43].

MMPs facilitate EC migration, promote EC sprouting, and induce growth of new vessels [45]. In response to occlusion and subsequent ischemia, diabetes may influence neoangiogenesis adversely [46, 47]. Some preclinical studies from rat hindlimb ischemia models indicated that in diabetics, angiogenesis is prevented by inhibition of MMP activity [48, 49]. Also, hyperglycemia ameliorated hepatocyte growth factor and AGE accumulation, leading to reduction of MMP activation [49]. If neovascularization takes place at atherosclerotic plaques, it may be crucial for continuous growth, intraplaque hemorrhage, and subsequent plaque rupture. MMPs are known as required components that can aid atherosclerotic plaque nutrition [50]. Although the role of MMPs in peripheral neoangiogenesis in diabetes remains obscure, angiogenesis in ischemic peripheral tissues might ameliorate the symptoms [50].

4.2 *Diabetic Macrovascular Complications and MMPs*

Atherosclerotic plaques and their stability are influenced by inflammatory activity in the vascular wall. Homeostasis of the fibrous cap also influences its stability. Results from previous studies suggest that degenerative proteases such as MMPs and cathepsins promote plaque instability [51]. A remarkable upregulation in intracellular MMP-9 levels was found in specimens obtained from patients with unstable angina compared to patients with stable angina [52]. Also, plaques extracted from patients undergoing carotid endarterectomy contain increased concentrations of MMP-9 [53]. Collagen present in the interstitial space is the chief contributor to the tensile strength of fibrous plaques [51]. Collagenases provide the degradation of collagen fragments by other MMPs. In the unstable plaques collagenase-1 (MMP-1) and collagenase-3 (MMP-13) colocalize with macrophages [54]. Neutrophil collagenase (MMP-8), which is secreted by several vascular cells, is also present in high concentration in these plaques [55]. Apart from the aforementioned MMPs, MMP-7 and MMP-12 are also present in carotid plaques [56]. The polymorphism present in stromelysin-1 (MMP-3), which is followed by subsequent increase of MMP-3 activity, is coupled with enhanced susceptibility to plaque rupture [57].

5 Matrix Metalloproteinases in Cardiovascular Complications of Diabetes

Various studies reported the role of metalloproteinases and their inhibitors in the pathophysiology of cardiovascular disease. They also influence the formation and destabilization of the atherosclerotic plaque. MMP-1, MMP-2, MMP-3, MMP-7, and MMP-9 activation within atherosclerotic arteries [58, 59] is an important process in this regard. There is evidence that one factor for occurrence of acute coronary events is polymorphism of the MMP-9 gene.

5.1 *MMPs and Atherosclerosis*

Compared to the general population, manifestation of atherosclerosis occurs earlier in diabetic patients. In diabetics it is more severe and has a more scattered character [60]. MMPs act by influencing the process of atherosclerotic lesion formation. One of the mechanisms by which MMPs act includes the increased migration of vascular smooth muscle cells through the internal elastic lamina into the intimal space, where they further increase plaque formation [61]. The fibrous cap is generally made up of extracellular matrix laid down by the smooth muscle cells, even though the matrix itself is acellular in nature. Various components of this include collagen, which provides strength, and elastin, which provides flexibility. In the normal remodeling process the fibrous cap is resorbed (destruction) or redeposited (construction). The destructive and constructive processes are balanced under stable conditions, and the plaque protects the lumen of the vessel from exposure to underlying thrombogenic material. However, under pro-inflammatory conditions, the remodeling process can be shifted in favor of resorption of the matrix, leading to a potentially dangerous weakness in the cap and resulting in plaque rupture. MMPs are often found associated with the presence of inflammatory cells, such as macrophages or T lymphocytes [21]. The MMPs are also associated with sites of potential weakness in the plaque where ruptures tend to occur or where extensive remodeling is taking place [62]. Diabetes and increased lipids lead to oxidative stress and release of free radicals in the intima and media of blood vessels, causing activation of MMPs. Additionally, cytokine release is stimulated by the presence of lipids (especially oxidized low density lipoprotein) from activated macrophages and T cells. This stimulation results in further increase in expression and activation of MMPs [62].

5.2 MMPs and Heart Failure

Studies have shown an increased level of MMPs in heart failure patients and increased level with progression of heart failure. The level of MMPs and their induction and activation systems are increased in pathological specimens of human heart failure [63]. Studies have also reported a significant rise in MMPs within hours of the precipitation of myocardial infarction, followed by local activation of cytokines and infiltration of inflammatory cells [64].

This increase in MMP concentrations decreases with the healing process, but then is followed by a second wave of activation that results in more ventricular dilation and progression toward heart failure [65].

5.3 MMPs and Acute Coronary Syndromes

Many authors have reported change in MMP levels in patients with acute myocardial infarction. In diabetic patients, worsening of outcome was observed, which leads to high mortality. When plasma levels of MMP-9 and TIMP-1 in patients with angiographically identified lesions in the left anterior descending artery were compared with normal subjects [66], it was found that plasma levels of MMP-9 and TIMP-1 increased during acute coronary syndromes [66]. Another study by Hojo et al. of subjects with acute myocardial infarction showed MMP-2 level was increased compared with healthy controls; they evaluated both MMP-2 and MMP-1 [67]. In another study concentrations of serum MMP-1 and TIMP-1 were examined in patients after their first myocardial infarction followed by successful reperfusion [68]. Results of this study showed a significant increase of MMP-1 and TIMP-1. Interestingly, concentrations of these enzymes peaked at the 14th day.

5.4 *MMPs and Aortic Aneurysms*

Studies both in humans and in animals have shown a relationship between elevated aortic tissue MMP-9 levels and abdominal aortic aneurysms. Histological findings of aneurysmal aortas suggest explicit changes in the extracellular matrix and the aortic wall. Histological study of such slides showed decrease in elastin content and an increase in collagen synthesis [69]. Proteolysis of elastin in pathological conditions also results in the release of elastin degradation product, release of MMP-1 and MMP-2, and smooth muscle cell proliferation [70].

6 Matrix Metalloproteinases 2 and 9 in Cardiovascular Complications of Diabetes

MMP-2 and MMP-9 greatly influence the extracellular matrix. By a different mechanism, hyperglycemia in diabetes influences MMP-2 and MMP-9 levels and acts on the ECM (Fig. 1). Earlier studies showed that MMP-2 and MMP-9, by acting on the ECM, may aggregate many pathological conditions. Because diabetes induces upregulation of MMP-2 and MMP-9, many cardiovascular events that are mediated by MMP-2 and MMP-9 are disturbed, resulting in cardiovascular complications of

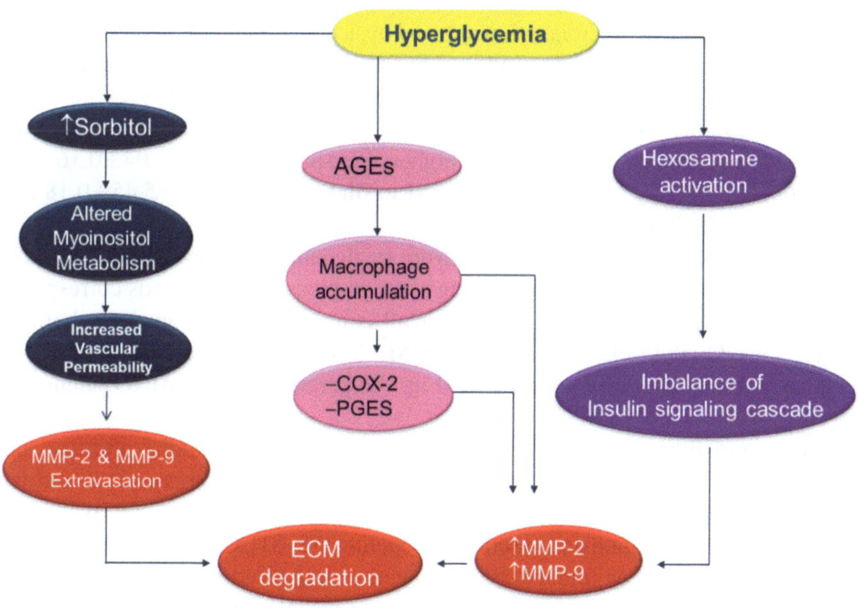

Fig. 1 Mechanism by which hyperglycemia induces matrix metalloproteinase (MMP)-2 and MMP-9 levels in diabetes

diabetes. In one study, hearts were obtained postmortem within 24 h from patients who died from causes other than coronary artery disease (CAD). Increased expression of MMP-2 and MMP-9 in expansively remodeled plaques was observed compared to constrictive remodeled segments of atherosclerotic coronary arteries [71]. Sun et al. suggested that an excess of tumor necrosis factor (TNF) in the myocardium is directly related to the increased production of local MMP-9 and MMP-2 [72]. Further, it was found that this increase in MMP-2 and MMP-9 is associated with changes in integrin isoform transition, which may lead to aggressive collagen dissolution that can cause acute myocardial rupture. Kai et al. demonstrated increased MMP-2 levels in 50 patients (22 with acute myocardial infarction, 11 with unstable angina, 17 with stable angina, and 17 normal volunteers) [73]. MMP-2 levels were increased by twofold in the unstable angina and acute myocardial infarction groups versus the stable angina and controls and were sustained over the 7-day period [73].

For degradation and reorganization of the ECM, vascular remodeling is needed, which could be either physiological or pathological. MMPs are important in this vascular remodeling [74]. In the progression of diabetes and arterial disease, high activity of MMP-2 and MMP-9 is widely reported [74]. The MMPs are secreted by inflammatory cells in the adventitia or smooth muscle cells in the media. The degraded elastic fiber induces calcium deposition, which in conjunction with altered vascular structure is associated with vessel stiffening [75].

In our previous studies we showed that long-term hyperglycemia induces upregulation of MMP-2 and MMP-9 levels in various tissues [76, 77]. This upregulation was responsible for thickening of the ECM in blood vessels, which then causes decreased vascular reactivity of the blood vessels [76]. Diabetes alters vascular reactivity, thus decreasing relaxation induced by acetylcholine (ACh). Endothelium-dependent relaxation responses induced by ACh was significantly decreased in a diabetic control group when compared with a normal group. Three-week treatment with minocycline (MINO), an inhibitor of MMP-2 and MMP-9, significantly increased relaxation and showed higher sensitivity ($pD2 = 6.04 \pm 0.16$ and $E_{max} = 64.31 \pm 2.537$, $P < 0.05$) as compared to diabetic rats ($pD2 = 5.45 \pm 0.18$ and $E_{max} = 40.36 \pm 2.99$) [76] (Fig. 2).

Protein levels in artery homogenate from treated and untreated diabetic rats and normoglycemic rats were assessed using zymography. Different bands corresponding with molecular weight were detected in all samples. A zymogram of the homogenate showed high levels of MMP-2 and MMP-9 protein in vehicle-treated diabetic animals compared to normal animals (Fig. 3). Treatment with MINO attenuated levels of MMP-2 and MMP-9 [76].

Increase in arterial stiffening in diabetes leads to increased cardiac afterload and resultant left ventricular hypertrophy. It also reduces coronary artery perfusion that may lead to myocardial ischemia, and increases pulse pressure, which finally promotes atheroma formation and vascular remodeling [78–81]. Circulating MMP-2 and MMP-9 levels have been found to be associated with large artery stiffness in humans [82]. In our studies we also evaluated the effect of diabetes on cardiovascular performance and correlated it with myocardial MMP-2 and MMP-9 levels. Streptozotocin (STZ)-induced diabetes mellitus produced decreased body and heart weights.

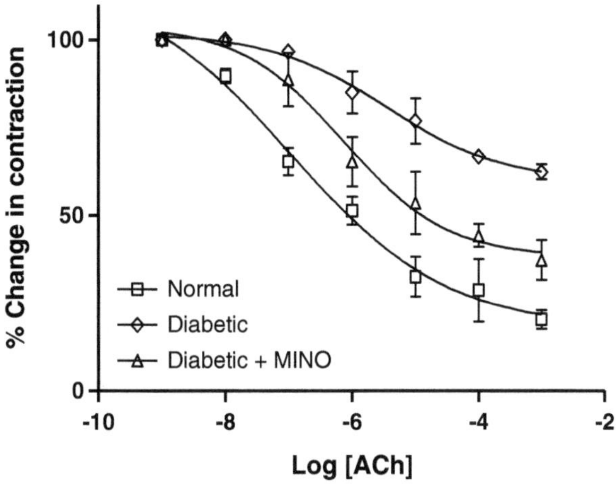

Fig. 2 Effect of hyperglycemia on vascular reactivity of rat aorta. Treatment with MMP-2 and MMP-9 inhibitor minocycline attenuated vascular reactivity assay. ACh concentration responses were taken after precontraction with PE on aorta from STZ-induced diabetic rats and age-matched controls

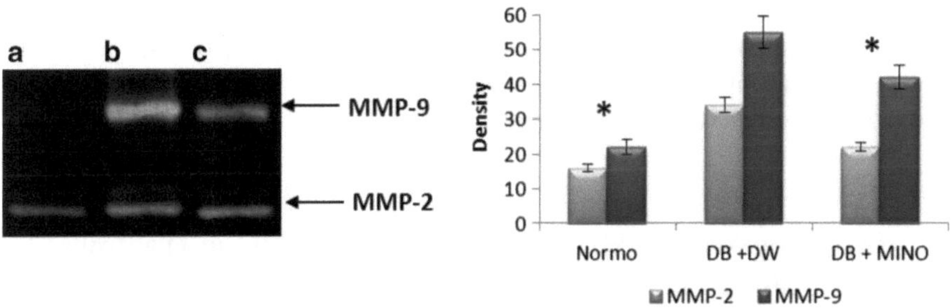

Fig. 3 Effect of hyperglycemia on MMP-2 and MMP-9 levels of rat aorta. Treatment with the MMP-2 and MMP-9 inhibitor minocycline attenuated MMP-2 and MMP-9 levels in rat aorta. (*a*) Normoglycemic. (*b*) Diabetic control. (*c*) Minocycline-treated diabetic. *$P < 0.05$ when compared with vehicle-treated diabetic group

The ratio of heart weight to body weight in hyperglycemic rats was increased compared with a normoglycemic group. Because cardiac hemodynamics is affected by changes in collagen content, total collagen content was analyzed in the hearts. Total collagen content was increased by twofold in vehicle-treated diabetic hearts compared with normoglycemic rats, whereas treatment with minocycline attenuated cardiac fibrosis significantly compared with vehicle-treated diabetic rats (Fig. 4).

Protein levels in heart homogenate from treated and untreated diabetic rats and normoglycemic rats were assessed using zymography. Different bands corresponding with molecular weight were detected in all samples. A zymogram of the homogenate showed high levels of MMP-2 and MMP-9 protein in vehicle-treated diabetic animals compared to normal animals. Treatment with minocycline attenuated levels of MMP-2 and MMP-9 protein. Hyperglycemia-induced induction of MMP-2 and MMP-9 in diabetics leads to cardiovascular dysfunctions. Treatment with the MMP-2 and MMP-9 inhibitor minocycline showed significant amelioration in

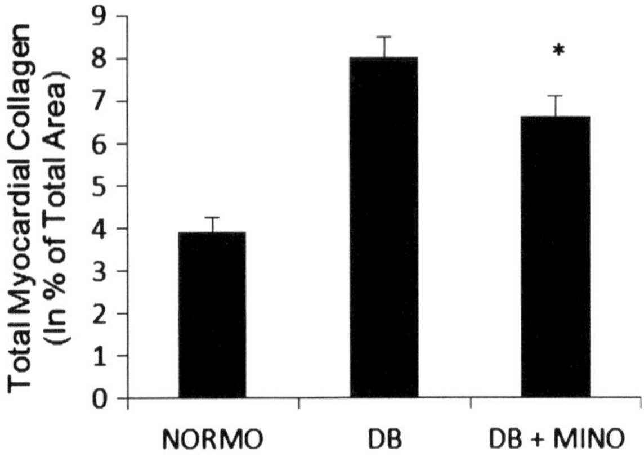

Fig. 4 Diabetes leads to significant increase in collagen levels. Treatment with the MMP-2 and MMP-9 inhibitor minocycline attenuated MMP-2 and MMP-9 levels in rat aorta.*$P < 0.05$ when compared with vehicle-treated diabetic group

cardiovascular dysfunction as evident from improvement in left ventricle performance, myocardial collagen, and MMP-2 and MMP-9 levels. Further, this treatment ameliorated vascular structure as evident from vascular reactivity assay and aortic collagen estimation. Thus, the study suggests a role of MMP-2 and MMP-9 in cardiovascular dysfunction in diabetic rats.

7 Targeting Matrix Metalloproteinase 2 and 9 for Treatment of Cardiovascular Dysfunction of Diabetes

MMPs may be effective as therapies commonly prescribed for patients with cardiovascular diseases. For example, use of nitroglycerin increases the expression and the activity of MMP-2 and MMP-9 and decreases TIMP-1 levels [83]. Studies reported the MMP-2 levels had been increased by heparin [84]. In cultured human vascular endothelial cells, calcium channel blockers were found to increase the activity of MMP-2 [85]. Losartan, an angiotensin-receptor blocker, has been shown to increase MMP-2 activity in human vascular smooth cells [86]. Earlier in this section the role of MMP-2 and MMP-9 is discussed in detail, and it is evident that in diabetes increased upregulation of MMP-2 and MMP-9 is responsible for cardiovascular complication of diabetes. Targeting MMP-2 and MMP-9 for the treatment of cardiovascular complication of diabetes can be a potential therapeutic approach.

8 Conclusion

In conclusion, matrix metalloproteinases have a very important role in the cardiovascular complications of diabetes. DM augments the risk for vascular events and worsens existing cardiovascular events. Various studies have shown how MMPs play a significant role in atherosclerosis and plaque rupture under hyperglycemia. Therefore, the MMP/TIMP system may provide a therapeutic approach to overcome macrovascular events in DM. MMP-2 and MMP-9, important factors in vascular remodeling and responsible for precipitation of cardiovascular events, can be targeted for relief. However, more clinical and mechanistic data are required for better understanding of the process by which MMP-2 and MMP-9 can become potential targets.

References

1. Rodriguez BL, Abbott RD, Fujimoto W et al (2002) The American Diabetes Association and World Health Organization classifications for diabetes: their impact on diabetes prevalence and total and cardiovascular disease mortality in elderly Japanese-American men. Diabetes Care 25:951–955
2. American Diabetes Association (2009) Diagnosis and classification of diabetes mellitus. Diabetes Care 32(Suppl 1):S62–S67
3. Zhang PZX, Brown JB, Vistisen D et al (2009) Economic impact of diabetes. In: Diabetes Atlas, International Diabetes Federation (IDF), 4th edn. http://www.idf.org/sites/default/files/Economic_impact_of_Diabetes.pdf
4. Leal J, Gray AM, Clarke PM (2009) Development of life-expectancy tables for people with type 2 diabetes. Eur Heart J 30:834–839
5. vanDieren S, Beulens JW, van der Schouw YT et al (2010) The global burden of diabetes and its complications: an emerging pandemic. EurJ Cardiovasc Prev Rehabil 17(Suppl 1):S3–S8
6. Brownlee M (2005) The pathobiology of diabetic complications: a unifying mechanism. Diabetes 54:1615–1625
7. Larkin ME, Lorenzi GM, Bayless M et al (2012) Evolution of the study coordinator role: the 28-year experience in Diabetes Control and Complications Trial/Epidemiology of Diabetes Interventions and Complications (DCCT/EDIC). Clin Trials 9:418–425
8. Genuth S (2008) The UKPDS and its global impact. Diabet Med 25(Suppl 2):57–62
9. Brownlee M (2001) Biochemistry and molecular cell biology of diabetic complications. Nature 414:813–820
10. Yabluchanskiy A, Li Y, Chilton RJ, Lindsey ML (2013) Matrix metalloproteinases: drug targets for myocardial infarction. Curr Drug Targets 14:276–286
11. Egeblad M, Werb Z (2002) New functions for the matrix metalloproteinases in cancer progression. Nat Rev Cancer 2:161–174
12. Van Wart HE, Birkedal-Hansen H (1990) The cysteine switch: a principle of regulation of metalloproteinase activity with potential applicability to the entire matrix metalloproteinase gene family. Proc Natl Acad Sci U S A 87:5578–5582
13. Kleifeld O, Van den Steen PE, Frenkel A et al (2002) Structural characterization of the catalytic active site in the latent and active natural gelatinase B from human neutrophils. J Biol Chem 275:34335–34343
14. Hu J, Van den Steen PE, Sang QX, Opdenakker G (2007) Matrix metalloproteinase inhibitors as therapy for inflammatory and vascular diseases. Nat Rev Drug Discov 6:480–498
15. Galis Z, Muszynski M, Sukhova GK et al (1995) Enhanced expression of vascular matrix metalloproteinases induced in vitro by cytokines and in regions of human atherosclerotic lesions. Ann N Y Acad Sci 748:501–507
16. Uzui H, Harpf A, Liu M et al (2002) Increased expression of membrane type 3-matrix metalloproteinase in human atherosclerotic plaque. Role of activated macrophages and inflammatory cytokines. Circulation 106:3024–3030

17. Jones CB, Sane DC, Herrington DM (2003) Matrix metalloproteinases: a review of their structure and role in acute coronary syndromes. Cardiovasc Res 59:812–823
18. Galis ZS, Khatri J (2002) Matrix metalloproteinases in vascular remodeling and atherogenesis. The good, the bad and the ugly. Circ Res 90:251–262
19. Negase H, Woessner JF Jr (1999) Matrix metalloproteinases. J Biol Chem 274:21491–21494
20. Visse R, Nagase H (2003) Matrix metalloproteinases and tissue inhibitors of metalloproteinases: structure, function and biochemistry. Circ Res 92:827–839
21. Knox JB, Sukhova GK, Whittermore AD et al (1997) Evidence for altered balance between matrix metalloproteinases and their inhibitors in human aortic diseases. Circulation 95: 205–212
22. Death A, Fisher EJ, McGrath KCY et al (2003) High glucose alters matrix metalloproteinase expression in two key vascular cells: potential impact on atherosclerosis in diabetes. Atherosclerosis 168:263–269
23. Uemura S, Matsushita H, Li W et al (2001) Diabetes mellitus enhances vascular matrix metalloproteinase activity. Role of oxidative stress. Circ Res 88:1291–1298
24. Benbow U, Brinckerhoff C (1997) The AP-1 site and MMP gene regulation: what is all the fuss about? Matrix Biol 15:519–526
25. Wilmer W, Cosio F (1996) DNA binding of activator protein-1 is increased in human mesangial cells cultured in high glucose concentrations. Kidney Int 53:1172–1181
26. Pascal M, Knott R, Forrester J (1996) Glucose mediated regulation of transforming growth factor beta in human retinal endothelial cells. Biochem Soc Trans 24:228S
27. Jormso S, Ye S, Moritz J et al (2000) Allele-specific regulation of matrix metalloproteinase-12 gene activity is associated with coronary artery luminal dimensions in diabetic patients with manifest coronary artery disease. Circ Res 86:998–1003
28. Yan SF, Ramasamy R, Naka Y et al (2003) Glycation, inflammation and RAGE. A scaffold for the macrovascular complications of diabetes and beyond. Circ Res 93:1159–1169
29. Virmani R, Kolodgie FD, Burke AP et al (2000) Lessons from sudden coronary death. A comprehensive morphological classification scheme for atherosclerotic lesions. Arterioscler Thromb Vasc Biol 20:1262–1275
30. Cipollone F, Pontera C, Pini B et al (2001) Overexpression of functionally coupled cyclooxygenase-2 and prostaglandin E synthase in symptomatic atherosclerotic plaques as a basis of prostaglandin E2-dependent plaque instability. Circulation 104:921–927
31. Ito H, Duxbury M, Benoit E et al (2004) Fibronectin-induced COX-2 mediates MMP-2 expression and invasiveness of rhabdomyosarcoma. Biochem Biophys Res Commun 318:594–600
32. Cha HS, Ahn KS, Jeon CH et al (2004) Inhibitory effect of cyclo-oxygenase-2 inhibitor on the production of matrix metalloproteinases in rheumatoid fibroblast-like synoviocytes. Rheumatol Int 24:207–211
33. Kalogeropoulou K, Mortzos G, Migdalis I et al (2002) Carotid atherosclerosis in type 2 diabetes mellitus: potential role of endothelin-1, lipoperoxides, and prostacyclin. Angiology 53: 279–285
34. Tsuruda T, Costello-Boerrigter LC, Burnett JC Jr (2004) Matrix metalloproteinases: pathways of induction by bioactive molecules. Heart Fail Rev 9:53–61
35. Roy-Beaudry M, Martel-Pelletier J, Pelletier JP et al (2003) Endothelin 1 promotes osteoarthritic cartilage degradation via matrix metalloprotease 1 and matrix metalloprotease 13 induction. Arthritis Rheum 48:2855–2864
36. Derosa G, Avanzini MA, Geroldi D et al (2004) Matrix metalloproteinase-2 may be a marker of microangiopathy in children and adolescents with type 1 diabetes mellitus. Diabetes Care 27:273–274
37. Das A, McGuire PG, Eriqat C et al (1999) Human diabetic neovascular membranes contain high levels of urokinase and metalloproteinase enzymes. Invest Ophthalmol Vis Sci 40:809–813
38. Saltzmann J, Limb GA, Khaw PT et al (2000) Matrix metalloproteinases and their natural inhibitors in fibrovascular membranes of proliferative retinopathy. Br J Ophthalmol 84: 1091–1096

39. McLennan SV, Death AK, Fisher EJ et al (1999) The role of the mesangial cell and its matrix in the pathogenesis of diabetic nephropathy. Cell Mol Biol 45:123–135
40. Friedman EA (1999) Advanced glycosylated end products and hyperglycemia in the pathogenesis of diabetic complications. Diabetes Care 22:B65–B71
41. McLennan SV, Fisher EJ, Martell SKY et al (2000) Effects of glucose on matrix metalloproteinases and plasmin activity in mesangial cells: possible role in diabetic nephropathy. Kidney Int 58(Suppl 77):S81–S87
42. Phillips AO, Steadman R, Morrissey K et al (1997) Exposure of human renal proximal cells to glucose leads to accumulation of type IV collagen and fibrinogen by decreased degradation. Kidney Int 52:973–984
43. McLennan SV, Martell SKY, Yue DK (2002) Effects of mesangium glycation on matrix metalloproteinase activities. Possible role in diabetic nephropathy. Diabetes 51:2612–2618
44. McLennan SV, Fisher EJ, Yue DK et al (1994) High glucose concentration causes a decrease in mesangium degradation. A factor in the pathogenesis of diabetic nephropathy. Diabetes 43:1041–1045
45. Vu TH, Shipley JM, Bergers G et al (1998) MMP-9/gelatinase B is a key regulator of growth plate angiogenesis and apoptosis of hypertrophic chondrocytes. Cell 93:411–422
46. Abaci A, Oguzhan A, Kahraman S et al (1999) Effect of diabetes mellitus on coronary collateral vessels. Circulation 99:2239–2242
47. Rivard A, Silver M, Chen D et al (1999) Rescue of diabetes-related impairment of angiogenesis by intramuscular gene therapy with adeno-VEGF. Am J Pathol 154:355–363
48. Tamarat R, Silvestre JS, Hujiberts M et al (2003) Blockade of advanced glycation end-product formation restores ischemia-induced angiogenesis in diabetic mice. Proc Natl Acad Sci U S A 100:8555–8560
49. Taniyama Y, Morishita R, Hiraoka K et al (2001) Therapeutic angiogenesis induced by human hepatocyte growth factor gene in rat diabetic hind limb ischemia model. Molecular mechanisms of delayed angiogenesis in diabetes. Circulation 104:2344–2350
50. McCarthy MJ, Loftus IM, Thompson MM et al (1999) Angiogenesis and the atherosclerotic carotid plaque: an association between symptomatology and plaque morphology. J Vasc Surg 30:261–268
51. Galis ZS, Sukhova GK, Lark MW et al (1994) Increased expression of matrix metalloproteinases and matrix degrading activity in vulnerable regions of human atherosclerotic plaques. J Clin Invest 90:775–778
52. Brown DL, Hibbs MS, Kearney M et al (1995) Identification of 92-kD gelatinase in human coronary atherosclerotic lesions. Circulation 91:2125–2131
53. Loftus IM, Naylor AR, Goodall S et al (2000) Increased matrix metalloproteinase-9 activity in unstable carotid plaques. A potential role in acute plaque disruption. Stroke 31:40–47
54. Sukhova GK, Schonbeck U, Rabkin E et al (1999) Evidence for increased collagenolysis from interstitial collagenases-1 and -3 in vulnerable human atheromatous plaques. Circulation 99:2503–2509
55. Herman MP, Sukhova GK, Libby P et al (2001) Expression of neutrophil collagenase (matrix metalloproteinase-8) in human atheroma. Circulation 104:1899–1904
56. Halpert I, Sire UI, Roby JD et al (1996) Matrilysin is expressed by lipid-laden macrophages at sites of potential rupture in atherosclerotic lesions and localizes to areas of versican deposition, a proteoglycan substrate for the enzyme. Proc Natl Acad Sci U S A 93:9748–9753
57. Hirashiki A, Yamada Y, Murase Y et al (2003) Association of gene polymorphisms with coronary artery disease in low- or high-risk subjects defined by conventional risk factors. J Am Coll Cardiol 42:1429–1437
58. Whalting C, McPheat W, Hurt-Camejo E (2004) Matrix management assigning different role of MMP-2 and MMP-9 in vascular remodeling. Arterioscler Thomb Vasc Biol 24:10–11
59. Lottus JM, Naylor AR, Bell PRF (2002) Matrix metalloproteinases and atherosclerotic plaque instability. Br J Surg 89:680–694
60. Lewandowski KC, Banach E, Bieńkiewicz M, Lewiński A (2011) Matrix metalloproteinases in type 2 diabetes and non-diabetic controls: effects of short-term and chronic hyperglycaemia. Arch Med Sci 7:294–303

61. Ravn HB, Falk E (1999) Histopathology of plaque rupture. Cardiol Clin 17:263–270
62. Galis ZS, Sukhova GK, Lark MW, Libby P (1994) Increased expression of matrix metalloproteinases and matrix degrading activity in vulnerable regions of human atherosclerotic plaques. J Clin Invest 94:2493–2503
63. Spinale FG, Coker ML, Heung LJ et al (2000) A matrix metalloproteinase induction/activation system exists in the human left ventricular myocardium and is upregulated in heart failure. Circulation 102:1944–1949
64. Tyagi SC, Campbell SE, Reddy HK et al (1996) Matrix metalloproteinase activity expression in infarcted, noninfarcted and dilated cardiomyopathic human hearts. Mol Cell Biochem 155:13–21
65. Sun M, Dawood F, Wen WH et al (2004) Excessive tumor necrosis factor activation after infarction contributes to susceptibility of myocardial rupture and left ventricular dysfunction. Circulation 110:3221–3228
66. Inokubo Y, Hanada H, Ishizaka H et al (2001) Plasma levels of matrix metalloproteinase-9 and tissue inhibitor of metalloproteinase-1 are increased in the coronary circulation in patients with acute coronary syndrome. Am Heart J 141:211–217
67. Hojo Y, Ikeda U, Ueno S et al (2001) Expression of matrix metalloproteinases in patients with acute myocardial infarction. Jpn Circ J 65:71–75
68. Hirohata S, Kusachi S, Murakami M et al (1997) Time dependent alterations of serum matrix metalloproteinase-1 and metalloproteinase-1 tissue inhibitor after successful reperfusion of acute myocardial infarction. Heart 78:278–284
69. Carmo M, Colombo L, Bruno A et al (2002) Alteration of elastin, collagen and their cross-links in abdominal aortic aneurysms. Eur J Vasc Endovasc Surg 23:543–549
70. Duca L, Floquet N, Alix AJ et al (2004) Elastin as a matrikine. Crit Rev Oncol Hematol 49:235–244
71. Pasterkamp G, Schoneveld AH, Hijnen DJ et al (2000) Atherosclerotic arterial remodeling and the localization of macrophages and matrix metalloproteases 1, 2, and 9 in the human coronary artery. Atherosclerosis 150:245–253
72. Sun M, Opavsky MA, Stewart DJ et al (2003) Temporal response and localization of integrins beta1 and beta3 in the heart after myocardial infarction: regulation by cytokines. Circulation 107:1046–1052
73. Kai H, Ikeda H, Yasukawa H et al (1998) Peripheral blood levels of matrix metalloproteinases-2 and -9 are elevated in patients with acute coronary syndromes. J Am Coll Cardiol 32:368–372
74. Kexin XU, Shukun HOU, Zhijun DU (2002) Prognostic value of matrix metalloproteinase-2 and tissue inhibitor of metalloproteinase-2 in bladder carcinoma. Chin Med J 115:743–745
75. Suzuki D (1998) Metalloproteinases in the pathogenesis of diabetic nephropathy. Nephron 80:125–133
76. Bhatt LK, Veeranjaneyulu A (2012) A therapeutic approach to treat cardiovascular dysfunction of diabetes. Exp Toxicol Pathol 64:847–853
77. Bhatt LK, Veeranjaneyulu A (2011) Minocycline with aspirin: an approach to attenuate diabetic nephropathy in rats. Ren Fail 33:72–78
78. Vlassara H, Striker LJ, Teichberg S (1994) Advanced glycation end products induce glomerular sclerosis and albuminuria in normal rats. Proc Natl Acad Sci U S A 91:11704–11708
79. Senatorski G, Paczek L, Sulowicz W (1998) Urine activity of cathepsin B, collagenase and urine excretion of TGF-beta 1 and fibronectin in membranous glomerulonephritis. Res Exp Med (Berl) 198:199–206
80. Zaoui P, Cantin JF, Alimardani-Bessette M (2000) Role of metalloproteases and inhibitors in the occurrence and progression of diabetic renal lesions. Diabetes Metab 26(Suppl 4):25–29
81. Fogo AB (1999) Current concepts in glomerulosclerosis. Am J Kidney Dis 34:LIV–LVI
82. Kurogi Y (2003) Mesangial cell proliferation inhibitors for the treatment of proliferative glomerular disease. Med Res Rev 23:15–31

83. Death AK, Nakhla S, McGrath KC et al (2002) Nitroglycerin upregulates matrix metalloproteinase expression by human macrophages. J Am Coll Cardiol 39:1943–1950
84. Tyagi SC, Kumar S, Katwa L (1997) Differential regulation of extracellular matrix metalloproteinase and tissue inhibitor by heparin and cholesterol in fibroblast cells. J Mol Cell Cardiol 29:391–404
85. Eickelberg O, Roth M, Mussmann R et al (1999) Calcium channel blockers activate the interleukin-6 gene via the transcription factors NF-IL-6 and NF-KB in primary human vascular smooth muscle cells. Circulation 99:2276–2282
86. Papakonstantinou E, Roth M, Kokkas B et al (2001) Losartan inhibits the angiotensin II-induced modifications on fibrinloysis and matrix deposition by primary human vascular smooth muscle cells. J Cardiovasc Pharmacol 38:715–728

Nutraceutical Approaches in the Management of Cardiovascular Dysfunctions Associated with Diabetes Mellitus

Saloni Daftardar, Ginpreet Kaur, and Veeranjaneyulu Addepalli

Abstract Diabetes is a prime risk factor in cardiovascular disease (CVD), which includes peripheral vascular disease, coronary artery disease, and cerebrovascular disease. Management of cardiovascular dysfunctions associated with diabetes has been a challenge for decades. The nutraceutical approach for prevention of diabetic cardiovascular complications has been a considerably newer trend. Nutraceuticals are medicinal foods that help in maintaining the health of individuals, thereby preventing or treating diseases. A literature search showed that several nutraceuticals used as dietary supplements have the ability to reduce cardiovascular risk factors. Thus, the nutraceutical approach can be very promising in the treatment of diabetic cardiovascular complications. In this review, we summarize the recent research findings on dietary fiber, antioxidants, prebiotics, and probiotics to highlight the benefits of using nutraceuticals in the management of diabetes-associated cardiovascular dysfunctions.

Keywords Insulin resistance • Inflammation • Antidiabetic • Nutraceuticals • Herbs

1 Introduction

Cardiovascular disease (CVD) is a major factor associated with diabetes that exacerbates disorders such as coronary artery disease and atherogenesis. The main factors associated with diabetes that lead to such diabetic cardiovascular complications are hyperglycemia, hyperlipidemia, and obesity. Management of cardiovascular complications associated with diabetes is often challenging. However, the use of

S. Daftardar • G. Kaur • V. Addepalli (✉)
Department of Pharmacology, SPP School of Pharmacy & Technology Management, SVKM's NMIMS, V. L. Mehta Road, Mumbai, Maharashtra 400056, India
e-mail: veeranjaneyulu.addepalli@nmims.edu

B. Turan and N.S. Dhalla (eds.), *Diabetic Cardiomyopathy: Biochemical and Molecular Mechanisms*, Advances in Biochemistry in Health and Disease 9,
DOI 10.1007/978-1-4614-9317-4_24,

nutraceuticals and dietary supplements has gained immense importance in clinical medicine recently because of patient compliance, rational and economic use, and acceptance by various authorities having policies that permit and encourage these practices. With increasing awareness of individuals in the light of management of diabetes using nutraceuticals, there is an increase in the demand for information on nutraceuticals. Nutraceuticals consisting of perplexing constituents offer a holistic approach to treatment of various diseases and disorders and also demonstrate the ability to treat them.

In spite of the enormous and rapid development in the field of allopathy during the past decade, medicinal plants remain one of the major sources of drugs in the modern as well as the traditional system of medicine globally. It has been projected that 80 % of the world's populations still use traditional medicine for their primary healthcare needs. According to the statistics given by WHO in 2008, about 80 % of the population in developing countries relies on traditional plant drugs [1]. Such traditional medicines comprise about 25 % of the market share of the entire pharmaceutical cache.

Nutraceuticals have been found to produce desirable effects with decreased side effects, thereby reducing the risks of diabetes. Nutraceuticals consist of natural health-promoting components that help in maintaining the normal physiological functions of individuals. Thus, the attainment of a maximum state of nutrition and health has become the ultimate goal using nutraceuticals. There has been a resurgence in the consumption and demand for medicinal plants in a complementary system of healthcare in the prevention and management of chronic lifestyle-related noncommunicable diseases. These plants are finding use as pharmaceuticals, nutraceuticals, cosmetics, and food supplements. They are available in the form of powders, pills, or capsules, as a single substance, or as combination preparations.

Nutritional supplementation along with diet and exercise are key for maintaining a healthy lifestyle and well-being, especially related to cardiovascular diseases, despite remarkable advances in medicine and pharmaceutical drug development [2, 3]. Therefore, there is a dire need to focus our research on nutraceuticals and their role in cardiovascular dysfunctions associated with diabetes mellitus. The term "nutraceutical" was coined in 1989 by Stephen De Felico, combining the word "nutrient" (a nourishing food or food component) with "pharmaceutical" (a medical drug). It has been used to describe medicinally or nutritionally functional foods or components, and may include a food, plant, or naturally occurring material that may have been purified or concentrated and is used for improvement of health by preventing or treating a disease [4].

Nutraceuticals are gaining popularity as people are relying on them for safeguarding their health and avoiding side effects associated with drugs as well. Nutraceuticals are destined to play an important role in future therapeutic developments and will continue to appeal to consumers because of better patient tolerance and decreased healthcare costs compared with conventional pharmaceuticals. They are found in a medley of products emerging from the food industry, the herbal and dietary supplement market, the pharmaceutical industry, and the newly merged pharmaceutical/agribusiness/nutrition firms. There is an urgent surge that pharmaceutical research is beginning to realize and to explore the unexplored role of natural products in the therapeutics and treatment of cardiovascular complication of diabetes mellitus, by strategizing alternative approaches and models to contribute safe and efficacious medicines with minimal adverse effects.

2 Classification of Nutraceuticals

Nutraceuticals can be grouped into the following three broad categories:

1. Substances with established nutritional functions, such as vitamins, minerals, amino acids and fatty acids: *Nutrients*
2. Herbs or botanical products as concentrates and extracts: *Herbals*
3. Products derived from other sources (e.g., pyruvate, steroid hormone precursors) and serving specific functions, such as sports nutrition, weight-loss supplements, and meal replacements: *Dietary supplements*

Thus, nutritional supplements cannot be termed “drugs”; drugs have potent therapeutic effects as well as side effects whereas nutraceuticals act more gently and maintain the physiological function in the body. They exert their action in several ways.

1. By reducing the level of LDL-cholesterol by modulating cholesterol production
2. By reducing reactive oxygen species (ROS) production
3. By reducing artery plaque formation and reducing blood pressure

3 Inflammation and Diabetic Complications

Thee Metabolic and immune systems of the body have evolved to be closely linked. Evidence has been accumulating that inflammation is a crucial factor in the onset of diabetes mellitus and its associated complications such as cardiovascular disease (CVD). Chronic inflammation (sometimes termed “silent” inflammation) is an abnormal process in which the body’s immune system continuously secretes inflammatory chemicals, precipitating diseases such as diabetes, heart disease, arthritis, neuropathy, and obesity. Recent studies suggest that pro-oxidative and pro-inflammatory processes play a significant role in the progression of atherosclerosis; thus, inflammatory markers are predictors of cardiovascular events and progression to type 2 diabetes in healthy individuals as well as those with metabolic syndrome, underscoring the link between inflammation, metabolic disorders, and cardiovascular disease. Diabetes has been linked to abnormalities in markers of systemic inflammation, and there has been hope that an understanding of this nexus could open the door to new therapeutic opportunities.

3.1 Obesity

Obesity associated with diabetes can be a driving force that causes inflammation. According to endocrinologist Jerrold Olefsky, fat cells grow in size as a result of weight gain; hence, they do not get sufficient oxygen from the blood and as a result begin to die. This cellular death recruits immune cells to the scene. Elevated plasma lipids precipitate peripheral tissue insulin resistance and atherosclerosis [5].

Obesity is associated with insulin resistance, hyperglycemia, dyslipidemia, hypertension, and pro-thrombotic and pro-inflammatory states. Adipose tissue releases a host of factors that include free fatty acids, interleukin 6, tumor necrosis factor-α, adiponectin, and plasminogen activator inhibitor 1: these apparently exacerbate the risk factors. Overexpression of tumor necrosis factor-α has been demonstrated in obese rodents to thus impair insulin signaling and has been linked to insulin resistance in rodents [6]. Loss of adiponectin, an anti-atherogenic protein secreted exclusively by adipose tissue, has been proposed to have a role in atherosclerosis, insulin resistance, and several components of metabolic syndrome, which increases the risk of cardiovascular diseases associated with diabetes mellitus [7]. Adipose tissue-derived proteins cause alteration in glucose and lipid metabolism. They also promote local and systemic inflammatory and thrombotic states. Elevated concentrations of free fatty acids result in reduction in hepatic and muscle insulin sensitivity with an increase in cholesterol production. Increase in free fatty acids leads to increases in triglyceride levels and in VLDL levels, thus increasing the risk of coronary heart disease.

3.2 Inflammation as a Major Pathogenic Mechanism in Development of Atherosclerosis

The stages involved in atherosclerosis, that is, initiation, growth, and complexity of atherosclerotic plaque, can be considered to be an inflammatory response. Various risk factors such as hypertension, dyslipidemia, and hyperglycemia are major injurious factors that lead to atherosclerosis via inflammation. These factors stimulate the secretion of leukocyte-soluble adhesion molecules, chemotactic factors, and other inflammatory mediators such as IL-1, IL-6, C-reactive protein (CRC), and a host of other acute-phase reactants. Metalloproteinases and other connective tissue enzymes cause rupture of the atherosclerotic plaque, which then exposes the core to arterial blood and induces thrombosis. Fibrinogen, plasminogen activator inhibitor 1 (PA-1), cytokines, and C-reactive protein are often seen in increased concentrations in patients with diabetes, resulting in pro-thrombotic and pro-inflammatory states; this also increases the risk of cardiovascular diseases.

3.3 *Insulin Resistance*

Another factor that is responsible for causing cardiovascular diseases associated with diabetes is insulin resistance [9]. Several studies have demonstrated that increases in plasma free fatty acids (FFA) cause insulin resistance in diabetic patients and in nondiabetic subjects [10, 11]. Increased levels of circulating FFA in obesity caused by diabetes enhance lipid accumulation in insulin target tissues and contribute to reducing their sensitivity to insulin. Insulin is an important antilipolytic hormone. Because insulin resistance leads to impaired fatty acid metabolism in adipose tissue, its development contributes to an increased release of fatty acids into the circulation. This, in turn, leads to multiple abnormalities in the lipoprotein profile, resulting in what is known as atherogenic dyslipidemia.

The prime molecular link between inflammation and diabetes is tumor necrosis factor-alpha (TNF-α), an inflammatory cytokine that is overexpressed in the adipose tissues of rodent models of diabetes. It has found to activate and increase production of numerous proteins that suppress the insulin-signaling pathways, thereby increasing the risk of insulin resistance [12]. The inhibition of signaling downstream of the insulin receptor is considered to be the main mechanism through which inflammation leads to insulin resistance. TNF-α stimulates inhibitory phosphorylation of insulin receptor substrates (IRS), which in turn reduces the tyrosine phosphorylation of IRS in response to insulin and the ability of IRS to associate with the insulin receptor and thereby inhibits downstream signaling, resulting in insulin resistance [13]. Hyperglycemia-induced oxidative stress, soluble advanced glycation end products, and products of lipid peroxidation serve as key activators of upstream kinases, leading to stimulation of inflammatory gene expression (Fig. 1). In a study where the monocyte cells were cultured and later incubated with high glucose (15 mmol/l for 18 h), there was an increase in the release of TNF-α mediated by reactive oxygen species via activation of transcription factors nuclear factor-κB (NF-κB) and activating protein 1.

It was demonstrated that the protein $FOXO_1$ instigates the expression of another inflammatory mediator called interleukin-1β, which interferes with insulin-signaling pathways and reduces the insulin produced from pancreatic beta cells [14]. Inflammatory signaling pathways can also become activated by metabolic stresses. The increased glucose metabolism can lead to higher mitochondrial production of ROS, which causes enhanced activation of inflammatory pathways.

4 Mechanisms Underlying Diabetes Mellitus with Cardiovascular Disease

Hyperglycemia is not the sole factor associated with CVD, although it instigates several potential mechanisms that increase the risk of atherosclerosis. Diabetic patients also have been found to show insulin resistance, obesity, and elevated blood

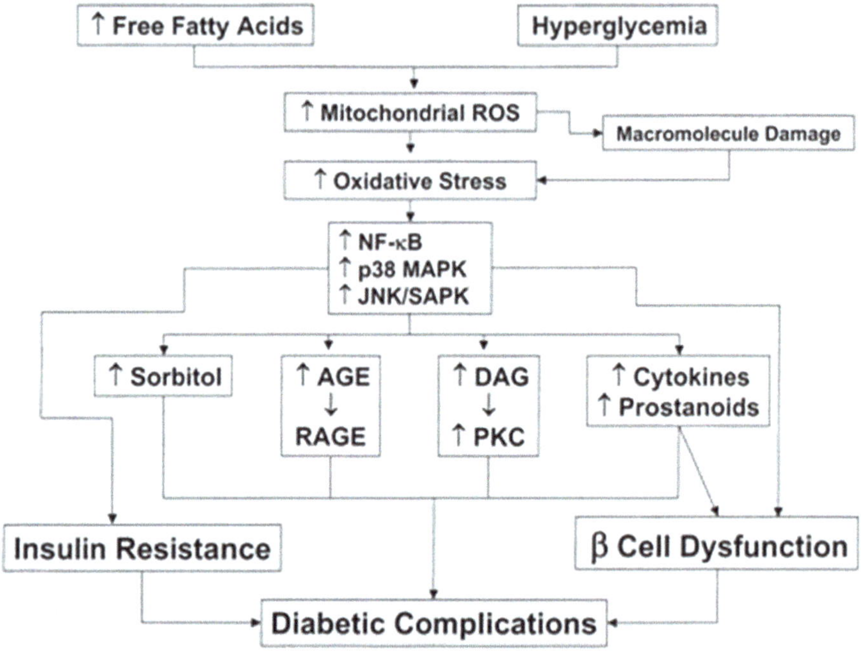

Fig. 1 General aspects related to development of diabetic complications via the activation of inflammatory signaling pathways and production of reactive oxygen species (*ROS*)

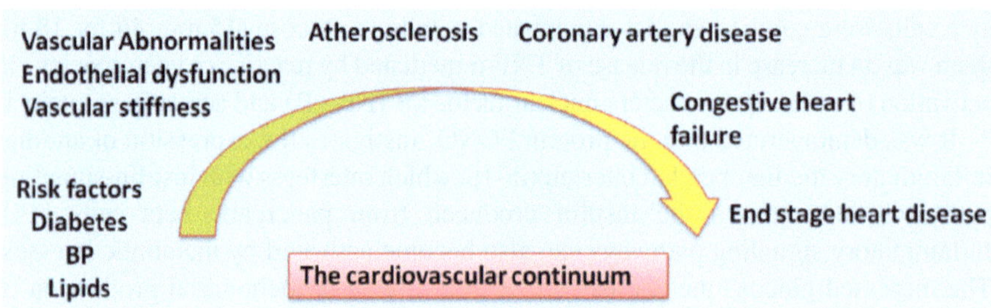

Fig. 2 Proposed cardiovascular continuum eliciting how the risk factors such as hypertension and elevated lipid levels culminate in microvascular abnormalities leading to coronary heart disease and congestive heart failure. (Adapted and modified from Dzau et al. 1991 [15])

pressure and triglyceride levels, which elevate the risk of acquiring diabetic complications such as heart disease.

Metabolic changes associated with cardiovascular risk factors of diabetes include vascular abnormalities such as endothelial dysfunction, atherosclerosis with thrombosis and fibrinolysis, and atherogenic dyslipidemia (Fig. 2).

Table 1 Abnormalities of endothelial dysfunction in diabetics

Characteristic feature	Parameter
Increased endothelial adhesiveness	Increased VCAM -1 and E-selectin
Impaired vasodilation	Decreased NO production, prostacyclin (PGI_2)
Increased coagulation	Decreased NO production, prostacyclin (PGI_2)
Increased tissue factor expression	–

4.1 *Development of Atherogenesis in Diabetes*

Atherosclerotic lesions that occur in the arterial system are the major factor leading to development of CVD. Lipids are deposited extracellularly, followed by infiltration of inflammatory mediators such as monocytes and T lymphocytes, resulting in the formation of fatty streaks. Monocytes transform into macrophages and scavenge lipids to form foam cells. The inflammatory mediators and foam cells produce reactive oxidative species, leading to migration of vascular smooth muscle cells. This is a continuous process that causes necrosis, further attracting the other inflammatory cells. There is also formation of fibrous tissue. A complex lesion is formed consisting of a fibrous tissue overlapping a mass of lipid and necrotic tissue. Expansion of this lesion can give rise to secondary symptoms such as reduction in arterial flow and thrombus formation, thereby leading to acute occlusion of the blood vessel [16, 17].

Multiple processes such as endothelial dysfunction, oxidative stress, and inflammation are thus involved in development of atherosclerosis.

4.2 *Endothelial Dysfunction*

Nitric oxide (NO) is an important component that mediates processes of vasodilation, vasoconstriction, and inflammatory and antiinflammatory activities. NO is a potent vasodilator that inhibits monocyte adhesion, platelet adhesion, and proliferation of vascular smooth muscle cells [18, 19]. In diabetics, there is a significant decrease in NO production because of insulin resistance (Table 1). This decrease is pivotal in the development of diabetic vascular complications with a complex array of abnormalities such as disruption of normal vasoregulation, increased adhesion and migration of leukocytes and platelets, and proliferation of vascular smooth muscle cells (Fig. 3). Attenuated NO-dependent vasodilation is the hallmark of endothelial dysfunction [20].

4.3 *Oxidative Stress*

Reactive oxygen species (ROS), a group of highly oxidative reactive molecules, are produced during normal cellular functions [22, 23]. For example, O_2^- is a principal

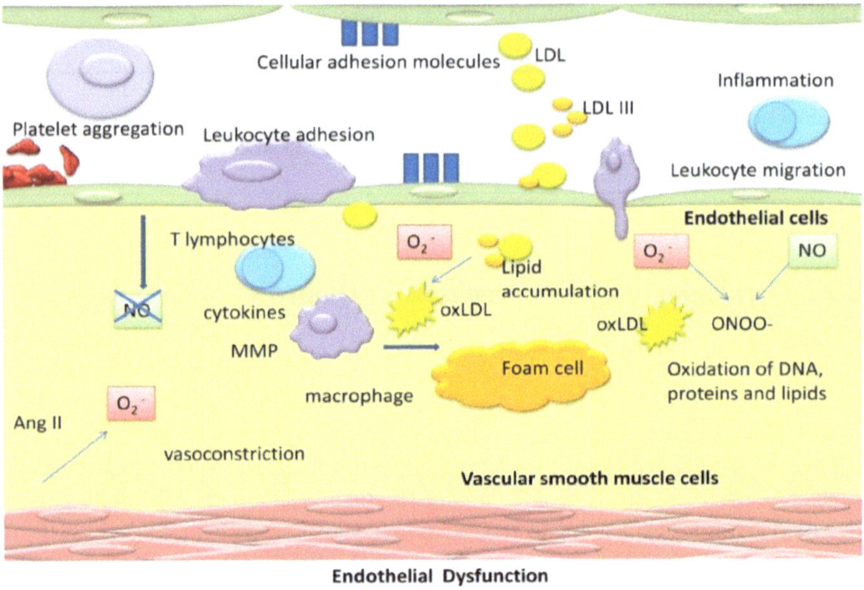

Fig. 3 Endothelial dysfunction: influence of nitric oxide (*NO*) on endothelial cell function and related inflammatory activities. (Adapted and modified from Hamilton et al. 2004 [21])

ROS produced by vasculature and inflammatory cells that further produces additional ROS through chain reactions.

A number of feedback mechanisms such as antioxidant scavenging systems maintain the equilibrium of the levels of ROS. There is a prevalence of imbalance between the production of ROS and antioxidant scavenging feedback mechanisms, resulting in increased oxidative stress. Hyperglycemia in diabetes is known to induce oxidative stress. Free radicals damage the cellular proteins and mitochondrial DNA. For example, increased oxidative stress reduces the bioavailability of NO and promotes leukocyte adhesion, inflammation, and endothelial dysfunction [24]. An increase in the levels of ROS potentiates the formation of oxidized LDL, which in turn further increases oxidative stress [25]. It plays a vital role in formation of atherosclerotic lesions.

4.4 *Atherogenic Dyslipidemia*

Lipid abnormalities in type 2 diabetics are associated mainly with insulin resistance. The important factor driving these abnormalities is hypertriglyceridemia. When the triglyceride levels increase in patients with diabetes, small dense LDL increases in proportion to the rise in triglycerides [26]; this is associated with hepatic lipase activity, which is increased in insulin-resistant patients. Hepatic lipase converts the triglyceride-enriched LDL to small dense LDL [27].

Diabetic patients having insulin resistance have higher levels of LDL and lower levels of HDL cholesterol [28]. Diabetics have a fundamental defect

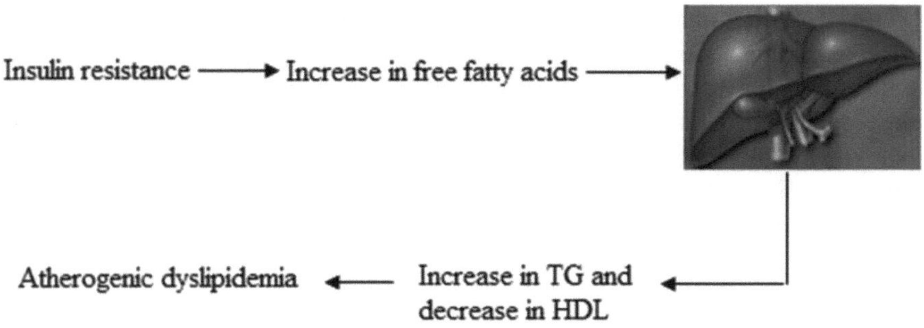

Fig. 4 Free fatty acids and atherogenic dyslipidemia

pertaining to lipid metabolism; there is overproduction of $VLDL_1$, which initiates the series of other changes in the lipoproteins such as small density LDL and oxidation of LDL particles. Reduced concentration of HDL and apolipoprotein A-1 results in the accumulation of cholesterol in the vessel wall and leads to atherosclerosis (Fig. 4). The increased glycation of apoprotein B is enhanced in diabetic patients and may contribute to the development of atherosclerosis. The small density LDL particles are associated with endothelial dysfunction. They enter the arterial wall rapidly and damage the endothelial cells. Also, they promote production of procoagulant factors.

Oxidized LDL has been found to increase oxidative stress. It increases superoxide production through xanthine oxidase, NADPH oxidase, and uncoupling of endothelial nitric oxide synthase (eNOS) [29]. The reduced bioavailability of NO results in impaired endothelium-dependent vasodilation.

4.5 Hypertension

The relationship between insulin resistance and hypertension is well established. Several different mechanisms are proposed. First, insulin is a vasodilator when given intravenously to people of normal weight, with secondary effects on sodium reabsorption in the kidney. Insulin resistance is also associated with increase in sodium reabsorption on kidney and decrease in vasodilation. These factors contribute to hypertension. Hyperinsulinemia may result in increased sympathetic nervous system (SNS) activity and contribute to the development of hypertension.

5 Management of Cardiovascular Diseases Associated with Diabetes Mellitus Using Nutraceuticals

Diabetes represents a clustering of risk factors related to an elevated risk of cardiovascular disease. Several nutraceuticals used in clinical practice represent a different approach to medicine, one based on nutrition and the health or wellness of the whole body, rather than treating the symptoms or effects of disease. New scientific evidence in the effectiveness of nutraceuticals together with the changing lifestyle of modern-day consumers has fueled more rapid development around the world today. Allopathic medicines cause side effects but nutraceutical products, being derived from natural sources, have no or minimal side effects. Nutraceuticals have the potential to play a role in prevention and treatment of diseases.

6 Several Nutraceuticals Target the Pathogenesis of CVD Associated with Diabetes Mellitus

1. Conjugated linoleic acids
2. Polyunsaturated fatty acids
3. Protective minerals
4. Antioxidants
5. α-Lipoic acid
6. Dietary fiber

6.1 Conjugated Linoleic Acid (CLA)

Conjugated linoleic acids (CLAs) are naturally occurring fatty acids with many health benefits that are a mixture of positional and geometric isomers of linoleic acid (*cis*-9, *cis*-12-octadecadienoic acid) that contain a conjugated double-bond system [30]. Studies have been proven its anticarcinogenic and antiatherogenic properties and antidiabetic effects [31–33]. CLAs can reduce body fats. The insulin-sensitizing effect of CLA has been defined by the activation of hepatic peroxisome proliferator-activated receptor (PPAR)-α and thereby increased fatty acid oxidation. Ryder et al. demonstrated that supplementing the diets of Zucker diabetic fatty (ZDF) rats with 1.5 % CLA reduced adiposity and improved glucose tolerance compared with control feedings. It also improved insulin-stimulated glucose transport in soleus skeletal muscles and insulin-stimulated glycogen synthase activity in soleus and extensor digitorum longus muscles [32]. The hypocholesterolemic and antiatherogenic properties of CLA were found by Lee et al. [31], who studied rabbits fed high-cholesterol diets with an appropriate amount of CLA per day. These rabbits had lower total cholesterol and reduced triacylglycerol concentrations with less atherosclerosis compared with rabbits on high cholesterol in the absence of CLA. The CLA lowers serum cholesterol; however, the exact reasons for that effect remain a mystery. It could act by inhibiting the secretion of apolipoprotein B or by enhancing the clearance rate of circulating LDL by way of increasing LDL-receptor activity [34].

6.2 Polyunsaturated Fatty Acids (PUFA)

Polyunsaturated fatty acids (PUFA) consist of a treasure of activities beneficial to health, including antiinflammatory and lipid-lowering effects and the prevention of coronary heart disease. PUFA are omega-3 fatty acids that are classified as essential fatty acids. Evidence suggests that PUFAs are significant in the prevention of coronary artery disease. Convincing studies have been performed in which fish meal or fish oil was given to a clinical group of patients suffering from CVD, including myocardial infarction and stroke, and it was proved that this supplementation significantly reduced these cardiovascular events [35, 36] α-Linolenic acid and docosahexanoic acid, types of omega-3 PUFA, are found to be effective at reducing blood pressure and heart rate [37]. These acids are used as supplementation in type 2 diabetes to lower triglycerides and VLDL-cholesterol and to reduce inflammatory markers [38]. Omega-3 fatty acids supplementation has no statistically significant effects on glycemic control or fasting insulin but may prevent or revert insulin resistance.

6.3 Protective Minerals

The following minerals play essential roles in activating the processes essential for the proper metabolism of sugars by stimulating the production of insulin and other digestive enzymes in the digestive tract.

(a) Chromium
Chromium regulates the uptake of glucose in cells in conjunction with insulin and thus plays a vital role in the glucose tolerance factor. Diabetics usually have deficient chromium levels. A few studies have cast light on the fact that chromium supplementation can lead to increased insulin sensitivity and improved glucose tolerance in patients with type 2 diabetes [39]. A recent study showed that supplemental intake of chromium (1,000 μg/day) caused significant improvements in glycosylated hemoglobin, glucose, insulin, and cholesterol levels when compared to placebo [40]. Further experimental and clinical studies have consolidated the efficacy of chromium in the management of type 2 diabetes and associated comorbidities such as insulin resistance and obesity [41].

(b) Magnesium
Diets rich in magnesium have proven to decrease risk for diabetes, suggesting an improvement in insulin sensitivity. Hence, supplemental nutrients should have adequate levels of magnesium with other nutrients and vitamins to restore the depleted levels of such nutrients in diabetic patients. In several rodent studies, magnesium helps to preserve adipocyte insulin sensitivity [42].

6.4 Antioxidants

The efficacy of antioxidants as prophylactic and therapeutic agents for cardiac complications of diabetes mellitus has encouraged the recent trend of antioxidant therapy in protecting against cardiovascular disease inflicted by accentuated oxidative stress caused by diabetes. It is evident from several preclinical studies that an adequate supply of dietary antioxidants may prevent or delay diabetes complications, including renal and neural dysfunction, by providing protection against oxidative stress. Epidemiologic data have proven that a strong link exists between intake of antioxidants and protection against cardiovascular diseases [43]. Several dietary foods that possess antioxidant properties are mentioned next.

(a) Vitamin C
Higher dietary intake of vitamin C lowers the risks of development of cardiovascular disease. It acts by scavenging ROS directly and preventing the propagation of chain reactions that would lead to a reduction in protein glycation. In animals, vitamin C is found to decrease diabetes-induced lipid peroxides in erythrocytes [44].

(b) Vitamin E
Vitamin E reacts with peroxyl and superoxide radicals and singlet oxygen and ensures protection to membranes from lipid peroxidation. In a study, patients with type 2 diabetes were given supplementation of vitamin E, and its effects on the risks for development of cardiovascular disease were assessed. Vitamin E treatment significantly reduced oxidation of low density lipoprotein and soluble cell-adhesion molecules [45]. Vitamin E is also found to improve renal function in patients with type 1 diabetes. The major CVD risk factors are microalbuminuria and hypertension. Microalbuminuria represents an advanced stage of CVD that may precipitate into cardiovascular morbidity and mortality [46]. Supplementation of patients with T2D with persistent micro/macroalbuminuria with both vitamins E and C significantly lowered their urinary albumin excretion rate, which suggests the use of vitamin E with other antioxidants in the treatment of microvascular complications of diabetes including CVD or nephropathy [47].

6.5 α-Lipoic Acid

α-Lipoic acid is a naturally occurring antioxidant with potent ROS-scavenging activity. It has the unusual property of being a ROS scavenger in its oxidized state; α-lipoic acid and dihydrolipoic acid work in a redox reaction and together have other antioxidant properties. Ischemic injury to the retina, considered to be one of the major causes of visual loss, occurs in diabetic retinopathy. α-Lipoic acid increases insulin sensitivity by approximately 18–20 % in patients with type 2 diabetes [48].

6.6 *Dietary Fiber*

Dietary fibers include polysaccharides, oligosaccharides, lignin, and associated plant substances. They can be classified into two groups:

(a) Water-soluble fibers: soluble fibers that dissolve in water are present in oats, dried beans, and legumes.
(b) Water insoluble fibers are present in brown rice, bananas, vegetables, and whole grains.
Soluble fibers are found to be more effective than insoluble fibers such as cellulose and lignins. Soluble fibers such as are found in oats, psyllium, pectin, flaxseed, and guar gum are associated with reduced cardiovascular risk. They have been found to increase insulin sensitivity and lower glucose levels. Soluble dietary fibers in the dose of 2–10 g/day cause a reduction in LDL cholesterol by 5–7 %; these bind to the bile acids, thus entrapping cholesterol and resulting in decreased cholesterol absorption, which leads to reduced cholesterol, increase in LDL receptors, and increase in LDL clearance.
(c) Psyllium as a dietary supplement.
Psyllium is derived from the husks of the seeds of *Plantago ovata*. It has been studied as a "non-systemic" cholesterol-lowering agent, as it is a soluble, indigestible, polysaccharide fiber that, when taken with water, slows gastric emptying and binds bile acids, causing increased excretion and synthesis of these acids. Because cholesterol is utilized in the synthesis of bile acids, the process results in a decrease in serum cholesterol.

7 Herbs Used in Treatment of Cardiovascular Complications Associated with Diabetes

Natural products identified from traditional medicinal plants present an exciting opportunity for the treatment of cardiovascular complications associated with diabetes because they contain a wide range of phytochemicals with diverse metabolic effects. The following plants have been screened for their hypolipidemic, antidiabetic, and anti-obesity activity.

7.1 Nigella sativa *Extract (Family: Ranunculaceae)*

According to Kanter et al., *Nigella sativa* or black cumin extracts were found to possess the property of lowering blood glucose [49]. Hypoglycemic effect of black cumin was observed by Meral et al. in which the blood glucose (194.41 mg/dl) was reduced significantly in diabetic rats treated with an extract of black cumin as compared to that of the control (340.43 mg/dl). The insulinotropic activity of black cumin oil was demonstrated by Farah et al. in a type 2 model [50]. The increased free radical production during diabetes decreases the enzyme activity and antioxidant status of the body. Because of its antioxidant properties, black cumin may reduce diabetes-induced oxidative stress. Lipid peroxidation often precipitates liver damage as there is a sharp increase in the level of different enzymes, which lowers the permeability of glucose precursors to liver, thereby leading to lower gluconeogenesis. Meral et al. found that black cumin prevents lipid peroxidation of biological membranes and also liver damage in diabetic rabbits by restoring the balance in the activities of liver enzymes [51].

Another study showed that black cumin fixed oil that is rich in sitosterol improved the lipid profile by decreasing cholesterol and triglycerides and elevating HDL levels [52]. A study carried out by Morikawa et al. demonstrated that nigellamine A-(5) decreased triglyceride levels in cultured mice hepatocytes and was found to be as effective as the synthetic drug clofibrate [53]. Black cumin fixed oil is rich in unsaturated fatty acids, linoleic acids, and eicosenoic acid, which help in lipid-lowering and antioxidant activity. It has been found to contain a reservoir of polyphenols, α-tocopherol, β-tocopherol, γ-tocopherol, selenium, and phytosterols such as β-sitosterol, 5-avenasterol, 7-avenasterol, the major ones, and stigmasterol and campesterol, present in minute quantities, which are responsible for hypoglycemic activity.

7.2 Momordica charantia *Extract (Family: Cucurbitaecae)*

Aqueous and alcoholic extracts of bitter melon have been traditionally used as a remedy for diabetes. Bitter melon contains several active constituents such as cucubitacin B, momordicin, and bioactive glycosides such as charantin, goyaglycosides, momordin, charantosides, and momordicosides. Bitter melon also includes cytotoxic proteins such as momordin and momorcharin and terpenoid compounds such as momordol, momordenol, momordicilin, and momordicinin. These organic ingredients are responsible for the hypoglycemic effect by activating 5'-AMP-activated protein kinase (AMPK), a protein molecule that regulates glucose uptake and suppresses excess appetite. The fresh juice of the unripe fruit was found to lower blood glucose levels in preclinical and human trials [54]. The alcoholic extract consists of mixed steroids that were found to be more potent than the synthetic hypoglycemic agent tolbutamide in an animal study [55]. Non-insulin-dependent diabetes mellitus animal models were found to respond to the effects of bitter melon whereas insulin-dependent animal models rarely showed any positive effects. Several studies performed in vitro have shown positive results of bitter melon treatment by production of an increased number of pancreatic beta cells and increased insulin-releasing activity in a few cases [56]. The seeds of *Momordica charantia* also have been determined to possess hypoglycemic and lipid-lowering properties. The hypoglycemic mechanism mimicked by *Momordica charantia* is the result of three effects:

1. Depression of gluconeogenic enzymes, increase in glucose transporters, and stimulation of glucose uptake in skeletal muscle cells.
2. Protection of islet beta cells and an increase in insulin secretory activity.
3. Reduction of oxidative stress.

7.3 Trigonella foenum-graecum *Extract (Family: Fabaceae)*

Fenugreek has been commonly for treatment of various diseases including diabetes. It has been popular for its blood glucose-lowering effects. Ethanolic extract of seed powder (50 mg/100 g body weight) decreased blood glucose, serum cholesterol, serum glutamic-oxaloacetic transaminase (SGOT), and serum glutamic-pyruvic transaminase (SGPT) levels in alloxan-induced diabetic rats. Hannan et al. evaluated the antidiabetic properties of the soluble dietary fiber fraction of a seed extract of fenugreek in type 1 and type 2 diabetic rats [57]. An increase in glucose tolerance and liver glycogen content with reduction in blood glucose, intestinal disaccharidase activity, and glucose absorption were observed. These findings indicate that the seed extracts showed antidiabetic effects mediated through inhibition of carbohydrate digestion and absorption and enhancement of peripheral insulin action [58]. Another study demonstrated that an amino acid extracted from fenugreek seeds, 4-hydroxyisoleucine, decreases glucose and insulin in diabetic rodents, thereby increasing glucose tolerance and inducing insulin secretion from isolated pancreatic islets. The components of fenugreek thought to be useful in diabetes were isolated and found to be trigonelline, coumarin (trigoforin), and nicotinic acid. An antihyperglycemic compound was isolated from water extract of the seeds of fenugreek and named GII [59]. Galactomannans, the polysaccharides found in fenugreek seeds, are composed of a 1,4-linked β-D-mannosyl backbone with single-unit galactoside side chains, α-linked at the O-6 oxygen. These compounds are found to be effective in treatment of type 2 diabetes patients by reducing hyperglycemia.

7.4 Allium *Species (*Allium cepa*) (Family: Amaryllidaceae)*

Various phytochemicals found in *Allium cepa* help reduce the risk of coronary heart disease. Onions possess vitamin C, niacin, folate, and potassium, fiber, flavonoids, fructans, sulfur-containing compounds, saponins, and flavonoids including quercetin and kaempferol. Several anthocyanins are found in onions, the predominant one being cyanidin-3-glycoside. These are potent antioxidants that are involved in different functions such as immune functions, platelet aggregation, and cholesterol metabolism. A new flavonoid called alliuocide A has been found to exhibit strong antioxidant potential for scavenging the free radicals of ROS and fatty acids [60].

Onion bulbs contain high contents of fructans (35–40 %) that are involved with reduction of blood glucose and serum cholesterol levels. Onion has found to contain organosulfur compounds such as cepaenes and thiosulfinates, which help to reduce the symptoms of cardiovascular dysfunctions associated with diabetes and inhibit platelet aggregation processes involved in thrombosis.

7.5 Nuts

Nuts, especially walnuts, are a treasure of mono- and polyunsaturated fatty acids, with small quantities of dietary fiber, phytosterols, and polyphenols. Intake of moderate quantities of walnuts is associated with reduced risk of CVD as it is found to decrease LDL-cholesterol. The constituents present in walnuts that may be important in reducing cardiovascular complications are polyunsaturated fatty acids (PUFA) and linoleic acid.

7.6 **Terminalia chebula** *Extract (Family: Combretaceae)*

Terminalia chebula, or *Kadukkai*, is one of the ancient traditional remedies for treatment of diabetes. It has been reported to possess numerous biological activities such as anticancer, antimutagenic, antibacterial, and antifungal. *Terminalia chebula* contains the triterpenes arjun glucoside 1, arjungenin, and chebulosides 1 and 2. Other constituents are chebulic acid 3–5 %, chebulinic acid 30 %, tannic acid 20–40 %, ellagic acid, gallic acid, ethyl gallate, terflavin A, anthraquinone, and flavonoids such as luteolin, rutins, and quercetin that may be responsible for the aforementioned therapeutic activities [61]. An experiment by Kaur et al. showed that the effect of *Terminalia chebula* extract on isolated rat pancreatic INS-1 β-cells and mouse 3 T3-L1 adipocytes demonstrated stimulatory action on the insulin secretion ability of pancreatic cells and an increase in the glucose consumption of adipocytes. Studies by Kim et al. [62] demonstrated that the extract of *Terminalia chebula* caused a significant reduction in blood glucose levels and lipids and improved serum biochemical values in diabetic rats. *Terminalia chebula* extract decreased advanced glycation end-products distribution in testis seminiferous tubules [63].

Through a literature search, the following nutraceuticals were found to be of significant importance, such as resveratrol from red grape products as an antioxidant, soluble dietary fiber products such as psyllium seed husk for reducing hypercholesterolemia, fenugreek as an antidiabetic, soy protein from soybeans for coronary heart disease, and garlic from *Allium sativum* Linn. for the treatment of cardiovascular diseases and many others. Herbs that have been screened for their antihyperlipidemic, antidiabetic, and antiobesity activity are listed in Table 2.

Table 2 List of herbs used in treatment of cardiovascular dysfunctions associated with diabetes mellitus

Herbs	Mechanism of action
Aloe vera (*Aloe barbadensis*)	Stimulates the release of insulin from beta cells in humans
Sadabhar (*Catharanthus roseus* Linn.)	Enhanced secretion of insulin and decreased blood glucose
Kali mirch (*Piper nigrum*)	Reverses oxidative damage and normalizes antioxidant enzyme levels in liver altered by diabetes
Turmeric (*Curcuma longa*)	Downregulates the inflammatory cytokines resistin and leptin and upregulates adiponectin and other associated proteins
Acacia arabica (Leguminosae)	Hypoglycemic effect by stimulating the release of insulin from pancreatic beta cells
Eugenia jambolana (Myrtaceae)	Enhances insulin secretion from pancreatic cells by inhibiting insulinase activity from liver and kidney
Caesalpinia bonducella (Leguminosae)	Hypoglycemic, hypolipidemic activity
Psyllium (soluble dietary fiber)	Decreases elevated blood pressure, improves glycemia and insulin sensitivity
Garlic	Inhibition of LDL oxidation
Flax seed	Lowers low density lipoprotein
Soy	Reduces cholesterol absorption
Glutathione	Increases reverse cholesterol transport

8 Conclusion

A multifactorial link exists between diabetes mellitus and cardiovascular complications including coronary artery disease and atherosclerosis [64]. Hyperglycemia, hypertension, obesity, and hyperlipidemia together culminate in the cardiovascular complications associated with diabetes. This review article has summarized all the factors causing the long-term complication of cardiovascular dysfunction of diabetes mellitus. Nutraceuticals now are gaining importance to treat such complications owing to their advantages such as cost-effective and economic treatment, wide and easy availability, and less or rare side effects. The botanicals discussed in this review possess hypoglycemic, hypolipidemic, antioxidant, and insulin-stimulating effects. Inflammation is the major factor leading to cardiovascular dysfunction [65, 66]. The nutraceuticals act through different mechanisms to control inflammatory signaling pathways and reactive oxygen species production, thereby preventing endothelial dysfunction [67]. Plants that treat dyslipidemia are used to prevent onset of cardiovascular disease by lowering the elevated lipid levels of LDL and triglycerides while increasing HDL-cholesterol.

Future scope lies in the investigation of the exact mechanism of action of the nutraceuticals and their constituents that are responsible for antidiabetic and insulin mimetic activity. Extensive research should be carried out exploiting the therapeutic properties and finding the toxic effects of nutraceuticals and dietary sources for treating diabetic complications in the light of development of new therapeutic strategies.

References

1. World Health Organization (2008) Traditional medicine. http://www.who.int/mediacentre/factsheets/fs134/en/
2. Zarraga I, Schwartz E (2006) Impact of dietary patterns and interventions on cardiovascular health. Circulation 114:961–973
3. Retelny VS, Neuendorf A, Roth JL (2008) Nutrition protocols for the prevention of cardiovascular disease. Nutr Clin Pract 23:468–476
4. Dureja H, Kaushik D, Kumar V (2003) Developments in nutraceuticals. Indian J Pharmacol 35:363–372
5. Jaap GN, Jerrold MO (2006) Inflamed fat: what starts the fire. J Clin Invest 116:33–35
6. Grundy SM, Brewer HB, Cleeman JI et al (2004) Definition of metabolic syndrome: report of the National Heart, Lung and Blood Institute/American Heart Association conference on scientific issues related to definition. Circulation 109:433–438
7. Jaswinder KS, Gokhan SH (1999) The role of TNF in adipocyte metabolism. Semin Cell Dev Biol 10:19–29
8. Stefan N, Vozarova B, Funahashi T et al (2002) Plasma adiponectin concentration is associated with skeletal muscle insulin receptor tyrosine phosphorylation and low plasma concentration precedes a decrease in whole-body insulin sensitivity in humans. Diabetes 51:1884–1888
9. He Z, Rask-Madsen C, King GL (2003) Managing heart disease: mechanisms of cardiovascular complications in diabetes and potential new pharmacological therapies. Eur Heart J Suppl 5(Suppl B):B51–B57
10. Perseghin G, Scifo P, Cobelli FD et al (1999) Intramyocellular triglyceride content is a determinant of in vivo insulin resistance in humans: a ^{1}H-^{13}C nuclear magnetic resonance spectroscopy assessment in offspring of type 2 diabetic parents. Diabetes 48:1600–1606
11. Tripathy D, Dandona P (2002) Acute elevation of plasma free fatty acids increases reactive oxygen species (ROS) generation by polymorphonuclear cells, induces nuclear factor-kB (NF-kB) and impairs brachial artery reactivity in healthy subjects. In: 62nd Scientific Sessions of the American Diabetes Association, San Francisco
12. Wellen KE, Hotamisligil GS (2005) Inflammation, stress and diabetes. J Clin Invest 115:1111–1119
13. Hotamisligil GS, Peraldi P, Budavari A et al (1996) IRS-1-mediated inhibition of insulin receptor tyrosine kinase activity in TNF-alpha- and obesity-induced insulin resistance. Science 271:665–668
14. Dong M, Su D, Coudriet GM, Hyun Kim D et al (2009) FoxO1 links insulin resistance to proinflammatory cytokine IL-1β production in macrophages. Diabetes 58(11):2624–2633
15. Dzau V, Braunwald E (1999) Resolved and unresolved issues in the prevention and treatment of coronary artery disease: a workshop consensus statement. Am Heart J 121:1244–1263
16. Ross R (1999) Atherosclerosis—an inflammatory disease. N Engl J Med 340:115–126
17. Woollard KJ, Geissmann F (2010) Monocytes in atherosclerosis: subsets and functions. Nat Rev Cardiol 7:77–86
18. Vallance P, Chan N (2001) Endothelial function and nitric oxide: clinical relevance. Heart 85:342–350
19. Landmesser U, Hornig B, Drexler H (2004) Endothelial function: a critical determinant in atherosclerosis. Circulation 109:11–27
20. Deanfield JE, Halcox JP, Rabelink TJ (2007) Endothelial function and dysfunction: testing and clinical relevance. Circulation 115:1285–1295
21. Hamilton CA, Miller WH, Brosnan MJ et al (2004) Strategies to reduce oxidative stress in cardiovascular disease. Clin Sci (Lond) 106:219–234
22. Taniyama Y, Griendling KK (2003) Reactive oxygen species in the vasculature: molecular and cellular mechanisms. Hypertension 42:1075–1081
23. Droge W (2002) Free radicals in the physiological control of cell function. Physiol Rev 82:47–95

24. Griendling KK, FitzGerald GA (2003) Oxidative stress and cardiovascular injury. Part II: Animal and human studies. Circulation 108:2034–2040
25. Jay D, Hitomi H, Griendling KK (2006) Oxidative stress and diabetic cardiovascular complications. Free Radic Biol Med 40:183–192
26. Packard CJ, Caslake M, Shepherd J (2000) The role of small, dense low density lipoprotein (LDL): a new look. Int J Cardiol 74:S17–S22
27. Ginsberg H, Zhang YL, Antonio H (2005) Regulation of plasma triglycerides in insulin resistance and diabetes. Arch Med Res 36:232–240
28. Reaven GM, Chen YD, Jeppesen J et al (1993) Insulin resistance and hyperinsulinemia in individuals with small, dense low density lipoprotein particles. J Clin Invest 92:141–146
29. Shaul PW (2003) Regulation of endothelial nitric oxide synthase. Annu Rev Physiol 64(749):774
30. Pariza MW, Park Y, Cook ME (2000) Mechanisms of action of conjugated linoleic acid: evidence and speculation. Proc Soc Exp Biol Med 223:8–13
31. Lee K, Kritchevsky D, Pariza M (1994) Conjugated linoleic acid and atherosclerosis in rabbits. Atherosclerosis 8:19–25
32. Ryder JW, Portocarrero CP, Song XM et al (2001) Isomer-specific antidiabetic properties of conjugated linoleic acid. Diabetes 50:1149–1157
33. Evans M, Brown J, McIntosh M (2002) Isomer-specific effects of conjugated linoleic acid (CLA) on adiposity and lipid metabolism. J Nutr Biochem 13:508–516
34. Grundy SM, Denke MA (1990) Dietary influences on serum lipids and lipoproteins. J Lipid Res 31:1149–1172
35. De Lorgeril M, Salen P, Martin JL et al (1999) Mediterranean diet, traditional risk factors, and the rate of cardiovascular complications after myocardial infarction: final report of the Lyon diet heart study. Circulation 99:779–785
36. Kris-Etherton PM, Taylor DS, Yu-Poth S et al (2000) Polyunsaturated fatty acids in the food chain in the United States. Am J Clin Nutr 71(Suppl 1):179S–188S
37. Mori TA, Bao DQ, Burke V et al (1999) Docosahexanoic acid but not eicosapentaenoic acid lowers ambulatory blood pressure and heart rate in humans. Hypertension 34:253–260
38. Hartweg J, Farmer AJ, Perera R et al (2007) Meta-analysis of the effects of n-3 polyunsaturated fatty acids on lipoproteins and other emerging lipid cardiovascular risk markers in patients with type 2 diabetes. Diabetologia 50:1593–1602
39. Lau FC, Bagchi M, Sen CK et al (2008) Nutrigenomic basis of beneficial effects of chromium (III) on obesity and diabetes. Mol Cell Biochem 317:1–10
40. Anderson RA, Cheng N, Bryden NA et al (1997) Elevated intakes of supplemental chromium improve glucose and insulin variables in individuals with type 2 diabetes. Diabetes 46(11):1786–1791
41. Broadhurst CL, Domenico P (2006) Clinical studies on chromium picolinate supplementation in diabetes mellitus: a review. Diabetes Technol Ther 8(6):677–687
42. Balon TW, Gu JL, Tokuyama Y et al (1995) Magnesium supplementation reduces development of diabetes in rat model of spontaneous non-insulin dependent diabetes mellitus. Am J Physiol 269:E745–E752
43. Kendall MJ, Nuttall SL, Martin U (1998) Antioxidant therapy—a new therapeutic option for reducing mortality from coronary artery disease. J Clin Pharm Ther 23:323–325
44. Riccioni G, Bucciarelli T, Mancini B et al (2007) Antioxidant vitamin supplementation in cardiovascular diseases. Ann Clin Lab Sci 37:89–95
45. Devaraj S, Jialal I (2000) Low-density lipoprotein postsecretory modification, monocyte function, and circulating adhesion molecules in type 2 diabetic patients with and without macrovascular complications: the effect of alpha-tocopherol supplementation. Circulation 102:191–196
46. Isomaa B, Almgren P, Tuomi T et al (2001) Cardiovascular morbidity and mortality associated with metabolic syndrome. Diabetes Care 24:683–689
47. Gaede P, Poulsen HE, Parving HH et al (2001) Double-blind, randomised study of the effect of combined treatment with vitamin C and E on albuminuria in type 2 diabetic patients. Diabet Med 18:756–760

48. Singh U, Jiala I (2008) Alpha-lipoic acid supplementation and diabetes. Nutr Rev 66: 646–657
49. Kanter M, Meral I, Yener Z et al (2003) Partial regeneration/proliferation of the β-cells in the islets of Langerhans by *Nigella sativa* L. in streptozotocin-induced diabetic rats. Tohoku J Exp Med 201:213–219
50. Fararh KM, Atoji Y, Shimizu Y et al (2002) Insulinotropic properties of *Nigella sativa* oil in streptozotocin plus nicotinamide diabetic hamster. Res Vet Sci 73:279–282
51. Meral I, Yener Z, Kahraman T et al (2001) Effect of *Nigella sativa* on glucose concentration, lipid peroxidation, antioxidant defense system and liver damage in experimentally-induced diabetic rabbits. J Vet Med A Physiol Pathol Clin Med 48:593–599
52. El-Dakhakhny M, Mady N, Lembert N et al (2002) The hypoglycaemic effect of *Nigella sativa* oil is mediated by extrapancreatic actions. Planta Med 68:463–464
53. Morikawa T, Xu F, Ninomiya K et al (2004) Nigellamines A3, A4, A5 and C, new dolabellane-type diterpene alkaloids, with lipid metabolism-promoting activities from the Egyptian medicinal food black cumin. Chem Pharm Bull 52:494–497
54. Sitasawad SL, Shewade Y, Bhonde R (2000) Role of bitter gourd fruit juice in STZ- induced diabetic state in vivo and in vitro. J Ethnopharmacol 73:71–79
55. Karunanayake EH, Jeevathayaparan S, Tennekoon KH (1990) Effect of *Momordica charantia* fruit juice on streptozotocin-induced diabetes in rats. J Ethnopharmacol 30:199–204
56. Welihinda J, Arvidson G, Gylfe E et al (1982) The insulin-releasing activity of the tropical plant *Momordica charantia*. Acta Biol Med Ger 41:1229–1240
57. Hannan JM, Ali L, Rokeya B et al (2007) Soluble dietary fibre fraction of *Trigonella foenum-graecum* (fenugreek) seed improves glucose homeostasis in animal models of type 1 and type 2 diabetes by delaying carbohydrate digestion and absorption and enhancing insulin action. Br J Nutr 97:514–521
58. Effect of fenugreek on blood sugar and insulin in diabetic humans. http://clinicaltrials.gov
59. Moorthy R, Prabhu KM, Murthy PS (2010) Anti-hyperglycemic compound (GII) from fenugreek (*Trigonella foenum-graecum* Linn.) seeds: purification and effect in diabetes mellitus. Indian J Exp Biol 48:1111–1118
60. Mohamed GA (2013) Alliuocide A: a new antioxidant flavonoid from *Allium cepa* L. Phytopharmacology 4:220–227
61. Cheng HY, Lin TC, Yu KH et al (2003) Antioxidant and free radical scavenging activities of *Terminalia chebula*. Biol Pharm Bull 26:1331–1335
62. Kim JH, Hong C, Koo YC et al (2011) Oral administration of ethyl acetate soluble portion of *Terminalia chebula* conferring protection from streptozotocin-induced diabetes mellitus and its complications. Biol Pharm Bull 34:1702–1709
63. Sschmidt AM, Yan SD, Wautier JL et al (1999) Activation of receptor for advanced glycation end products: a mechanism for chronic vascular dysfunction in diabetic vasculopathy and atherosclerosis. Circ Res 84:489–497
64. Hartge M, Unger T, Kintscher U (2007) The endothelium and vascular inflammation in diabetes. Diab Vasc Dis Res 4:84–88
65. Navarro J, Mora C (2005) Role of inflammation in diabetic complications. Nephrol Dial Transplant 20:2601–2604
66. Festa A, Haffner S (2005) Inflammation and cardiovascular disease in patients with diabetes: lessons from the Diabetes Control and Complications Trial. Circulation 111:2414–2415
67. Omara E, Kama A, Alqahtani A et al (2010) Herbal medicines and nutraceuticals for diabetic vascular complications: mechanisms of action and bioactive phytochemicals. Curr Pharm Des 16:3776–3807

Nutritional Management of Cardiovascular Complications Caused by Diabetes

Adriana Adameova, Paramjit S. Tappia, Yan-Jun Xu, and Naranjan S. Dhalla

Abstract Several lines of epidemiological, experimental, and clinical evidence have demonstrated a correlation between diet and increased risk of developing diabetes-induced cardiovascular complications. The increased consumption of refined and simple carbohydrates, fats, red meats, and low fibre as well as low intake of specific minerals and vitamins have been shown to impair insulin response and increase plasma glucose levels and thus produce damage to the macro- and micro-vasculature. On the other hand, healthy dietary choices, specific nutrients, and some herbal supplements are known to control the glycemic index in subjects with diabetes and thus suppress the progression of diabetic cardiomyopathy. In addition, the nutritional experiences of the developing fetus and maternal anemia have been reported to alter glucose metabolism and insulin signaling in the offspring. This chapter is focused on reviewing the influence of nutrition on the risk of developing diabetes and the dietary approaches that are undertaken for the management of diabetes and prevention of cardiovascular complications. Furthermore, it is intended to highlight the impact of the nutritional experience of the developing fetus and describe the role of epigenetics and maternal nutrition on the risk of developing diabetes and cardiovascular complications.

A. Adameova
Department of Pharmacology and Toxicology, Faculty of Pharmacy,
Comenius University, Bratislava, Slovakia

P.S. Tappia (✉)
Asper Clinical Research Institute, St. Boniface Hospital Research Centre,
CR3129-369 Tache Avenue, Winnipeg, MB R2H 2A6, Canada
e-mail: ptappia@sbrc.ca

Y.-J. Xu • N.S. Dhalla
Institute of Cardiovascular Sciences & Department of Physiology,
Faculty of Medicine, University of Manitoba, Winnipeg, MB, Canada

B. Turan and N.S. Dhalla (eds.), *Diabetic Cardiomyopathy: Biochemical and Molecular Mechanisms*, Advances in Biochemistry in Health and Disease 9,
DOI 10.1007/978-1-4614-9317-4_25,

Keywords Nutrition • Diabetes • Heart dysfunction • Metabolic abnormalities • Maternal nutrition • Epigenetics

1 Introduction

It has been estimated that around 171 million people are affected by diabetes around the world, and it is predicted to affect 366 million people by 2030. Moreover, it is generally thought that about 30 % of the population with diabetes remain undiagnosed in industrialized nations [1]. Up to 10 % of the diabetic population suffers from type 1 diabetes whereas 85–90 % of diabetics are affected by type 2 diabetes, making this disease the world's most prevalent metabolic disease [2]. The onset of type 1 diabetes occurs in children and type 2 diabetes occurs in adulthood; however, because of unbalanced diet and early obesity, type 1 diabetes (insulin-dependent) is also being identified in adults, whereas type 2 diabetes (insulin-non-dependent) has been detected in children and teenagers. Regardless of the type of diabetes, increased glucose in the plasma produces damage to the small and large vasculature, which underlies the main factor for the increased mortality in diabetic populations. In fact, heart disease is the leading cause of death among diabetics and accounts for most of the deaths.

Approximately 70–80 % of diabetic patients die of cardiovascular complications, such as ischemic heart disease, atherosclerosis, hypertension, arrhythmias, and congestive heart failure [1]. In addition, diabetic cardiomyopathy has been shown to occur in the absence of coronary heart disease in diabetic patients [3]. Diabetic cardiomyopathy is characterized by structural, cellular, and molecular abnormalities leading to diastolic dysfunction, which progresses to left ventricular hypertrophy followed by systolic dysfunction [4, 5]. Figure 1 shows that diabetes is associated with the activation of the sympathetic nervous system and the renin-angiotensin system (RAS) [6, 7]. In addition, there is a metabolic shift as well as an increase in plasma glucose levels that result in increased formation of oxyradicals. The activation of RAS, which results in the depletion of tissue antioxidant levels as well as impaired endothelial function, may also provide conditions favoring oxidative stress. Furthermore, oxidative stress induces subcellular remodeling that results in Ca^{2+}-handling abnormalities and finally in contractile dysfunction in diabetic heart [4, 6, 7].

Several pharmacological approaches are used to treat and prevent the progression of diabetic cardiomyopathy [8]. In general, the treatment of cardiac dysfunction in diabetic subjects is similar to that for individuals without diabetes, and the clinical response to these drugs is similar or better. The beneficial effects of RAS blockade by angiotensin-converting enzyme inhibitors (ACEI), angiotensin receptor, and β-adrenoceptor blockers as well as antioxidants on cardiovascular outcomes in diabetes has been well documented [9–15]. In addition, pharmacological approaches to lower blood glucose levels and improve insulin action are used as first-line therapy for the management of diabetes [8, 16, 17].

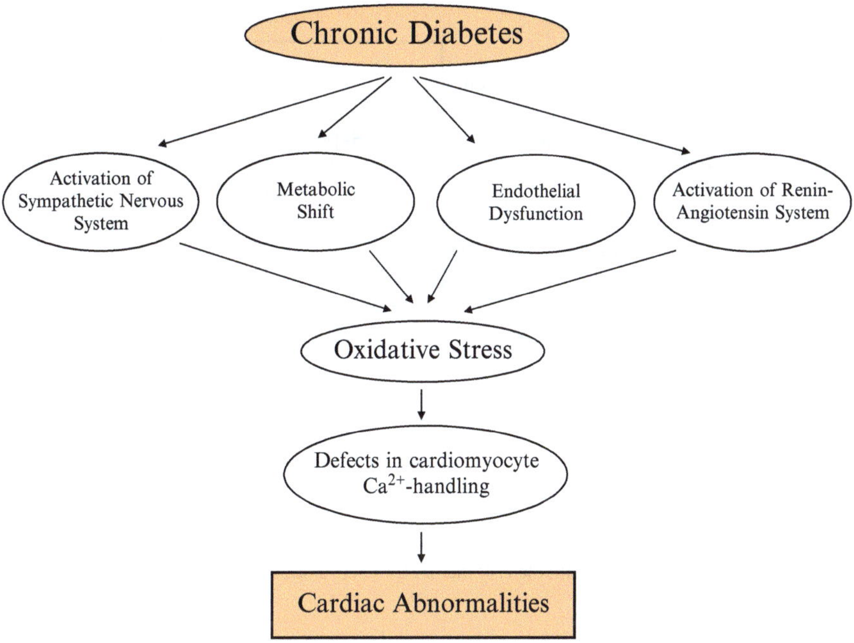

Fig. 1 Major diabetes-induced defects leading to cardiac abnormalities

Although pharmacological interventions are of benefit, following a recommended diet, regular physical activity, self-monitoring of blood glucose, and smoking cessation [18, 19] are now also regarded as key contributors to a healthy lifestyle that can even prevent cardiovascular complications resulting from diabetes. It can be seen from Fig. 2 that there are a number of nonmodifiable (age, gender, genetic) and modifiable factors that are linked to increased risk for heart disease. Clearly, these modifiable factors are interrelated, and nutrition is a major determinant of heart disease. Nutritional interventions have been evaluated as cost-effective and highly efficient components of a strategy to reduce the growing disease burden [20]. In fact, an increase in public awareness for lifestyle changes for diabetes management and prevention has been conducted through educational programs, public forums, and brochures. In this chapter, we focus only on diet and review some effects of nutrition on the development and prevention of diabetes and diabetes-induced defects in heart function. Furthermore, the influence of specific types of food, minerals, and vitamins, as well as decoctions of some herbal products, on diabetes management and prevention of diabetes-induced cardiovascular complications are also briefly mentioned. In addition, the impact of an adverse nutritional environment of the developing fetus is briefly discussed to further increase the awareness that adult disease may have its origins during fetal development.

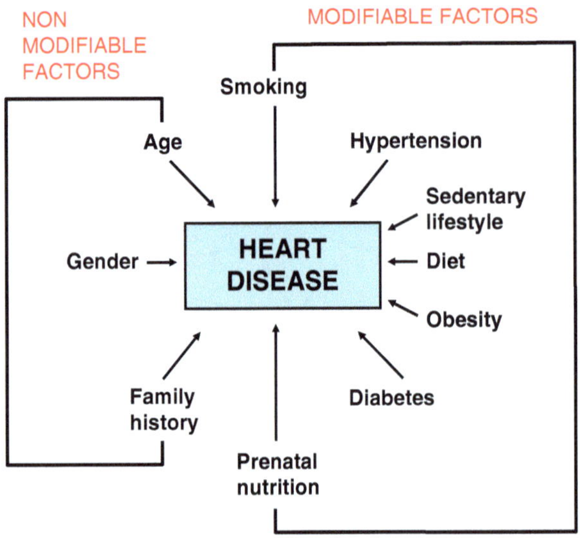

Fig. 2 Nonmodifiable and modifiable risk factors for heart disease

2 Beneficial Effects of Nutrients on the Development and Prevention of Diabetes

Diet is considered to be the cornerstone in the treatment of diabetes and associated heart dysfunction. Indeed, dietary advice can be tailored to the individual management needs. Although, as reviewed here, a low-fat diet in general is recommended for patients with diabetes, particular emphasis is put on nutritional support, encouraging food fortification, full-fat products, and high-energy snacks in those patients with diabetes who are identified as being at risk of malnutrition [21]. Prevention and attenuation of the diabetes-induced deleterious effects on heart function are based on modulation of the diet. These dietary recommendations are very similar to those given to the general population for a healthy diet and include increased fibre and fruit and vegetable intake and reduced fat ingestion [22, 23].

Dietary fibre has been, for several years, a healthy component in popular nutrition; however, with changes in dietary habits, the intake of fibre is actually decreasing. Lack of fibre in the diet has been reported as a major risk factor for the development of diabetes and heart disease, although strong inverse relationships are seen between whole grain intake and incidence of type 2 diabetes [24, 25]. In fact, high fibre intake has been shown to improve glucose tolerance and lipid metabolism in subjects with type 2 diabetes [26] and to decrease systolic blood pressure without any changes in increased diastolic blood pressure [27]. There are a number of potential mechanisms by which dietary fibre may modify insulin sensitivity and glucose tolerance. For instance, fibre can influence intestinal tract time, absorption of macronutrients, alteration of the action of digestive enzymes, and secretion of pancreatic and gastrointestinal hormones [28]. Soluble fibre delays gastric emptying, slowing the absorption and digestion of carbohydrates and potentially delaying the insulin response, whereas insoluble fibre decreases intestinal tract time and thereby reduces time for the absorption of carbohydrates in the jejunum [29]. In addition, nondigestive fibre (resistant starch) can be fermented by the microflora of the colon, leading into the production of short-chain fatty acids that can also influence carbohydrate metabolism [30]. Furthermore, resistant starch has potential application in weight management, in the improvement of blood lipid profile, glucose tolerance, and insulin sensitivity [31, 32]. Although dietary guidelines recommend a daily resistant starch intake of 15–20 g, many populations consume significantly lower amounts of resistant starch/day. Thus, increasing resistant starch intake in the diet can be seen as a major step in diabetes management and prevention of cardiovascular complications.

Current dietary guidelines for the prevention of type 2 diabetes recommend consuming at least 400 g fruit and vegetables per day [33]; however, data from these studies are less conclusive probably because of low quality and heterogeneity of the subjects [34]. In spite of this, the association between greater fruit and vegetable intake and reduced risk of cardiovascular disease and cancer has been reported [35, 36]. It has been estimated that up to 2.7 million lives worldwide could be saved with sufficient fruit and vegetable consumption [37]. Recently, it has been reported that plasma vitamin C, a biomarker for fruit and vegetable intake, is inversely associated with glycated hemoglobin, as well as blood glucose level [38]. However, the role of fruit and vegetable intake, independent of each other, need to be further investigated. In developing countries, dietary habits have progressively changed. In fact, a shift from a healthy traditional high-fibre, low-fat, and low-calorie diet toward calorie-dense foods containing refined carbohydrates (fructose and sucrose), saturated fats, red meats, and low fibre has occurred [39, 40]. These changes in dietary habits are seen to contribute to increases in obesity, type 2 diabetes mellitus, metabolic syndrome, and cardiac dysfunction [41–43].

Reduced consumption of saturated and *trans*-unsaturated fatty acids is generally advised for diabetic patients because a negative relationship between palmitic acid and insulin sensitivity has been reported [44]. In contrast to saturated fatty acids, polyunsaturated fatty acids have been shown to be inversely related to incidence of diabetes [45]. It should be noted that the incorporation of different fatty acids into cell-membrane phospholipids alters the microenvironment of membrane-associated proteins such as cation channels, enzymes, and receptors, including the insulin receptor, and thus may modify their function [46, 47]. Consumption of low-cost unhealthy vegetable oils is not cardioprotective because it increases dietary intake of *trans*-fatty acids, which are known to modify the balance between free fatty acid oxidation and lipogenesis and promote insulin resistance [40].

During diabetes, because glucose transport and oxidation are defective, cardiac energy production is almost exclusively by breakdown of fatty acids that are supplied in excess to the heart [48]. Overload of fatty acids, in particular saturated fatty acids such as palmitic acid, has been suggested to be linked with apoptosis, which seems to occur independently of hyperglycemia [49, 50]. Becuase diabetic cardiomyopathy is associated with the occurrence of programmed cell death [51, 52], which in turn is associated with an impaired heart function [14], it is likely that such diabetes-associated complications in the accumulation of fatty acids may contribute to heart disease. In addition to unhealthy fat intake, there is evidence that increased consumption of red meat may increase the risk of type 2 diabetes. In fact, a 19 % higher risk of the disease was noted with each serving of processed meat consumed per day [53, 54]. In contrast, vegetarian food and a prudent diet allowing small amounts of red meat have been found to exert favorable effects on metabolic parameters, blood pressure, and reduced risk of type 2 diabetes and cardiovascular diseases [55, 56].

3 Amino Acids and Minerals in Diabetes

Amino acids are considered to be essential nutrients for maintaining the physiological function of the heart, and inadequate levels of some amino acids under different pathophysiological conditions have been associated with cardiac dysfunction [5, 57]. Three amino acids in particular have been studied. Taurine is the most abundant intracellular sulfur-containing amino acid [58]. Although it can be synthesized from methionine and cysteine in the presence of vitamin B_6 [58], taurine can be obtained from the diet, predominantly through eggs, meat, and seafood. Cysteine availability is considered as a major rate-limiting factor for the production of the major cellular antioxidant, glutathione. Sources of cysteine in humans are dietary protein, endogenous proteolysis, and conversion via methionine. Carnitine and its derivatives are natural substances involved in both carbohydrate and lipid metabolism. The main food sources of carnitine are red meat and dairy products as well as fish and poultry.

Deficiencies of these amino acids are known to occur in diabetes patients. In diabetes, intracellular accumulation of sorbitol resulting from the high extracellular levels of glucose leads to the depletion of intracellular taurine levels and is associated with the development of diabetic cardiomyopathy [59]. Experimental and clinical studies have reported the beneficial actions of taurine in both type 1 and type 2 diabetes. Indeed, the beneficial actions of taurine in diabetes have been attributed to the inhibition of platelet aggregation [60], improved lipid profile [61], attenuation of oxidative stress [62], and reduced myocardial cell death [63]. Carnitine deficiency is often observed in diabetes [64, 65], and exogenous carnitine has been shown to improve cardiac metabolism and function [66]. The mechanisms of carnitine action include an increase in glucose metabolism and lipid-lowering effects [66–68], increase in insulin sensitivity [69], and antioxidant effects [70] as well as reduction of pro-atherogenic lipoprotein A [71]. Cysteine, as a major rate-limiting factor for gluthathione production, has been suggested to play an important role in the development of cardiac defects caused by diabetes as well as other diseases associated with oxidative stress [57]. In experimental studies in diabetic rats, cysteine was found to prevent cardiac dysfunction due to diabetes [72]. It was suggested that the beneficial effects of cysteine may result from the direct scavenging of free radicals or improved antioxidant capacity through glutathione preservation [73].

With respect to minerals, chromium has been proposed to be an essential trace element for the improvement of insulin sensitivity [74]. Although earlier studies have indicated the insulin-sensitizing action of chromium picolinate in experimental models of type 1 and type 2 diabetes [75], recent animal studies have revealed that a diet with low chromium content has no effect on body composition, glucose metabolism, or insulin sensitivity compared with a chromium-sufficient diet, and thus it was suggested that chromium may not be an essential trace element [76]. Although the notion that chromium is an essential element for carbohydrate and lipid metabolism that may be questionable [77], in a recent placebo-controlled single-blind prospective study conducted in newly-onset patients with type 2 diabetes, chromium supplementation (9 g brewers yeast, 42 μg chromium) for 3 months was found to reduce fasting blood glucose, glycated hemoglobin (HbA1c), total cholesterol, triglycerides, and LDL levels. These results suggest that chromium exerts beneficial effects on glycemic control and blood lipid profile [78]. Chromium supplementation has also been reported to shorten QTc interval, a pro-arrhythmic predictor, in diabetic patients [79]. Despite these lines of evidence, further work is warranted in determining the benefits of chromium.

Experimental studies have reported that vanadate treatment normalizes both cardiac function and hyperglycemia [80]. Indeed, both vanadate (V5+) and vanadyl (V4+) forms exhibit antidiabetic activity [81]. Interestingly, following a withdrawal of vanadyl treatment of up to 30 weeks, diabetic animals were reported to remain normoglycemic and to have normalized glucose tolerance and improved plasma insulin levels [81]. In addition, treatment of streptozotocin (STZ)-induced diabetic animals with bis(maltolato)oxovanadium (IV) has been shown to enhance insulin-mediated GLUT4 translocation in cardiac tissue and improve cardiac function [82, 83]. Improved cardiomyocyte Ca^{2+} transients have also been reported in STZ-induced diabetic rats treated with vanadate administered in a black tea extract for an 8-week period [84]. Furthermore, T/V treatment resulted in a contractile response that was not different from cardiomyocytes of control animals. In a recent phase IIa trial, treatment of type 2 diabetic individuals with bis(ethylmaltolato)oxovanadium (IV) (20 mg, daily for 28 days) resulted in a reduction in fasting blood glucose and HbA1c and improved responses to oral glucose tolerance testing [85]. Thus, it is evident that vanadate may be of benefit for diabetes and prevention of cardiovascular complications.

Selenium, as an antioxidant, has been considered as part of a strategy to control oxidative stress and antagonize cardiovascular complications from diabetes [86, 87]. Treatment of STZ-induced diabetic rats with sodium selenite has been reported to reverse the altered mechanical and electrical activities caused by restoration of the diminished K^+ currents. In addition, selenium also restored the cell glutathione redox cycle [88]. In another study, sodium selenite (5 μmol/kg body weight) of STZ-induced diabetic rats for 4 weeks prevented the loss of myofibrils and reduction of cardiomyocyte diameter [89]. Furthermore, in the sodium selenite-treated diabetic rat heart, alterations of the discus intercalaris and nucleus were corrected and degeneration in myofilaments and Z-lines was reversed [89]. Thus, it would appear that selenium can protect against diabetes-induced structural alterations of the heart. From the evidence provided, selenium may therefore be of value for the prevention of cardiovascular complications caused by diabetes.

4 Herbal Medicine and Diabetes

In traditional medicine, antidiabetic effects of certain botanicals have been investigated in the diabetic population. Interestingly, metformin, an antidiabetic agent, was developed from the herb French lilac (*Galega officinalis*) [90]. Numerous other herbs have been suggested to be targeted for antidiabetic drug development. Several lines of evidence indicate that ginseng (*Panax* spp.), ivy gourd (*Coccinia indica*), garlic (*Allium sativum* and *Allium cepa*), holy basil (*Ocimum sanctum*), fenugreek (*Trigonella foenum graecum*), prickly pear cactus or nopal (*Opuntia streptacantha*), milk thistle (*Silibum marianum*), fig leaf (*Ficus carica*), gurmar (*Gymnema sylvestre*), bitter melon (*Momordica charantia*), *Aloe vera*, and *Ginkgo biloba* improve glycemic control. Although the mechanisms responsible for hypoglycemic effects of these herbs are not completely defined, it has been suggested that they may delay glucose absorption in the gut (*A. vera*, prickly pear cactus), increase glucose uptake/disposal (fig leaf, ivy gourd), and glucose-stimulated insulin secretion, or may exert two or more of these effects simultaneously [91–93].

In spite of these potential beneficial effects, antihyperglycemic efficacy remains inconclusive for many of these herbs, which may be the result of the small number of clinical trials, small sample size, lack of randomization, absence of blinding, and inadequate reporting of dropouts. In addition, the safety of long-term use of herbal products as well as possible interactions with concomitant medications remains to be fully understood for many of these products, and thus caution must be exercised. In fact, recently we have reviewed some of the experimental and clinical evidence for hypoglycemic and lipid-lowering effects of different functional foods as well as some herbal products as possible adjunctive/alternative therapies for diabetes and cardiovascular complications [94]. Figure 3 summarizes the possible mechanisms for the hypoglycemic effects of herbal products and dietary interventions.

5 Diet During Fetal Development and Epigenetic Changes

Several studies have shown that the risk of developing diabetes and associated heart diseases in adulthood is determined not only by the Western diet, but also by nutritional environment of fetus and infant. In fact, poor growth in utero has been found to be associated with increased risk of developing type 2 diabetes and cardiovascular disease in late-life period [95, 96]. The detrimental effects of poor fetal growth on metabolic conditions appear to be amplified, if it is followed by accelerated postnatal growth and/or obesity. In this regard, children with low birth weights

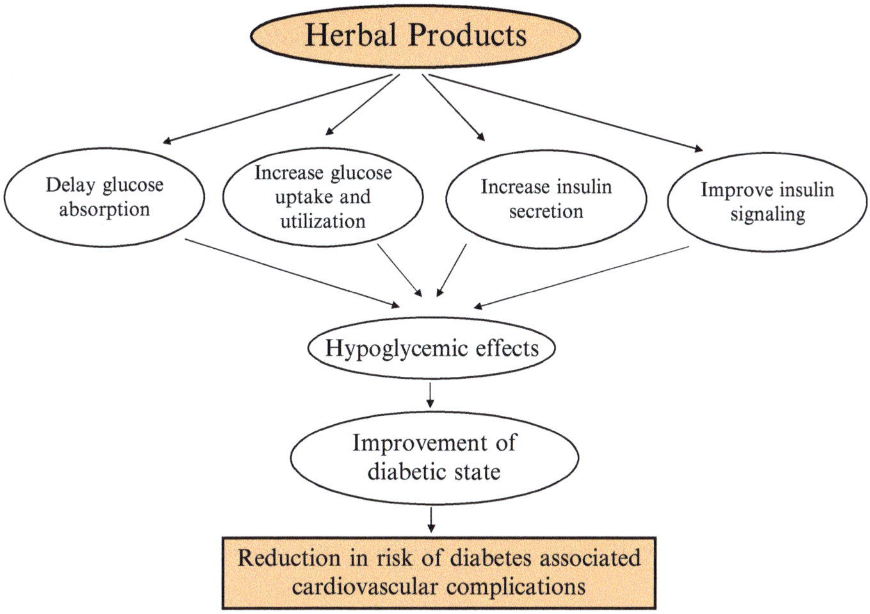

Fig. 3 Putative mechanisms of beneficial action of herbs for reduced risk of diabetes and associated cardiovascular complications

(≤2.5 kg) who underwent rapid childhood weight gain had the worst glucose tolerance and increased risk of cardiovascular disease [97]. Similarly, the highest death rate from coronary heart disease occurred in men who were thin at birth but later their body mass index was average or above average from the age of 7-year children [98]. Interestingly, studies with both monozygotic and dizygotic twin pairs who were discordant for diabetes have found that the diabetic twin has a lower birth weight than the normoglycemic co-twin [99]; this provides further evidence for a link between diabetes and low birth weight.

The impact of maternal nutrition on the health of offspring has been intensively examined and several mechanisms underlying the fetal origins of adult disease or programming have been suggested. One of the proposed mechanisms is related to suboptimal nutrition, in utero. There is now accumulating evidence that epigenetic regulation of transcription is a mechanism for inducing changes in phenotype [100]. Such epigenetic changes are defined as reversible changes that occur as a result of heritable changes without the involvement of primary DNA sequence alterations. Early nutritional deficits have been shown to modify DNA and histones, proteins that regulate gene activity of the developing organism [101, 102]. The main epigenetic modulations caused by adverse environment include DNA methylation and histone modifications, which have been shown to maintain specific gene regions in a transcriptionally silent state (Fig. 4). Such epigenetic marks, which are particularly vulnerable during early stages of embryonic development, can persist long after the duration of the initiating factor or condition [103–105] and thus promote defects in later age.

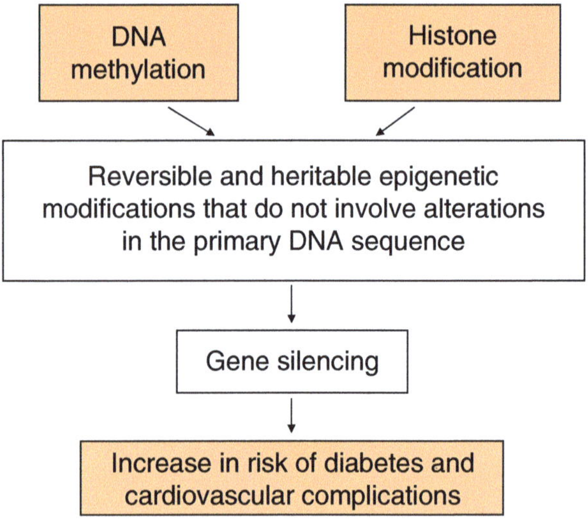

Fig. 4 Reversible epigenetic modifications of the primary DNA sequence leading to gene silencing

The main environmental conditions that adversely influence the developing fetus include low-protein maternal diets, maternal calorie restriction, and maternal anemia. In fact, a maternal low-protein diet has been found to alter the methylation status of peroxisome proliferator-activated receptor (PPAR)-α [102], a transcription factor that is involved in the regulation of numerous metabolic processes, including metabolism of carbohydrates [106]. These alterations lead to a large change in PPAR-α expression [102]. Other studies in offspring have also shown upregulation of PPAR-α transcripts because of postnatal maternal protein restriction; however, there were no significant changes in PPAR-α promoter methylation [107]. In addition to these changes, in offspring of female rats fed a protein-deficient diet, modifications in the promoters of angiotensin receptor [108] and glucocorticoid receptors [101] have also been reported. Since PPAR and glucocorticoid receptors regulate triglyceride and glucose metabolism these posttranslational modifications are likely to account for the development of diabetes and associated cardiovascular disease. In addition, histone modifications from caloric restriction during the second half of pregnancy have been found to reduce GLUT4 expression [106, 109]. As histone modifications by demethylases is dependent on flavin adenine dinucleotide or α-ketoglutarate [110], it is very likely that this process is also responsive to cellular energy status. Therefore, it seems that nutritional status may not only exert specific effects, but also exhibit global effects on histone modifications and DNA methylation.

In addition to modifications in DNA and histones, upregulation of the gene expression as well as protein levels of key components of glucose metabolism and insulin signal transduction pathway, such as phosphatidylinositol-3-kinase, insulin receptor β-subunit and insulin receptor substrate-1, have been suggested to account, at least in part, for the accelerated energy supply, demand, and utilization in offspring of female rats fed a low-protein (casein) diet. These effects were associated with an increase in left ventricular internal diameters and wall thickness [96], indicating a direct relationship between adverse environment during development in utero, diabetes, and subsequent myocardial abnormalities. The other suggested mechanisms by which maternal protein restriction can lead to type 2 diabetes development later in life are oxidative stress and impaired oxidative defenses as evidenced by lipid peroxidation as well as the reduced levels of antioxidant enzymes [111].

6 Conclusions

From the aforementioned discussion as well as extensive literature review by others [112–115], it is evident that cardiovascular disease is a major cause of mortality in the diabetic population. Furthermore, it is becoming clear that prenatal and postnatal dietary practices could promote the development of diabetes and increased risk of cardiovascular complications. Inadequate or unhealthy diets can be reversed for disease prevention and management. Increased intake of fibre, fruit, vegetables, and a reduced consumption of saturated and *trans*-fats as well as carbohydrates are recommended by dietary guidelines. Although some herbs and herbal products exhibit hypoglycemic effects and could serve as adjunct therapies, their safety and efficacy remain to be fully determined. It is also apparent that the health benefits of food and food products in the prevention of cardiovascular disease have been recognized in recent years. The market for "food for health" is being driven by an increasing consumer understanding of the link between diet and disease, aging populations, rising healthcare costs, and advances in food technology and nutrition research. Although this chapter is not meant to deemphasize pharmacological therapies and the role of physical activity, it is our contention that healthy nutrition may attenuate diabetes as well as diabetes-induced cardiovascular complications.

Acknowledgments The research in this article was supported by a grant from the Slovak Scientific Grant Agency (VEGA) 2/0054/11, 1/0638/12 and APVV-0523-10, APVV-0102-11. Infrastructural support for the work reported in this article was provided by the St. Boniface Hospital Research Foundation.

References

1. Dhalla NS (2005) Cardiovascular dysfunction in chronic diabetes. Bus Brief US Endocr Rev:35–38
2. Kaneto H, Katakami N, Matsuhisa M, Matsuoka TA (2010) Role of reactive oxygen species in the progression of type 2 diabetes and atherosclerosis. Mediators Inflamm 2010:453892
3. Rubler S, Dlugash J, Yuceoglu YZ et al (1972) New type of cardiomyopathy associated with diabetic glomerulosclerosis. Am J Cardiol 30:595–602

4. Dhalla NS, Liu X, Panagia V, Takeda N (1998) Subcellular remodeling and heart dysfunction in chronic diabetes. Cardiovasc Res 40:239–247
5. Vecchini A, Del Rosso F, Binaglia L et al (2000) Molecular defects in sarcolemmal glycerophospholipid subclasses in diabetic cardiomyopathy. J Mol Cell Cardiol 32:1061–1074
6. Dhalla NS, Pierce GN, Innes IR, Beamish RE (1985) Pathogenesis of cardiac dysfunction in diabetes mellitus. Can J Cardiol 1:263–281
7. Dhalla NS, Rangi S, Zieroth S, Xu YJ (2012) Alterations in sarcoplasmic reticulum and mitochondrial functions in diabetic cardiomyopathy. Exp Clin Cardiol 17:115–120
8. Coccheri S (2007) Approaches to prevention of cardiovascular complications and events in diabetes mellitus. Drugs 67:997–1026
9. Machackova J, Liu X, Lukas A, Dhalla NS (2004) Renin-angiotensin blockade attenuates cardiac myofibrillar remodelling in chronic diabetes. Mol Cell Biochem 261:271–278
10. Machackova J, Sanganalmath SK, Barta J et al (2010) Amelioration of cardiac remodeling in congestive heart failure by β-adrenoceptor blockade is associated with depression in sympathetic activity. Cardiovasc Toxicol 10:9–16
11. Machackova J, Sanganalmath SK, Elimban V, Dhalla NS (2011) β-Adrenergic blockade attenuates cardiac dysfunction and myofibrillar remodelling in congestive heart failure. J Cell Mol Med 15:545–554
12. Grimm D, Jabusch HC, Kossmehl P et al (2002) Experimental diabetes and left ventricular hypertrophy: effects of β receptor blockade. Cardiovasc Pathol 11:229–237
13. Sugawara T, Fujii S, Zaman TA et al (2001) Coronary capillary remodeling in non-insulin-dependent diabetic rats: amelioration by inhibition of angiotensin-converting enzyme and its potential clinical implications. Hypertens Res 24:75–81
14. Tsutsui H, Matsushima S, Kinugawa S et al (2007) Angiotensin II type 1 receptor blocker attenuates myocardial remodeling and preserves diastolic function in diabetic heart. Hypertens Res 30:439–449
15. Xu Y-J, Arneja AS, Tappia PS, Dhalla NS (2008) The potential health benefits of taurine in cardiovascular disease. Exp Clin Cardiol 13:57–65
16. Lavine SJ, Gellman SD (2002) Treatment of heart failure in patients with diabetes mellitus. Drugs 62:285–307
17. O'Keefe JH, Abuannadi M, Lavie CJ, Bell DS (2011) Strategies for optimizing glycemic control and cardiovascular prognosis in patients with type 2 diabetes mellitus. Mayo Clin Proc 86:128–138
18. Minet L, Moller S, Vach W et al (2010) Mediating the effect of self-care management intervention in type 2 diabetes: controlled trials. Patient Educ Couns 80:29–41
19. Steyn NP, Mann J, Bennett PH et al (2004) Diet, nutrition and the prevention of type 2 diabetes. Public Health Nutr 7:147–165
20. Dalziel K, Segal L (2007) Time to give nutrition interventions a higher profile: cost-effectiveness of 10 nutrition interventions. Health Promot Int 22:271–283
21. Hughes S (2012) Diabetes: support for those at risk of malnutrition in the community. Br J Community Nurs 17:529–530
22. Pan XR, Li GW, Hu YH et al (1997) Effects of diet and exercise in preventing NIDDM in people with impaired glucose tolerance: the Da Qing IGT and diabetes study. Diabetes Care 20:537–544
23. Knowler WC, Barrett-Connor E, Fowler SE et al; Diabetes Prevention Program Research Group (2002) Reduction in the incidence of type 2 diabetes with lifestyle intervention or metformin. N Engl J Med 346:393–403
24. Liu S, Manson JE, Stampfer MJ et al (2000) A prospective study of whole-grain intake and risk of type 2 diabetes mellitus in US women. Am J Public Health 90:1409–1415
25. Meyer KA, Kushi LH, Jacobs DR et al (2000) Carbohydrates, dietary fibre, and incident type 2 diabetes in older women. Am J Clin Nutr 71:921–930
26. Li J, Kaneko T, Qin LQ et al (2003) Effects of barley intake on the glucose tolerance, lipid metabolism, and bowel function in women. Nutrition 19:926–929

27. Li J, Wang J, Kaneko T et al (2004) Effects of fibre intake on the blood pressure, lipids, and heart rate in Goto Kakizaki rats. Nutrition 20:1003–1007
28. Anderson JW (1986) Fibre and health: an overview. Am J Gastroenterol 81:892–897
29. Montonen J, Knekt P, J¨arvinen R et al (2003) Whole-grain and fibre intake and the incidence of type 2 diabetes. Am J Clin Nutr 77:622–629
30. Thorburn A, Muir J, Proietto J (1993) Carbohydrate fermentation decreases hepatic glucose output in healthy subjects. Metabolism 42:780–785
31. Raben A, Tagliabue A, Christensen NJ et al (1994) Resistant starch: the effect on postprandial glycemia, hormonal response, and satiety. Am J Clin Nutr 60:544–551
32. Yamada Y, Hosoya S, Nishimura S et al (2005) Effect of bread containing resistant starch on postprandiol blood glucose levels in humans. Biosci Biotechnol Biochem 69:559–566
33. World Health Organization (2003) Diet, nutrition and the prevention of chronic diseases. Technical report 916. http://www.who.int/dietphysicalactivity/publications/trs916/en/
34. Carter P, Gray LJ, Troughton J et al (2010) Fruit and vegetable intake and incidence of type 2 diabetes mellitus: systematic review and meta-analysis. Br Med J 341:c4229
35. Van't Veer P, Jansen CJF, Klerk M, Kok FJ (2000) Fruits and vegetables in the prevention of cancer and cardiovascular disease. Public Health Nutr 3:103–107
36. Dauchet L, Amouyel P, Hercberg S, Dallongeville J (2006) Fruit and vegetable consumption and risk of coronary heart disease: a meta-analysis of cohort studies. J Nutr 136:2588–2593
37. World Health Organization and Food and Agriculture Organization of the United Nations (2004) Fruit and vegetables for health 2004. http://www.who.int/dietphysicalactivity/publications/fruit_vegetables_report.pdf
38. Carter P, Gray LJ, Talbot D et al (2013) Fruit and vegetable intake and the association with glucose parameters: a cross-sectional analysis of the Let's Prevent Diabetes Study. Eur J Clin Nutr 67:12–17
39. Popkin BM (1994) The nutrition transition in low-income countries: an emerging crisis. Nutr Rev 52:285–298
40. Maire B, Lioret S, Gartner A, Delpeuch F (2002) Nutritional transition and non-communicable diet-related chronic diseases in developing countries. Sante 12:45–55
41. Tucker KL, Buranapin S (2001) Nutrition and aging in developing countries. J Nutr 131:2417S–2423S
42. Lovejoy J, DiGirolamo M (1992) Habitual dietary intake and insulin sensitivity in lean and obese adults. Am J Clin Nutr 55:1174–1179
43. Marshall JA, Hamman RF, Baxter J (1991) High-fat, low-carbohydrate diet and the etiology of non-insulin-dependent diabetes mellitus: the San Luis Valley diabetes study. Am J Epidemiol 134:590–603
44. Vessby B, Tengblad S, Lithell H (1994) Insulin sensitivity is related to the fatty acid composition of serum lipids and skeletal muscle phospholipids in 70-year-old men. Diabetologia 37:1044–1050
45. Salmerón J, Hu FB, Manson JE et al (2001) Dietary fat intake and risk of type 2 diabetes in women. Am J Clin Nutr 73:1019–1026
46. Pelikanova T, Kohout M, Valek J et al (1989) Insulin secretion and insulin action related to the serum phospholipid fatty acid pattern in healthy men. Metabolism 38:188–192
47. Haag M, Dippenaar NG (2005) Dietary fats, fatty acids and insulin resistance: short review of a multifaceted connection. Med Sci Monit 11:RA359–RA367
48. Rodrigues B, Cam MC, Jian K et al (1997) Differential effects of streptozotocin-induced diabetes on cardiac lipoprotein lipase activity. Diabetes 46:1346–1353
49. Sparagna GC, Hickson-Bick DL, Buja LM, McMillin JB (2000) A metabolic role for mitochondria in palmitate-induced cardiac myocyte apoptosis. Am J Physiol Heart Circ Physiol 279:H2124–H2132
50. Dyntar D, Eppenberger-Eberhardt M, Maedler K et al (2001) Glucose and palmitic acid induce degeneration of myofibrils and modulate apoptosis in rat adult cardiomyocytes. Diabetes 50:2105–2113

51. Baraka A, AbdelGawad H (2010) Targeting apoptosis in the heart of streptozotocin-induced diabetic rats. J Cardiovasc Pharmacol Ther 15:175–181
52. Yang ZH, Peng XD (2010) Insulin resistance and heart injury in rats with insulin resistance or type 2 diabetes mellitus. Acta Cardiol 65:329–335
53. Aune D, Ursin G, Veierød MB (2009) Meat consumption and the risk of type 2 diabetes: a systematic review and meta-analysis of cohort studies. Diabetologia 52:2277–2287
54. Micha R, Wallace SK, Mozaffarian D (2010) Red and processed meat consumption and risk of incident coronary heart disease, stroke, and diabetes mellitus: a systematic review and meta-analysis. Circulation 121:2271–2283
55. McEvoy CT, Temple N, Woodside JV (2012) Vegetarian diets, low-meat diets and health: a review. Public Health Nutr 15:2287–2294
56. Yang SY, Li XJ, Zhang W et al (2012) Chinese lacto-vegetarian diet exerts favorable effects on metabolic parameters, intima-media thickness, and cardiovascular risks in healthy men. Nutr Clin Pract 27:392–398
57. Kendler BS (2006) Supplemental conditionally essential nutrients in cardiovascular disease therapy. J Cardiovasc Nurs 21:9–16
58. Bouckenooghe T, Remacle C, Reusens B (2006) Is taurine a functional nutrient? Curr Opin Clin Nutr Metab Care 9:728–733
59. Hansen SH (2001) The role of taurine in diabetes and the development of diabetic complications. Diabetes Metab Res Rev 17:330–346
60. Franconi F, Bennardini F, Mattana A et al (1995) Plasma and platelet taurine are reduced in subjects with insulin-dependent diabetes mellitus: effects of taurine supplementation. Am J Clin Nutr 61:1115–1119
61. You J, Chang KJ (1998) Effects of taurine supplementation on lipid peroxidation, blood glucose and blood lipid metabolism in streptozotocin-induced diabetic rats. Adv Exp Med Biol 442:163–168
62. Lim E, Park S, Kim H (1998) Effect of taurine supplementation on the lipid peroxide formation and the activities of glutathione-related enzymes in the liver and islet of type I and II diabetic model mice. Adv Exp Med Biol 442:99–103
63. Tappia PS, Thliveris J, Xu YJ et al (2011) Effects of amino acid supplementation on myocardial cell damage and cardiac function in diabetes. Exp Clin Cardiol 16:e17–e25
64. Evangeliou A, Gourgiotis D, Karagianni C et al (2010) Carnitine status and lactate increase in patients with type I juvenile diabetes. Minerva Pediatr 62:551–557
65. Mamoulakis D, Galanakis E, Dionyssopoulou E et al (2004) Carnitine deficiency in children and adolescents with type 1 diabetes. J Diabetes Complications 18:271–274
66. Lango R, Smolenski RT, Narkiewicz M et al (2001) Influence of L-carnitine and its derivatives on myocardial metabolism and function in ischemic heart disease and during cardiopulmonary bypass. Cardiovasc Res 51:21–29
67. Rodrigues B, Xiang H, McNeill JH (1988) Effect of L-carnitine treatment on lipid metabolism and cardiac performance in chronically diabetic rats. Diabetes 37:1358–1364
68. Rodrigues B, Ross JR, Farahbakshian S, McNeill JH (1990) Effects of in vivo and in vitro treatment with L-carnitine on isolated hearts from chronically diabetic rats. Can J Physiol Pharmacol 68:1085–1092
69. González-Ortiz M, Hernández-Gonzalez SO, Hernández-Salazar E, Martinez-Abundis E (2008) Effect of oral L-carnitine administration on insulin sensitivity and lipid profile in type 2 diabetes mellitus patients. Ann Nutr Metab 52:335–338
70. Malaguarnera M, Vacante M, Avitabile T et al (2009) L-Carnitine supplementation reduces oxidized LDL cholesterol in patients with diabetes. Am J Clin Nutr 89:71–76
71. Chandni R, Ramamoorthy KP (2012) Lipoprotein(a) in type 2 diabetic subjects and its relationship to diabetic microvascular complications. World J Diabetes 3:105–109
72. Tappia PS, Xu YJ, Rodriguez-Leyva D et al (2013) Cardioprotective effects of cysteine alone or in combination with taurine in diabetes. Physiol Res 62:171–178

73. Shackebaei D, King N, Shukla B, Suleiman MS (2005) Mechanisms underlying the cardioprotective effect of l-cysteine. Mol Cell Biochem 277:27–31
74. Yoshimoto S, Sakamoto K, Wakabayashi I, Masui H (1992) Effect of chromium administration on glucose tolerance in stroke-prone spontaneously hypertensive rats with streptozotocin-induced diabetes. Metabolism 41:636–642
75. Shinde UA, Goyal RK (2003) Effect of chromium picolinate on histopathological alterations in STZ and neonatal STZ diabetic rats. J Cell Mol Med 7:322–329
76. Di Bona KR, Love S, Rhodes NR et al (2011) Chromium is not an essential trace element for mammals: effects of a "low-chromium" diet. J Biol Inorg Chem 16:381–390
77. Shindea UA, Sharma G, Xu YJ et al (2004) Insulin sensitising action of chromium picolinate in various experimental models of diabetes mellitus. J Trace Elem Med Biol 18:23–32
78. Sharma S, Agrawal RP, Choudhary M et al (2011) Beneficial effect of chromium supplementation on glucose, HbA1C and lipid variables in individuals with newly onset type-2 diabetes. J Trace Elem Med Biol 25:149–153
79. Vrtovec M, Vrtovec B, Briski A et al (2005) Chromium supplementation shortens QTc interval duration in patients with type 2 diabetes mellitus. Am Heart J 149:632–636
80. Tahiliani AG, McNeill JH (1986) Diabetes-induced abnormalities in the myocardium. Life Sci 38:959–974
81. Cros GH, Cam MC, Serrano JJ et al (1995) Long-term antidiabetic activity of vanadyl after treatment withdrawal: restoration of insulin secretion? Mol Cell Biochem 153:191–195
82. Li SH, McNeill JH (2001) In vivo effects of vanadium on GLUT4 translocation in cardiac tissue of STZ-diabetic rats. Mol Cell Biochem 217:121–129
83. Yuen VG, Orvig C, Thompson KH, McNeill JH (1993) Improvement in cardiac dysfunction in streptozotocin-induced diabetic rats following chronic oral administration of bis(maltolato) oxovanadium(IV). Can J Physiol Pharmacol 71:270–276
84. Clark TA, Maddaford TG, Tappia PS et al (2010) Restoration of cardiomyocyte function in streptozotocin-induced diabetic rats after treatment with vanadate in a tea decoction. Curr Pharm Biotechnol 11:906–910
85. Thompson KH, Lichter J, LeBel C et al (2009) Vanadium treatment of type 2 diabetes: a view to the future. J Inorg Biochem 103:554–558
86. Turan B (2010) Role of antioxidants in redox regulation of diabetic cardiovascular complications. Curr Pharm Biotechnol 11:819–836
87. Ayaz M, Turan B (2006) Selenium prevents diabetes-induced alterations in $[Zn^{2+}]i$ and metallothionein level of rat heart via restoration of cell redox cycle. Am J Physiol Heart Circ Physiol 290:H1071–H1080
88. Ayaz M, Ozdemir S, Ugur M et al (2004) Effects of selenium on altered mechanical and electrical cardiac activities of diabetic rat. Arch Biochem Biophys 426:83–90
89. Ayaz M, Can B, Ozdemir S, Turan B (2002) Protective effect of selenium treatment on diabetes-induced myocardial structural alterations. Biol Trace Elem Res 89:215–226
90. Marles RJ, Farnsworth NR (1995) Antidiabetic plants and their active constituents. Phytomedicine 2:137–189
91. Yeh GY, Eisenberg DM, Kaptchuk TJ, Phillips RS (2003) Systematic review of herbs and dietary supplements for glycemic control in diabetes. Diabetes Care 26:1277–1294
92. Vuksan V, Sievenpiper JL, Xu Z et al (2001) Ginseng and diabetes: a new way to use an old medicine. Can J Diab Care 25:111–120
93. Vuksan V, Sung M, Sievenpiper JL et al (2008) Korean red ginseng (*Panax ginseng*) improves glucose and insulin regulation in well-controlled, type 2 diabetes: results of a randomized, double-blind, placebo-controlled study of efficacy and safety. Nutr Metab Cardiovas Dis 18:46–56
94. Tappia PS, Xu YJ, Dhalla NS (2013) Reduction of cholesterol and other cardiovascular risk factors by alternative therapies. Clin Lipidol 8:345–359
95. Gluckman PD, Hanson MA, Beedle AS, Raubenheimer D (2008) Fetal and neonatal pathways to obesity. Front Horm Res 36:61–72

96. Tappia PS, Guzman C, Dunn L, Aroutiounova N (2013) Adverse cardiac remodeling due to maternal low protein diet is associated with alterations in expression of genes regulating glucose metabolism. Nutr Metab Cardiovasc Dis 23:130–135
97. Crowther NJ, Cameron N, Trusler J, Gray IP (1998) Association between poor glucose tolerance and rapid post natal weight gain in seven-year-old children. Diabetologia 41:1163–1167
98. Hales CN, Barker DJ, Clark PM et al (1991) Fetal and infant growth and impaired glucose tolerance at age 64. Br Med J 303:1019–1022
99. Poulsen P, Vaag AA, Kyvik KO et al (1997) Low birth weight is associated with NIDDM in discordant monozygotic and dizygotic twin pairs. Diabetologia 40:439–446
100. Le Clair C, Abbi T, Sandhu H, Tappia PS (2009) Impact of maternal undernutrition on diabetes and cardiovascular disease risk in adult offspring. Can J Physiol Pharmacol 87:161–179
101. Lillycrop KA, Phillips ES, Jackson AA et al (2005) Dietary protein restriction of pregnant rats induces and folic acid supplementation prevents epigenetic modification of hepatic gene expression in the offspring. J Nutr 135:1382–1386
102. Lillycrop KA, Phillips ES, Torrens C et al (2008) Feeding pregnant rats a protein-restricted diet persistently alters the methylation of specific cytosines in the hepatic PPAR alpha promoter of the offspring. Br J Nutr 100:278–282
103. Bird A (2002) DNA methylation patterns and epigenetic memory. Genes Dev 16:6–21
104. Gabory A, Attig L, Junien C (2011) Developmental programming and epigenetics. Am J Clin Nutr 94:1943S–1952S
105. Gabory A, Attig L, Junien C (2011) Epigenetic mechanisms involved in developmental nutritional programming. World J Diabetes 2:164–175
106. Jay MA, Ren J (2007) Peroxisome proliferator-activated receptor (PPAR) in metabolic syndrome and type 2 diabetes mellitus. Curr Diabetes Rev 3:33–39
107. Ivanova E, Chen JH, Segonds-Pichon A et al (2012) DNA methylation at differentially methylated regions of imprinted genes is resistant to developmental programming by maternal nutrition. Epigenetics 7:1200–1210
108. Bogdarina I, Welham S, King PJ et al (2007) Epigenetic modification of the renin-angiotensin system in the fetal programming of hypertension. Circ Res 100:520–526
109. Raychaudhuri N, Raychaudhuri S, Thamotharan M, Devaskar SU (2008) Histone code modifications repress glucose transporter 4 expression in the intrauterine growth-restricted offspring. J Biol Chem 283:13611–13626
110. Teperino R, Schoonjans K, Auwerx J (2010) Histone methyl transferases and demethylases; can they link metabolism and transcription? Cell Metab 12:321–327
111. Tarry-Adkins JL, Chen JH, Jones RH et al (2010) Poor maternal nutrition leads to alterations in oxidative stress, antioxidant defense capacity, and markers of fibrosis in rat islets: potential underlying mechanisms for development of the diabetic phenotype in later life. FASEB J 24:2762–2771
112. Deedwania P, Srikanth S (2008) Diabetes and vascular disease. Exp Rev Cardiovasc Ther 6:127–138
113. Furie K, Inzucchi SE (2008) Diabetes mellitus, insulin resistance, hyperglycemia, and stroke. Curr Neurol Neurosci Rep 8:12–29
114. Haffner SM, Lehto S, Rönnemaa T et al (1998) Mortality from coronary heart disease in subjects with type 2 diabetes and in nondiabetic subjects with and without prior myocardial infarction. N Engl J Med 339:229–234
115. Kannel WB, McGee DL (1979) Diabetes and cardiovascular disease. The Framingham study. J Am Med Assoc 241:2035–2038

Index

B. Turan and N.S. Dhalla (eds.), *Diabetic Cardiomyopathy: Biochemical and Molecular Mechanisms*, Advances in Biochemistry in Health and Disease 9, DOI 10.1007/978-1-4614-9317-4, © Springer Science+Business Media New York 2014

MIX
Papier aus verantwortungsvollen Quellen
Paper from responsible sources
FSC® C105338

If you have any concerns about our products,
you can contact us on
ProductSafety@springernature.com

In case Publisher is established outside the EU,
the EU authorized representative is:
Springer Nature Customer Service Center GmbH
Europaplatz 3, 69115 Heidelberg, Germany

Printed by Libri Plureos GmbH
in Hamburg, Germany